I0057353

Cell and Molecular Biology

Cell and Molecular Biology

Edited by **Samantha Granger**

New York

Published by Callisto Reference,
106 Park Avenue, Suite 200,
New York, NY 10016, USA
www.callistoreference.com

Cell and Molecular Biology
Edited by Samantha Granger

International Standard Book Number: 978-1-63239-750-8 (Hardback)

Printed in the United States of America.

Contents

Preface

This book was inspired by the evolution of our times; to answer the curiosity of inquisitive minds. Many developments have occurred across the globe in the recent past which has transformed the progress in the field.

Cell is the fundamental building block of all the existing life forms. Conducting studies related to cells and its various functions and activities are crucial for understanding all the essential life sustaining processes. Cell biology tries to explain the structure, physiological properties, metabolic processes, signaling pathways and life cycles, and other processes of the cells. Molecular biology studies the interaction of biomolecules with different systems of the cell including proteins, DNA, RNA and their synthesis. It is a sub-field of cell biology. This book traces the progress of this field and highlights some of its key concepts and applications. The topics covered in this extensive book deal with the core topics related this area of study. It unfolds the innovative aspects of cell biology which will be crucial for the progress of this field in the future. It aims to assist cell biologists, molecular biologists, researchers and students associated with this field.

This book was developed from a mere concept to drafts to chapters and finally compiled together as a complete text to benefit the readers across all nations. To ensure the quality of the content we instilled two significant steps in our procedure. The first was to appoint an editorial team that would verify the data and statistics provided in the book and also select the most appropriate and valuable contributions from the plentiful contributions we received from authors worldwide. The next step was to appoint an expert of the topic as the Editor-in-Chief, who would head the project and finally make the necessary amendments and modifications to make the text reader-friendly. I was then commissioned to examine all the material to present the topics in the most comprehensible and productive format.

I would like to take this opportunity to thank all the contributing authors who were supportive enough to contribute their time and knowledge to this project. I also wish to convey my regards to my family who have been extremely supportive during the entire project.

Editor

1

Expression and Localization of NANOS1 in Spermatogenic Cells during Spermatogenesis in Rat

Sadaki Yokota, Yuko Onohara
Section of Functional Morphology, Faculty of Pharmaceutical Sciences, Nagasaki International University, Nagasaki, Japan
Email: syokota@niu.ac.jp

ABSTRACT

Expression and localization of NANOS1 in the spermatogenic cells of rat testis were studied by immunofluorescence and immunoelectron microscopy. Using immunofluorescence techniques, NANOS1 was localized in the cytoplasm and discrete granules of spermatocytes and spermatids. The staining intensity of NANOS1 signal varied depending upon the stage of the cycle of seminiferous epithelium. Double immunofluorescence staining with antibodies against NANOS1 and DDX4 showed that several DDX4-positive compartments of nuage were also stained for NONOS1. Immunoelectron microscopy revealed that the major subcellular localization sites for NANOS1 were the chromatoid body (CB) and satellite body (SB), and the minor sites were the rest of the nuage compartments, including the irregularly-shaped perinuclear granules (ISPG), and intermitochondrial cement (IMC). Non-nuage structures such as mitochondria-associated granules (MAG), granulated body (GB), and reticulated body (RB) were also labeled by theNANOS1 antibody. In addition, NANOS1 was found in the outer dense fibers of flagella of spermatids at steps 12-19, and in the head cap of late spermatids after step 15. These results suggest that NANOS1 is one of the nuage proteins and functions mainly in the CBs as well as in the cytoplasm. NANOS1 may also have additional functions in non-nuage structures such as MAGs, GBs and RBs.

Keywords: NANOS1; Nuage; Chromatoid Body; Satellite Body; Non-Nuage Compartments; Immunocytochemistry

1. Introduction

The nanos gene has been extensively studied in *Drosophila*, and has been shown to play multiple roles in the development and maintenance of the germ line [1-3], embryonic patterning [4-7], migration of the primordial germ cells to the embryonic gonad [1,8-10], and inhibition of apoptosis of germ line cells in embryo [11]. There are three homologous genes, NANOS1, NANOS2 and NANOS3, in mouse [10,12]. NANOS1 is not expressed in the primordial germ cells (PGCs) but its expression can be detected in the spermatogenic cells of mature testis. During embryogenesis, the level of maternally expressed NANOS1 is found to be rapidly reduced after fertilization. The expression of NANOS1 can be detected in the central nervous system, and continues to occur in the adult brain [12]. NANOS2 is predominantly expressed in the male germ cells, and the elimination of this gene causes the complete loss of spermatogonia [10]. NANOS2 can substitute NANOS3 during early PGC development but by contrast NANOS3 does not rescue the defects in *NANOS*2-null mice [13]. In addition, NANOS2 is known to suppress meiosis [14]. In human, NANOS3 is expressed highly in germ cell nuclei where the protein co-localizes with chromosomal DNA during mitosis/meiosis [15]. NANOS3 plays a role in the maintenance of PGC migration to the primary gonads [1,16]. *NANOS*3-null-mice are characterized by a complete absence of germ cells in both sexes [10]. NANOS3 interacts with PUMILIO2 and delays the cell cycle progression of spermatogonial cells [17].

Human NANOS1 forms a stable complex with PUMILIO2 through interaction of highly conserved domains [18]. PIMILIO2 interacts with the DEAD-box RNA helicase GEMIN3, a microRNA biogenesis factor. GEMIN3 has been shown to co-precipitate with NANOS 1 and PUMILIO2 in transfected mammalian cells. By double immunofluorescence staining, protein complexes composed of NANOS1, PUMILIO2 and GEMIN3 are localized to the cytoplasmic granules which are also stained for DDX4, a marker protein of chromatoid body (CB) [19]. Moreover, PUMILIO2 and NANOS1 interact with SNARE-associated component SNAPIN to form a complex in human spermatogenic cells [20]. Nanos1 forms a complex with PUMILIO2 and functions to regulate the translation of selective mRNAs, specifically via association with the 3'-UTR of its mRNA targets. Thus, the roles of NANOS in mammalian spermatogenic cells are gradually being elucidated. Despite of the colocalization of NANOS1 with DDX4 in the CB, the interaction of the two proteins has not been shown in other nuage compartments such as intermitochondrial cement (IMC) and irregularly-shaped perinuclear granules (ISPG). Moreover, it has been reported that several nuage-resident proteins are localized to the non-nuage structures, including mitochondria-associated granules (MAG), granulated body (GB), reticulated body (RB) and ribosome aggregates (RA) [21,22]. It is unclear whether NANOS1 is localized to the nuage or non-nuage compartments. In the present work, we investigated the subcellular localization of NANOS1 in rat testis using immunofluorescence and immunoelectron microscopy. Our data show that NANOS1 is co-localized with DDX4 in all nuage compartments and non-nuage structures.

2. Materials and Methods

2.1. Animals

Male guinea pigs (320 - 450 g) and Male Wistar rats (180 - 220 g) were obtained from Kyudo Co. (Tosu, Japan) and fed appropriate standard diets and water ad libitum. The animal experiments were performed in accordance with the guidance for animal experiments issued by Nagasaki International University.

2.2. Antibodies

Polypeptide (MEAFPWAPRSPRRARAPAP) consisting of 19 amino acids at the N-terminus of mouse NANOS1 was synthesized and at the C-terminus cysteine was added to facilitate chemical binding to the carrier protein, The polypeptide (1mg) was conjugated with egg albumin (50 mg) by m-maleimidobenzoyl-N-hydrosuccinimide ester. The conjugate was emulsified with Freund's complete adjuvant. Guinea pigs were intracutaneously injected with 50 μg of peptide. The injections were carried out 4 times with 2-weeks interval, followed by collection of blood samples 2 weeks after the last injection. The specific antibody was affinity-purified by the peptide-coupled column. Rabbit antibody against mouse DDX4 was prepared as previously [23]. Horseradish peroxidase (HRP)-labeled swine anti-guinea pig IgG was obtained from DAKO Japan (Tokyo). Rabbit anti-mitochondrial isocitrate dehydrogenase (ICD1) was described previously [24]. Rabbit anti-histone H2B, rabbit anti-protein disulfide isomerase (PDI), and rabbit anti-tubulin were obtained from Cell Signaling Technology (Boston, MA), Sigma-Aldrich Japan (Tokyo), and Thermo Fisher Scientific (Fremont, CA), respectively. Alexa 568® or Alexa 488® conjugated goat anti-rabbit IgG or goat anti-rat IgG were obtained from Molecular Probes (Eugene, OR). HRP-labeled goat antibody to rabbit IgG was purchased from DAKO Japan (Tokyo) and Invitrogen Japan, respectively. Protein A-gold probe (15 nm gold) was prepared by the method of de Roe *et al.* [25].

2.3. Biochemical Analysis

2.3.1. Western Blotting

Rat testes were homogenized in 5 mM MOPS-KOH buffer (pH 7.4) containing 0.25 M sucrose, 1 mM ethylenediaminetetraacetic acid, 1 mM phenylmethylsulfonyl fluoride and a cocktail of protease inhibitors (1 μg/ml), including leupeptin, pepstatin, aprotinin, and antipain (medium A), using a Potter-Elvehjem homogenizer. Ten percent homogenate (w/v) was centrifuged at 800 × g for 10 min. The resulting supernatant was centrifuged at 10,000 × g for 20 min and the pellet (mitochondrial fraction) was suspended in a small volume of medium A. The supernatant was centrifuged at 100,000 × g for 60 min in a Beckman ultracentrifuge using a SW 41 swing rotor. The resulting pellets were suspended in medium A and used as the microsomal fraction, while the supernatant was used as the cytosol fraction. The cell fractions were stored at −70°C. Protein concentrations were determined by the bicinchoninic acid method (Pierce Chemical, Rockford, IL) using bovine serum albumin (BSA) as a standard. The protein concentrations of the fractions were adjusted to 1 mg/ml, mixed with one volume of sample buffer for SDS-PAGE, and heated in boiling water for 5 min. Ten micrograms of each sample were analyzed by western blotting. The molecular mass of NANOS1 was estimated by prestained protein maker (Nippon Genetics Europe, Dueren, Germany).

2.3.2. Dot Blot Analysis of NANOS1 Content in Cell Fractions of Mouse Testis

The mitochondrial, microsomal, and cytosolic cell fractions were isolated from rat testes homogenate by differential centrifugation as described above. Nuclei were isolated by the method of Rickwood and Ford [26].

Briefly, the crude nuclear fraction (800 g pellet) was suspended in medium A containing 1% Triton X-100 and centrifuged at 800 g for 10 min. The resulting pellet was suspended in 2 ml of medium A, mixed with 72.5% (wt/vol) metrizamide (Sigma-Aldrich) in medium A without 0.25 M sucrose and centrifuged at 10,000 g for 20 min at 5°C The nuclei-rich pellicles at the surface of the metrizamide solution were collected, suspended in a 10-times volume of medium A, and then centrifuged at 6000 g for 10 min. The pellet was suspended in 1.6 ml of medium A. Each fraction was diluted by medium A. The diluted fractions (90 μl) were mixed with 90 μl of SDS-PAGE sample buffer and 20 μl of 0.3 M iodoacetamide, treated in boiling water for 5 min, and centrifuged at 10,000 g for 10 min. The supernatant fractions were diluted 1000-fold and 1 - 5 μl was loaded onto polyvinylidene difluoride (PVDF) membranes for immunoblotting. NANOS1 was visualized by combination of HRP-labeled swine anti-guinea pig IgG and 3,3'-diaminobenzidine (DAB) reaction. Internal standard proteins, histone H2A (H2A, nuclei), mitochondrial isocitrate dehydrogenase (ICD1, mitochondria), protein disulfide isomerase (PDI, microsomes), and α-tubulin (α-Tub, cytosol) were visualized using a combination of rabbit antibodies against each protein, HRP-labeled goat anti-rabbit IgG, and DAB reaction. The staining intensity was measured by densitometer. The total amount of NANOS1 in each cell fraction was calculated as follows: the densitometric values obtained were multiplied by the final volume of each fraction. The data were obtained from three measurements and the average values and stand ard deviations were plotted.

2.4. Immunocytochemical Analysis

2.4.1. Immunofluorescence Staining of Testis Tissue Sections

Rat testes were embedded in Tissue-Tek (SAKURA Fintec, Tokyo, Japan) and frozen in isopentane cooled by liquid nitrogen. Frozen sections (6 μm thickness) were cut with a Leitz cryostat (Leica Instruments, Nussloch, Germany) and picked up on silicone-coated glass slides. Sections were fixed in a fixative consisting of 4% paraformaldehyde, 0.01% $CaCl_2$, and 0.1 M HEPES-KOH buffer (pH 7.4) for 15 min at room temperature (RT). Sections were treated with 0.2% Triton X-100 and 0.2% Saponin for 15 min and then incubated in blocking solution containing 2% fish gelatin and 15 mM NaN_3 in PBS for 30 min. The sections were subsequently incubated with guinea pig anti-NANOS1 antibody (1 μg/ml) overnight at 4˚C. After washing in PBS, they were incubated with Alexa 488 or 568-labeled goat anti-guinea pig IgG solutions containing 3 μM DAPI (Hoechst, Tokyo) for 60 min at RT. For the immunofluorescence control, preimmune guinea pig sera were used for the primary reaction, followed by Alexa 488-labeled or 568-labeled secondary antibodies. For double staining, sections were incubated separately with mixtures of rabbit anti-DDX4/guinea pig anti-NANOS1, followed by incubation with Alexa 488- or 568-labeled secondary antibodies. The sections were examined with a Nikon Eclipse E600 fluorescence microscope (Nikon, Tokyo).

The images were merged using Adobe Photoshop 7.0 to visualize cell contours. The stage of seminiferous tubules was judged from the size and shape of spermatocytes and spermatids stained by DAPI with reference to the stages of the cycle illustrated by Russell and coworkers [27].

2.4.2. Immunoelectron Microscopic Staining of Rat Testis

Rat testes were dissected out, cut into small tissue blocks, and fixed in 4% paraformaldehyde + 0.2% glutaraldehyde in HEPES-KOH buffer (pH 7.4) for 1h at 4˚C. The tissue blocks were dehydrated in graded ethanol series at −20˚C and embedded in LR White, which was polymerized under UV light at −20˚C. Thin sections of LR White-embedded testis tissues were cut with a diamond knife equipped with a Reichert Ultracut R, mounted on nickel grids, and incubated with affinity purified guinea pig anti-NANOS1 (1 μg/ml) overnight at 4˚C. Preimmune guinea pig sera were used instead of the primary specific antibody for the control experiments. Sections were contrasted and then examined with a Hitachi H7650 electron microscope (Hitachi, Tokyo, Japan) at acceleration voltage of 80 kV. The stage of seminiferous tubules and step of spermatids were judged according to stages of the cycle illustrated by Russell *et al.* [27].

3. Results

3.1. Western Blotting

Our peptide antibody reacted with two protein bands with molecular masses of 29 and 40 kDa, respectively. The former band was detected in all cell fractions, whereas the latter was present only in the mitochondrial and cytosolic fractions (**Figure 1**). Dot blot analysis of NANOS1 contents in cell fractions showed that approximately 61%, 27%, 8%, and 4% of the protein was detected in the cytosolic, mitochondrial, nuclear, and microsomal fractions, respectively. Marker proteins, H2A, ICD1, PDI, and α-Tub, were detected prominently in the nuclear, mitochondrial, microsomal, and cytosol fractions, respectively (**Figure 1(b)**, Dot blot).

3.2. Immunofluorescence Staining of NANOS1 Protein (NANOS1) in Rat Testis

Two staining patterns were noted in spermatogenic cells after immunofluorescence staining of NANOS1, one being diffusive in the cytoplasm and the other with a discrete granular distribution. The former staining pattern was observed in spermatocytes and spermatids at all steps with various staining intensities **Figures 2(a)**, **(d)** and **(g)**. The latter was seen in early to late spermatids and the size and shape of stained granules varied depending on their differentiation steps. In spermatids at steps 1 to 8, irregularly shaped granules were strongly stained, which were also positive for DDX4, a marker for chromatoid bodies (CB) (**Figures 2(a)-(c)**). Small granules located in the cytoplasmic lobe of spermatids at steps 10-13 were also stained for NANOS1 but not for DDX4 (**Figures 2(d)-(i)**). In pachytene spermatocytes, irregularly-shaped perinuclear granules (ISPG) were stained weakly for NANOS1 but strongly for DDX4 (**Figures 2(d)-(i)**). Frequently, the granules stained only either for NANOS1 or DDX4 were noted (**Figures 2(d)-(i)**). The nuclei of spermatocytes were almost negative for NANOS1 but those of elongated spermatids at steps 9-10 were slightly

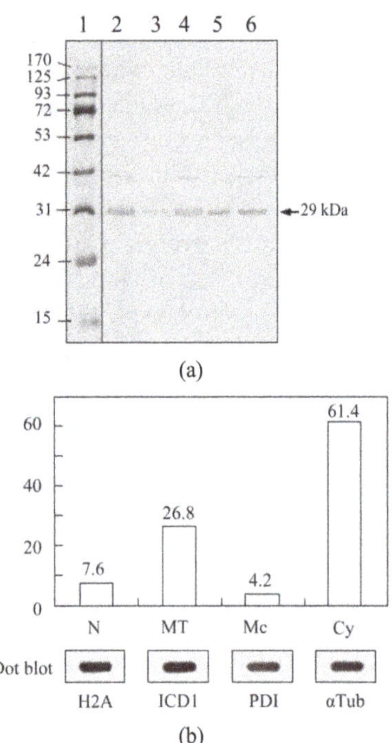

Figure 1. (a) Western blotting analysis of cell fractions of rat testis. Lane 1: molecular weight marker, lane 2: homogenate, lane 3: nuclear fraction, lane 4: mitochondrial fraction, lane 5: microsomal fraction, lane 6: cytosol fraction. A single band with a molecular mass of 29 kDa is seen (arrows); (b) Dot blotting analysis of cell fractions isolated from rat testis homogenate. N: nuclear fraction, Mt: mitochondrial fraction, Mc: microsomal fraction, Cy: cytosol fraction. Dot blot shows marker protein of each cell fraction. H2A: histone H2A, ICD1: mitochondrial isocitrate dehydrogenase, PDI: protein disulfide isomerase, α-Tub: α-tubulin.

stained (**Figure 2(d)**). Areas surrounding clusters of elongated spermatid heads were stained for NANOS1 (**Figure 2(a)**). Flagella of spermatids at steps 12-19 were weakly stained (**Figure 2(g)**). At higher magnification, staining of CB for NANOS1 and DDX4 was almost completely overlapped in early spermatid (**Figure 3(a)**) but partly stuck out each other in spermatids at steps 5 and 6 (**Figure 3(b)**). ISPG in pachytene spermatocytes at stage IX-XII were stained partially for NANOS1 and DDX4 (**Figure 3(c)**). All the staining for NANOS1 and DDX4 was not observed in the control sections incubated in preimmune sera, followed by the secondary antibodies conjugated with Alexa 568® or Alexa 488® (not shown).

3.3. Immunoelectron Microscopic Localization of NANOS1 in Nuage Compartments

NANOS1 signals detected in all nuage compartments as classified by Russell and Frank [28], including 70 - 90 nm particles, cluster of 60 - 90 nm particles, intermitochondrial cement (IMC), chromatoid body (CB) and satellite body (SB) with various labeling intensities. The staining intensity of NANOS1 in 70 - 90 nm particles was very weak, as compared with the staining for DDX4 (data not shown).The clusters of 60 - 90 nm particles corresponding to ISPG identified by immunofluorescence staining were intermediately stained for NANOS1 (**Figure 4(a)**). The labeling intensity for NANOS1 was weaker than that for DDX4 (**Figure 4(a)**, inset). Many of gold-labeled signals were found to be associated with the surface of the dense material of the cluster (**Figure 4(a)**).

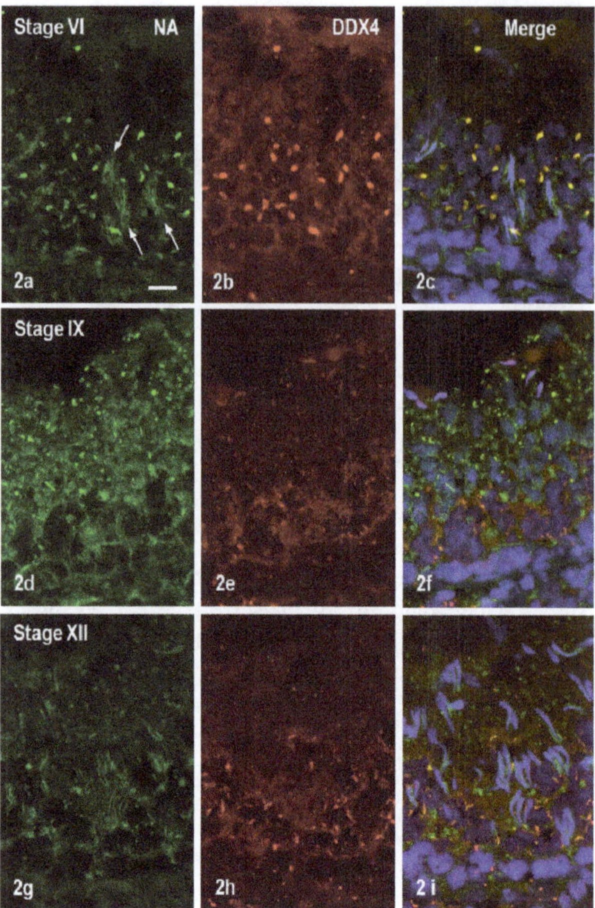

Figure 2. IF staining of rat testis. (a)-(c). Seminiferous tubule at stage IV. (a) Staining for NANOS1; (b) Staining for DDX4; (c) Merged (a) and (b) and staining with DAPI for nuclei. Relatively large granules are stained for both NANOS1 and DDX4 but staining around clusters of spermatid heads is for solely for NANOS1. Bar = 50 µm; (d)-(f) Seminiferous tubule at stage IX. (d) Staining for NANOS1; (e) DDX staining; (f) Merged (d) and (e) and stained with DAPI. Note that NANOS1-stained granules in step 10 spermatids are almost negative for DDX4 staining, whereas DDX4-positive small granules in pachytene spermatocytes are weakly stained for NANOS1. (g)-(i) Tubule at stage XII. (g) Staining for NANOS1; (h) Staining for DDX4; (i) Merged (g) and (h) and stained with DAPI. Granules in step 12 spermatids some in pachytene spermatocytes are stained only for NANOS1, whereas irregularly shaped perinuclear granules are stained for both NANOS1 and DDX4 with diverse intensities.

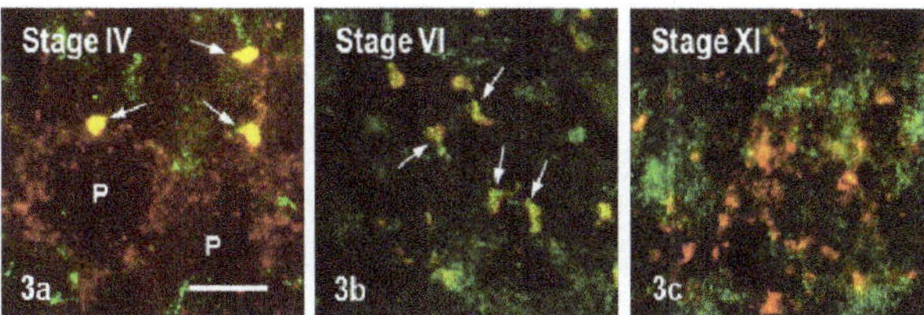

Figure 3. High power view of double IF staining for NANOS1 (green) and DDX4 (red). (a) Granules in step 4 spermatids are stained for both proteins. (a) Stage IV. Large granules of spermatid at step 4 are stained for both proteins (arrows). Many small granules in pachytene spermatocytes (P) are stained for DDX4 and small numbers of granules are stained for NANOS1; (b) Stage VI. Spermatids are shown. Irregularly-shaped granules are stained for both proteins with various staining intensities (arrows); (c) Stage XI. Pachytene spermatocytes are shown. NANOS1 is stained in numerous small granules whereas DDX4 is in ISPG, some of which are positive for NANOS1. Bar = 10 µm.

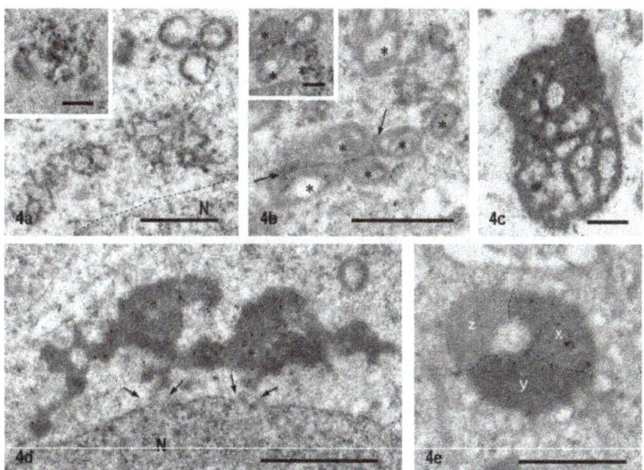

Figure 4. Immunoelectron Microscopic staining for NANOS1 and DDX4. (a) Cluster of 60-90nm particles corresponding to ISPG by IF staining is weakly labeled for NANOS1, whereas the same structure is stained strongly for DDX4 (Inset). Dashed line shows nuclear membrane. N: nucleus; (b) IMC of pachytene spermatocyte at stage IX. It is weakly labeled for NANOS1 (arrows) but very strongly labeled for DDX4 (Inset). *: mitochondria; (c) Typical CB in step 5 spermatid. A half upper dashed line is darker than lower half and labeled stronger than lower part; (d) CB of step 1 spermatid stained for NANOS1. It locates closely to the nucleus. Between CB and nuclear membrane vague particles are recognized (arrows). N: nucleus; (e) Late CB in step 9 spermatid. Three different areas (x, y and z) which are bounded with dashed lines are seen. Labeling intensity is highest in the area x, decreases in the y and no label in the z. Bar = 1 µm for all pictures and 0.5 µm for inset.

NANOS1 labeling in IMC was also low, compared with DDX4 labeling (**Figure 4(b)**). Frequently, the structure of IMC was connected to the cluster of 60 - 90 nm particles. There were two types of CBs, one having homogeneous matrix and the other being heterogeneous. Both types reacted positively for NANOS1 with various staining intensities. The former type of CB was observed generally in early round spermatids, located closely to the nuclear envelope and seemed to communicate with the nucleus (**Figure 4(d)**). The latter type of CB was seen in spermatids at steps 5-10 and had the areas stained heterogeneously from negative to positive (**Figure 4(c)**). This heterogeneous labeling pattern increased as differentiation of spermatids progressed. In the CBs of late elongated spermatids, even the completely unlabeled area appeared (**Figure 4(e)**). SB appeared in pachytene spermatocytes at stages IX-XI and was relatively heavily labeled for NANOS1 (**Figure 5(a)**). Gold-labeled signals were associated with the dense cord-like structure. In the meiotic cells, NANOS1 was localized in small particles, which were scattered in the cytoplasm or closely associated with mitochondria (**Figure 5(b)**). All the gold signals were not observed in the immunoelectron microscopic control sections incubated with preimmune sera, followed by incubation with protein A-gold probe (**Figures 6(a)-(e)**).

3.4. Immunoelectron Microscopic Localization of NANOS1 in Non-Nuage Structures

Next we focused on non-nuage structures, in (MAG), granulated body (GB), reticulated body (RB), ribosome aggregate (RA), head cap of late spermatids, manchette, and dense fiber of flagella.

MAG was 1 - 2 µm in diameter, composed of a fine granular material, and appeared in the elongated cytop-

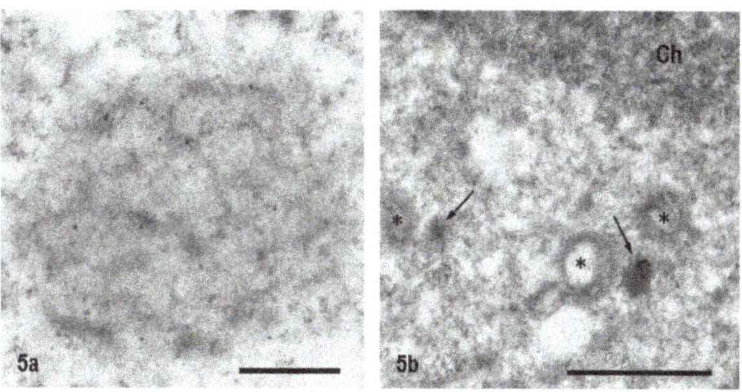

Figure 5. (a) SB seen in pachytene spermatocyte at stage IX. Gold particles showing NANOS1 are associated closely with dense reticular network; (b) Dividing spermatocyte. NANOS1 signals are seen in small dense particles which located closely to mitochondria (*) (arrows). Ch: chromosome. Bar = 0.5 µm.

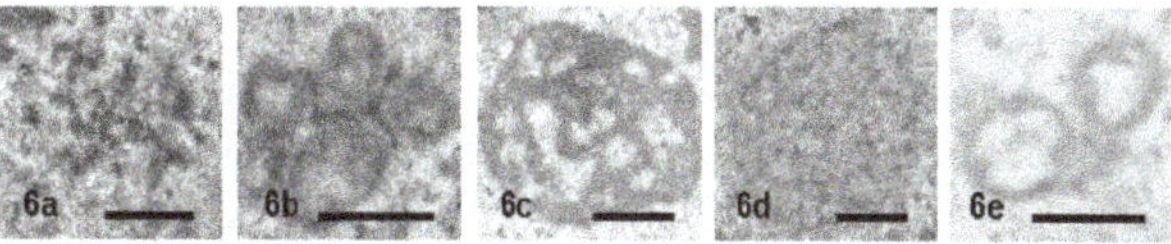

Figure 6. Nuage structures in immunoelectron microscopic control sections incubated with preimmune sera, followed by protein A-gold probe. (a) ISPG; (b) IMC; (c) CB; (d) SB. and (e) MAG appeared in dividing spermatocyte. No gold labeling is observed. Bar = 0.5 µm.

lasmic lobes of steps 9-17 spermatids. Its marked feature was the close association with mitochondria [29]. MAG was relatively strongly labeled for NANOS1 (**Figure 7(a)**) GB appeared in the cytoplasm lobes of step 14 spermatids, was most abundant in step 17 spermatids, and decreased in number during subsequent stages of sper matogenesis. The dense matrix of GB was labeled for NANOS1 but not less dense particles closely associated with GB (**Figure 7(b)**). RB first appeared in the cytoplasm of step 14 spermatids and completely disappeared in step 18 spermatids. It was characterized by several dense, anastomosed cords with a width of 80 - 100 nm. Gold signals showing NANOS1 were closely associated with the surface of RB but not with the densecord (**Figure 7(c)**). RA was observed in the residual body of step 19 spermatids and was composed of massive ribosome particles and other unknown materials. NANOS1 signals were scattered in RA but not in the cytoplasm surrounding it (**Figure 7(d)**). In spermatids after step 15, NANOS1 signals were noted in the ventral head cap but the dorsal head cap was weakly or not labeled (**Figure 7(e)**). The acrosome itself was almost negative. In the flagellum of spermatids at steps 15-17, outer dense fibers of the middle piece and principal piece were stained for NANOS1 but not microtubules of the end piece (**Figures 7(f)** and **(h)**). In immunoelectron microscopic control sections incubated with preimmune sera, followed by protein A-gold probe, gold labeling in aforementioned structures was abolished (**Figures 8(a)-(h)**).

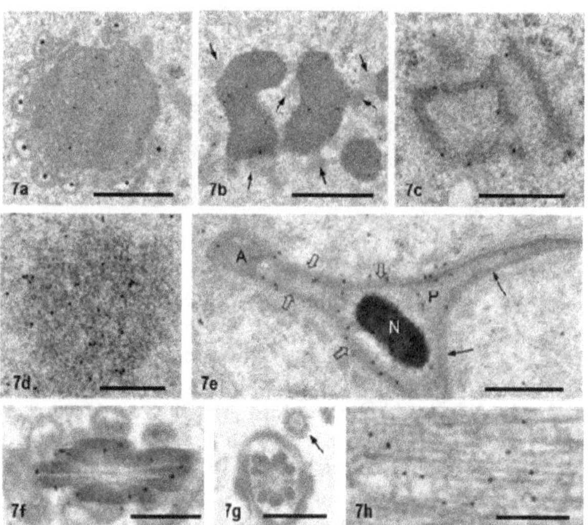

Figure 7. NANOS1 localization in non-nuage compartments. (a) MAG seen in elongated cytoplasmic lobe of step 15 spermatid. It is surrounded by many mitochondria (*). NANOS1 signals are present in the matrix; (b) GB in spermatids at step 15. Small less dense particles attach to it (arrows). Gold particles showing NANOS1 are in the fine granular matrix; (c) RB in spermatid at step 15. NANOS1 signals are associated with rod-like structures; (d) RA in the residual body of step-19 spermatid; (e) Cross-section of sickle-shaped nucleus of step 17 spermatid. Head cap is labeled for NANOS1 (open arrows) but the cap at dorsal side is not (arrows). N: nucleus; (f) Cross section of flagellum of step16 spermatid. Outer dense fibers are labeled for NANOS1; (g) Cross-section of flagellum of step16 spermatid. Outer dense fibers are labeled for NANOS1 but microtubules in the end piece are not (arrow); (h) Manchette of step 10 spermatid. Gold particles are associated with microtubules. Bar = 1 μm for (a) and (b). Bar = 0.5 μm for the other pictures.

3.5. NANOS1 Expression in Spermatogenic Cells during Spermatogenesis in Rat Testis

Figure 9 summarizes the results of this experiment. Data from immunofluorescence and immunoelectron microscopy were combined and the expression of NANOS1 in nuage and non-nuage compartments was evaluated by staining intensity. In spermatocytes, weak cytosolic staining for NANOS1 was observed during all stages. NANOS1 appeared in ISPG and IMC of pachytene spermatocytes at stage V, the strongest staining seen in stages IX-X, decreased gradually in subsequent stages. NANOS1 started to appear in SB of pachytene spermatocytes at stage VIII, and then it rapidly increased, reached the peak level at stage IX, then started to decrease, and disappeared in stage XI. NANOS1 in CB appeared in round spermatids at step 1, increased rapidly to reach the peak level at steps 2-3. This state was maintained until step 8, following the signals decreased gradually and eventually disappeared in step 11. As described above, NANOS1 was localized in four kinds of non-nuage structures, MAG, RB, GB and RA located in the elongated cytoplasmic lobe of late spermatids. MAG was detected in spermatids at steps 10-17 while

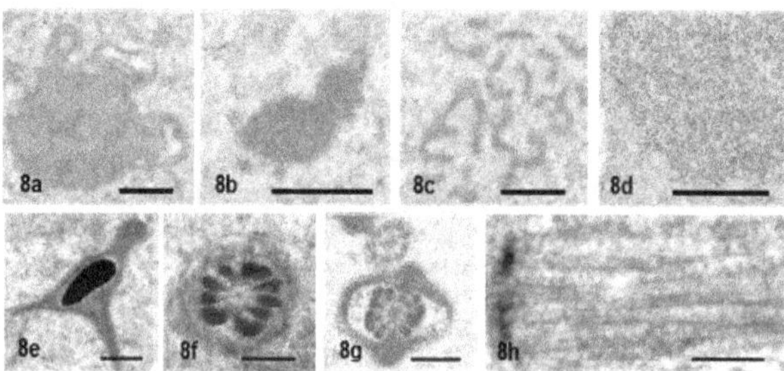

Figure 8. Non-nuage structures in Immuno-electron microscopy control sections. (a) MAG; (b) GB; (c) RB; (d) RA; (e) Head cap; (f) Cross section of flagellum at the middle piece; (g) Cross section of flagellum at the principal piece; (h) Manchette of step 10 spermatid. Bar = 0.5 μm for all.

NANOS1 was detected in all steps. RB appeared in late spermatids at steps 14-16 but the NANOS1 signals were seen in RB only at steps 15 and 16. GB occurred in spermatids at steps 15-16 and the signals for NANOS1 were associated with GB in all steps. RA was observed in the elongated cytoplasmic lobe of the step-18 spermatids

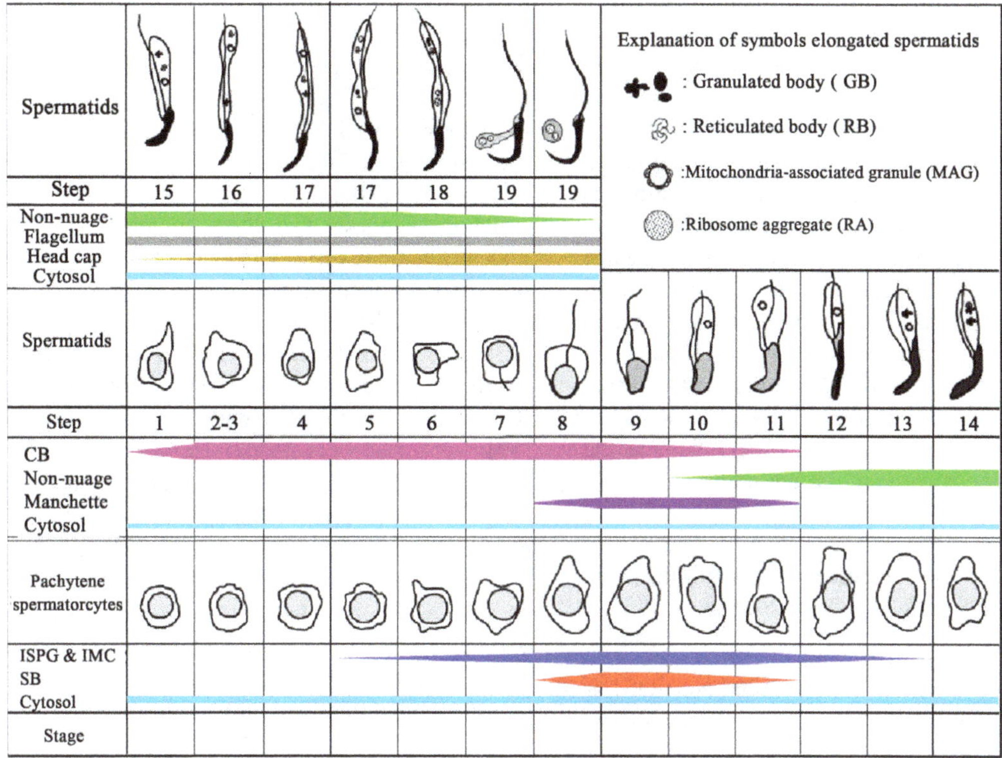

Figure 9. Expression of NANOS1 in nuage and non-nuage structures during spermatogenesis of rat testis. The width of colored bars indicates the number and size of nuage and non-nuage granules and cytosol staining, which are estimated arbitrarily. The position of colored bars indicates the stage (spermatocytes) and step (spermatids) where the staining of the granules and cytosol appears and disappears. ISPG and IMC are indicated by blue color, SB by red, CB by pink, non-nuage including GB, RB, MAG and RA, by green, cytosol by cyan, outer dense fiber by gray, manchette by violet, and head cap by ocher. Illustration showing developing cells is modified from that of Russell *et al.* [27].

and in the residual bodies of spermatids at step 19 and the NANOS1 signals were found to be stronger in step 19 than in step 18. Manchette appeared in spermatids at step 7 and continued to be observed until step 14. However, the signals for NANOS1 were detected in manchette in spermatids at steps 8-11. Flagella started to form in spermatids at step 7 and the NANOS1 signals appeared first in the dense fiber of spermatids at step 12. Afterward, the signals were consistently observed until step 19. The NANOS1 signals in the head cap appeared in spermatids at step 15, gradually increased and reached the peak level at step 19. Weak cytosolic staining of NANOS1 in spermatids was seen in all steps.

4. Discussion

4.1. Subcellular Distribution of NANOS1 in Rat Testis Cell Fractions

Our peptide antibody against NANOS1 reacted with a main 29 KDa band and an additional 40 KDa band on by Western blotting analysis. The molecular mass of the main band is almost consistent with that calculated from amino acid sequence reported in SWISS-PROT database, indicating that the antibody reacts with NANOS1. Although the latter band was observed in the mitochondrial and cytosol fractions but not in the nuclear and microsomal fractions, it is unclear whether it shows the band larger molecular species of NANOS1 or the other protein. The larger molecules of NANOS1 were reported [15,30]. Dot blot analysis showed that approximately 61% of NANOS1 distributed in the cytosol fraction and about 27% in the mitochondrial fraction. The results are almost consistent with immunofluorescence staining. Immunoelectron microscopic observations show no labeling in the mitochondria, so the dot blot value for the mitochondria indicates NANOS1 containing IMC, CB and cluster of 60 - 90 nm particles, which are co-precipitated with the mitochondria after differential centrifugation.

4.2. Immunofluorescence Localization of NANOS1 in Rat Testis

Our immunofluorescence staining results showed that NANOS1 is distributed in the cytosol and granular components of spermatogenic cells, but is absent in the nucleus. Double staining for NANOS1 andDDX4, a marker for nuage, showed that the granular components were mainly nuage compartments in which CB and SB were major site and ISPG and IMC minor sites. The distribution of NANOS1 in CB has previously been reported [19]. The present study demonstrates that NANOS1 is present in all nuage compartments. Our double immunofluorescence staining results suggest that there is a dynamic development process of nuage compartments, in which the NANOS1 negative nuage compartments continue to fuse with NANOS1-positive ones. Moreover, cytosolic NANOS1 might be incorporated into the nuage compartments. On the other hand, the double staining results also revealed that there appear to have many NANOS1-containing granules that are not stained for DDX4 in spermatogenic cells. Most of them are distributed in the elongated cytoplasmic lobe of late spermatids. Immunoelectron microscopic observations reveal that these DDX4-negative granules are non-nuage structures.

4.3. Immunoelectron Microscopic Localization of NANOS1 in Nuage and Non-Nuage Compartments

In the present study, our data showed that NANOS1 is localized to five kinds of nuage compartments as described by Russell and Frank [28]. The NANOS1 signals were weaker than those observed for other nuage proteins such as DDX4 [23], DDX25 [21] and BRU-NOL2 [22]. As shown by immunofluorescence staining and dot blotting analysis, NANOS1 was abundantly present in the cytosol fraction. It is possible that NA-NOS1 is synthesized in the cytosol and transported to the nuage compartments. It has been shown that human NANOS1 interacts with PUMILIO2 through highly conserved domains to form a stable complex [18]. Human PIMILIO2 has been shown to interact with the DEAD-box RNA helicase GEMIN3, a microRNA biogenesis factor, which is co-precipitated with NANOS1 and PU-MILIO2 in transfected mammalian cells [19]. By double immunofluorescence staining, the complexes of NA-NOS1, PUMILIO2 and GEMIN3 have been localized to CB [19]. Thus, it is suggested that CB is the main subcellular site where NANOS1 forms complexes with PUMILIO2 and GEMIN3 in the spermatids. The present results showed that NANOS1 is also localized to other nuage compartments such as 70 - 90 nm particles, cluster of 60 - 90 nm particles, IMC and SB, which occurred mainly in pachytene spermatocytes. Thus, NANOS1-PUMILIO2-GEMIN3 complexes might be one of the building components of these nuage compartments. The staining heterogeneity within the CB, in which the NANOS1 staining of one area was stronger than the other, was noticeable in spermatids at steps 8-11. It is likely that the observed staining heterogeneity may represent degradation levels of CB contents as suggested previously [31].

The present study showed that NANOS1 is also localized to the non-nuage structures, such as MAG, GB, RB, RA, Head cap of the step-17 spermatids, and outer dense fibers of flagella. Although the functions of MAG, GB, RB, and RA remain unknown, our previous studies have shown that DDX25 and BRUNOL2 are present in these structures [21,23]. The existence of nuage proteins such as BRUNOL2, DDX25 and NANOS1 in these structures suggests that these proteins may have biological functions there. However, it is not clear whether they perform their primary functions, such as the regulation of translation and mRNA silencing by BRUNOL2 [32], the participation of DDX25 in the nuclear export of RNAs [33], and the involvement of NANOS1 in microRNA biogenesis together with PUMILIO and GEMIN3 [19].

In conclusion, NANOS1 is a protein component of all nuage compartments, including CB, IMC, ISPG, SB and cluster of 60 - 90 nm particles, as well as several non-nuage compartments such as MAG, GB and RB. NANOS1 may exert its biological function in the nuage compartments via formation of protein complexes with PUMILIO and GEMIN3. How NANOS1 performs its function in the non-nuage structures is unclear.

5. Acknowledgements

This work was supported by the University research fund, in part by a grant-in-aid (17570158) from the Ministry of Education, Science, Culture and Sport, and by Science Research Promotion Fund from the Promotion and Mutual Aid Corporation for Private Schools of Japan.

REFERENCES

[1] S. Kobayashi, M. Yamada, M. Asaoka and T. Kitamura, "Essential Role of the Posterior Morphogen Nanos for Germline Development in *Drosophila*," *Nature*, Vol. 380, No. 6576, 1996, pp. 708-711. doi:10.1038/380708a0

[2] K. Subramaniam and G. Seydoux, "Nos-1 and Nos-2, Two Genes Related to *Drosophila* Nanos, Regulate Primordial Germ Cell Development and Survival in *Caenorhabditis elegans*," *Development*, Vol. 126, No. 21, 1999, pp. 4861-4871.

[3] Z. Wang and H. Lin, "Nanos Maintains Germline Cell Self-Renewal by Preventing Differentiation," *Science*, Vol. 303, No. 5666, 2004, pp. 2016-2019. doi:10.1126/science.1093983

[4] D. Curtis, J. Apfeld and L. Lehmann, "Nanos Is an Evolutionarily Conserved Organizer of Anterior-Posterior Polarity," *Development*, Vol. 121, No. 6, 1995, pp. 1899-1910.

[5] R. Lehmann and C. Nüsslein-Volhard, "The Maternal Gene Nanos Has a Central Role in Posterior Pattern Formation of the *Drosophila embryo*," *Development*, Vol. 112, No. 3, 1991, pp. 679-691.

[6] C. Wang, K. K. Dickinson and R. Lehman, "Genetics of

nanos Localization in *Drosophila*," *Developmental Dynamics*, Vol. 199, No. 2, 1999, pp. 103-115. doi:10.1002/aja.1001990204

[7] Y. Wang, R. M. Zayas T. Guo and P. A. Newmark, "*Nanos* Function Is Essential for Development and Regeneration of Planarian Germ Cells," *Proceedings of National Academy of Sciences of the United States of America*, Vol. 104, No. 14, 2007, pp. 5901-5906. doi:10.1073/pnas.0609708104

[8] G. Deshpande, G. Calhoun, J. L. Yanowitzand and P. D. Schedl, "Novel Functions of *nanos* in Down Regulating Mitosis and Transcription during the Development of the *Drosophila* Germline," *Cell*, Vol. 99, No. 3, 1999, pp. 271-181. doi:10.1016/S0092-8674(00)81658-X

[9] Y. Hayashi, M. Hayashi and S. Kobayashi, "Nanos Suppresses Somatic Cell Fate in *Drosophila* Germ Line," *Proceedings of National Academy of Sciences of the United States of America*, Vol. 101, No. 28, 2004, pp. 10338-10342. doi:10.1073/pnas.0401647101

[10] M.Tsuda, Y. Sasaoka, M. Kiso, K. Abe, S. Haraguchi, S. Kobayashi and Y. Saga, "Conserved Role of *nanos* Proteins in Germ Cell Development," *Science*, Vol. 301, No. 5637, 2003, pp. 1239-1241. doi:10.1126/science.1085222

[11] K. Sato, Y. Hayashi, Y. Ninomiya, S. Shigenobu, K. Arita, M. Mukai and S. Kobayashi, "Maternal Nanos Represses *hid*/*skl*-Dependent Apoptosis to Maintain the Germ Line in *Drosophila* Embryos," *Proceedings of National Academy of Sciences of the United States of America*, Vol. 104, No. 18, 2007, pp. 7455-7460. doi:10.1073/pnas.0610052104

[12] S. Haraguchi, M. Tsuda, S. Kitajima, Y. Sasaoka, A. Nomura-Kitabayashi, K. Kurokawa and Y. Saga, "*nanos*1: A Mouse *nanos* Gene Expressed in the Central Nervous System Is Dispensable for Normal Development," *Mechanism of Development*, Vol. 120, No. 6, 2003, pp. 721-731. doi:10.1016/S0925-4773(03)00043-1

[13] A. Suzuki, M. Tsuda and Y. Saga, "Functional Redundancy among Nanos Proteins and a Distinct Role of Nanos2 during Male Germ Cell Development," *Development*, Vol. 134, No. 1, 2007, pp. 77-83. doi:10.1242/dev.02697

[14] A. Suzuki and Y. Saga, "Nanos2 Suppresses Meiosis and Promotes Male Germ Cell Differentiation," *Genes & Development*, Vol. 22, No. 4, 2008, pp. 430-435. doi:10.1101/gad.1612708

[15] V. T. A. Julaton and R. A. R. Pera, "*NANOS*3 Function in Human Germ Cell Development," *Human Molecular Genetics*, Vol. 20, No. 11, 2011, pp. 2238-2250. doi:10.1093/hmg/ddr114

[16] A. Forbes and R. Lehmann, "Nanos and Pumilio Have critical Roles in the Development and Function of *Drosophila* Germline Stem Cells," *Development*, Vol. 125, 1998, pp. 679-690.

[17] F. Lolicato, R. Marino, M. P. Paronetto, M. Pellegrini, S. Dolci, R. Geremia and P. Grimaldi, "Potential Role Nanos3 in Maintaining the Undifferentiated Spermatogonia Population," *Developmental Biology*, Vol. 313, No. 2, 2008, pp. 725-738. doi:10.1016/j.ydbio.2007.11.011

[18] J. Jaruzelska, M. Kotecki, K. Kusz, A. Spik, M. Firpo and R. A. R. Pera, "Conservation of a Pumilio-Nanos Complex from *Drosophila* Germ Plasm to Human Germ Cells," *Development Genes and Evolution*, Vol. 213, No. 3, 2003, pp. 120-126.

[19] B. Ginter-Matuszewska, K. Kusz, A. Spik, D. Grzeszkowiak, A. Rembiszewska, J. Kupryjanczyk and J. Jaruzelska, "NANOS1 and PUMILIO2 Bind MicroRNA Biogenesis Factor GEMIN3, within Chromatoid Body in Human Germ Cells," *Histochemistry and Cell Biology*, Vol. 136, No. 3, 2011, pp. 279-287. doi:10.1007/s00418-011-0842-y

[20] B. Ginter-Matuszewska, A. Spik, A. Rembiszewska, C. Koyias, J. Kupryjańczyk and J. Jaruzelska, "The SNARE-Associated Component SNAPIN Binds PUMILIO2 and NANOS1 Proteins in Human Male Germ cells," *Molecular Human Reproduction*, Vol. 15, No. 3, 2009, pp. 173-179. doi:10.1093/molehr/gap004

[21] Y. Onohara and S. Yokota, "Expression of DDX25 in Nuage Components of Mammalian Spermatogenic Cells: Immunofluorescence and Immunoelectron Microscopic Study," *Histochemistry and Cell Biology*, Vol. 137, No. 1, 2012, pp. 37-51. doi:10.1007/s00418-011-0875-2

[22] H. Yonetamari, Y. Onohara and S. Yokota, "Localization of BRUNOL2 in Rat Spermatogenic Cells Revealed by Immunofluorescence and Immunoelectron Microscopic Techniques," *Open Journal of Cell Biology*, Vol. 2, No. 2, 2012, pp. 11-20. doi:10.4236/ojcb.2012.22002

[23] Y. Onohara, T. Fujiwara, T. Yasukochi, M. Himeno and S. Yokota, "Localization of Mouse Vasa Homolog Protein in Chromatoid Body and Related Nuage Structures of Mammalian Spermatogenic Cells during Spermatogenesis," *Histochemistry and Cell Biology*, Vol. 133, No. 6, 2010, pp. 627-639. doi:10.1007/s00418-010-0699-5

[24] C. M. Haraguchi, T. Mabuchi and S. Yokota, "Localization of a Mitochondrial Type of NADP-Dependent Isocitrate Dehydrogenase in Kidney and Heart of Rat: An Immunocytochemical and Biochemical Study," *Journal of Histochemistry and Cytochemistry*, Vol. 51, No. 2, 2003, pp. 215-226. doi:10.1177/002215540305100210

[25] C. de Roe, P. J. Courtoy and P. Baudhuin, "A Model of Protein-Colloidal Gold Interactions," *Journal of Histochemistry and Cytochemistry*, Vol. 35, No. 11, 1987, pp. 1191-1198. doi:10.1177/35.11.3655323

[26] D. Rickwood and T. C. Ford, "Preparation and Fractionation of Nuclei, Nucleoli and Deoxyribonucleoproteins," In: D. Rickwood Ed., *Iodinated Density Gradient Media, Practical Approach*, IRL Press, Oxford, Washington DC, 1983, pp. 69-89.

[27] L. D. Russell, R. A. Ettlin, A. S. P. Hikimand and E. D. Clegg, "Histological and Histopathological Evaluation of Testis," Cache River Press, Clearwater, 1990.

[28] L. Russell and B. Frank, "Ultrastructural Characterization of Nuage in Spermatocytes of the Rat Testis," *Anatomical Record*, Vol. 190, No. 1, 1978, pp. 79-97. doi:10.1002/ar.1091900108

[29] Y. Clermont, R. Oko and L. Hermo, "Immunocytochemical Localization of Proteins Utilized in the Formation of Outer Dense Fibers and Fibrous Sheath in Rat Spermatids: An Electron Microscope Study," *Anatomical*

Record, Vol. 227, No. 4, 1990, pp. 447-457. doi:10.1002/ar.1092270408

[30] K. M. Kusz, L. Tomczyk, M. Sajek, A. Spik, A. Latos-Bielenska, P. Jedrzejczak, L. Pawelczyk and J. Jaruzelska, "The Highly Conserved NANOS2 Protein: Testis-Specific Expression and Significance for the Human Male Reproductionm," *Molecular Human Reproduction*, Vol. 15, No. 3, 2009, pp. 165-171. doi:10.1093/molehr/gap003

[31] C. M. Haraguchi, T. Mabuchi, S. Hirata, T. Shoda, K. Hoshi, K. Akasaki and S. Yokota, "Chromatoid Bodies: Aggresome-Like Characteristics and Degradation Sites for Organelles of Spermiogeniccells," *Journal of Histochemistry and Cytochemistry*, Vol. 53, No. 4, 2000, pp. 455-465. doi:10.1369/jhc.4A6520.2005

[32] C. Barreau, L. Paillard, A. Méreau and H. B. Osborne, "Mammalian CELF/Bruno-Like RNA-Binding Proteins: Molecular Characteristics and Biological Functions," *Biochemie*, Vol. 88, No. 5, 2006, pp. 515-525. doi:10.1016/j.biochi.2005.10.011

[33] Y. Sheng, C.-H. Tsai-Morris, R. Gutti, Y. Maeda and M. L. Dufau, "Gonadotropin-Regulated Testicular RNA Helicase (GRTH/Ddx25) Is a Transport Protein Involved in Gene-Specific mRNA Export and Protein Translation during Spermatogenesis," *Journal of Biological Chemistry*, Vol. 281, No. 46, 2006, pp. 35048-35056. doi:10.1074/jbc.M605086200

2

Regeneration from a Cell Biological Perspective—Fascinating New Insights and Paradigms*

Bor Luen Tang
Department of Biochemistry, Yong Loo Lin School of Medicine, National University Health System,
NUS Graduate School for Integrative Sciences and Engineering, National University of Singapore, Singapore, Singapore
Email: bchtbl@nus.edu.sg

ABSTRACT

Regeneration research is more focused on translational values. However, lying at its very foundation is an understanding of how tissues and organs repair and renew themselves at the cellular level. The past decade has witnessed paradigm changing advances in regenerative biology, many of these stems from novel insights into stemness, pluripotency, cell death and their related intra- and inter-cellular biochemical and molecular processes. Some of these new insights are highlighted in the paragraphs that follow. We now have a much better understanding of how regeneration occurs in lower organisms. We have also discovered tools and means of nuclear reprogramming to generate induced pluripotency and changes in cell fate in mammalian models. With further research, there is reasonable hope that various obstacles of regeneration in humans can be better understood and tackled. As regeneration research enters a new era, *CellBio* welcomes timely review articles and original papers on the theme of "The Cell Biology of Regeneration".

Keywords: Inflammation; Induced Pluripotent Stem (iPS) Cells; Progenitor/Stem Cells; Regeneration; Reprogramming; Wnt

1. Introduction

The ability to regenerate injured tissues or organs, as well as rejuvenation of the senesced or aged, has been an elusive goal of ancient alchemy and modern biomedicine alike. Biologists have marveled at the ability of plants and lower animals to regenerate. Planarians and cnidarians could regenerate entire organism from small body fragments, or even dissociated single cells. However, for more complex animals, this regenerative capacity is apparently attenuated, or completely loss. While organ and limb regeneration are still readily observed in fishes, reptiles and amphibians, this almost never occurs to any significant extent in mammals. Even at the cellular level, one resigns to the vast amount of data demonstrating that whole tissues aside, most terminally differentiated cell types, such as brain neurons and skeletal muscle fibers, simply do not regenerate. While this latter notion remains accurate, the past few years have witnessed multiple advances that are paradigm changing in terms of our understanding of regeneration from a cell biological perspective. The following paragraphs highlight a few aspects of the novel insights associated with adult animal regeneration that have become clear after the turn of the century.

2. From cNeoblasts to Blastema Stem Cells-Endogenous Pluripotent and Multipotent Stem Cells Enable Regeneration

Whether complex tissues could be regenerated appears to depend primarily in the availability of stem cells, their relative lineage differentiation potency, and their state of quiescence (and how this latter state could be changed when the need for regeneration arises). At least in theory, stem/progenitor cells required for regeneration could exist as an ever present pool, or dedifferentiated from differentiated cells. To be able to account for their regenerative capacity at the organismal level, pluripotent, if not totipotent, stem cell types must exist in adult planarians and cnidarians, and for that matter widely distributed throughout the adult organism, Indeed, a population of undifferentiated adult dividing cells, the neoblasts, has been identified to be responsible for planarian regenerative capacity. Using a clonal analysis approach of lethal ionizing radiation followed by single-cell transplan-

*A preface to CellBio's thematic review series on "The Cell Biology of Regeneration".

tation in Schmidtea mediterranea, planarian clonogenic neoblasts (cNeoblasts) was shown to be able to differentiated to almost all known postmitotic cell types throughout the body. Intriguingly, single transplanted cNeoblasts could restore regeneration in a lethally irradiated worm [1]. On the other hand, tissue pluripotency in the cnidarian Hydra involves three independent cell lineages form the body of the polyp, namely epithelial stem cells from the ectodermal and endodermal layers respectively, as well as interstitial stem cells [2]. The epithelial stem cells are pluripotent [3] but the interstitial cells at best multipotent.

At a level of more modest regenerative capacity, reptiles and amphibians are able to generate severed limbs or other appendages. This is no mean feat as vertebrate appendages are composed of a mixture of tissue types from multiple germ layers. Regeneration in this regard is also dependent on resident stem cells [4]. The process begins with the formation of a blastema at the site if injury or amputation, which is a collection of progenitor cells that appear to be homogenous, but these are at best multipotent, with a good degree of lineage potential restriction [5]. The equivalent of a pluripotent planarian eNeoblast is most likely either completely absent in adult vertebrate tissues, or is not available in any significant numbers that would enable regeneration at a more massive scale. In mammals, limb regeneration is further reduced to the ability to regenerate digit tips, and this was recently shown to occur via ectodermal and mesodermal fate-restricted progenitors that regenerate their own lineages within the digit tip [6]. It is speculative at the moment as to why regenerative capacity reduces with complexity, or that a phenotype of having pluripotent stem cells at stock was selected against in higher vertebrates. One reason could be the difficulty in the maintenance of a large amount of pluripotent stem cells quiescent and the increase probability of malignant transformation. Understanding more about how lower organisms use their endogenous stem cells to regenerate may provide clues as to how endogenous stem cells in various niches of the adult human could be harness (or activated) to aid regeneration.

3. Rising from the Ashes of the Dead

Injury often causes massive cell death. Attraction of immune cells to the site of injury underlies the associated inflammatory responses, which together create a non-conducive post-injury environment that is conventionally viewed to be hostile, impairing the survival of spared cells as well as anti-proliferative against regenerating cells. This view may be overtly oversimplified, as recent work suggest that both apoptosis (or more accurately, programmed cell death) and inflammation play important roles in triggering regeneration from cnidarian to vertebrates. Midgastric bisection of Hydra precipitates a rapid wave of apoptosis and transient release of Wnt3 among interstitial cells at the head regenerating end, and the latter activates the canonical Wnt/β-catenin pathway in neighbouring cycling cells to enhance cell cycle progression [7]. This sort of apoptotic cell-induced compensatory cell proliferation has also been documented in regeneration models of higher organisms, including Drosophila wing disc regeneration [8] and tail regeneration in the tadpoles of Xenopus laevis [9].

The adult mammalian brain is a well-known organ where regeneration is particularly restrictive. In fact, it was believed for a long time prior to the identification and characterization of adult neurogenic regions that adult neurogenesis (*i.e.* the formation of new neurons from progenitors) [10] does not occur in the mammalian brain. The much simpler fish brain, on the other hand, could regenerate to a significant degree. Neuroinflammation characterizing cases of acute ischemic or traumatic injuries, as well as more chronic neurodegenerative diseases in human brain pathology, is widely recognized as a major barrier to regeneration of any kind. Interestingly, recent findings points to inflammation as being required and sufficient for enhancing the proliferation of neural progenitors and their subsequent neurogenesis in the adult zebra fish brain [11]. In connection with apoptosis-driven regeneration discussed above, Wnt signalling appears to be a key pathway in balancing brain damage and repair. Exogenous Wnt3a injected into mouse striatum was recently shown to enhance neurogenesis and significantly functional recovery after ischemic injury [12]. Wnt signalling components are only present in immune cells as well as brain glia cells in adult mammals, and the crosstalk between these cells in a post-injury inflammatory setting, particularly in influencing neurogenesis [13,14], could be exploited for therapeutic intervention purposes. Regenerative capacities are not only conserved between lower vertebrates and mammals in terms of signalling. It is worth noting that both neurogenic adult neural progenitors in fish and mammals have a similar morphological phenotype and niche—they all appear to be derived from ventricular radial glia [15,16].

4. Starting Over-Nuclear Reprogramming to Pluripotency, Multipotency or Alternative Fates

Erasure of epigenetic markings of differentiation and aging, as well as induction of pluripotency, occur naturally during reproduction, be it in the case of a budded Saccharomyces. cerevisiae daughter cell or after the fusion of a spermatozoa and an ovum in humans. The ability of an enucleated ovum to reprogram somatic cell nuclei to a state of pluripotency underlies the promise of somatic cell nuclear transfer (SCNT) for the generation of embryonic stem cells, and thus materials for autologous transplantation (or for cloning). Just as the community begins to feel that perhaps efficient SCNT-based reprogramming is for some reason unachievable for primates and humans, the discovery of induced pluripotency [17] literally changed overnight the way many approach the subject. The technology is based on a deceptively simple concept that nuclear reprogramming could be achieved by the introduction and expression of the four Yamanaka factors (Oct3/4, Sox2, Klf4 and c-Myc), or a subset of these in combination with others genes/compounds, into easily sampled somatic cells such as fibroblasts or keratinocytes. These genes initiate a cascade of changes in genetic and epigenetic profiles, converting differentiated somatic cells these over a period of time into pluripotent stem cells [18]. Work on or related to induced pluripotent stem (iPS) cells has now amassed more than 4500 PUBMED entries, and related new findings are being made at an unprecedentedly fast pace.

Of particular therapeutic interest is the potential of iPS methods to generate individual-specific autologous cells or tissues that are safe for grafting. In accordance to the generalized notion that grafting differentiated cells runs a lower risk of tumorigenesis, researchers quickly develop methods of direct reprogramming of fibroblast into differentiated cell types of other lineages, such as neurons [19], cardiomyocytes [20] or endothelial cells [21] without passage through the undifferentiated pluripotent iPS stage. Modifications of factors and culture methods allowed the generation of multipotent neural progenitors [22-24] and hematopoietic progenitors [25]. Beyond providing therapeutic materials, the seemingly limitless lineage conversion to either fully differentiated cell types or more immediate progenitors from clinically accessible cells like fibroblasts will also greatly advance studies on disease etiology and development. Granted that nuclear reprogramming may be incomplete in the case of iPS cells and residual epigenetic memories of the cell of origin may limit their usefulness, the paradigm shift in terms of research approach using iPS-based methods has clearly revolutionize regenerative biology.

5. Acknowledgements

The author is supported by NUS Graduate School for Integrative Sciences and Engineering (NGS), and declares no conflict of interest.

REFERENCES

[1] D. E. Wagner, I. E. Wang and P. W. Reddien, "Clonogenic Neoblasts Are Pluripotent Adult Stem Cells That Underlie Planarian Regeneration," *Science*, Vol. 332, No. 6031, 2011, pp. 811-816. doi:10.1126/science.1203983

[2] T. C. G. Bosch, F. Anton-Erxleben, G. Hemmrich and K. Khalturin, "The Hydra Polyp: Nothing but an Active Stem Cell Community," *Development, Growth & Differentiation*, Vol. 52, No. 1, 2010, pp. 15-25. doi:10.1111/j.1440-169X.2009.01143.x

[3] J. Wittlieb, K. Khalturin, J. U. Lohmann, F. Anton-Erxleben and T. C. G. Bosch, "Transgenic Hydra Allow *in Vivo* Tracking of Individual Stem Cells during Morphogenesis," *Proceedings of the National Academy of Sciences of USA*, Vol. 103, No. 16, 2006, pp. 6208-6211. doi:10.1073/pnas.0510163103

[4] J. I. Morrison, S. Lööf, P. He and A. Simon, "Salamander Limb Regeneration Involves the Activation of a Multipotent Skeletal Muscle Satellite Cell Population," *Journal of Cell Biology*, Vol. 172, No. 3, 2006, pp. 433-440. doi:10.1083/jcb.200509011

[5] K. Tamura, S. Ohgo and H. Yokoyama, "Limb Blastema Cell: A Stem Cell for Morphological Regeneration," *Development, Growth & Differentiation*, Vol. 52, No. 1, 2010, pp. 89-99. doi:10.1111/j.1440-169X.2009.01144.x

[6] J. A. Lehoczky, B. Robert and C. J. Tabin, "Mouse Digit Tip Regeneration Is Mediated by Fate-Restricted Progenitor Cells," *Proceedings of the National Academy of Sciences of USA*, Vol. 108, No. 51, 2011, pp. 20609-20614. doi:10.1073/pnas.1118017108

[7] S. Chera, L. Ghila, K. Dobretz, Y. Wenger, C. Bauer, W. Buzgariu, J. C. Martinou and B. Galliot, "Apoptotic Cells Provide an Unexpected Source of Wnt3 Signaling to Drive Hydra Head Regeneration," *Developmental Cell*, Vol. 17, No. 2, 2009, pp. 279-289. doi:10.1016/j.devcel.2009.07.014

[8] H. D. Ryoo, T. Gorenc and H. Steller, "Apoptotic Cells Can Induce Compensatory Cell Proliferation through the JNK and the Wingless Signaling Pathways," *Developmental Cell*, Vol. 7, No. 4, 2004, pp. 491-501. doi:10.1016/j.devcel.2004.08.019

[9] A. S. Tseng, D. S. Adams, D. Qiu, P. Koustubhan and M. Levin, "Apoptosis Is Required during Early Stages of Tail Regeneration in *Xenopus laevis*," *Developmental Biology*, Vol. 301, No. 1, 2007, pp. 62-69. doi:10.1016/j.ydbio.2006.10.048

[10] C. Zhao, W. Deng and F. H. Gage, "Mechanisms and Functional Implications of Adult Neurogenesis," *Cell*, Vol. 132, No. 4, 2008, pp. 645-660. doi:10.1016/j.cell.2008.01.033

[11] N. Kyritsis, C. Kizil, S. Zocher, V. Kroehne, J. Kaslin, D. Freudenreich, A. Iltzsche and M. Brand. "Acute Inflammation Initiates the Regenerative Response in the Adult Zebrafish Brain," *Science*, Vol. 338, No. 6112, 2012, pp. 1353-1356. doi:10.1126/science.1228773

[12] A. Shruster, T. Ben-Zur, E. Melamed and D. Offen, "Wnt Signaling Enhances Neurogenesis and Improves Neurological Function after Focal Ischemic Injury," *PloS One*, Vol. 7, No. 7, 2012, e40843. doi:10.1371/journal.pone.0040843

[13] T. Kuwabara, J. Hsieh, A. Muotri, G. Yeo, M. Warashina, D. C. Lie, L. Moore, K. Nakashima, M. Asashima and F. H. Gage, "Wnt-Mediated Activation of NeuroD1 and Retro-Elements during Adult Neurogenesis," *Nature Neuroscience*, Vol. 12, No. 9, 2009, pp. 1097-1105.

doi:10.1038/nn.2360

[14] N. C. Inestrosa and E. Arenas, "Emerging Roles of Wnts in the Adult Nervous System," *Nature Reviews Neuroscience*, Vol. 11, No. 2, 2010, pp. 77-86. doi:10.1038/nrn2755

[15] F. T. Merkle, A. D. Tramontin, J. M. García-Verdugo and A. Alvarez-Buylla, "Radial Glia Give Rise to Adult Neural Stem Cells in the Subventricular Zone," *Proceedings of the National Academy of Sciences of USA*, Vol. 101, No. 50, 2004, pp. 17528-17532. doi:10.1073/pnas.0407893101

[16] V. Kroehne, D. Freudenreich, S. Hans, J. Kaslin and M. Brand, "Regeneration of the Adult Zebrafish Brain from Neurogenic Radial Glia-Type Progenitors," *Development*, Vol. 138, No. 22, 2011, pp. 4831-4841. doi:10.1242/dev.072587

[17] K. Takahashi and S. Yamanaka, "Induction of Pluripotent Stem Cells from Mouse Embryonic and Adult Fibroblast Cultures by Defined Factors," *Cell*, Vol. 126, No. 4, 2006, pp. 663-676. doi:10.1016/j.cell.2006.07.024

[18] J. M. Polo, E. Anderssen, R. M. Walsh, B. A. Schwarz, C. M. Nefzger, S. M. Lim, M. Borkent, E. Apostolou, S. Alaei, J. Cloutier, O. Bar-Nur, S. Cheloufi, M. Stadtfeld, M. E. Figueroa, D. Robinton, S. Natesan, A. Melnick, J. Zhu, S. Ramaswamy and K. Hochedlinger, "A Molecular Roadmap of Reprogramming Somatic Cells into iPS Cells," *Cell*, Vol. 151, No. 7, 2012, pp. 1617-1632. doi:10.1016/j.cell.2012.11.039

[19] T. Vierbuchen, A. Ostermeier, Z. P. Pang, Y. Kokubu, T. C. Südhof and M. Wernig, "Direct Conversion of Fibroblasts to Functional Neurons by Defined Factors," *Nature*, Vol. 463, No. 7284, 2010, pp. 1035-1041. doi:10.1038/nature08797

[20] M. Ieda, J. D. Fu, P. Delgado-Olguin, V. Vedantham, Y. Hayashi, B. G. Bruneau and D. Srivastava, "Direct Reprogramming of Fibroblasts into Functional Cardiomyocytes by Defined Factors," *Cell*, Vol. 142, No. 3, 2010, pp. 375-386. doi:10.1016/j.cell.2010.07.002

[21] A. Margariti, B. Winkler, E. Karamariti, A. Zampetaki, T. N. Tsai, D. Baban, J. Ragoussis, Y. Huang, J. D. J. Han, L. Zeng, Y. Hu and Q. Xu, "Direct Reprogramming of Fibroblasts into Endothelial Cells Capable of Angiogenesis and Reendothelialization in Tissue-Engineered Vessels," *Proceedings of the National Academy of Sciences of USA*, Vol. 109, No. 34, 2012, pp. 13793-13798. doi:10.1073/pnas.1205526109

[22] M. Thier, P. Wörsdörfer, Y. B. Lakes, R. Gorris, S. Herms, T. Opitz, D. Seiferling, T. Quandel, P. Hoffmann, M. M. Nöthen, O. Brüstle and F. Edenhofer, "Direct Conversion of Fibroblasts into Stably Expandable Neural Stem Cells," *Cell Stem Cell*, Vol. 10, No. 4, 2012, pp. 473-479. doi:10.1016/j.stem.2012.03.003

[23] E. Lujan, S. Chanda, H. Ahlenius, T. C. Südhof and M. Wernig. "Direct Conversion of Mouse Fibroblasts to Self-Renewing, Tripotent Neural Precursor Cells," *Proceedings of the National Academy of Sciences of USA*, Vol. 109, No. 7, 2012, pp. 2527-2532. doi:10.1073/pnas.1121003109

[24] K. L. Ring, L. M. Tong, M. E. Balestra, R. Javier, Y. Andrews-Zwilling, G. Li, D. Walker, W. R. Zhang, A. C. Kreitzer and Y. Huang, "Direct Reprogramming of Mouse and Human Fibroblasts into Multipotent Neural Stem Cells with a Single Factor," *Cell Stem Cell*, Vol. 11, No. 1, 2012, pp. 100-109. doi:10.1016/j.stem.2012.05.018

[25] E. Szabo, S. Rampalli, R. M. Risueño, A. Schnerch, R. Mitchell, A. Fiebig-Comyn, M. Levadoux-Martin and M. Bhatia, "Direct Conversion of Human Fibroblasts to multilineage Blood Progenitors," *Nature*, Vol. 468, No. 7323, 2010, pp. 521-526. doi:10.1038/nature09591

3

Intracellular Notch1 May Induce a Conformational Change in CSL/DNA, without Forming ICN1/CSL/DNA Molecular Complex, *in Vitro*

Alexei A. Stortchevoi*
Department of Pathology, Yale University, New Haven, USA
Email: alexei.stortchevoi@yale.edu

ABSTRACT

Intracellular Notch (ICN) initiates DNA transcription in cooperation with CSL that acts as repressor in the absence of ICN. The ICN mediates recruitment of MAML protein, leading to the formation of minimal transcriptional complex, MAML/ICN/CSL/DNA. Crystal structure reveals that different conformations exist between the free (CSL/DNA) and bound (ICN/MAML/CSL/DNA) forms. The significance of this modulation of the CSL/DNA molecular complex can be better understood by experimental approaches that aim to elucidate the cause and timing of these events. There are four orthologues of human ICN (ICN1-4). We studied interactions between human full-length ICN1 and CSL/DNA without involvement of MAML, *in vitro*, and found that 1) the EMSA profile of CSL/DNA is altered in the presence of ICN1 as a consequence of an intrinsic change(s) in CSL/DNA, and not due to the formation of an ICN/CSL/DNA molecular complex; 2) ICN1 destabilizes CSL/DNA. These findings indicate that human ICN1 functions to modulate the CSL/DNA molecular complex for subsequent recruitment of MAML, and that modulated CSL/DNA cannot accommodate ICN1 in the absence of MAML. The latter in turn, implies that the formation of the MAML/ICN1/CSL/DNA is likely to be a collective event, wherein preassembly of MAML and ICN1 as a binary complex co-localizes at the CSL/DNA promoter site, or the MAML/ICN1/CSL complex is pre-assembled prior to binding to the promoter, rather than ICN1 arriving at CSL/DNA ahead of MAML and/or other associated transcription factors. The novel finding that ICN1 destabilizes the CSL/DNA complex opens new possibilities of transcriptional regulation by Notch.

Keywords: CBF1; CSL; Hes1 Promoter; Notch; RbpJ; Su(H)

1. Introduction

Notch is a transmembrane receptor protein that controls numerous developmental decisions through binding of ligands displayed on the surface of adjacent cells. Notch signaling pathway regulates cell fate decisions in neurogenesis [1], T cell development [2] and hypoxia [3] and has been implicated in multiple cross talks with major pathways that influence cell proliferation, differentiation, survival and migration [4]. Aberrations in Notch receptors have been linked to malignancies [5-9], and lately, the components of Notch signal transduction have been targeted in anti-cancer drug development [10-12].

Upon ligand binding, the Notch receptor is proteolytically cleaved at the membrane site, and intracellular Notch (ICN) migrates to the nucleus, where it interacts with various promoters to stimulate transcription in cooperation with transcription factor CSL (also known as RBP-J, CBF1, Su(H) and LAG-1), reviewed [13-15]. In the absence of ICN, CSL acts as a transcriptional repressor, bound at the promoter to "GTGGGAAA" CSL specific sequence. The ICN has not been found to bind directly to DNA but its RAM domain (RAM = RBP-J Associating Molecule) and Ankyrin domain (ANK) interact with CSL [16]. Transcriptional repression by CSL is reportedly achieved through recruitment of histone deacetylase complexes, SMRT/mSin3A/HDAC-1 or NCor/nSin3A/HDAC-1 [17], CIR/SAP30/HADC-2 [18] or SHARP/CtIP [19] to a gene promoter site and compacting of the chromatin. It is believed that ICN1 plays an important role in displacing the co-repressor complexes, allowing the chromatinized DNA to unfold and thus fa-

*Current address: Department of Genetics, Harvard Medical School, Center for Human Genetic Research, Massachusetts General Hospital's Richard B. Simches Research Center, Boston, USA.

cilitating transcription. This ICN function is supported by Mastermind-like, a human homologue of Drosophila protein (MAML) [20-22], Ski-interacting protein (SKIP) that binds both, CSL and ICN [23] and histone acetylase CBP/p300 [20]. MAML is also thought to control withdrawal of ICN1 from the transcription complex [20,24]. Other proteins involved in ICN-mediated transcription include negative regulator KyoT2 [25], transcriptional co-activator p300 [26], hypoxia-inducible fator 1 alfa (Hif-1a) [3,27], mediator protein Med220, cell-cycle depended kinase 7 (CDK7) and CDK8 [28], elongation factors FACT, Spt6, P-TEFb, reviewed [13]. While the number of transcription factors involved is extensive, and continues to grow, the sequence of the events leading to transcription and its turnover is far from been clearly understood.

One of the lingering questions surrounding ICN mediated transcription events is whether ICN binds independently to CSL/DNA, or is dependent upon other factors to form the ICN/CSL/DNA complex. Some data point to a higher probability of an assisted rather than independent ICN1 binding to CSL/DNA in mammals: Timed ChIP/PCR study shows that the arrival and the departure of ICN1 at the CSL-occupied *HES*1 promoter occurs within the same 30 min. interval that is associated with MAML1 and SKIP binding to CSL/DNA [28]. Crystal structure of the CSL/*HES*1 promoter fragment with the ANK domain of murine ICN1 and N-terminal peptide of MAML1 shows that mutual dependence of these proteins is required to maintain stability of the association [29]. When crystal structures are compared between CSL/DNA [30] and MAML/ICN1/CSL/DNA (harboring the RAMANK domain of mouse ICN1) [31], it reveals conformational change occurs in CSL/DNA; this observation raises the important question of whether the change in CSL/DNA structure is induced by ICN1 only, or via a synergistic interaction between ICN1 and MAML.

Human ICN1 is ~90% homologous to the mouse counterpart, with most of the discrepancies located within the C-terminal "Trans-Activation Domain" that is absent in the crystal structures, leaving the impact of interspecies differences open to speculation.

We analyzed the impact of full-length human recombinant ICN1 on CSL/DNA binding. The study was designed to include only the initial trio: human *HES*1 promoter DNA, full-length human CSL and full-length human ICN1. We show that ICN1 does not form stable association with CSL/DNA in the absence of MAML (at the conditions used) but induces a change in the CSL/DNA molecular complex that alters its electromobility in EMSA. Another effect of the ICN1is that it destabilizes the bond between the CSL protein and *Hes*1 promoter, and that the degree of CSL dissociation is dependent on the length of DNA fragment that is used for the EMSA.

The results of these experiments suggest that human ICN1 can induce a conformational change(s) in CSL/DNA, and recruit additional DNA sequence(s) that maintain CSL binding to the core promoter during the interaction. Our data complement and expand the existing knowledge of ICN1 interaction with CSL/DNA; we demonstrate that use of the full-length human ICN1 (and not the RAMANK truncated form of murine ICN), is capable of conformational change(s) in CSL/DNA. In addition, we show that ICN1 alone is capable of inducing the change (likely, to accommodate MAML binding to CSL/DNA) but cannot form stable association with CSL/DNA without MAML (likely because the new conformation of CSL/DNA favors binding of ICN1/MAML but not ICN1). The results imply that in human species binding of MAML and ICN1 to CSL/DNA at *Hes*1 promoter is a simultaneous, mutually assisted event. Another interesting finding of this work is destabilization of the CSL/DNA complex by Notch, which potentially may play a role in negative regulation of transcription.

2. Materials and Methods

2.1. Antibodies and Transfection Reagents

Anti-Flag M2 antibody (F1804), anti-Flag M2 antibody HRP conjugate (A8592), anti-Flag M2 affinity gel (A2220), anti-HA monoclonal antibody, clone HA-7, HRP conjugate (H6533), anti-HA monoclonal antibody (H9658), anti-polyHistidine monoclonal antibody (H1029) were obtained from Sigma (Sigma Chemicals, St. Louis, MO, USA); anti-6xHis tag antibody HRP conjugate (ab1187) was obtained from Abcam (Abcam, Cambridge, UK), anti-ICN1 polyclonal serum (goat) was prepared in the lab. DMEM and RPMI media, Lipofectamine 2000 and Lipofectamine TM were purchased from Invitrogen (Invitrogen, Carisbad, CA, USA).

2.2. DNA Sequences

For EMSA probes, we used fragments of *HES1* promoter (ENSEMBL genome ID: ENST00000232424, human chromosome 3: 195,853,833 to 195,853,892). The basic 59 bp probe imitated −101 to −42 fragment of the transcription start (−337 to −278 before the ATG translation initiation codon) of that sequence. The purified DNA duplexes were purchased from Eurofins MWG (Huntsville, AL, USA, oligo-us@eurofins.com).

2.3. CSL and ICN1 Protein Expression in *E. coli*

Recombinant ICN1 and CSL genes and cloning sites can be viewed in the supplement to this article.

The genes were cloned in pET28a vector (Novagen, Madison, WI, USA) and expressed in BL21.DE3 strain of *E. coli*. Expression of the ICN1 and 6His-ICN1-6His recombinant proteins: 0.4 mM IPTG at OD600 = 0.8, 3 hours, 37˚C after induction; Expression of CSL, CSL-6His Flag-CSL and Flag-CSL-6His recombinant proteins: 0.4 mM IPTG at OD600 = 0.8 followed by overnight growth at 16˚C: at such conditions ~80% of the recombinant CSL protein was soluble. The 6His- tagged proteins were purified using Ni-NTA Agarose (Invitrogen R901-01 or Qiagen 30230) according to manufacturers' instructions at native conditions. After purification, proteins were dialyzed in phosphate buffer (50 mM NaH_2PO_4 pH 8.0, 500 mM NaCl) snap-frozen and stored at −80˚C or stored at −20˚C in if dialyzed in phosphate buffer in presence of 40% glycerol.

2.4. CSL and ICN1 Protein Expression in HEK293T Cells

HEK293T cells were transiently transfected in a 10 cm tissue culture dish with 2 - 10 µg of a recombinant vector DNA, using Lipofectamin 2000 or Lipofectamin TM reagents. Cells were harvested for protein extract preparation in 24 - 72 hours: washed with an ice-cold PBS, then scraped from the plate and transferred to a 15 ml Falcon polypropylene tube. After pelleting by centrifugation, the cells were lysed by osmotic shock on ice for 10 min. in 1 ml/10^7 cells of the Hypotonic lysis buffer (10 mM TrisHCl, pH 7.4, 10 mM NaCl, 2 mM EDTA, 0.5% Triton X-100) in the presence of 1 mM PMSF and a 1:50 dilution of protease inhibitor cocktail (Sigma P8340). After lysis, the NaCl concentration was adjusted to 150 mM and the cell extract was cleared of debris by centrifugation at 4˚C for 15 minutes at maximum speed on a microcentrifuge, the extracts were taken for EMSA analysis immediately or snap-frozen in aliquots and stored at −80˚C prior to single use. Protein expression in the extracts was confirmed by Western blotting.

2.5. ChIP/PCR Assay

Chromatin immunoprecipitation was adapted from the published protocol [32] with adjustment for antibody-agarose conjugates (precipitation with protein A was not necessary). Small fractions of cell culture were scraped prior to cross-linking with formaldehyde for analysis of protein expression by SDS-PAGE/Western blotting. PCR Primers: CCCTGGCTCCAAAAGAAATAGAC and GAAGTTTCACACGAGCCGTTCG. After running PCR products in 1.5% TAE agarose gel, quantification of the bands was done with GeneTools Version 4.01 software (Syngene (A Division of Synoptics Ltd), Beacon House, Cambridge, CB41TF, England).

2.6. EMSA

The EMSA Accessory Kit (Novagen 71282-3) was used for sample preparation according to manufacturer's instructions. (4× EMSA binding buffer = 400 mM KCl, 8 mM HEPES, 2 mM DTT, 0.8 mM EDTA, 80% glycerol, pH 8.0; 1× EMSA binding buffer was supplemented with sonicated salmon sperm DNA (125 - 200 ng) and Poly (dI-dC) (dI-dC) (0.01U)). EMSA sample (20 ml) contained, unless specified otherwise, 300 nM of DNA probe (10,000 - 100,000 cpm of ^{32}P activity) 2 - 10 mg of total protein extract or 100 - 400 nM (100 - 300 ng)/reaction of purified CSL or ICN1. The DNA/CSL mix was pre-incubated on ice for 10 minutes prior to loading in gel. After addition of ICN1, the sample was loaded in gel and ran immediately or pre-incubated on ice for up to 30 minutes, and then ran, at 7.5 V/cm for 2 - 8 hours (depending on the required resolution) at room temperature. The 15 × 20 cm 1 mm - 1.5 mm thick 4% - 6% TBE acrylamide:bis (29:1) gel was pre-run for 30 - 40 min prior to loading the samples. After the run was finished, the gel was taken off the glass plates, wrapped in plastic film and exposed to autoradiography film or to a phosphorimaging screen.

2.7. EMSA/Western Blotting

The DNA and proteins from EMSA gel were transferred on to a PVDF membrane (Immobilon-P Cat. No: IPVH 00010 Millipore, Billerica, MA, USA), using the optimized protocol for semi-dry transfer: Three transfer buffers, containing 10% methanol were: anode buffer I (0.3 M Tris pH 10.4), anode buffer II (25 mM Tris pH 10.4), and cathode buffer (25 mM Tris and 40 mM e-amino-caproic acid, pH 9.4). The transfer membrane, sandwiched between transparent plastic sheets, was then exposed to a film (2 hours at −80˚C with enhancer screen), or a phosphorimaging screen, with reference points marked at the membrane. After radioactivity exposure, Western blotting analysis was done in a single step, using an HRP-conjugated (anti-Flag, anti-HA or anti-6His) antibody.

2.8. Phosphorimaging Analysis

Special care was taken to ensure equal distribution of CSL and ^{32}P-labeled DNA probe between the "No ICN1" and "+ICN1" samples prior to loading. After each run, the wet EMSA gel was placed between plastic sheets and exposed to ^{32}P sensitized phosphorimaging screen for 2 - 6 hours. Image reading and quantification was done with Image Quant software (Molecular Dynamics, Foster City, CA).

2.9. DNase Footprinting Analysis

20 µl of EMSA sample in EMSA buffer containing 10,000 - 20,000 cpm ^{32}P-labeled DNA probe was treated with 1 - 10 u of DNase I for 5 - 10 seconds prior to ex-

traction once with phenol (volume adjusted to 100 µl), once-with phenol/chloroform, and once-with chloroform. DNA was ethanol precipitated with added 4 µg glycogen (as carrier). Pellets were re-dissolved in 20 µl of sequencing buffer and 2 - 4 µl per lane were taken for analysis in a sequencing gel.

3. Results

3.1. Testing of Recombinant CSL and ICN1 in EMSA Shows That ICN1 Has Supershifting and Weakening Effects on CSL/DNA Band

DNA probes for Electromobility Shift Assay (EMSA), used in this work, represented the fragments of human *HES*1 promoter, where it contains two consensus CSL binding sites [33]. The 59 bp wild type (WT) probe, "59 bp CPS", referred to as CPS for CSL Paired Sites [34] had the two original sites: one site—GTGGGAAA—was closer to 5'-end of the fragment and designated "5'CS" (CS = CSL Site)," and the other—TTTCACAC (antiparallel)—was closer to 3'-end of the fragment and designated "3'CS". Derived from the "59 bp CPS" probe were two shorter WT probes: "18 bp 5'CS" contained only one (5'CS) CSL site [29] and "21 bp No CS" contained no CSL sites. There were also three full-length mutant probes: "59 bp 5'CS" and "59 bp 3'CS" had only one alternative CSL site-active and another-inactivated by mutations, and "59 bp No CS", where both sites were inactivated (**Figure 1(a)**).

Using the WT and the mutant 59 bp probes, we confirmed CSL site-specificity of DNA shifts in EMSA with HEK293T cell extract overexpressing untagged full-length recombinant human CSL (**Figure 1(b)**, right panel), and titrated the amount per reaction of bacterially expressed purified CSL-6His to comparable results (**Figure 1(b)**, left panel). Consistent with the number of active CSL sites per probe, "59 bp CPS" produced two bands in the shift, corresponding to one and two CSL molecules bound to a single DNA probe (**Figure 1(b)** lane 1); "59 bp 5'CS" and "59 bp 3'CS" each produced single band (**Figure 1(b)** lanes 2 and 4); and "59 bp No CS"-produced no EMSA shifts at all (**Figure 1(b)** lane 3). Having confirmed experimentally that the mutations eliminated CSL specificities of the CSL sites, we no longer used "59 bp No CS" probe as negative control but used the "21 bp No CS", instead.

After confirming the binding of the recombinant CSL to DNA in EMSA, we tested "ICN1+DNA" mixtures. The ICN1 produced no specific EMSA shifts with any of the CS-containing probes (not shown) but when added to pre-incubated CSL/DNA, it created visible impacts: 1) the primary CSL/DNA shift was retarded (**Figure 1(c)** lane 2) and 2) the shift was either weakened (as with "59

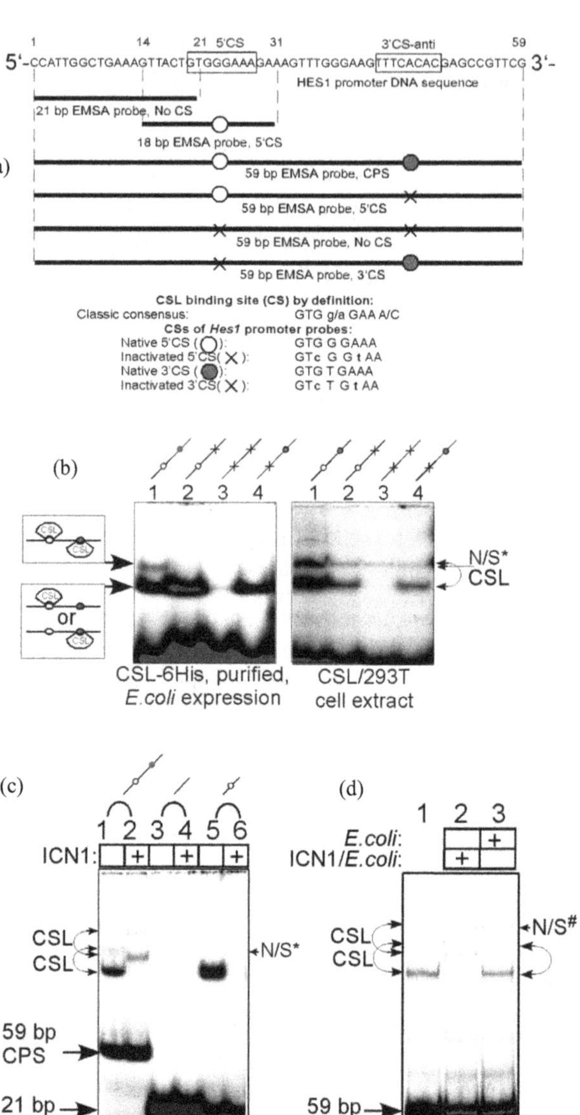

Figure 1. Initial EMSA with *Hes*1 promoter probes. (a) Schematic EMSA probes are linked by dotted lines with the 59 bp reference sequence. The CSL binding sites (CSs)-boxed in the sequence-are depicted as circular shapes in the diagram: Clear-for active 5'CS; filled-for active 3'CS. Inactivated CSs are depicted as X-shapes instead of circular. The original CS sequences and the inactivating mutations (substituted nucleotides—in small caps) are shown below, referred to consensus sequence [33]; (b) Testing the 59 bp probes (shown schematically above the EMSA lanes) with 100 nM purified CSL (left panel) and with 8 mg of CSL/293T extract (right panel). Diagram on the left of EMSA picture depicts putative CSL/DNA complexes. N/S*-"Non-specific" DNA-binding activity of 293T extract; (c) EMSA of "DNA + 8 mg of CSL/293T extract" and "DNA/8 mg of CSL/293T extract + 200 nM of 6His-ICN1-6His" sample pairs with different DNA probes, depicted above; (d) EMSA of 59 bp CPS probe with soluble bacterial extracts (5 mg) expressing untagged CSL and ICN1. N/S#-Non-specific DNA-binding activity of *E.coli* BL21.DE3 extract.

bp CPS" in **Figure 1(c)** lane 2) or completely disap peared (as with "18 bp 5'CS" in **Figure 1(c)** lane 6). To exclude artifact, potentially brought by 6Histag of recombinant ICN1, we replaced the purified 6His-ICN1-6His in the reaction by soluble *E. coli* extract, overexpressing untagged ICN1 (ICN1/*E.coli*): the "ICN1/*E.coli"* extract produced the same effects as the purified 6His-ICN1-6His protein (**Figure 1(d)** lane 2) whereas *E.coli* extract without ICN1 expression (*E. coli*) had no impact (**Figure 1(d)** lane 3).

3.2. Supershifting of CSL/DNA Band in EMSA Is Not Caused by Formation of ICN1/CSL/DNA Complex, Because ICN1 Is Absent from the Shift

The retardation of primary EMSA shift in "+ICN1" sample could reflect the formation of ICN1/CSL/DNA molecular complex with lower electromobility mobility due to its larger size than CSL/DNA, but could also be a consequence of an allosteric modification of CSL/DNA by ICN1, where modified form of CSL/DNA would have lower electromobility in EMSA. To find out what was really the case, we took three approaches to detect whether ICN1 was present in the secondary shift.

Firstly, we tried supershifting with monoclonal anti-His and polyclonal anti-ICN1 antibodies. Both antibodies were proven in Western blotting but neither of them produced any effect on the secondary CSL/DNA shift in EMSA (**Figure 2(a)** lanes 5 and 7), which would be the expected if ICN1 formed the ICN1/CSL/DNA molecular complex. At the same time, supershifting worked efficiently with anti-Flag antibody, confirming the presence of Flag-CSL in the primary and the secondary shifts (**Figure 2(a)** lanes 3 and 6).

Secondly, we used sequentially decreasing concentrations of ICN1 with constant concentration of CSL, in EMSA reactions. The expected outcome was coexistence in EMSA of two bands (one for CSL/DNA and one for ICN1/CSL/DNA) when CSL was in excess to ICN1 and the amount of ICN1 was sufficient to produce a visible secondary shift. Instead, only the secondary shift was present and no primary shifts, with four sequential dilutions of ICN1 (**Figure 2(b)** lanes 2 - 5); The last sample in the range, where ICN1 concentration was the lowest, contained only the primary shift but no secondary shifts (**Figure 2** lane 6), presumably because the dilution of ICN1 reached the point where supershifting effect could no longer be produced at given conditions. Such result was more consistent with a conformational change in the CSL/DNA molecular complex than with ICN1 joining the complex.

Finally, we used EMSA/Western blotting technique to

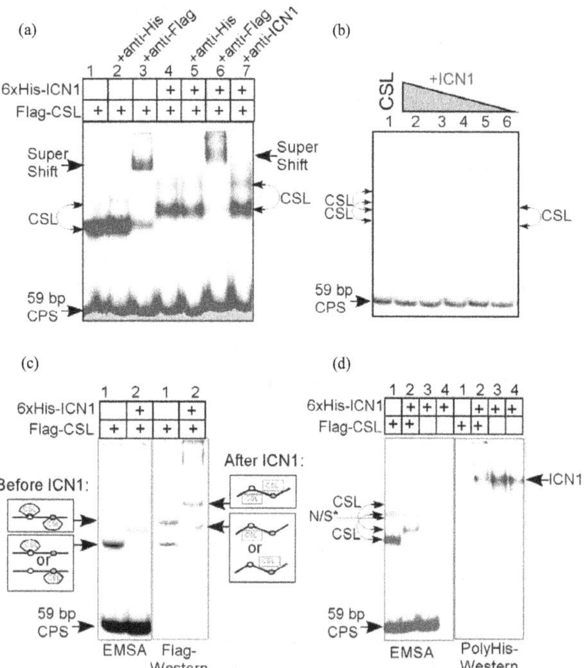

Figure 2. Analysis of EMSA bands. (a) Primary and secondary EMSA shifts were supershifted with 100 ng anti-Flag monoclonal antibody (lanes 3 and 6), but not with either100 ng anti-6×His monoclonal antibody (lanes 2 and 5) or 100 ng anti-ICN1 polyclonal serum (lane 7). The samples contained 300 nM of 5'-end-labeled [^{32}P]59 bp CPS DNA probe, 100 nM purified Flag-CSL and 200 nM purified 6His-ICN1-6His; (b) Titration of decreasing concentrations of ICN1 fails to induce a co-existance of "CSL/DNA" and "ICN1/CSL/DNA" EMSA bands (ICN1: lane 1 = 0 nM; lane 2 = 200 nM; lane 3 = 150 nM; lane 4 = 100 nM; lane 5 = 50 nM; lane 6 = 10 nM. CSL: 100 nM; DNA: 300 nM; (c) Flag-CSL was detected by Western blotting (right panel) in primary and secondary (with ICN1) shifts. The diagrams at the sides of the picture panels show the putative CSL/DNA molecular complexes that form the shifts; (d) The recombinant 6×His- ICN1-6xHis protein was detected by anti-His Western blotting (right panel) not in the DNA shift but much higher in the gel,. The 6His-ICN1-6His protein (200 nM) was present in the samples 2 - 4, including "No DNA" in lane 4. N/S* in lanes 1 and 2-"Non-specific" DNA-binding activity of 293T extract, located in gel very closely to 2 XCSL/DNA band.

establish the exact locations of 6His-ICN1-6His and Flag-CSL proteins in EMSA gel. Once again, the presence of Flag-CSL in the primary and the secondary shifts was confirmed (**Figure 2(c)** right panel, lanes 1 and 2), however 6His-ICN1-6His was not in the shift but detected at a much higher position in the gel, not co-localizing with DNA-bound CSL (**Figure 2(d)** right panel, lanes 2 - 4).

Based on these findings we concluded that the secondary EMSA shift of "+ICN1" sample was not ICN1/CSL/DNA molecular complex but an alternative conformation of CSL/DNA molecular complex, induced by ICN1.

3.3. Weakening of EMSA Shifts in "+ICN1" Samples is Due to CSL/DNA Dissociation

The second effect, weakening of EMSA shift in "+ICN1" samples, indicated dissociation of DNA from CSL/DNA molecular complex. We measured the effect quantitatively and confirmed that when the amount of DNA in the shift was decreasing, the amount of free (unbound) DNA was complementary increasing in the same sample (**Figure 3(a)** The total DNA loads of "–ICN1" and "+ICN1" samples varied negligibly (raw measurements prior to converting to "% of total" are provided in **Table S1**), excluding loss of DNA by degradation. The CSL/DNA was dissociated about 50% when 59 bp CPS probe was used (**Figure 3(a)** and almost complete when the 18 bp 5'CS probe was used.

Since the two probes had different length (59 bp vs 18 bp) and different number of CSL sites per probe (5'CS + 3'CS vs 5'CS only), we investigated what had greater influence on the extent of CSL/DNA dissociation in presence of ICN1, and found that it was the length of the probe. In **Figure 3(b)**, EMSA was performed alongside for 59 bp 3'CS, 59 bp 5'CS (each having one but alternative CS) and 59 bp CPS (having two CSs). All three probes shared the same sequence (except CSL sites) and equal length (59 bp): the dissociating/retardation effect in presence ICN1 was similar for all three probes (**Figure 3(b)** lanes 2, 4 and 8). However between the 18 bp 5'CS and the 59 bp 5'CS probes, having the same CSL binding site but different length, the difference was sharp (**Figure 3(b)** lane 12 vs 14).

3.4. DNase Footprinting Analysis Supports Conformational Change in DNA/CSL in Presence of ICN1

Since the length of the DNA probe played a role in the extent of DNA dissociation from CSL/DNA in presence of ICN1 whereas CSL binding sites did not seem to have influence, we were seeking an additional evidence of involvement of sequence(s) other than CSL binding sites of the "59 bp CPS" during the DNA/CSL-ICN1 interaction. Our method of choice was DNase footprinting analysis. We kept the CSL concentration at minimum (to enhance the sensitivity to ICN1) and added ICN1, trying to identify the changes in the degenerate DNA ladder. The gel image in left panel of **Figure 3(c)** shows titration of CSL protein with 59 bp CPS, in footprinting (lanes 4 - 8): the footprints left by CSL in DNA degenerate ladder are indicated by arrows on the left and the vertical bars at the right side of the picture. In the right panel of **Figure 3(c)**, 2 mg and 4 mg of 6His-ICN1-6His (lanes 4 and 5) were added to 1 mg of CSL-6His (lanes 3-5) compared to controls where soluble *E.coli* extract, not expressing CSL or ICN1 was used in place of CSL (lanes 6-8) and where 6His-ICN1-6His was used alone (lane 9). As seen from the figure, the "CSL + ICN1" samples (lanes 4 and 5) showed difference from all other samples at positions 34 - 36 (TTG) of the degenerate DNA ladder, implicating a structural change, consistent with EMSA results.

3.5. The Results of ChIP/PCR of Hes1 Promoter from HEK293T Cells, Transfected with CSL and ICN1 Genes Are Consistent with Partial CSL/DNA Dissociation in Presence of Overexpressed ICN1 inside the Cell

In the previous section, we showed that full-length human ICN1 caused partial dissociation of CSL from the *Hes*1 promoter *in vitro* in the absence of stabilizing transcription factors. To find out if similar effect of ICN1 takes place *in vivo*, we used chromatin immunoprecipitation (ChIP) [32] from the cells, transiently transfected with CSL and ICN1 genes. Modeling the interaction between CSL/DNA and ICN1 *in vivo* without interference of other Notch-related factors presented a challenge. We minimized such interference by using the non Notch-expressing transfected host cell line, HEK293T, which is not expected to express Notch-specific factors in stoichiometric amounts to the overexpressed recombinant ICN1 and CSL. Robust overexpression of ICN1 and CSL genes in HEK293T would dwarf any background activity of the endogenous CSL, present in HEK293T [35], but to clear the background even further we used Flag-tagged human CSL expression (Flag-CSL) for anti-Flag rather than anti-CSL ChIP assay.

Anti-Flag ChIP/PCR from Flag-CSL-transfected cells confirmed that the recombinant Flag-CSL protein is bound to *Hes1* promoter (**Figure 4(a)** lane 7). In contrary, anti-ICN1 ChIP/PCR from the cells co-transfected with CSL and ICN1 genes together was negative, disregarding the type of the antibody, used (**Figure 4(a)** lanes 4-6 as shown). The result suggested that ICN1 was either unbound to CSL/DNA or the immunoprecipitation of ICN1/CSL/DNA complexes by ICN1 epitops was inefficient. We targeted Flag-epitope after "Flag-CSL + Flag-ICN1" co-transfection-and such ChIP/PCR was positive but *HES*1 promoter was under-represented in "CSL + ICN1"-transfected cells compared to "CSL only"-transfected cells (**Figure 4(b)** upper panel lanes 1 and 2). Serial testing of anti-Flag ChIP/PCR after co-transfections of cells with Flag-CSL plus either ICN1-HA, 6His-ICN1 or untagged ICN1 suggested that the target epitope that helped bring down the *HES*1 promoter belonged to Flag-CSL and not to ICN1 (**Figure 4(b)** lanes 1 and 2, all panels). Anti-ICN1 ChIP was negative not only for "CSL+ICN1" transfected cells (as described) but also for "ICN1 only" transfections (**Figure 4(b)** lane 3, all panels). Prior to each ChIP experiment, the expression of CSL and ICN1 proteins was confirmed by Western blotting (**Figure 4(c)**, for Flag-CSL and ICN1-HA pro-

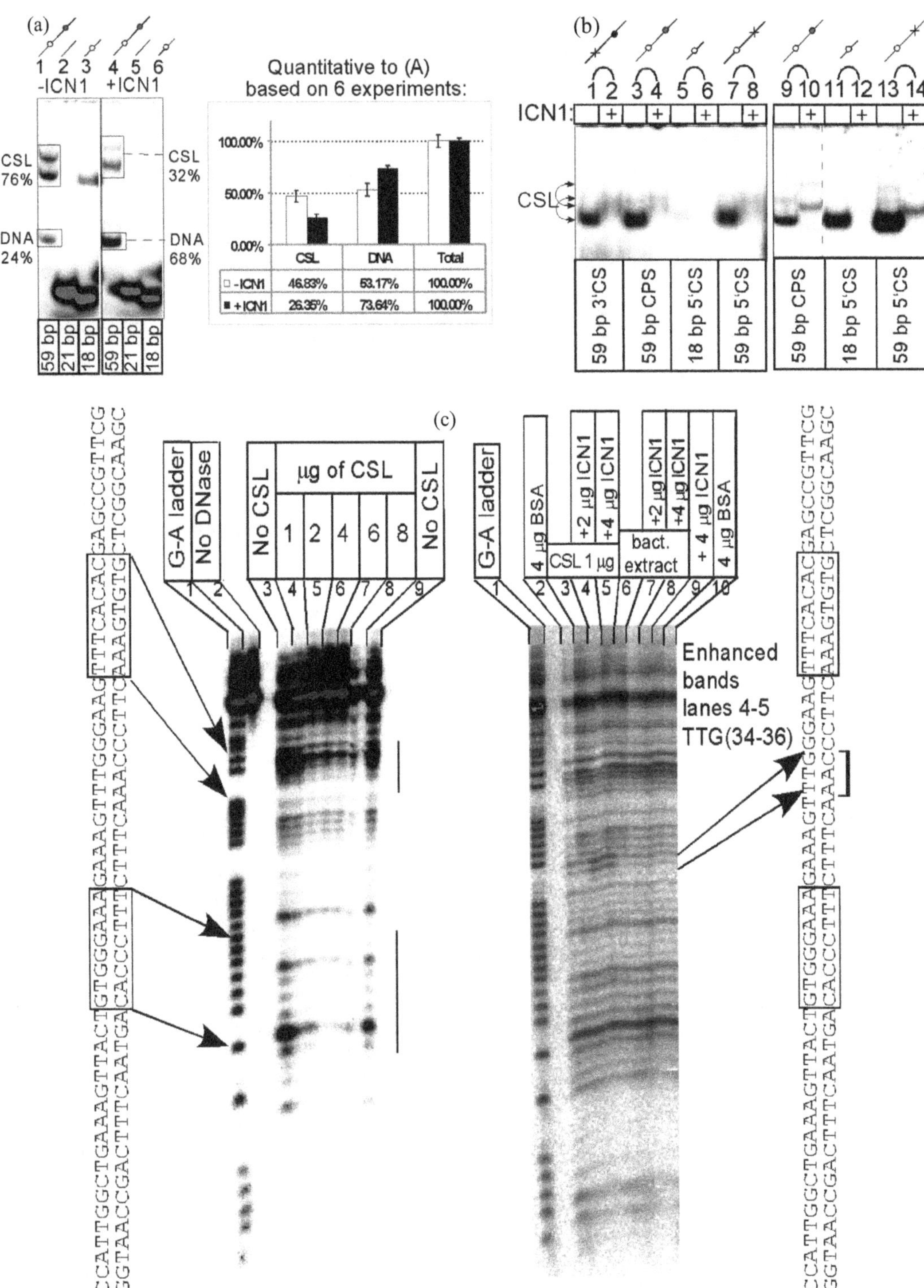

Figure 3. DNA quantification and footprinting analysis. (a) Left panel-EMSA gel of 59 bp CPS probe: lane 1, "DNA/CSL" shift of the "-ICN1" sample contained 76% of total DNA; lane 4, "DNA/CSL shift of the "+ICN1" sample contained only 32% of total DNA, indicating dissociation. Right panel: quantification results obtained from six EMSA gels (error bars reflect variations between the samples). Raw data are shown in Table 1 of the Supplement; (b) All 59 bp probes had similar CSL/DNA band shift pattern in presence of ICN1 (CSL/DNA not dissociated completely), different from 18 bp 5'CS probe (complete CSL/DNA dissociation); (c) Left panel-footprinting analysis with recombinant CSL: the footprints at CSL-binding sites of the 59 bp CPS probe develop increased intensity as CSL concentration rises from 1 to 8 mg (lanes 4 - 8); Right panel - footprinting analysis with CSL and ICN1: local disturbance in DNA structure is detected exactly in the middle of CSL paired site, in presence of CSL and ICN1 (lanes 4 and 5). The effect is not observed with ICN1 alone (lane 9), or ICN1 with total protein extract of *E. coli* (lanes 7 and 8), or without ICN1 in presence of CSL (lane 3) or total protein extract of *E.coli* (lane 6).

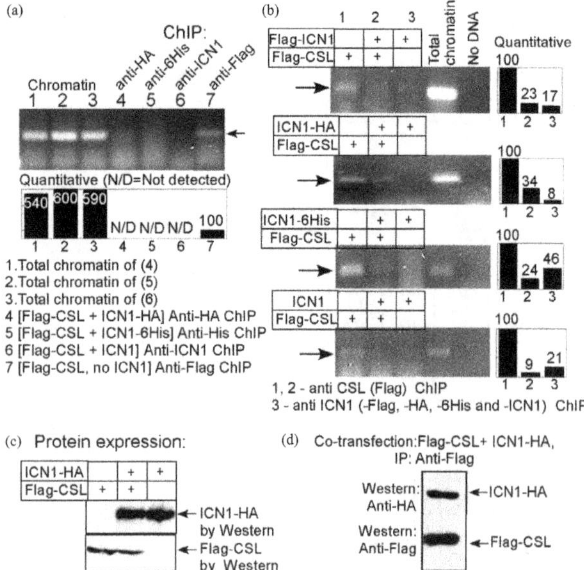

Figure 4. ChIP/PCR of *Hes1* promoter. (a) No immunoprecipitation of *Hes1* promoter was detected with anti-ICN1 antibodies (lanes 4-6 as marked) from [CSL + ICN1] transfected cells; The immunoprecipitation worked with anti-CSL (anti-Flag) antibody from "CSL only" transfected cells; (b) The anti-Flag ChIP/PCR of *Hes1* promoter was much more efficient from "Flag-CSL only" transfected cells (lane 1) than from "Flag-CSL + ICN1" co-transfected cells (lane 2) and anti-ICN1 ChIP/PCR was inefficient from "ICN1 only" transfected cells (lane 3). Antibodies and ICN1 species as marked; (c) Expression of the recombinant CSL and ICN1 was confirmed by Western blotting prior to ChIP/PCR; (d) The recombinant ICN1-HA co-immunoprecipitated from "Flag-CSL + ICN1-HA" transfected cells by anti-Flag IP, confirming accessibility of Flag-epitope of Flag-CSL/ICN1 molecular complex.

teins), excluding the possibility that under-representation of *HES*1 promoter was caused by the lack of Flag-CSL. Interesting to note that Flag-CSL and ICN1-HA could be co-immunoprecipitated together from the extracts of co-transfected HEK293T, as detected by Western blotting (**Figure 4(d)**), indicating that Flag-tag epitope of the Flag-CSL was accessible in the Flag-CSL/ICN1-HA protein pair complex, at least when unbound to DNA.

In summary to the *in vivo* results, we concluded that in the transfected HEK293T cells 1) ICN1 was not associated with *HES*1 promoter, 2) ICN1 was not associated with CSL/DNA complex at *HES*1 promoter, and 3) ICN1 caused partial dissociation of CSL from *HES*1 promoter. To further substantiate these conclusions, we analyzed the transfected HEK293T extracts in EMSA.

3.6. EMSA with Transfected HEK293T Extracts Is Similar to the One with Purified ICN1, Except Larger Magnitude of Retardation of CSL/DNA Band

CSL-expressing HEK293T extract (CSL/293T) was already tested in EMSA alongside bacterially expressed purified CSL-6His (*i.e.*, **Figure 1(b)**, etc.). Here again, CSL/293T produced primary EMSA shift (**Figure 5(a)** lane 1). But neither untransfected cell extract (HEK293T) no cell extract transfected with ICN1 gene only (ICN1/293T) had any specific effect on the DNA** (**Figure 5(a)**

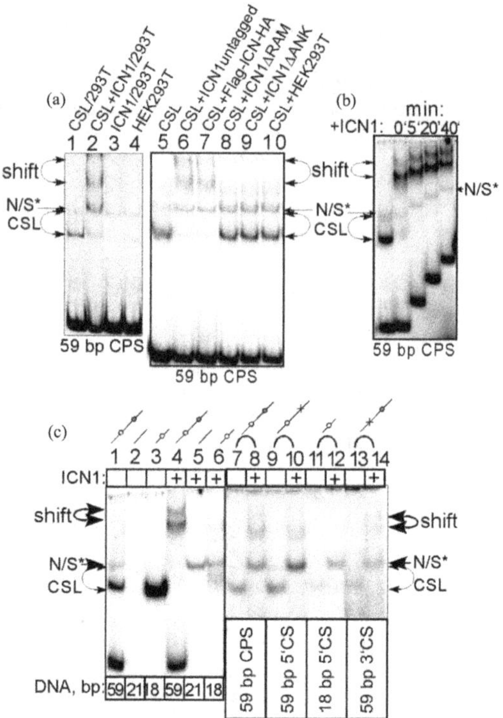

Figure 5. EMSA with HEK293T extracts. (a) "CSL only" transfected HEK293T cell extract produced primary CSL shift in EMSA (lanes 1 and 5) while "ICN1 only" (lane 3) and untransfected HEK293T (lane 4 and 10) extracts produced no specific shifts; Cellular extract of HEK293T co-transfected with "CSL+ICN1" produced primary and secondary EMSA shifts (lane 2); Similar primary and secondary shifts were also observed with CSL/293T extract combined with ICN1/293T extract (lanes 6 and 7); mutant DRAM (lane 8) and DANK (lane 9) ICN1 species caused no secondary shifts of CSL/DNA; (b) Time course: Addition of ICN1/293T extract to pre-incubated CSL/DNA caused secondary EMSA shift when loaded in EMSA gel at "0" time and was unchanged in 40 min (as marked); (c) Testing of different DNA probes with CSL/293T-for primary and CSL/293T + ICN1/293T-for secondary shifts: The 59 bp CPS probe had double-banded primary and double-banded secondary shift (lanes 1 and 4, 7 and 8); 18 bp 5'CS probe had a single-banded primary and no secondary shift (lanes 3 and 6, 11 and 12) and the 59 bp 5'CS and 59 bp 3'CS both had a single-banded primary but double-banded secondary shift (lanes 9, 10 and 13, 14). N/S*-"Non-specific" DNA-binding activity of 293T extract, located very closely to 2XCSL/DNA band.

**All HEK293T extracts produced a non-specific (N/S) shift in EMSA, running close to 2 x CSL/DNA band. This N/S shift was also seen with 7 random DNA probes not containing CSs (in separate testing) and therefore was ignored as a general background.

lanes 3 and 4). When the extract of the cells, co-transfected with CSL+ICN1 genes (CSL+ICN1)/293T (**Figure 5(a)** lane 2) or the mixture of ICN1/293T + CSL/293T extracts (**Figure 5(a)** lanes 6 and 7 for tagged and untagged modifications of ICN1) were used in EMSA, CSL/DNA band was supershifted much higher in the gel: the effect resembling that of bacterially expressed/purified ICN1 except the larger magnitude of retardation. RAM- and Ankyrin-deletions of ICN1 (ICN1DRAM/293T and ICN1DANK/293T) abolished the effect of WT ICN1/293T (**Figure 5(a)** lanes 8 and 9), pointing to a direct involvement of ICN1 in the supershifting. Shorter or longer exposure of pre-formed CSL/ 59 bp CPS to WT ICN1/293T had no difference: the full effect took place at "zero time" to the same extent as after 40 minutes of pre-incubation (**Figure 5(b)**).

We compared the effects of the "59 bp CPS", "21 bp No CS" and "18 bp 5'CS" probes- and the results were similar to bacterially expressed/purified CSL and ICN1 (except the depth of retardation of the CSL/DNA shift): there was a primary CSL/DNA shift formed with the "59 bp CPS" and "18 bp 5'CS" probes, and not with the "21 bp No CS" probe (**Figure 5(c)** lanes 1-3); two characteristic bands were seen in the shift with the "59 bp CPS" (**Figure 5(c)** lane 1) and only one band-with the "18 bp CS" (**Figure 5(c)** lane 3); The addition of ICN1/293T caused retardation of CSL/DNA band in case of CSL/59 bp CPS probe (**Figure 5(c)** lane 4) and disappearance of CSL/DNA band in case of CSL/18 bp 5'CS band (**Figure 5(c)** lane 6).

As inferred from ChIP/PCR data (in Section 2.5), ICN1/293T extract induced obvious dissociation of CSL/DNA molecular complex in case of the 18 bp 5'CS probe, but the super-retardation of the 59 bp CPS without an apparent reduction of DNA material in the shift had to be explained.

We compared the 18 bp 5'CS probe with other CS probes in EMSA with ICN1/293T (**Figure 5(c)** right panel): Unlike the 18 bp 5'CS probe, the 59 bp 5'CS and 59 bp 3'CS probes did not show obvious signs of dissociation (**Figure 5(c)** lanes 10 and 14 vs lane 12), implying the involvement of extra sequence of 59 bp probe in maintaining stability during the interaction with ICN1. Surprisingly, both 59 bp CS probes performed identically to the 59 bp CPS probe by producing the double-banded pattern of the super-retarded shift (**Figure 5(c)** lanes 8, 10 and 14). In EMSA with purified CSL and ICN1 (as described in Section 2.2) , the double-banded pattern of the CSL/DNA shift was due to two CSL binding sites present in one probe (the exclusive feature of the CPS probe), but that did not seem to be the case for the double-banded pattern of the super-retarded shift in EMSA with HEK293T extracts. Rather, they represented either ICN1/CSL/DNA + X/ICN1/CSL/DNA or X/CSL/DNA + X/DNA molecular complexes, where "X"-unidentified factor(s) from HEK293T extract. We further elucidated the possibilities using EMSA/Western blotting and DNA quantification.

3.7. EMSA/Western Blotting Reveals Presence of CSL but Not ICN1 in the Retarded Shift; CSL Is Detected in Only One of the Two Bands of the Shift

We identified the positions of ICN1-HA and Flag-CSL proteins in EMSA gel using EMSA/Western blotting technique. As seen in the **Figure 6(a)** (anti-HA Western: lane 1 = ICN1 without DNA, lane 2 = ICN1+DNA, lane 3 = ICN1+CSL/DNA), the ICN1-HA protein was not co-localized in the DNA shifts (lane 3), indicating that molecular complexes were unlikely to contain ICN1 (*i.e.*, not ICN1/CSL/DNA or X/ICN1/CSL/DNA, etc.).

Flag-CSL protein, in contrary, was detected in the super-retarded shift of "ICN1/293T sample (**Figure 6(a)** anti-Flag Western, lane 3; **Figure 6(b)** anti-Flag Western, lane 3) as well as in the primary shifts (**Figure 6(b)** anti-Flag Western lane 2). To closer examine the two bands of the shift we resolved them by extended EMSA run in 4% polyacrylamide gel and subjected to anti-Flag Western analysis (**Figure 6(b)** lanes 4 and 5): interestingly, Flag-CSL protein was detected only in the upper but not in the lower band (**Figure 6(b)** lane 5, right panel), consistent with the presumed CSL-heterogeneity of the shift (*i.e.*, CSL/X /DNA and X /DNA molecular complexes, as depicted in the right-side diagram in **Figure 6(b)**).

3.8. DNA Quantification Provides Evidence of Partial CSL Dissociation from CSL/X/DNA in Presence of ICN1/293T

Quantitative analysis revealed higher contents of ^{32}P-labeled DNA probe in the super-retarded shift of the "+ICN1/293T" sample than in the primary CSL/DNA shift of "No ICN1" sample (40% of the total in lane 2 compared 34% in **Figure 6(c)**). Since the amounts of CSL protein were strictly equalized between the samples in lanes 1 and 2, at least some of the DNA in the super-retarded shift (6% or more of the total in **Figure 6(c)**) had to be bound in the protein complexes without CSL: Subsequently, the CSL-heterogeneity of the shift, detected by Western blotting, was supported by quantification of DNA in EMSA bands. Presuming that the lower band of the super-retarded shift is X/DNA (no CSL, no ICN1) molecular complex, and the upper band-CSL/X/DNA molecular complex, the amount of CSL in the DNA-bound molecular complex of "+ICN1/293T" sample (CSL/X/DNA) can be compared with the amount of CSL in the "No ICN1" sample (CSL/DNA) by quanti-

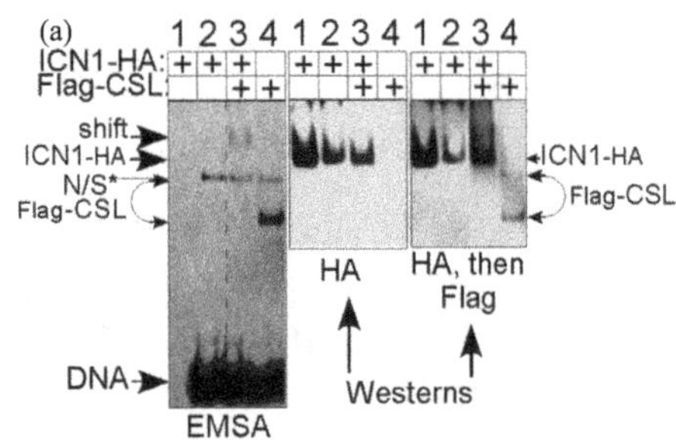

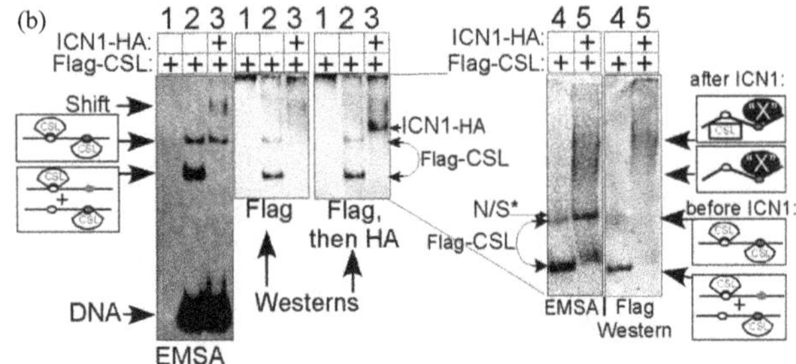

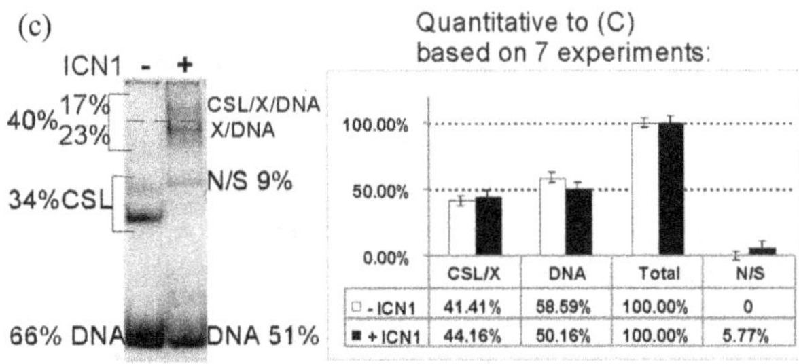

Figure 6. Localization of ICN1 and CSL by Western blotting. (a) Localization of ICN1-HA protein of ICN1-HA/293T extract: Left panel-^{32}P exposure of EMSA gel, middle panel-anti-HA Western blotting of the corresponding transfer on the PVDF membrane, right panel-anti-HA and then anti-Flag Western of the same membrane. Lanes: 1, ICN1-HA/293T without DNA; 2, ICN1-HA/293T + DNA; 3, ICN1-HA/293T + CSL/293T/DNA; 4, CSL/293T/DNA; (b) Localization of Flag-CSL protein of Flag-CSL/293T extract: Panel (from the left)-EMSA gel; panel 2, anti-Flag Western blotting; panel 3, anti-Flag then anti-HA Western of the same protein transfer; panel 4, extended EMSA run; panel 5, anti-Flag Western of the extended EMSA. Lanes: 1, Flag-CSL/293T without DNA; 2, Flag-CSL/293T+DNA; 3, Flag-CSL/293T+DNA + ICN1-HA/293T; 4, Flag-CSL/293T + DNA (extended EMSA run); 5, Flag-CSL/293T + DNA + ICN1-HA/293T (extended EMSA run); (c) Quantification of ^{32}P-activity in EMSA bands, produced by HEK293T extracts. In the left panel-a selected picture of EMSA gel used in quantification: Lanes: 1, CSL/293T + DNA with no ICN1; 2, CSL/293T+DNA + ICN1/293T. In the right panel-combined results of seven independent experiments in one plot with error bars representing variability between the samples. N/S*—"Non-specific" DNA-binding activity of 293T extract.

fication the equamolar amounts of the DNA in the correspondent EMSA bands. Interestingly, with such presumption only ~half of the CSL is still bound to the DNA in the "+ICN1/293T" sample compared to "No ICN1" sample (17% of CSL/X/DNA in lane 2 vs 34% of CSL/DNA in lane 1, **Figure 6(c)**), meaning that CSL dissociated in presence of ICN1: consistent with ChIP/PCR data and remarkably similar to the EMSA results obtained with purified ICN1 and CSL.

4. Discussion

The goal of this study was to evaluate the impact of full-length human ICN1 protein on CSL/DNA molecular complex. The experimental results reveal that ICN1 may function alone to allosterically modulate CSL/DNA complex, at *HES*1 promoter. Specifically, the data here suggest that full length human ICN1 does not form a stable association with CSL/DNA but induces a conformational change in CSL/DNA; the ICN1-induced CSL-DNA conformation is unstable and prone to dissociation; the bond between the CSL and DNA in the new conformation may involve *HES*1 promoter sites other than the consensus CSL-binding sites. Transient dissociation of the CSL/DNA complex may sponsor the recruitment of additional factors and DNA sequences.

We acknowledge that several research groups have reported detectable ICN/CSL/DNA shifts in EMSA [20, 29], although the lack of binding of ICN1 to DNA/CSL was also reported: [22]. A number of reasons may account for these differences, amongst them:

- Human ICN1 and non-human ICN1 may possess different binding affinities/stoichiometry for the CSL/DNA;
- In the most studies, the ICN1/CSL/DNA molecular complex was reportedly assembled with truncated ICN1 (ANK, RAM or RAMANK domains, etc.): The use of a non-truncated, full-length human ICN1 could alter protein-protein interaction and binding with CSL/DNA;
- The *in vitro* setup and DNA probes are not identical and often differ between laboratories and this can lead to discrepancies in data outcomes;
- A conformational change in CSL/DNA could decrease electrophoretic mobility, and this may be falsely identified as the ICN1/CSL/DNA molecular complex in EMSA: the presence of ICN1 in the shift must be verified;

An excellent illustration of the interspecies differences is provided in the recent publication by Kovall and coworkers [36] wherein they report a 50-fold difference in affinity of the RAM domain to CSL between the worm and mouse species. In the same study, the reported ability of RAM domain to induce modulation of CSL/DNA is only 50% compared to the extent of modulation found in the ternary complex (MAML/ICN/CSL + DNA) with RAMANK-can serve as an illustration of change that is initially influenced by the size of the truncated ICN1. Of course, in the latter case there is also unknown contribution of MAML protein.

The binding of MAML to CSL/DNA is dependent upon the participation of ICN1, because it requires opening the loop in N-terminal domain of CSL that remains closed in the free CSL/DNA form, as presumed from crystallization studies [30,37]. The instability of CSL/DNA, detected in our experiments, implies that CSL conformation in the MAML/ICN1/CSL/DNA complex (as seen in crystal structure) is different from the conformation induced by ICN1 in the absence MAML, because the ternary complex is not characterized by low affinity to DNA. Instead, CSL prints in DNA footprinting are enhanced with MAML [20]. Thus, it is likely that MAML, in turn, re-configures the CSL/DNA complex that is modulated by ICN1, and stabilizes the bond of CSL with DNA. Our model presumes that highest level of cooperation between ICN1 and MAML is required to produce maximal affinity for binding to CSL/DNA. The interaction of ICN1 with DNA/CSL lasts seconds, taking into account that the mixed (ICN1 + DNA/CSL) samples were applied immediately to the EMSA [the interacting molecules are "fixed" together by "caging effect" after entering the gel [38] and the time of entry into the gel lasted less than 1 minute. ICN1 opens CSL conformation and compromises the CSL bond with the promoter DNA. The allosteric modulation of CSL/DNA that is induced in the presence of ICN1 may prove to be of higher significance than physical association of ICN1 with the CSL/DNA molecular complex. Jones and co-workers have discussed that "transactivation" by Notch is achieved through recruitment of available non-DNA-binding co activators such as MAML, SKIP, Deltex, and not by functionality of ICNs own domains [20]. There is a growing family of factors, known to interact with Notch, whose function is not completely understood. MAML proteins interact with partners other than Notch and CSL, and the Notch pathway that competes for MAML1, titrates this protein molecule from interacting with the other transcription factors [39]. SKIP, binding to CSL and ICN, may well prove to be another "stabilizer" of ICN1-CSL/DNA, alternative to MAML, but crystal structure of SKIP/ICN1/CSL/DNA is yet to be determined. Deltex does not bind to CSL [40], but may act with ICN to modulate other enhancer-binding proteins. Many genes that contain CSL-binding sites in their promoters are not always responsive to ICN, despite the presence of CSL in the nuclei of the cells [41,42]; The ability of Notch to modify CSL/DNA, and potentially, "Another Effector/DNA", opens an intriguing possibility that activation of Notch-regulated promoters is achieved through recruitment of multiple different, promoter-specific co-activators (MAML, SKIP, Deltex, etc.), which may take advantage of the universal function of ICN1 as "modulator" of enhancer-binding protein/DNA complexes (including the CSL/DNA). In some cases, "opening" of an enhancer-binding protein (such as CSL) may be the only necessary and sufficient role of ICN1; In the case of *HES*1 transcription, ICN1 actively modulates CSL into a configuration that accepts the transcription co-activator MAML, plus ICN1: if MAML is available and appropriately positioned at the time of ICN1-CSL/DNA interaction, the "open" conformation is re-adjusted and a stable MAML/ICN1/CSL/DNA transcription complex is formed; if MAML is unavailable—CSL may dissociate or revert to the "free DNA-bound" conformation, and thereby inhibit transcription. Recent studies reveal that the consensus CSL binding sequences are present in hundreds if not thousands promoters of different genes, and CSL occupancy of those promoters is quite high [43,44]; Thus, the differential expression of promoter-specific CSL/DNA-stabilizing factors may even form the basis of precise response(s) to Notch signaling.

The Notch-induced instability/modulation of CSL/DNA promoter complexes as demonstrated in this study contributes to our understanding of how Notch, when expressed in small amounts within the nucleus, can displace CSL-bound co-repressor proteins that exist at much higher concentrations [2]. The current theory is that SKIP protein assists Notch in the displacement of co-repressors from CSL [45]. Another possibility is that allosteric modulation of CSL/DNA can render a conformation not favorable to binding of co-repressor complexes, thus causing their dissociation. Or even that the CSL may dissociate from the promoter itself upon ICN1 impact, vacating the place to pre-assembled MAML/ICN1/CSL molecular complexes, whose existence is in discussion in the alternative model of Notch activation [36,46,47], different from the model where CSL is constantly statically bound to DNA.

5. Acknowledgements

This work was supported by NCI (NIH) grant 5 R01 CA 06245011.

REFERENCES

[1] G. Weinmaster and C. Kintner, "Modulation of Notch Signaling during Somitogenesis," *Annual Review of Cell and Developmental Biology*, Vol. 19, 2003, pp. 367-395. doi:10.1146/annurev.cellbio.19.111301.115434

[2] J. S. Mumm and R. Kopan, "Notch Signaling: From the Outside in," *Developmental Biology*, Vol. 228, No. 2, 2000, pp. 151-165. doi:10.1006/dbio.2000.9960

[3] M. V. Gustafsson, *et al.*, "Hypoxia Requires Notch Signaling to Maintain the Undifferentiated Cell State," *Developmental Cell*, Vol. 9, No. 5, 2005, pp. 617-628. doi:10.1016/j.devcel.2005.09.010

[4] G. D. Hurlbut, *et al.*, "Crossing paths with Notch in the

Hyper-Network," *Current Opinion in Cell Biology*, Vol. 19, No. 2, 2007, pp. 166-175. doi:10.1016/j.ceb.2007.02.012

[5] F. Jundt, R. Schwarzer and B. Dorken, "Notch Signaling in Leukemias and Lymphomas," *Current Molecular Medicine*, Vol. 8, No. 1, 2008, pp. 51-59. doi:10.2174/156652408783565540

[6] R. M. Demarest, F. Ratti and A. J. Capobianco, "It's T-ALL about Notch," *Oncogene*, Vol. 27, No. 38, 2008, pp. 5082-5091. doi:10.1038/onc.2008.222

[7] F. Wu, A. Stutzman and Y. Y. Mo, "Notch Signaling and Its Role in Breast Cancer," *Frontiers in Bioscience*, Vol. 12, 2007, pp. 4370-4383. doi:10.2741/2394

[8] M. Roy, W. S. Pear and J. C. Aster, "The Multifaceted Role of Notch in Cancer," *Current Opinion in Genetics & Development*, Vol. 17, No. 1, 2007, pp. 52-59. doi:10.1016/j.gde.2006.12.001

[9] U. Koch and F. Radtke, "Notch and Cancer: A Double-Edged Sword," *Cellular and Molecular Life Sciences*, Vol. 64, No. 21, 2007. pp. 2746-2762. doi:10.1007/s00018-007-7164-1

[10] Z. Wang, *et al.*, "Exploitation of the Notch Signaling Pathway as a Novel Target for Cancer Therapy," *Anticancer Research*, Vol. 28, No. 6A, 2008, pp. 3621-3630.

[11] M. A. Villaronga, C. L. Bevan and B. Belandia, "Notch Signaling: A Potential Therapeutic Target in Prostate Cancer," *Current Cancer Drug Targets*, Vol. 8, No. 7, 2008, pp. 566-580. doi:10.2174/156800908786241096

[12] P. Rizzo, *et al.*, "Rational Targeting of Notch Signaling in Cancer," *Oncogene*, Vol. 27, No. 38, 2008, pp. 5124-5131. doi:10.1038/onc.2008.226

[13] R. Kopan and M. X. Ilagan, "The Canonical Notch Signaling Pathway: Unfolding the Activation Mechanism," *Cell*, Vol. 137, No. 2, 2009, pp. 216-233. doi:10.1016/j.cell.2009.03.045

[14] M. E. Fortini, "Notch Signaling: The Core Pathway and Its Posttranslational Regulation," *Developmental Cell*, Vol. 16, No. 5, 2009, pp. 633-647. doi:10.1016/j.devcel.2009.03.010

[15] T. Borggrefe and F. Oswald, "The Notch Signaling Pathway: Transcriptional Regulation at Notch Target Genes," *Cellular and Molecular Life Sciences*, Vol. 66, No. 10, 2009, pp. 1631-1646. doi:10.1007/s00018-009-8668-7

[16] M. Le Gall and E. Giniger, "Identification of Two Binding Regions for the Suppressor of Hairless Protein within the Intracellular Domain of Drosophila Notch," *The Journal of Biological Chemistry*, Vol. 279, No. 28, 2004, pp. 29418-29426. doi:10.1074/jbc.M404589200

[17] H. Y. Kao, *et al.*, "A Histone Deacetylase Corepressor Complex Regulates the Notch Signal Transduction Pathway," *Genes & Development*, Vol. 12, No. 15, 1998, pp. 2269-2277. doi:10.1101/gad.12.15.2269

[18] J. J. Hsieh, *et al.*, "CIR, a Corepressor Linking the DNA Binding Factor CBF1 to the Histone Deacetylase Complex," *Proceedings of the National Academy of Sciences of USA*, Vol. 96, No. 1, 1999, pp. 23-28. doi:10.1073/pnas.96.1.23

[19] F. Oswald, *et al.*, "RBP-Jkappa/SHARP Recruits CtIP/CtBP Corepressors to Silence Notch Target Genes," *Molecular and Cellular Biology*, Vol. 25, No. 23, 2005, pp. 10379-10390. doi:10.1128/MCB.25.23.10379-10390.2005

[20] C. J. Fryer, *et al.*, "Mastermind Mediates Chromatin-Specific Transcription and Turnover of the Notch Enhancer Complex," *Genes & Development*, Vol. 16, No. 11, 2002, pp. 1397-1411. doi:10.1101/gad.991602

[21] A. G. Petcherski and J. Kimble, "Mastermind Is a Putative Activator for Notch," *Current Biology*, Vol. 10, No. 13, 2000, pp. R471-473. doi:10.1016/S0960-9822(00)00577-7

[22] L. Wu, *et al.*, "MAML1, a Human Homologue of Drosophila Mastermind, Is a Transcriptional Co-Activator for NOTCH Receptors," *Nature Genet*, Vol. 26, No. 4, 2000, pp. 484-489. doi:10.1038/82644

[23] S. Zhou, *et al.*, "A Role for SKIP in EBNA2 Activation of CBF1-Repressed Promoters," *Journal of Virology*, Vol. 74, No. 4, 2000, pp. 1939-1947. doi:10.1128/JVI.74.4.1939-1947.2000

[24] L. Espinosa, *et al.*, "Phosphorylation by Glycogen Synthase Kinase-3 Beta Down-Regulates Notch Activity, a Link for Notch and Wnt Pathways," *Journal of Biological Chemistry*, Vol. 278, No. 34, 2003, pp. 32227-32235. doi:10.1074/jbc.M304001200

[25] H. Qin, *et al.*, "RING1 Inhibits Transactivation of RBP-J by Notch through Interaction with LIM Protein KyoT2," *Nucleic Acids Research*, Vol. 32, No. 4, 2004, pp. 1492-1501. doi:10.1093/nar/gkh295

[26] F. Oswald, *et al.*, "p300 Acts as a Transcriptional Coactivator for Mammalian Notch-1," *Molecular and Cellular Biology*, Vol. 21, No. 22, 2001, pp. 7761-7774. doi:10.1128/MCB.21.22.7761-7774.2001

[27] E. A. Johnson, "HIF Takes It Up a Notch," *Science Signal*, Vol. 4, No. 181, 2011, p. pe33. doi:10.1126/scisignal.2002277

[28] C. J. Fryer, J. B. White and K. A. Jones, Mastermind Recruits CycC:CDK8 to Phosphorylate the Notch ICD and Coordinate Activation with Turnover," *Molecular Cell*, Vol. 16, No. 4, 2004, pp. 509-520. doi:10.1016/j.molcel.2004.10.014

[29] Y. Nam, *et al.*, "Structural Basis for Cooperativity in Recruitment of MAML Coactivators to Notch Transcription Complexes," *Cell*, Vol. 124, No. 5, 2006, pp. 973-983. doi:10.1016/j.cell.2005.12.037

[30] R. A. Kovall and W. A. Hendrickson, "Crystal Structure of the Nuclear Effector of Notch Signaling, CSL, Bound to DNA," *The EMBO Journal*, Vol. 23, No. 17, 2004, pp. 3441-3451. doi:10.1038/sj.emboj.7600349

[31] J. J. Wilson and R. A. Kovall, "Crystal Structure of the CSL-Notch-Mastermind Ternary Complex Bound to DNA," *Cell*, Vol. 124, No. 5, 2006, pp. 985-996. doi:10.1016/j.cell.2006.01.035

[32] A. S. Weinmann, *et al.*, "Use of Chromatin Immunoprecipitation to Clone Novel E2F Target Promoters," *Molecular and Cellular Biology*, Vol. 21, No. 20, 2001, pp. 6820-6832. doi:10.1128/MCB.21.20.6820-6832.2001

[33] D. T. Nellesen, E.C. Lai and J.W. Posakony, "Discrete Enhancer Elements Mediate Selective Responsiveness of Enhancer of Split Complex Genes to Common Transcriptional Activators," *Developmental Biology*, Vol. 213, No. 1, 1999, pp. 33-53. doi:10.1006/dbio.1999.9324

[34] A. M. Bailey and J. W. Posakony, "Suppressor of Hairless Directly Activates Transcription of Enhancer of Split Complex Genes in Response to Notch Receptor Activity," *Genes and Development*, Vol. 9, No. 21, 1995, pp. 2609-2622. doi:10.1101/gad.9.21.2609

[35] S. Jeffries, D. J. Robbins and A. J. Capobianco, "Characterization of a High-Molecular-Weight Notch Complex in the Nucleus of Notch(ic)-Transformed RKE Cells and in a Human T-Cell Leukemia Cell Line," *Molecular and Cellular Biology*, Vol. 22, No. 11, 2002, pp. 3927-3241. doi:10.1128/MCB.22.11.3927-3941.2002

[36] D. R. Friedmann, J. J. Wilson and R. A. Kovall, "RAM-Induced Allostery Facilitates Assembly of a Notch Pathway Active Transcription Complex," *The Journal of Biological Chemistry*, Vol. 283, No. 21, 2008, pp. 14781-14791. doi:10.1074/jbc.M709501200

[37] R. A. Kovall, "Structures of CSL, Notch and Mastermind Proteins: Piecing Together an Active Transcription Complex," *Current Opinion in Structural Biology*, Vol. 17, No. 1, 2007, pp. 117-127. doi:10.1016/j.sbi.2006.11.004

[38] M. Fried and D. M. Crothers, "Equilibria and Kinetics of Lac Repressor-Operator Interactions by Polyacrylamide Gel Electrophoresis," *Nucleic Acids Research*, Vol. 9, No. 23, 1981, pp. 6505-6525. doi:10.1093/nar/9.23.6505

[39] A. S. McElhinny, J. L. Li and L. Wu, "Mastermind-Like Transcriptional Co-Activators: Emerging Roles in Regulating Cross Talk among Multiple Signaling Pathways," *Oncogene*, Vol. 27, No. 38, 2008, pp. 5138-5147. doi:10.1038/onc.2008.228

[40] N. Yamamoto, *et al.*, "Role of Deltex-1 as a Transcriptional Regulator Downstream of the Notch Receptor," *The Journal of Biological Chemistry*, Vol. 276, No. 48, 2001, pp. 45031-45040. doi:10.1074/jbc.M105245200

[41] J. W. Cave, *et al.*, "A DNA Transcription Code for Cell-Specific Gene Activation by Notch Signaling," *Current Biology*, Vol. 15, No. 2, 2005, pp. 94-104. doi:10.1016/j.cub.2004.12.070

[42] A. Neves and J. R. Priess, "The REF-1 Family of bHLH Transcription Factors Pattern *C. elegans* Embryos through Notch-Dependent and Notch-Independent Pathways," *Developmental Cell*, Vol. 8, No. 6, 2005, pp. 867-879. doi:10.1016/j.devcel.2005.03.012

[43] L. M. Persson and A. C. Wilson, "Wide-Scale Use of Notch Signaling Factor CSL/RBP-Jkappa in RTA-Mediated Activation of Kaposi's Sarcoma-Associated Herpesvirus Lytic Genes," *Journal of Virology*, Vol. 84, No. 3, 2009, pp. 1334-1347. doi:10.1128/JVI.01301-09

[44] H. Hamidi, *et al.*, "Identification of Novel Targets of CSL-Dependent Notch Signaling in Hematopoiesis," *PLoS ONE*, Vol. 6, No. 5, 2011, p. e20022. doi:10.1371/journal.pone.0020022

[45] S. Zhou, *et al.*, "SKIP, a CBF1-Associated Protein, Interacts with the Ankyrin Repeat Domain of NotchIC to Facilitate NotchIC Function," *Molecular and Cellular Biology*, Vol. 20, No. 7, 2000, pp. 2400-2410. doi:10.1128/MCB.20.7.2400-2410.2000

[46] A. Krejci and S. Bray, "Notch Activation Stimulates Transient and Selective Binding of Su(H)/CSL to Target Enhancers," *Genes & Development*, Vol. 21, No. 11, 2007, pp. 1322-1327. doi:10.1101/gad.424607

[47] Y. Nam, *et al.*, "Cooperative Assembly of Higher-Order Notch Complexes Functions as a Switch to Induce Transcription," *Proceedings of the National Academy of Sciences of the United States of America*, Vol. 104, No. 7, 2007, pp. 2103-2108. doi:10.1073/pnas.0611092104

Supplement

1) ICN1 and CSL recombinant gene sequences used for expression in HEK293T cells and *E. coli* strain BL21DE3.
N-terminal ICN1 sequence (Flag- epitope) added when cloned in pcDNA3.1:

```
HindIII                           BamHI
AAGCTTCCACCATGGATTACAAGGATGACGACGATAAGGGATCC
          M  D  Y  K  D  D  D  D  K  G  S
```

N-terminal ICN1 sequence (with 6xHis epitope) added when cloned in pET28a:

```
3/1                                     33/11
atg ggc agc agc cat cat cat cat cat cac agc agc ggc ctg gtg ccg cgc ggc agc cat
Met gly ser ser his his his his his his ser ser gly leu val pro arg gly ser his
63/21                                   93/31    BamHI
atg gct agc atg act ggt gga cag caa atg ggt cgc gga tcc
met ala ser met thr gly gly gln gln met ggy arg gly ser
```

Sequence Range: 1 to 2482

```
        10        20        30        40        50
GGATCCGGAGGCATGGGCTCCCGCAAGCGCCGGCGGCAGCATGGCCAGCT
CCTAGGCCTCCGTACCCGAGGGCGTTCGCGGCCGCCGTCGTACCGGTCGA
 G  S  G  G  M  G  S  R  K  R  R  R  Q  H  G  Q  L>
______________TRANSLATION OF ICN1 [A]________________>

        60        70        80        90       100
CTGGTTCCCTGAGGGCTTCAAAGTGTCTGAGGCCAGCAAGAAGAAGCGGC
GACCAAGGGACTCCCGAAGTTTCACAGACTCCGGTCGTTCTTCTTCGCCG
  W  F  P  E  G  F  K  V  S  E  A  S  K  K  K  R>
______________TRANSLATION OF ICN1 [A]________________>

       110       120       130       140       150
GGGAGCCCCTCGGCGAGGACTCCGTGGGCCTCAAGCCCCTGAAGAACGCT
CCCTCGGGGAGCCGCTCCTGAGGCACCCGGAGTTCGGGGACTTCTTGCGA
R  E  P  L  G  E  D  S  V  G  L  K  P  L  K  N  A>
______________TRANSLATION OF ICN1 [A]________________>

       160       170       180       190       200
TCAGACGGTGCCCTCATGGACGACAACCAGAATGAGTGGGGGGACGAGGA
AGTCTGCCACGGGAGTACCTGCTGTTGGTCTTACTCACCCCCCTGCTCCT
 S  D  G  A  L  M  D  D  N  Q  N  E  W  G  D  E  D>
______________TRANSLATION OF ICN1 [A]________________>

       210       220       230       240       250
CCTGGAGACCAAGAAGTTCCGGTTCGAGGAGCCCGTGGTTCTGCCTGACC
GGACCTCTGGTTCTTCAAGGCCAAGCTCCTCGGGCACCAAGACGGACTGG
  L  E  T  K  K  F  R  F  E  E  P  V  V  L  P  D>
______________TRANSLATION OF ICN1 [A]________________>

       260       270       280       290       300
TGGACGACCAGACAGACCACCGGCAGTGGACTCAGCAGCACCTGGATGCC
ACCTGCTGGTCTGTCTGGTGGCCGTCACCTGAGTCGTCGTGGACCTACGG
L  D  D  Q  T  D  H  R  Q  W  T  Q  Q  H  L  D  A>
______________TRANSLATION OF ICN1 [A]________________>

       310       320       330       340       350
GCTGACCTGCGCATGTCTGCCATGGCCCCCACACCGCCCCAGGGTGAGGT
```

```
CGACTGGACGCGTACAGACGGTACCGGGGGTGTGGCGGGGTCCCACTCCA
 A  D  L  R  M  S  A  M  A  P  T  P  P  Q  G  E  V>
_______________TRANSLATION OF ICN1 [A]________________>

       360       370       380       390       400
TGACGCCGACTGCATGGACGTCAATGTCCGCGGCCTGATGGCTTCACCC
ACTGCGGCTGACGTACCTGCAGTTACAGGCGCCCGGACTACCGAAGTGGG
  D  A  D  C  M  D  V  N  V  R  G  P  D  G  F  T>
_______________TRANSLATION OF ICN1 [A]________________>

       410       420       430       440       450
CGCTCATGATCGCCTCCTGCAGCGGGGGCGGCCTGGAGACGGGCAACAGC
GCGAGTACTAGCGGAGGACGTCGCCCCCGCCGGACCTCTGCCCGTTGTCG
P  L  M  I  A  S  C  S  G  G  G  L  E  T  G  N  S>
_______________TRANSLATION OF ICN1 [A]________________>

       460       470       480       490       500
GAGGAAGAGGAGGACGCGCCGGCCGTCATCTCCGACTTCATCTACCAGGG
CTCCTTCTCCTCCTGCGCGGCCGGCAGTAGAGGCTGAAGTAGATGGTCCC
 E  E  E  E  D  A  P  A  V  I  S  D  F  I  Y  Q  G>
_______________TRANSLATION OF ICN1 [A]________________>

       510       520       530       540       550
CGCCAGCCTGCACAACCAGACAGACCGCACGGGCGAGACCGCCTTGCACC
GCGGTCGGACGTGTTGGTCTGTCTGGCGTGCCCGCTCTGGCGGAACGTGG
  A  S  L  H  N  Q  T  D  R  T  G  E  T  A  L  H>
_______________TRANSLATION OF ICN1 [A]________________>

       560       570       580       590       600
TGGCCGCCCGCTACTCACGCTCTGATGCCGCCAAGCGCCTGCTGGAGGCC
ACCGGCGGGCGATGAGTGCGAGACTACGGCGGTTCGCGGACGACCTCCGG
L  A  A  R  Y  S  R  S  D  A  A  K  R  L  L  E  A>
_______________TRANSLATION OF ICN1 [A]________________>

       610       620       630       640       650
AGCGCAGATGCCAACATCCAGGACAACATGGGCCGCACCCCGCTGCATGC
TCGCGTCTACGGTTGTAGGTCCTGTTGTACCCGGCGTGGGGCGACGTACG
 S  A  D  A  N  I  Q  D  N  M  G  R  T  P  L  H  A>
_______________TRANSLATION OF ICN1 [A]________________>

       660       670       680       690       700
GGCTGTGTCTGCCGACGCACAAGGTGTCTTCCAGATCCTGATCCGGAACC
CCGACACAGACGGCTGCGTGTTCCACAGAAGGTCTAGGACTAGGCCTTGG
  A  V  S  A  D  A  Q  G  V  F  Q  I  L  I  R  N>
_______________TRANSLATION OF ICN1 [A]________________>

       710       720       730       740       750
GAGCCACAGACCTGGATGCCCGCATGCATGATGGCACGACGCCACTGATC
CTCGGTGTCTGGACCTACGGGCGTACGTACTACCGTGCTGCGGTGACTAG
R  A  T  D  L  D  A  R  M  H  D  G  T  T  P  L  I>
_______________TRANSLATION OF ICN1 [A]________________>

       760       770       780       790       800
CTGGCTGCCCGCCTGGCCGTGGAGGGCATGCTGGAGGACCTCATCAACTC
```

```
GACCGACGGGCGGACCGGCACCTCCCGTACGACCTCCTGGAGTAGTTGAG
 L  A  A  R  L  A  V  E  G  M  L  E  D  L  I  N  S>
______________TRANSLATION OF ICN1 [A]_______________>

      810       820       830       840       850
ACACGCCGACGTCAACGCCGTAGATGACCTGGGCAAGTCCGCCCTGCACT
TGTGCGGCTGCAGTTGCGGCATCTACTGGACCCGTTCAGGCGGGACGTGA
  H  A  D  V  N  A  V  D  D  L  G  K  S  A  L  H>
______________TRANSLATION OF ICN1 [A]_______________>

      860       870       880       890       900
GGGCCGCCGCCGTGAACAATGTGGATGCCGCAGTTGTGCTCCTGAAGAAC
CCCGGCGGCGGCACTTGTTACACCTACGGCGTCAACACGAGGACTTCTTG
W  A  A  A  V  N  N  V  D  A  A  V  V  L  L  K  N>
______________TRANSLATION OF ICN1 [A]_______________>

      910       920       930       940       950
GGGGCTAACAAAGATATGCAGAACAACAGGGAGGAGACACCCCTGTTTCT
CCCCGATTGTTTCTATACGTCTTGTTGTCCCTCCTCTGTGGGGACAAAGA
 G  A  N  K  D  M  Q  N  N  R  E  E  T  P  L  F  L>
______________TRANSLATION OF ICN1 [A]_______________>

      960       970       980       990      1000
GGCCGCCCGGGAGGGCAGCTACGAGACCGCCAAGGTGCTGCTGGACCACT
CCGGCGGGCCCTCCCGTCGATGCTCTGGCGGTTCCACGACGACCTGGTGA
  A  A  R  E  G  S  Y  E  T  A  K  V  L  L  D  H>
______________TRANSLATION OF ICN1 [A]_______________>

     1010      1020      1030      1040      1050
TTGCCAACCGGGACATCACGGATCATATGGACCGCCTGCCGCGCGACATC
AACGGTTGGCCCTGTAGTGCCTAGTATACCTGGCGGACGGCGCGCTGTAG
F  A  N  R  D  I  T  D  H  M  D  R  L  P  R  D  I>
______________TRANSLATION OF ICN1 [A]_______________>

     1060      1070      1080      1090      1100
GCACAGGAGCGCATGCATCACGACATCGTGAGGCTGCTGGACGAGTACAA
CGTGTCCTCGCGTACGTAGTGCTGTAGCACTCCGACGACCTGCTCATGTT
 A  Q  E  R  M  H  H  D  I  V  R  L  L  D  E  Y  N>
______________TRANSLATION OF ICN1 [A]_______________>

     1110      1120      1130      1140      1150
CCTGGTGCGCAGCCCGCAGCTGCACGGAGCCCCGCTGGGGGGCACGCCCA
GGACCACGCGTCGGGCGTCGACGTGCCTCGGGGCGACCCCCCGTGCGGGT
  L  V  R  S  P  Q  L  H  G  A  P  L  G  G  T  P>
______________TRANSLATION OF ICN1 [A]_______________>

     1160      1170      1180      1190      1200
CCCTGTCGCCCCCGCTCTGCTCGCCCAACGGCTACCTGGGCAGCCTCAAG
GGGACAGCGGGGGCGAGACGAGCGGGTTGCCGATGGACCCGTCGGAGTTC
T  L  S  P  P  L  C  S  P  N  G  Y  L  G  S  L  K>
______________TRANSLATION OF ICN1 [A]_______________>

     1210      1220      1230      1240      1250
CCCGGCGTGCAGGGCAAGAAGGTCCGCAAGCCCAGCAGCAAAGGCCTGGC
```

```
GGGCCGCACGTCCCGTTCTTCCAGGCGTTCGGGTCGTCGTTTCCGGACCG
 P  G  V  Q  G  K  K  V  R  K  P  S  S  K  G  L  A>
_______________TRANSLATION OF ICN1 [A]_______________>

     1260      1270      1280      1290      1300
CTGTGGAAGCAAGGAGGCCAAGGACCTCAAGGCACGGAGGAAGAAGTCCC
GACACCTTCGTTCCTCCGGTTCCTGGAGTTCCGTGCCTCCTTCTTCAGGG
  C  G  S  K  E  A  K  D  L  K  A  R  R  K  K  S>
_______________TRANSLATION OF ICN1 [A]_______________>

     1310      1320      1330      1340      1350
AGGATGGCAAGGGCTGCCTGCTGGACAGCTCCGGCATGCTCTCGCCCGTG
TCCTACCGTTCCCGACGGACGACCTGTCGAGGCCGTACGAGAGCGGGCAC
Q  D  G  K  G  C  L  L  D  S  S  G  M  L  S  P  V>
_______________TRANSLATION OF ICN1 [A]_______________>

     1360      1370      1380      1390      1400
GACTCCCTGGAGTCACCCCATGGCTACCTGTCAGACGTGGCCTCGCCGCC
CTGAGGGACCTCAGTGGGGTACCGATGGACAGTCTGCACCGGAGCGGCGG
 D  S  L  E  S  P  H  G  Y  L  S  D  V  A  S  P  P>
_______________TRANSLATION OF ICN1 [A]_______________>

     1410      1420      1430      1440      1450
ACTGCTGCCCTCCCCGTTCCAGCAGTCTCCGTCCGTGCCCCTCAACCACC
TGACGACGGGAGGGGCAAGGTCGTCAGAGGCAGGCACGGGGAGTTGGTGG
  L  L  P  S  P  F  Q  Q  S  P  S  V  P  L  N  H>
_______________TRANSLATION OF ICN1 [A]_______________>

     1460      1470      1480      1490      1500
TGCCTGGGATGCCCGACACCCACCTGGGCATCGGGCACCTGAACGTGGCG
ACGGACCCTACGGGCTGTGGGTGGACCCGTAGCCCGTGGACTTGCACCGC
L  P  G  M  P  D  T  H  L  G  I  G  H  L  N  V  A>
_______________TRANSLATION OF ICN1 [A]_______________>

     1510      1520      1530      1540      1550
GCCAAGCCCGAGATGGCGGCGCTGGGTGGGGGCGGCCGGCTGGCCTTTGA
CGGTTCGGGCTCTACCGCCGCGACCCACCCCCGCCGGCCGACCGGAAACT
 A  K  P  E  M  A  A  L  G  G  G  G  R  L  A  F  E>
_______________TRANSLATION OF ICN1 [A]_______________>

     1560      1570      1580      1590      1600
GACTGGCCCACCTCGTCTCTCCCACCTGCCTGTGGCCTCTGGCACCAGCA
CTGACCGGGTGGAGCAGAGAGGGTGGACGGACACCGGAGACCGTGGTCGT
  T  G  P  P  R  L  S  H  L  P  V  A  S  G  T  S>
_______________TRANSLATION OF ICN1 [A]_______________>

     1610      1620      1630      1640      1650
CCGTCCTGGGCTCCAGCAGCGGAGGGGCCCTGAATTTCACTGTGGGCGGG
GGCAGGACCCGAGGTCGTCGCCTCCCCGGGACTTAAAGTGACACCCGCCC
T  V  L  G  S  S  S  G  G  A  L  N  F  T  V  G  G>
_______________TRANSLATION OF ICN1 [A]_______________>

     1660      1670      1680      1690      1700
TCCACCAGTTTGAATGGTCAATGCGAGTGGCTGTCCCGGCTGCAGAGCGG
```

```
AGGTGGTCAAACTTACCAGTTACGCTCACCGACAGGGCCGACGTCTCGCC
 S  T  S  L  N  G  Q  C  E  W  L  S  R  L  Q  S  G>
______________TRANSLATION OF ICN1 [A]________________>

      1710      1720      1730      1740      1750
CATGGTGCCGAACCAATACAACCCTCTGCGGGGGAGTGTGGCACCAGGCC
GTACCACGGCTTGGTTATGTTGGGAGACGCCCCCTCACACCGTGGTCCGG
  M  V  P  N  Q  Y  N  P  L  R  G  S  V  A  P  G>
______________TRANSLATION OF ICN1 [A]________________>

      1760      1770      1780      1790      1800
CCCTGAGCACACAGGCCCCCTCCCTGCAGCATGGCATGGTAGGCCCGCTG
GGGACTCGTGTGTCCGGGGGAGGGACGTCGTACCGTACCATCCGGGCGAC
P  L  S  T  Q  A  P  S  L  Q  H  G  M  V  G  P  L>
______________TRANSLATION OF ICN1 [A]________________>

      1810      1820      1830      1840      1850
CACAGTAGCCTTGCTGCCAGCGCCCTGTCCCAGATGATGAGCTACCAGGG
GTGTCATCGGAACGACGGTCGCGGGACAGGGTCTACTACTCGATGGTCCC
 H  S  S  L  A  A  S  A  L  S  Q  M  M  S  Y  Q  G>
______________TRANSLATION OF ICN1 [A]________________>

      1860      1870      1880      1890      1900
CCTGCCCAGCACCCGGCTGGCCACCCAGCCTCACCTGGTGCAGACCCAGC
GGACGGGTCGTGGGCCGACCGGTGGGTCGGAGTGGACCACGTCTGGGTCG
  L  P  S  T  R  L  A  T  Q  P  H  L  V  Q  T  Q>
______________TRANSLATION OF ICN1 [A]________________>

      1910      1920      1930      1940      1950
AGGTGCAGCCACAAAACTTACAGATGCAGCAGCAGAACCTGCAGCCAGCA
TCCACGTCGGTGTTTTGAATGTCTACGTCGTCGTCTTGGACGTCGGTCGT
Q  V  Q  P  Q  N  L  Q  M  Q  Q  Q  N  L  Q  P  A>
______________TRANSLATION OF ICN1 [A]________________>

      1960      1970      1980      1990      2000
AACATCCAGCAGCAGCAAAGCCTGCAGCCGCCACCACCACCACCACAGCC
TTGTAGGTCGTCGTCGTTTCGGACGTCGGCGGTGGTGGTGGTGGTGTCGG
 N  I  Q  Q  Q  Q  S  L  Q  P  P  P  P  P  P  Q  P>
______________TRANSLATION OF ICN1 [A]________________>

      2010      2020      2030      2040      2050
GCACCTTGGCGTGAGCTCAGCAGCCAGCGGCCACCTGGGCCGGAGCTTCC
CGTGGAACCGCACTCGAGTCGTCGGTCGCCGGTGGACCCGGCCTCGAAGG
  H  L  G  V  S  S  A  A  S  G  H  L  G  R  S  F>
______________TRANSLATION OF ICN1 [A]________________>

      2060      2070      2080      2090      2100
TGAGTGGAGAGCCGAGCCAGGCAGACGTGCAGCCACTGGGCCCCAGCAGC
ACTCACCTCTCGGCTCGGTCCGTCTGCACGTCGGTGACCCGGGGTCGTCG
L  S  G  E  P  S  Q  A  D  V  Q  P  L  G  P  S  S>
______________TRANSLATION OF ICN1 [A]________________>

      2110      2120      2130      2140      2150
CTGGCGGTGCACACTATTCTGCCCCAGGAGAGCCCCGCCCTGCCCACGTC
```

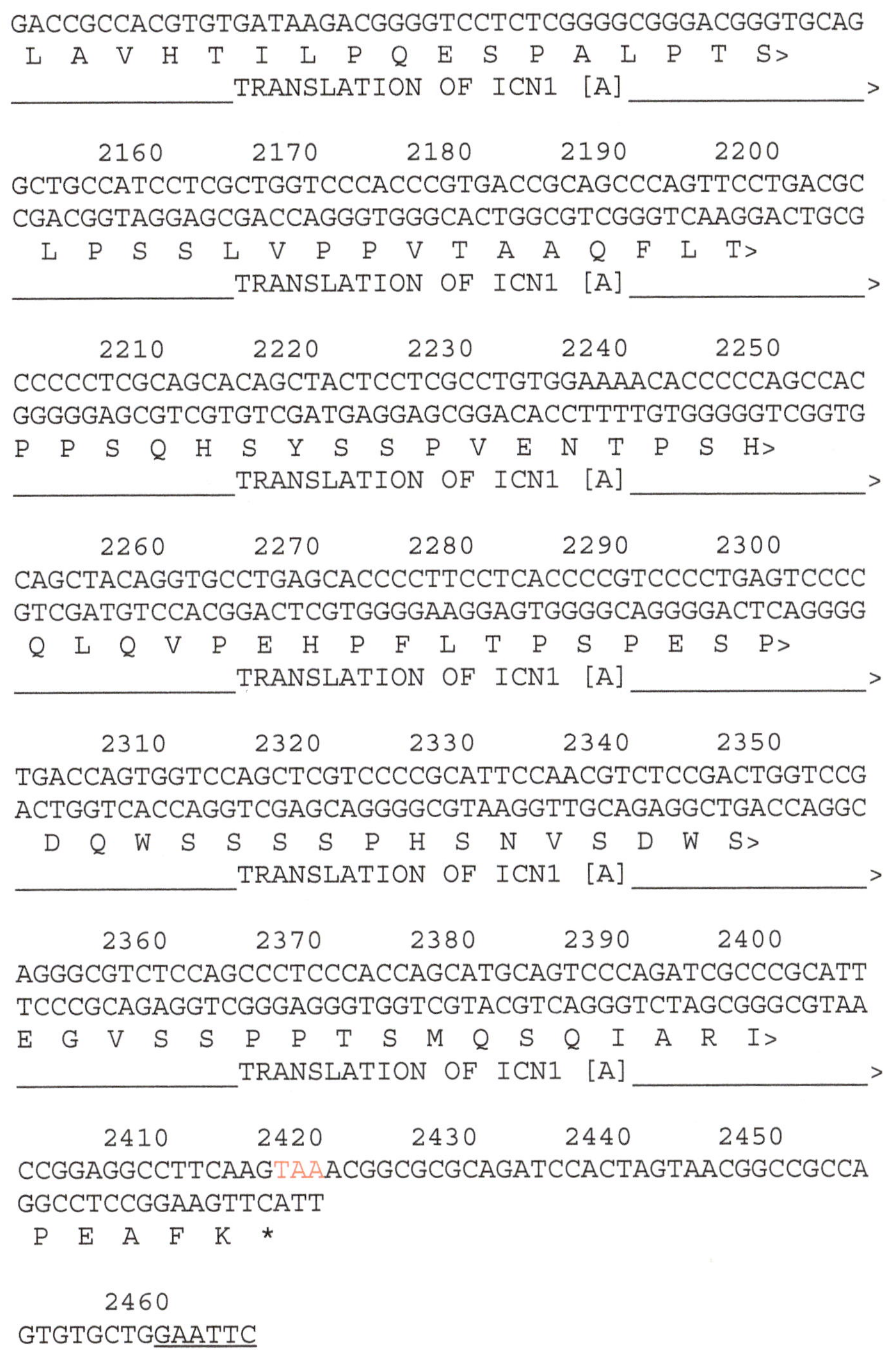

```
GACCGCCACGTGTGATAAGACGGGGTCCTCTCGGGGCGGGACGGGTGCAG
 L  A  V  H  T  I  L  P  Q  E  S  P  A  L  P  T  S>
_______________TRANSLATION OF ICN1 [A]________________>

      2160      2170      2180      2190      2200
GCTGCCATCCTCGCTGGTCCCACCCGTGACCGCAGCCCAGTTCCTGACGC
CGACGGTAGGAGCGACCAGGGTGGGCACTGGCGTCGGGTCAAGGACTGCG
  L  P  S  S  L  V  P  P  V  T  A  A  Q  F  L  T>
_______________TRANSLATION OF ICN1 [A]________________>

      2210      2220      2230      2240      2250
CCCCCTCGCAGCACAGCTACTCCTCGCCTGTGGAAAACACCCCCAGCCAC
GGGGGAGCGTCGTGTCGATGAGGAGCGGACACCTTTTGTGGGGGTCGGTG
P  P  S  Q  H  S  Y  S  S  P  V  E  N  T  P  S  H>
_______________TRANSLATION OF ICN1 [A]________________>

      2260      2270      2280      2290      2300
CAGCTACAGGTGCCTGAGCACCCCTTCCTCACCCCGTCCCCTGAGTCCCC
GTCGATGTCCACGGACTCGTGGGGAAGGAGTGGGGCAGGGGACTCAGGGG
 Q  L  Q  V  P  E  H  P  F  L  T  P  S  P  E  S  P>
_______________TRANSLATION OF ICN1 [A]________________>

      2310      2320      2330      2340      2350
TGACCAGTGGTCCAGCTCGTCCCCGCATTCCAACGTCTCCGACTGGTCCG
ACTGGTCACCAGGTCGAGCAGGGGCGTAAGGTTGCAGAGGCTGACCAGGC
  D  Q  W  S  S  S  S  P  H  S  N  V  S  D  W  S>
_______________TRANSLATION OF ICN1 [A]________________>

      2360      2370      2380      2390      2400
AGGGCGTCTCCAGCCCTCCCACCAGCATGCAGTCCCAGATCGCCCGCATT
TCCCGCAGAGGTCGGGAGGGTGGTCGTACGTCAGGGTCTAGCGGGCGTAA
E  G  V  S  S  P  P  T  S  M  Q  S  Q  I  A  R  I>
_______________TRANSLATION OF ICN1 [A]________________>

      2410      2420      2430      2440      2450
CCGGAGGCCTTCAAGTAAACGGCGCGCAGATCCACTAGTAACGGCCGCCA
GGCCTCCGGAAGTTCATT
 P  E  A  F  K  *

      2460
GTGTGCTGGAATTC
        EcoRI
```

For C-terminal tags 3'- end of ICN1 was modified by deletion of a single "C" nucleotide to create an AflII site next to TAA – STOP codone: ...CTTcAAGtaa -> CTTAAGtaa. Then –HA or - 6xHis sequences were added as follows:

-HA sequence:

```
AflII                          XhoI
CTTAAGTTACCCATACGATGTTCCTGACTATGCGTAACTCGAG
 L  S  Y  P  Y  D  V  P  D  Y  A  stop
```

-6xHis sequence:

```
AflII                 XhoI
CTTAAGTCACCACCACCACCACCACTGACTCGAG
 L  S  H  H  H  H  H  H  Stop
```

2) Recombinant CSL gene.

```
        10         20         30         40         50
ATGGACTACAAAGACGATGACGACAAGCTTATGGACCACACGGAGGGC
TACCTGATGTTTCTGCTACTGCTGTTCGAATACCTGGTGTGCCTCCCG
M  D  Y  K  D  D  D  D  K  L  M  D  H  T  E  G>
______TRANSLATION OF FLAGCSLRECOMBINANT [A]_______>

          60         70         80         90        100
TTGCCCGCGGAGGAGCCGCCTGCGCATGCTCCATCGCCTGGGAAATTTGG
AACGGGCGCCTCCTCGGCGGACGCGTACGAGGTAGCGGACCCTTTAAACC
 L  P  A  E  E  P  P  A  H  A  P  S  P  G  K  F  G>
_______TRANSLATION OF FLAGCSLRECOMBINANT [A]________>

         110        120        130        140        150
TGAGCGGCCTCCACCTAAACGACTTACTAGGGAAGCTATGCGAAATTATT
ACTCGCCGGAGGTGGATTTGCTGAATGATCCCTTCGATACGCTTTAATAA
  E  R  P  P  P  K  R  L  T  R  E  A  M  R  N  Y>
_______TRANSLATION OF FLAGCSLRECOMBINANT [A]________>

         160        170        180        190        200
TAAAAGAGCGAGGGGATCAAACAGTACTTATTCTTCATGCAAAAGTTGCA
ATTTTCTCGCTCCCCTAGTTTGTCATGAATAAGAAGTACGTTTTCAACGT
L  K  E  R  G  D  Q  T  V  L  I  L  H  A  K  V  A>
_______TRANSLATION OF FLAGCSLRECOMBINANT [A]________>

         210        220        230        240        250
CAGAAGTCATATGGAAATGAAAAAAGGTTTTTTTGCCCACCTCCTTGTGT
GTCTTCAGTATACCTTTACTTTTTTCCAAAAAAACGGGTGGAGGAACACA
 Q  K  S  Y  G  N  E  K  R  F  F  C  P  P  P  C  V>
_______TRANSLATION OF FLAGCSLRECOMBINANT [A]________>

         260        270        280        290        300
ATATCTTATGGGCAGCGGATGGAAGAAAAAAAAAAGAACAAATGGAACGCG
TATAGAATACCCGTCGCCTACCTTCTTTTTTTTTTCTTGTTTACCTTGCGC
  Y  L  M  G  S  G  W  K  K  K  K  E  Q  M  E  R>
_______TRANSLATION OF FLAGCSLRECOMBINANT [A]________>

         310        320        330        340        350
ATGGTTGTTCTGAACAAGAGTCTCAACCGTGTGCATTTATTGGGATAGGA
TACCAACAAGACTTGTTCTCAGAGTTGGCACACGTAAATAACCCTATCCT
D  G  C  S  E  Q  E  S  Q  P  C  A  F  I  G  I  G>
_______TRANSLATION OF FLAGCSLRECOMBINANT [A]________>

         360        370        380        390        400
AATAGTGACCAAGAAATGCAGCAGCTAAACTTGGAAGGAAAGAACTATTG
TTATCACTGGTTCTTTACGTCGTCGATTTGAACCTTCCTTTCTTGATAAC
 N  S  D  Q  E  M  Q  Q  L  N  L  E  G  K  N  Y  C>
_______TRANSLATION OF FLAGCSLRECOMBINANT [A]________>

         410        420        430        440        450
CACAGCCAAAACATTGTATATATCTGACTCAGACAAGCGAAAGCACTTCA
GTGTCGGTTTTGTAACATATATAGACTGAGTCTGTTCGCTTTCGTGAAGT
  T  A  K  T  L  Y  I  S  D  S  D  K  R  K  H  F>
_______TRANSLATION OF FLAGCSLRECOMBINANT [A]________>
```

```
        460       470       480       490       500
TTTTTTCTGTAAAGATGTTCTATGGCAACAGTGATGACATTGGTGTGTTC
AAAAAAGACATTTCTACAAGATACCGTTGTCACTACTGTAACCACACAAG
I  F  S  V  K  M  F  Y  G  N  S  D  D  I  G  V  F>
________TRANSLATION OF FLAGCSLRECOMBINANT [A]_________>

        510       520       530       540       550
CTCAGCAAGCGGATAAAAGTCATCTCCAAACCTTCCAAAAAGAAGCAGTC
 L  S  K  R  I  K  V  I  S  K  P  S  K  K  K  Q  S>
GAGTCGTTCGCCTATTTTCAGTAGAGGTTTGGAAGGTTTTTCTTCGTCAG
________TRANSLATION OF FLAGCSLRECOMBINANT [A]_________>

        560       570       580       590       600
ATTGAAAAATGCTGACTTATGCATTGCCTCAGGAACAAAGGTGGCTCTGT
TAACTTTTTACGACTGAATACGTAACGGAGTCCTTGTTTCCACCGAGACA
  L  K  N  A  D  L  C  I  A  S  G  T  K  V  A  L>
________TRANSLATION OF FLAGCSLRECOMBINANT [A]_________>

        610       620       630       640       650
TTAATCGACTACGATCCCAGACAGTTAGTACCAGATACTTGCATGTAGAA
AATTAGCTGATGCTAGGGTCTGTCAATCATGGTCTATGAACGTACATCTT
F  N  R  L  R  S  Q  T  V  S  T  R  Y  L  H  V  E>
________TRANSLATION OF FLAGCSLRECOMBINANT [A]_________>

        660       670       680       690       700
GGAGGTAATTTTCATGCCAGTTCACAGCAGTGGGGAGCCTTTTTTATTCA
CCTCCATTAAAAGTACGGTCAAGTGTCGTCACCCCTCGGAAAAAATAAGT
 G  G  N  F  H  A  S  S  Q  Q  W  G  A  F  F  I  H>
________TRANSLATION OF FLAGCSLRECOMBINANT [A]_________>

        710       720       730       740       750
TCTCTTGGATGATGATGAATCAGAAGGAGAAGAATTCACAGTCCGAGATG
AGAGAACCTACTACTACTTAGTCTTCCTCTTCTTAAGTGTCAGGCTCTAC
  L  L  D  D  D  E  S  E  G  E  E  F  T  V  R  D>
________TRANSLATION OF FLAGCSLRECOMBINANT [A]_________>

        760       770       780       790       800
TCTACATCCATTATGGACAAACATGCAAACTTGTGTGCTCAGTTACTGGC
AGATGTAGGTAATACCTGTTTGTACGTTTGAACACACGAGTCAATGACCG
V  Y  I  H  Y  G  Q  T  C  K  L  V  C  S  V  T  G>
________TRANSLATION OF FLAGCSLRECOMBINANT [A]_________>

        810       820       830       840       850
ATGGCACTCCCAAGATTGATAATTATGAAAGTTGATAAGCATACCGCATT
TACCGTGAGGGTTCTAACTATTAATACTTTCAACTATTCGTATGGCGTAA
 M  A  L  P  R  L  I  I  M  K  V  D  K  H  T  A  L>
________TRANSLATION OF FLAGCSLRECOMBINANT [A]_________>

        860       870       880       890       900
ATTGGATGCAGATGATCCTGTGTCACAACTCCATAAATGTGCATTTTACC
TAACCTACGTCTACTAGGACACAGTGTTGAGGTATTTACACGTAAAATGG
  L  D  A  D  D  P  V  S  Q  L  H  K  C  A  F  Y>
________TRANSLATION OF FLAGCSLRECOMBINANT [A]_________>
```

```
       910       920       930       940       950
TTAAGGATACAGAAAGAATGTATTTGTGCCTTTCTCAAGAAAGAATAATT
AATTCCTATGTCTTTCTTACATAAACACGGAAAGAGTTCTTTCTTATTAA
L  K  D  T  E  R  M  Y  L  C  L  S  Q  E  R  I  I>
______TRANSLATION OF FLAGCSLRECOMBINANT [A]_______>

       960       970       980       990      1000
CAATTTCAGGCCACTCCATGTCCAAAAGAACCAAATAAAGAGATGATAAA
GTTAAAGTCCGGTGAGGTACAGGTTTTCTTGGTTTATTTCTCTACTATTT
 Q  F  Q  A  T  P  C  P  K  E  P  N  K  E  M  I  N>
______TRANSLATION OF FLAGCSLRECOMBINANT [A]_______>

      1010      1020      1030      1040      1050
TGATGGCGCTTCCTGGACAATCATTAGCACAGATAAGGCAGAGTATACAT
ACTACCGCGAAGGACCTGTTAGTAATCGTGTCTATTCCGTCTCATATGTA
  D  G  A  S  W  T  I  I  S  T  D  K  A  E  Y  T>
______TRANSLATION OF FLAGCSLRECOMBINANT [A]_______>

      1060      1070      1080      1090      1100
TTTATGAGGGAATGGGCCCTGTCCTTGCCCCAGTCACTCCTGTGCCTGTG
AAATACTCCCTTACCCGGGACAGGAACGGGGTCAGTGAGGACACGGACAC
F  Y  E  G  M  G  P  V  L  A  P  V  T  P  V  P  V>
______TRANSLATION OF FLAGCSLRECOMBINANT [A]_______>

      1110      1120      1130      1140      1150
GTAGAGAGCCTTCAGTTGAATGGCGGTGGGGACGTAGCAATGCTTGAACT
CATCTCTCGGAAGTCAACTTACCGCCACCCCTGCATCGTTACGAACTTGA
 V  E  S  L  Q  L  N  G  G  G  D  V  A  M  L  E  L>
______TRANSLATION OF FLAGCSLRECOMBINANT [A]_______>

      1160      1170      1180      1190      1200
TACAGGACAGAATTTCACTCCAAATTTACGAGTGTGGTTTGGGGATGTAG
ATGTCCTGTCTTAAAGTGAGGTTTAAATGCTCACACCAAACCCCTACATC
  T  G  Q  N  F  T  P  N  L  R  V  W  F  G  D  V>
______TRANSLATION OF FLAGCSLRECOMBINANT [A]_______>

      1210      1220      1230      1240      1250
AAGCTGAAACTATGTACAGGTGTGGAGAGAGTATGCTCTGTGTCGTCCCA
TTCGACTTTGATACATGTCCACACCTCTCTCATACGAGACACAGCAGGGT
E  A  E  T  M  Y  R  C  G  E  S  M  L  C  V  V  P>
______TRANSLATION OF FLAGCSLRECOMBINANT [A]_______>

      1260      1270      1280      1290      1300
GACATTTCTGCATTCCGAGAAGGTTGGAGATGGGTCCGGCAACCAGTCCA
CTGTAAAGACGTAAGGCTCTTCCAACCTCTACCCAGGCCGTTGGTCAGGT
 D  I  S  A  F  R  E  G  W  R  W  V  R  Q  P  V  Q>
______TRANSLATION OF FLAGCSLRECOMBINANT [A]_______>

      1310      1320      1330      1340      1350
GGTTCCAGTAACTTTGGTCCGAAATGATGGAATCATTTATTCCACCAGCC
CCAAGGTCATTGAAACCAGGCTTTACTACCTTAGTAAATAAGGTGGTCGG
  V  P  V  T  L  V  R  N  D  G  I  I  Y  S  T  S>
______TRANSLATION OF FLAGCSLRECOMBINANT [A]_______>
```

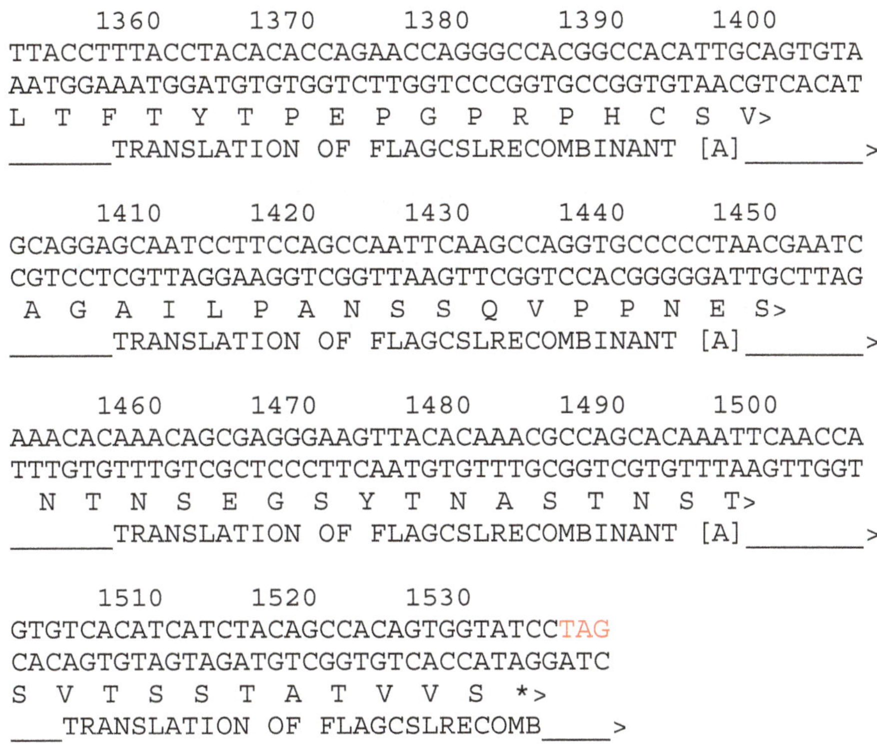

```
       1360      1370      1380      1390      1400
TTACCTTTACCTACACACCAGAACCAGGGCCACGGCCACATTGCAGTGTA
AATGGAAATGGATGTGTGGTCTTGGTCCCGGTGCCGGTGTAACGTCACAT
L  T  F  T  Y  T  P  E  P  G  P  R  P  H  C  S  V>
______TRANSLATION OF FLAGCSLRECOMBINANT [A]_______>

       1410      1420      1430      1440      1450
GCAGGAGCAATCCTTCCAGCCAATTCAAGCCAGGTGCCCCCTAACGAATC
CGTCCTCGTTAGGAAGGTCGGTTAAGTTCGGTCCACGGGGGATTGCTTAG
 A  G  A  I  L  P  A  N  S  S  Q  V  P  P  N  E  S>
______TRANSLATION OF FLAGCSLRECOMBINANT [A]_______>

       1460      1470      1480      1490      1500
AAACACAAACAGCGAGGGAAGTTACACAAACGCCAGCACAAATTCAACCA
TTTGTGTTTGTCGCTCCCTTCAATGTGTTTGCGGTCGTGTTTAAGTTGGT
  N  T  N  S  E  G  S  Y  T  N  A  S  T  N  S  T>
______TRANSLATION OF FLAGCSLRECOMBINANT [A]_______>

       1510      1520      1530
GTGTCACATCATCTACAGCCACAGTGGTATCCTAG
CACAGTGTAGTAGATGTCGGTGTCACCATAGGATC
S  V  T  S  S  T  A  T  V  V  S  *>
___TRANSLATION OF FLAGCSLRECOMB____>
```

For 6xHis-tag expression the last six nucleotides of CSL sequence were modified by mutagenesis to create XhoI site: TCCTAG -> CTCGAG. The modification allowed cloning directly in pET28a vector between 5' - NcoI and 3' - XhoI sites in frame with 3'- 6xHis-tag sequence, provided in the vector.

3) Quantification **Tables 1** and **2** for EMSA.

Table 1. Quantification of ^{32}P-activity in EMSA bands with bacterially expressed/purified CSL and ICN1, 6 independent experiments. In bold–data correspondent to EMSA in Figure 3.

59 bp CPS probe, No ICN1, pxl[1]			59 bp CPS probe, + ICN, pxl		
DNA-CSL	DNA	Total	DNA-CSL	DNA	Total
306758	**95866**	**402624**	**146782**	**319045**	**465827**
148573	434698	583271	64652	598300	662952
136546	160695	297241	92470	196675	289145
24636	17709	42345	7829	29633	37462
39090	76385	115475	32326	82371	114697
39745	56339	96084	34521	61949	96570

[1]Pxl units = amount of radioactivity converted to pixels by phosphorimaging.

Table 2. Quantification of ^{32}P-activity in EMSA bands with CSL/HEK293T and ICN1/HEK293T extracts, 7 independent experiments. In bold–data correspondent to EMSA in Figure 6.

No ICN1/HEK293T, pxl			+ ICN1/HEK293T, pxl			
DNA-CSL	DNA	Total	Shift	DNA	Non-specific	Total
24636	17709	42345	25512	14187	3205	42904
39090	**76385**	**115475**	**47863**	**60489**	**11107**	**119459**
39745	56339	96084	33145	52297	9245	94687
53628	117939	171567	53964	102001	9136	165101
157099	268372	425471	160484	228974	32693	422151
167115	255395	422510	182137	232023	1053	414160
120387	126638	247025	170213	114081	978	284294

4

The Impact of Seasonal Fluctuations on Rat Liver Mitochondria Response to Tested Compounds—A Comparison between Autumn and Spring. New Insight into Collecting and Interpretation of Experimental Data Originating from Different Seasons

Magdalena Labieniec-Watala[1*], Karolina Siewiera[2]
[1]Department of Thermobiology, Faculty of Biology and Environmental Protection, University of Lodz, Lodz, Poland
[2]Department of Haemostasis and Haemostatic Disorders, Medical University of Lodz, University Clinical Hospital No. 2, Lodz, Poland
Email: [*]magdalab@biol.uni.lodz.pl, ksiewiera@gmail.com

ABSTRACT

Seasonal variations play an essential role in the metabolism, behavior and activity of the laboratory animals. This study was aimed to examine whether mitochondrial function can be influenced by the seasonal changes and how large is the impact of these fluctuations on experiments with using animal models and further results interpretation. Liver mitochondria were isolated from male Wistar rats and exposed to calcium ions, PAMAM dendrimers G2.5 or their combination: (Ca^{2+}) and dendrimer. The scientific hypothesis assumed that dendrimer G2.5 is able to limit the detrimental effect of Ca^{2+} ions on mitochondria function, possibly through affecting the following parameters: calcium transport, mitochondrial potential and membrane fluidity. The activity of mitochondria was monitored using fluorescent labels. The changes in calcium transport were detected using Calcium Green 5-N, the mitochondrial membrane potential and membrane fluidity were elucidated using JC-1 and DPH, respectively. The experiments were carried out during the autumn (October/November) or during the spring (May/June). The obtained data emphasize the effect of seasonal differences on liver mitochondria originating from laboratory animals and outline the importance of planning the experiments during the same seasonal period in order to receive objective and reliable results in the future. Finally, it was revealed the neutral effect of G2.5 dendrimer on mitochondria and its inability to protect mitochondria against overload of calcium ions regardless of seasonality. It was also evidenced that liver mitochondria isolated from autumn-derived animals were more sensitive to calcium and/or dendrimer exposure in comparison with mitochondria isolated from animals investigated during the spring.

Keywords: Calcium Ions; Fluorescence Measurements; Mitochondrial Function; PAMAM Dendrimer G2.5; Seasonal Variations

1. Introduction

Mitochondria are important organelles for every cell as the powerhouse to provide energy for a multitude of cellular processes. They are also considered as the hub of metabolic pathways, primary sources of reactive oxygen species, regulators of apoptosis, and buffers of intracellular calcium. Calcium is a universal intracellular signaling molecule that regulates many pathways critical to cell survival. Under physiological conditions, mitochondrial calcium levels are maintained through calcium cycling between cytosol and matrix through the uniporter and Na/Ca exchanger activities. In addition, large amounts of calcium can be found in the form of precipitates in the matrix. Calcium overload occurs when mitochondria are exposed to calcium concentrations exceeding the matrix capacity, and may result in membrane de-energization, ROS production, cytochrome c release, permeability transition phenomenon (mPT) and apoptosis activation [1]. Then, the mitochondrial dysfunction leads to many human maladies, including cardiovascular diseases, neurodegenerative disease, and cancer. Therefore, it is very important to protect mitochondria against detrimental impact of calcium overload.

Polyamidoamine (PAMAM) dendrimers are, by far,

[*]Corresponding author.

the best studied of the commercialised and divergently synthesised dendrimers. Typically, these dendrimers are available in full generations (amine terminated) and half-generations (carboxylic acid terminated) that are representative of both their size (*i.e.*, diameter in angstroms) and molecular weight. Even at low concentrations, the peripheral amine groups of cationic dendrimers damage cell membranes and lead to cell toxicity. In contrast, anionic dendrimers are much less toxic or even non-toxic and have wider application in medical sciences. Therefore, the attention was focused on using PAMAM G2.5 (**Scheme 1**) as an agent limiting the detrimental Ca^{2+} effect on mitochondria.

In 2008 we used for the first time the anionic PAMAM dendrimer of the higher generation (G3.5) to protect mitochondria against calcium overload. Unfortunately, we observed that dendrimer G3.5 used at the concentration above 10 μM strongly affected the functionality of rat liver mitochondria. On the other hand, the decreased level of calcium ions added to mitochondria externally was revealed after exposure to G3.5 [2]. This promising observation helped to decide to continue these experiments and to use lower dendrimer's generation at the reduced concentration (5 μM). We believed that these "improvements" would lead us to the positive verification of our hypothesis stating that chelating properties of anionic PAMAM dendrimers could be applied also against diseases involving mitochondrial dysfunctions.

Recently, some papers reported that PAMAM dendrimers were able to bind metal ions [3-5]. It was evidenced using different methods that ions can be chelated not only buy dendrimers' terminal groups but also by their internal parts, *i.e.* by ethylenediamine core [4]. It was also revealed that anionic dendrimers could form complexes with metals which interact much less strongly with cell biomacromolecules than pure dendrimer or pure metal [5].

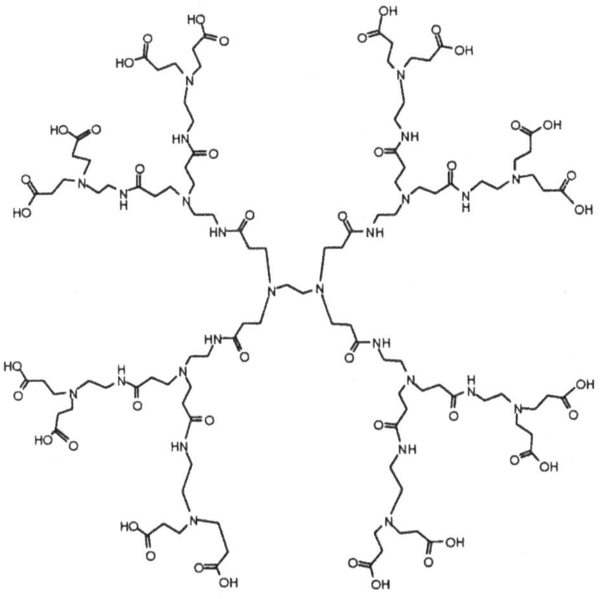

Scheme 1. The structure of PAMAM dendrimer G2.5.

Considering the above reports, as well as the other (non referred to herein), the aim of this study was to evaluate the ability of PAMAM G2.5 to bind calcium ions used at the concentrations toxic for rat liver mitochondria. We have chosen to this study only liver mitochondria. First of all, this research is a continuation of our earlier studies on liver mitochondria and anionic PAMAM dendrimer [2]. Secondly, mitochondria isolated from different tissues strongly differ between themselves. Therefore we focused our attention only on one type of mitochondria. Thirdly, liver mitochondria are less sensitive to "environemental changes", *i.e.* various calcium concentrations in comparison to brain or heart mitochondria. Therefore, in our opinion, the liver mitochondria are a very good model for our study, the main goal of which was to test the hypothesis that PAMAM G2.5 protects the mitochondria against detrimental overload with calcium ions. The calcium-induced mitochondrial changes in some parameters (calcium transport, membrane potential and membrane fluidity) were examined upon exposure to the low concentration of PAMAM dendrimer G2.5, using liver mitochondria isolated from Wistar rats. The exposure of mitochondria to PAMAM dendrimers was conducted in two approaches: 1) without the prior incubation with G2.5 or 2) with 10 min pre-incubation with the dendrimer. The study was performed in four steps. First, we selected the calcium concentration with the highest toxic activity towards the selected mitochondrial parameter, the calcium movement. Second, we evaluated the concentration of PAMAM dendrimer, which appeared neutral (not detrimental) towards another mitochondrial parameter, the transmembrane potential. Third, the activity of G2.5 against the detrimental influence of calcium ions was assessed. Fourth, the impact of seasonal variations (autumn and spring) on the collected data was analyzed.

2. Materials and Methods

2.1. Chemicals

PAMAM G2.5 dendrimers, 10 wt% solution in methanol (ethylenediamine core; molecular weight of this dendrimer and the number of terminal sodium carboxylate groups are 6265.60 and 32, respectively), succinate, rotenone, $CaCl_2$, carbonylcyanide m-chlorophenylhydrazone (CCCP), mannitol, EGTA, Tris-HCl, sucrose, Calcium Green 5-N (CG5-N), 5,5',6,6'-tetrachloro-1,1',3,3'-tetraethyl benzimidazolyl carbocyanine iodide (JC-1) and 1,6-diphenyl-1,3,5-hexatriene (DPH) were all purchased from Sigma-Aldrich. All other reagents and solvents used in this study were of the highest analytical reagent grade.

2.2. Biological Material

Mitochondria were isolated from livers of Wistar rats (male, 200 - 250 g) during autumn (October/November) and spring months (May/June). All animals used for experiment were at the same age and were maintained on the same diet throughout the whole experimental period(s). Before the experiment, the animals were housed under standard environmental conditions for 2 weeks (25°C, with a light/dark cycle of 12 h/12 h). The experiments were conducted in accordance with the Guide for the Care and Use of Laboratory Animals published by the US National Institute of Health (NIH Publication No. 85-23, revised 1985), as well as with the guidelines formulated by the European Community for the Use of Experimental Animals (L358-86/609/EEC) and the Guiding Principles in the Use of Animals in Toxicology (1989).

2.3. Isolation of Liver Mitochondria

Liver mitochondria were isolated from male Wistar rats according to the procedure described previously [2]. Briefly, liver was removed from the rat killed by decapitation and immediately cooled down to 4°C in the homogenization medium (75 mM sucrose, 225 mM mannitol, 5 mM Tris-HCl, pH 7.4, 0.5% BSA and 0.1 M EGTA). The liver tissue was washed to remove blood and connective tissue, cut into small pieces, homogenized using glass homogenizer and centrifuged for 5 min at 740 × g. Then, the pellet was discarded, and the supernatant was centrifuged again for 10 min at 9000 × g. The resulting supernatant was removed together with the top layer of the pellet, which contained damaged mitochondria. The dark-brown, "dense" mitochondrial pellet was resuspended in isolation buffer II (225 mM mannitol, 75 mM sucrose, 5 mM Tris-HCl, pH 7.4, 0.5% BSA) and centrifuged again for 10 min at 10,000 × g. For the final spin (10,000 × g for 10 min), the pellet was resuspended in the isolation buffer III (225 mM mannitol, 75 mM sucrose, 5 mM Tris-HCl, pH 7.4). The mitochondrial pellet was gently resuspended in the isolation buffer (containing 250 mM sucrose, 0.5 mM EDTA, 10 mM Tris and 1 g/l bovine serum albumin, pH 7.4) and stored on ice prior to the experiments.

All isolation procedures were carried out at 0°C - 4°C. During experimentation, mitochondria were stored on ice in the isolation medium until use. The experiments were performed up to 4 h after preparation. The protein concentration in each sample was determined by the Lowry method.

2.4. Transport of Ca^{2+} Ions across Mitochondrial Membranes

Transmembrane transport of calcium ions following the treatment with PAMAM G2.5 was monitored according to the procedure reported previously [6]. Briefly, the fluorescent dye Calcium Green 5-N was used to record the fluctuations in the extramitochondrial level of calcium ions upon addition of dendrimer G2.5. The whole experiment was divided into 2 stages: 1) the evaluation of EC_{50} and EC_{100} for Ca^{2+} ions (study carried out in autumn) and 2) the estimation of EC_{50} for dendrimer's activity (the study carried out in spring). Mitochondria (0.3 mg/ml protein) were suspended in 1.5 ml of buffer containing 100 mM KCl, 10 mM HEPES, 25 μM EGTA, 10 μM rotenone, 5 mM succinate, and 200 nM Calcium Green 5-N (K_D = 190 nM), pH 7.2. The following variants were used: 1) 5 μM PAMAM G2.5; 2) 10 min preincubation with 5 μM PAMAM G2.5; 3) 10 min preincubation with methanol (solvent for dendrimer); and 4) without PAMAM G2.5 (control). The Ca^{2+} uptake by mitochondria was monitored as the decreased fluorescence of Calcium Green 5-N along with the portions of external Ca^{2+} added to the medium. A mitochondrial uncoupler, *i.e.*, CCCP (5 μM), was used for internal control, to obtain maximum depolarization (data not shown), and massive release of Ca^{2+} by mitochondria in the presence of this protonophore. The titration of the suspension of mitochondria with external calcium led to stepwise Ca^{2+} accumulation in the medium until calcium overload was achieved. The measurements were conducted with a constant stirring at 37°C, using a PerkinElmer luminescence spectrometer (Model LS55), and emission continuously recorded at 531 nm with an excitation wavelength set at 506 nm. The relationships between the concentration of external calcium (x) and the fluorescence intensity of Calcium Green 5-N (y) were fitted to the four parametric regression curves: $y = a + \{[b - a]/[1 + (x/c)^d]\}$, and the values of EC_{50} (the effective dose of concentration for which the 50% effect occurs) for the tested samples were calculated by resolving of this function for $f(x) = y_{max}/2$. The coefficients of the above regression curves were iterated with the use of the quasi-Newton algorithm, using the convergence criterion of 0.0001 for max 500 iterations with the initial step of 0.5 for all the coefficients (GraphPad Prism ver. 5 and STATISTICA for Windows, ver. 10).

2.5. Evaluation of Transmembrane Potential with JC-1

Mitochondrial transmembrane potential following the treatment with Ca^{2+} ions and/or PAMAM G2.5 was monitored with the fluorescent cationic dye JC-1 according to the procedure reported by Feeney *et al.* [7]. The whole experiment was divided into 2 stages: 1) the evaluation of the safe (neutral) concentration of PAMAM dendrimer G2.5 (tested concentrations: 1, 2.5, 5, 10, 20 and 50 μM; the study carried out in autumn) and 2) the assessment of the activity of the concentration of dendrimer selected in the stage 1 (non- and 10-min pre-incubated) with and without Ca^{2+} ions (used at EC_{50} and EC_{100}), (the study carried out in spring). CCCP (5 μM) was used as a positive control (uncoupling compound).

Mitochondria (0.3 mg/ml protein) were suspended in 1.5 ml of buffer containing 125 mM KCl, 10 mM Tris-HCl, 2.5 mM KH_2PO_4, 10 mM EGTA, 10 μM rotenone, 5 mM succinate, and 1 μM JC-1, pH 7.4. Fluorescence measurements were performed in a Perkin-Elmer luminescence spectrometer (Model LS55). The excitation wavelength for JC-1 was 490 nm and the emission wavelengths were observed at 535 nm (fluorescence characterised for JC-1 monomers) and at 595 nm (fluorescence characterized for JC-1 aggregates). Then, the ratio of monomers to aggregates (535/595) was evaluated. The higher membrane depolaryzation, the higher value of this ratio was observed.

2.6. Assessment of Membrane Fluidity with DPH

The fluorescence probe DPH (1,6-diphenyl-1,3,5-hexatriene) was used to assess the so-called "fluidity" of the mitochondrial membrane in rat liver mitochondria after the treatment with PAMAM dendrimer G2.5 (5 μM) and Ca^{2+} ions (at EC_{50} and EC_{100}, 37°C) according to the procedure reported previously [6]. Mitochondria were resuspended at a final protein concentration of 0.75 mg/ml in the "analysis buffer", containing 60 mM KCl, 10 mM KH_2PO_4, 60 mM Tris-HCl, 0.5 mM EDTA, 5 mM succinate, and 10 μM rotenone, pH 7.4. Samples were incubated with DPH for 15 min to allow complete incorporation of the probe into the membranes. Fluorescence measurements were performed in a Perkin Elmer luminescence spectrometer (Model LS55). The excitation and emission wavelengths for DPH were selected with monochromators set to 360 nm (5 nm slit width) and 450 nm (5 nm slit width), respectively. The degree of fluorescence anisotropy was calculated according to Shinitzky and Barenholz [8].

2.7. Statistical Analysis

All measurements of mitochondrial parameters were performed in triplicates. Due to occasional data asymmetry in some variables and groups all data were expressed as median and interquartile lower-upper range (25% - 75%). Data normality was checked using the Shapiro-Wilk's test and variance homogeneity was verified with Levene's test. Then, data with evidenced normality were analyzed with parametric tests and these with non-proved normality were analyzed with non-parametric tests (Mann-Whitney U test with Bonferroni's correction for multiple comparisons). For heterogeneous variances, the non-parametric Kruskal-Wallis test and non-parametric Conover-Inman post-hoc test for multiple comparisons were used. The statistical significance between homogenous groups was estimated using one-way ANOVA or two-way ANOVA and post hoc Tukey tests. For all experiments, the number of sample size was estimated for type I and II statistical errors of 0.05 and 0.8, respectively. Furthermore, the power of used tests was also checked for each (parametric) analysis. The power test below 80% was considered as unbelievable outcome and the constructive conclusions were not formulated. All statistical calculations were made with the use of STATISTICA.PL v.9 or 10 (StatSoft) and StatsDirect (Stats-Direct Limited). EC_{50} parameters were calculated using GraphPad Prism (ver. 5).

3. Results

All presented results were collected during two different seasonal stages: autumn and spring. We do not show the data from winter or summer. Results obtained in autumn were compared to these ones received in spring and based on these comparisons we have drawn the conclusions presented below. In this paper we report only the part of our study performed on liver mitochondria.

3.1. Calcium Movements across Mitochondrial Membranes

The sensitivity of rat liver mitochondria on calcium ions was assessed by fluorescence method, using Calcium Green 5-N for the monitoring of calcium concentration outside the organelles in the suspension of mitochondria. The design of this experiment was to titrate mitochondria suspension with small aliquots of Ca^{2+} ions (up to 2 μM) and to determine the external concentration of Ca^{2+} ions, up to which calcium gradually accumulates inside mitochondria until the abrupt depolarization of mitochondrial membrane.

3.1.1. Effect of Ca^{2+} Ions on Mitochondria—The Evaluation of EC_{50} and EC_{100} during the Autumn Measurements

Based on the obtained titration curves (9 curves from 9 animals), we were able to evaluate the effect of calcium on the "iterated" values of the EC_{50} and EC_{100} parameters. The example of such a curve, characteristic for this stage of the study (autumn measurements), was presented in **Figure 1**. The value of EC_{50} was 18.5 ± 1.019 μM and EC_{100} was 37 ± 2.040 μM. Data were expressed as mean ± SEM.

3.1.2. Effect of PAMAM G2.5 on Ca^{2+} Mobilization across Mitochondrial Membranes—The Evaluation of EC_{50} during the Spring Measurements

In this experiment, the dendrimer G2.5 was used at the

concentration of 5 μM (selected based on the measurements of membrane potential with JC-1). The effect of the dendrimer was tested in two approaches: 1) without the prior pre-incubation and 2) with 10 min pre-incubation. In **Table 1** the values of EC_{50} are presented, calculated for the autumn and spring measurements.

Based on statistical calculations it was revealed that mitochondria isolated from animals investigated in autumn were more sensitive to calcium ions compared to the organelles isolated from spring animals. As shown in **Figure 2**, the higher concentration of calcium ions (33.3 μM) was needed to reach 50% calcium-induced mitochondria depolarization isolated from spring animals compared to autumn rats (18.5 μM), $P < 0.05$.

The effect of PAMAM dendrimer G2.5 on calcium ions movement across mitochondrial membranes was not observed, neither in the mitochondria not pre-incubated nor in those pre-incubated with PAMAM G2.5. This suggests that dendrimer used at this concentration (5 μM) was not able to limit the mitochondrial depolarization caused by Ca^{2+} ions. On the other hand, the dendrimer did not contribute to the increased sensitivity of mitochondria to calcium, as we evidenced its neutral impact on the tested parameter.

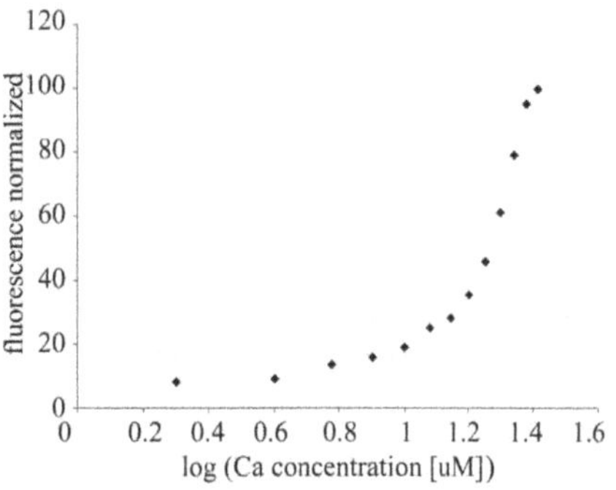

Figure 1. Titration of rat liver mitochondria with calcium ions. The EC_{50} and EC_{100} were calculated based on the resolving of the four-parametric curve, $y = a + \{[b - a]/[1 + (x/c)^d]\}$, using GraphPad Prism software, ver. 5 for non-linear regression curves. For further experimental and data analysis details see Section 2.

Table 1. Values of EC_{50} evaluated in different seasons for rat liver mitochondria exposed to Ca^{2+} ions and PAMAM G2.5.

Autumn	Spring			P
Control	Control	+G2.5	+G2.5 pre-incubated	
18.5 ± 1.0	33.3 ± 5.5	26.3 ± 4.6	38.0 ± 4.5	*0.0382

Data are expressed as mean ± SEM, n = 9 animals (measurements done in triplicates for each animal); EC_{50} given in μM. Statistical significance (*P = 0.0382) refers to the comparison of "control" measured in autumn months vs. "control" measured in spring months. The differences for samples measured in spring were not statistically significant: control vs. G2.5, P = 0.6707; control vs. G2.5 (pre-incubated), P = 0.9058 and G2.5 vs. G2.5 (pre-incubated), P = 0.1556. For further details see Section 2.

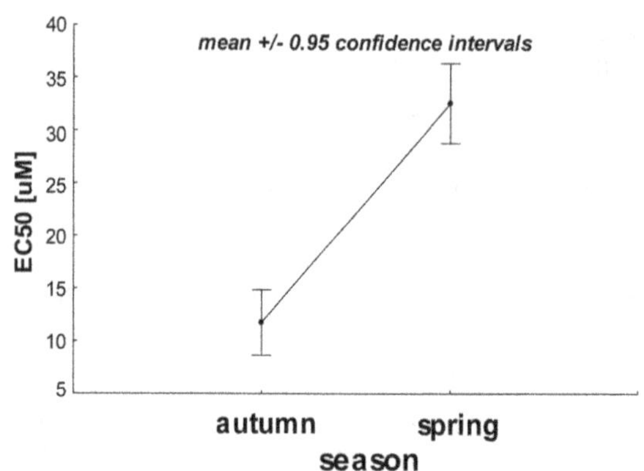

Figure 2. The effect of seasonality on the calcium-mediated rat liver mitochondrial depolarization. Data presented as mean ± 95% confidence intervals. Statistical significance the between tested samples (using two-way ANOVA) was P = 0.0382 (STATISTICA ver. 10, StatSoft). For further experimental and statistical details see Section 2.

3.2. Transmembrane Potential Evaluation

Rat liver mitochondrial potential was measured spectrofluorimetrically using JC-1. JC-1 was applied at the concentration of 1 μM. CCCP (5 μM) was used as uncoupler (positive control, data not shown). The interpretation of the mitochondrial potential was based on the changes in the ratio between the fluorescence of JC-1 monomers and JC-1 aggregates, depending on the changes in mitochondrial potential.

3.2.1. Effect of PAMAM Dendrimer G2.5 on Mitochondrial Potential—Evaluation of the Neutral Concentration of the Dendrimer—The Autumn Measurements

The "neutral" concentration of PAMAM G2.5 was chosen based on the changes in mitochondrial potential measured during the autumn period. As shown in **Figure 3**, none of the tested G2.5 concentrations (0, 1, 2.5, 5, 10, 20 and 50 μM without preincubation) was shown to significantly affect the mitochondrial potential.

Based on these and other observations (non included in this paper) the concentration of 5 μM was selected to be used in further investigations, as completely non-toxic. The significant reduction in mitochondrial potential was observed only for the mitochondria treated with 5 μM CCCP (data not shown).

3.2.2. Effect of PAMAM G2.5 and Ca^{2+} Ions on Mitochondrial Potential—The Spring Measurements

The results obtained in this part of the study revealed that the mitochondria studied during the spring period are not sensitive to the tested compounds. Some different approaches were employed: 1) calcium ions used at EC_{50} (18.5 μM) or EC_{100} (37 μM); 2) dendrimer G2.5 used at

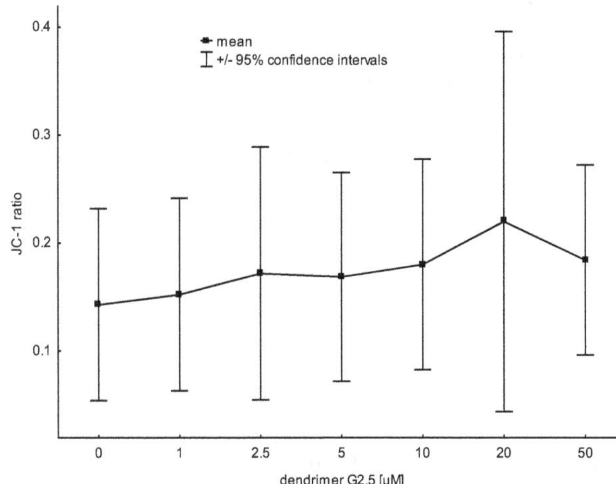

Figure 3. The impact of PAMAM dendrimer G2.5 on rat liver mitochondrial potential, measured with the use of the fluorescence probe, JC-1 (autumn measurements). Data are expressed as mean ± 95% confidence intervals, n = 6 animals (for each animal 3 mitochondrial preparations were tested). No statistically significant differences were observed between the tested concentrations of the dendrimer. For further experimental and statistical details see Section 2.

5 μM (mitochondria not pre-incubated and pre-incubated for 10 min with the dendrimer); and 3) the combinations of G2.5 and calcium ions used at both the concentrations. As shown in **Table 2**, calcium used at EC_{50} and EC_{100} did not cause any changes in the mitochondrial potential.

No alterations were also observed for dendrimers and the combinations of dendrimer + Ca^{2+} with the respect to control samples (mitochondria alone). It means that all the tested variants of both the agents had no impact on mitochondrial potential. However, the statistically significant differences were revealed between: 1) the higher (37 μM) and the lower (18.5 μM) calcium concentration ($P < 0.05$), as well as between 2) the combination of "5 μM G2.5 + 37 μM Ca^{2+}", and "5 μM G2.5" ($P < 0.05$). These data suggest that there was a tendency to reduce the mitochondrial potential by the higher Ca^{2+} concentrations, although we did not observe the statistically significant depolarization of mitochondria exposed to 37 μM Ca^{2+} (non-significant versus control mitochondria).

Interestingly, it was noticed that there was a significant difference in the responses to G2.5 (5 μM) exposure between the "autumn" mitochondria and the "spring" mitochondria. As shown in **Figure 4**, the "autumn" mitochondria appeared more sensitive (the higher ratio of JC-1 means the reduction in the mitochondrial potential) to G2.5 activity in comparison with the "spring" mitochondria ($P = 0.0094$). Nevertheless, the changes in the potential between the control "autumn" mitochondria and the control "spring" mitochondria were not revealed ($P = 0.0681$).

Table 2. The effect of PAMAM G2.5 and Ca^{2+} ions on rat liver mitochondrial potential—the spring evaluation.

Sample	JC-1 ratio	P value
Control	0.036 ± 0.003	
18 μM Ca^{2+}	0.025 ± 0.002	ns vs. control
37 μM Ca^{2+}	0.046 ± 0.006	ns vs. control 0.011 vs. 18 μM Ca^{2+}
5 μM G2.5	0.027 ± 0.002	ns vs. control
5 μM G2.5 (10 min pre-incubated with mitochondria)	0.029 ± 0.002	ns vs. control
5 μM G2.5 + 18 μM Ca^{2+}	0.028 ± 0.002	ns vs. control
5 μM G2.5 (pre-incubated) + 18 μM Ca^{2+}	0.032 ± 0.002	ns vs. control
5 μM G2.5 + 37 μM Ca^{2+}	0.060 ± 0.013	ns vs. control 0.027 vs. 5 μM G2.5

Data (JC-1 ratio of monomers to aggregates) are presented as mean ± SEM, n = 6 animals (the measurements were done in triplicates for each animal); ns-non statistically significant. For further experimental and statistical details see Section 2.

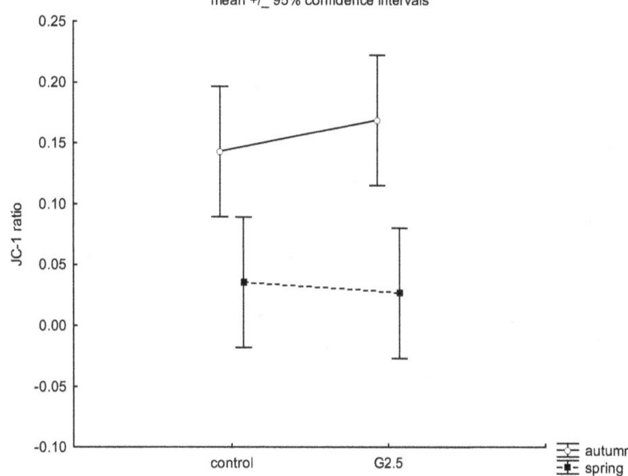

Figure 4. The effect of seasonality (autumn and spring) on PAMAM dendrimer G2.5 (5 μM) activity. Data (JC-1 ratio of monomers to aggregates) are presented as mean ± 95% confidence intervals, n = 6 animals (measurements were done in triplicates for each animal). The statistical significance (two-way ANOVA, StatSoft, ver. 10) was evidenced only between the treatment with PAMAM G2.5 during autumn and spring ($P < 0.005$). For further experimental and statistical details see Section 2.

Moreover, it was evidenced that the time plays an important role in the mitochondrial potential studies. All tested variants of samples measured after 30 min incubation (room temperature in dark) showed depolarization of mitochondrial membrane at the level of statistically significant, *i.e.* the JC-1 ratio for control mitochondria in "0 time" and control mitochondria after "30 min incubation" increased from 0.036 ± 0.003 to 0.094 ± 0.006 (mean ± SEM), $P < 0.0005$ (the complete data not shown). These results suggest that mitochondria are aging rapidly and it is important to take into account the "passage of time" as a meaningful factor, when measuring mitochondrial potential.

3.2.3. Mitochondrial Membrane Fluidity Evaluation—The Spring Measurements

Rat liver membrane fluidity was evaluated based on the anisotropy fluorescence of DPH. The tested compounds were studied according to the design presented in **Table 3**. The obtained results revealed that none of the applied compounds (or their combinations) had an impact on membrane fluidity. Nevertheless, the power of the performed comparisons was relatively low (below 70% or lower). Therefore, we have to consider the results of these experiments with a caution. The presented data notwithstanding, are intriguing and certainly deserve to be further explored in the future. Similar phenomena were evidenced for the same samples measured after 30 min of incubation at room temperature: membrane fluidity remained unchanged upon the treatment, and the power of test was below 40%.

4. Discussion

In the present study, the hypothesis that PAMAM dendrimer G2.5 protects mitochondria against the overload with calcium ions was tested. It is widely considered that the activity of Ca^{2+} ions towards mitochondria is ambiguous. On one hand, there are known advantageous effects of calcium, *i.e.* stimulation of ATP production,

Table 3. The effect of PAMAM G2.5 and Ca^{2+} ions on rat liver mitochondrial fluidity—the spring evaluation.

Sample	DPH anisotropy (a.u.)	P value	Power of test [%]
control	0.384 ± 0.004	-	-
18 μM Ca^{2+}	0.395 ± 0.002	ns	66
37 μM Ca^{2+}	0.393 ± 0.003	ns	10
5 μM G2.5	0.382 ± 0.004	ns	60
5 μM G2.5 (10 min pre-incubated with mitochondria)	0.391 ± 0.005	ns	14
5 μM G2.5 + 18 μM Ca^{2+}	0.389 ± 0.006	ns	12
5 μM G2.5 (pre-incubated) + 18 μM Ca^{2+}	0.383 ± 0.005	ns	5
5 μM G2.5 + 37 μM Ca^{2+}	0.381 ± 0.005	ns	60

Data (DPH fluorescence anisotropy) are presented as mean ± SEM, n = 6 animals (measurements were done in triplicates for each animal); ns-difference non statistically significant. Statistical analysis and power of test were calculated using STATISTICA ver. 9 (StatSoft). Power of test was measured for the comparisons in relation to control. For further experimental and statistical details see Section 2.

coordination of cell metabolism or activation of mitochondrial enzymes. On the other hand, in contrast to this beneficial activity, the pathological effect of calcium on mitochondria function has been well evidenced [9]. It cannot be ignored that Ca^{2+}, particularly at high concentrations, appears to have several negative impacts on mitochondria, *i.e.* induction of mPT pores, stimulation of mitochondrial uncoupling, activation of apoptosis via mitochondria or interruption of cellular signaling. Therefore, it has become important to find compounds/agents, which would work as efficient modulators of the excessive calcium in the various cellular compartments, including mitochondria.

For a few years our attention has been focused on anionic PAMAM dendrimers. It has been well demonstrated that PAMAM dendrimers are successfully used for many *in vitro* and *in vivo* biomedical applications [10]. On the other side, PAMAM dendrimers were tested for their effects on cellular cytotoxicity, lysosomal pH, and mitochondria dependent apoptosis. Some results suggest that full generations of PAMAM dendrimers are endocytosed into the cells through a lysosomal pathway, leading to lysosomal alkalinization and induction of mitochondria-mediated apoptosis [11]. Other experimental data also prove the detrimental modulatory effects of cationic dendrimres (especially G4 and higher) on mitochondria function [12-14]. Nevertheless, there are only few literature data concerning the negative influence of half-generations of PAMAMs on mitochondria. Our previous study [2] suggested that, on one hand, PAMAM G3.5 used above 10 μM affected the selected rat liver mitochondrial parameters (*i.e.* caused membrane depolarization), but on the other effectively hindered the influx of Ca^{2+} to mitochondria. Thus, PAMAM dendrimer G3.5 did not have the desired biological activity, however, our other findings, based upon still unpublished data, suggest that smaller generations and/or smaller concentrations of these molecules should be more effective in the protection of mitochondria against the overload with calcium ions.

Our present data show that PAMAM G2.5, used in a concentration of 5 μM, does not affect the rat liver mitochondrial parameters, such as membrane potential, membrane fluidity and calcium transport across the membrane. On the other hand, the protective role of this dendrimer (towards mitochondria) was not observed as it was expected at the stage of the formulation of research hypothesis. Maybe, the used concentration of dendrimer was too low, but some data (not shown here) revealed that the concentration of G2.5 above 7.5 μM significantly caused the disturbances in mitochondrial bioenergetics (measurements done using the high-fidelity Oroboros-2k oxygraph). Therefore, it was safer to use lower G2.5 concentration when testing its activity. The main assumption of this study was based on the proven chelating properties of half-generations PAMAM dendrimers [3-5]. We believed that G2.5 is able to successfully bind excess of Ca^{2+} ions and thus limits their input to mitochondria. However, it was important to meet two experimental conditions in order to confirm these assumptions. The first one concerned the choice of the neutral concentration of G2.5 towards mitochondrial function and the second one concerned the ability of binding calcium ions by this dendrimer. Neutral concentration of G2.5 was assessed experimentally in our laboratory, but we did not check its chelating activity towards Ca^{2+}. Based only on the literature data, we hoped that even low generations of PAMAMs can effectively bind metal ions, such as: Ag^{+}, Cu^{2+}, Zn^{2+}, Fe^{2+}, and Fe^{3+} [5,15]. Nevertheless, so far, the investigation of binding Ca^{2+} by PAMAM dendrimers was not proven using appropriate experimental techniques (*i.e.* mass spectrometry). Based on the above information, we need to emphasize that this study has been designed simple examination of the relationship of the "concentration-effect", rather than the proving of Ca^{2+}-chelating properties of G2.5.

Interestingly, we have noticed, although not for the first time [16,17], that the response of mitochondria to the treatment with various compounds depends on the seasonality. The results presented in this paper also indicate that there is a statistically significant difference between the activities of the mitochondria isolated from rats investigated in autumn (October-November) and the mitochondria isolated from animals studied in spring (May-June). Based on the activity of mitochondrial parameters, such as mitochondrial potential and Ca^{2+} movement across the mitochondrial transmembranes, it appears that "autumn" mitochondria are more sensitive in comparison with "spring" mitochondria. The mitochondrial potential decrease (the higher depolarization) and the resistance to the input of calcium ions overload were observed. Not many laboratories, which conduct the studies with using rats as their animal model, distinguished the impact of seasonal variations on the changes observed in the cell or tissue at the molecular level. So far, some reports relate to the changes in the breeding activity and animal metabolism [18], others refer to the differences in immune system in birds [19], food-catching activity [20] or reproduction behavior [21]. Nevertheless, the recent study of Konior *et al.*, demonstrated that guinea-pig and rat hearts generate more super oxide anion, and human subjects excrete more 8-isoprostane in the summer compared to other seasons [22]. In contrary to our observations, the conclusions mentioned above indicate that the increase of oxidative stress was characteristic for animals studied in summer in comparison to changes observed for animals tested in winter months. On the other hand, it was evidenced in young rats that their arteries were protected from thrombogenesis and lipid accumulation during summer months vs. winter period [23]. The study of Mock and Frankel showed that older rats (5 - 6 months old) might be more sensitive to seasonal influences than younger rats (4 months old). Their results indicate that male laboratory rats exhibited seasonality in several reproductive parameters, even though the animals were maintained under constant laboratory conditions [24]. There are a lot of similar examples of studies demonstrating the role of seasonality, which makes us to believe that the influence of the seasons should always be considered when designing of the study. Otherwise, the consequences of carrying out the studies during the whole year without blocking of the obtained data according to the employed laboratory conditions can deliver misleading outcomes. Unfortunately, at this stage of our research, we are not able to explain at biochemical level (*i.e.* the monitoring of liver glycogen content) why mitochondria differ physiologically, when studied in different seasons. Until now there is a lack of information concerning seasonal influence on the laboratory animals, and thus, more studies are needed to answer questions raised also in this paper.

Despite the strong evidence linking the calcium ions and mitochondria damage, the precise concentration of Ca^{2+} needed to cause mitochondrial changes remains unknown. It seems likely to be dependent on the source of mitochondria, cellular metabolism and other cellular conditions. Some data showed that Ca^{2+} used at 10 μM caused irreversible and very fast membrane depolarization of rat liver mitochondria [25]. Also the study of Vergun and Reynolds revealed that depolarization of rat liver mitochondria required the use of calcium at 20 μM after ca. 10 min incubation with mitochondria [26]. The same concentration of Ca^{2+} ions was used to reduce the effectiveness of oxidative phosphorylation in rat liver mitochondria [27]. Taking into account these data it seems that our result—18.5 μM (EC_{50} for Ca^{2+}, assessed during autumn) is in agreement with other literature findings. Surprisingly however, we were not able to obtain the similar results for the same experiment carried out a few months later (spring period). The EC_{50} for calcium ions was 33.3 μM, which indicated that the tested mitochondria were much less sensitive to detrimental effects of calcium ions. Probably, the main reason of this differentiation in the received data lies in the distinct seasons (autumn/spring), during which these measurements were done. Consequently, when conducting of the experiments in spring, and using the values of Ca^{2+} and G2.5 evaluated in autumn, we might have received the data, which could not properly verify the hypothesis. It was very hard to check the protective activity of G2.5 towards mitochondria, which were treated by calcium ions used at the concentration much lower than that needed to observe the desired changes. Therefore, in fact, we could only test the impact of PAMAM dendrimer on mitochondrial parameters. Nevertheless, without repeating of the study related to the finding of both non-cytotoxic and cytotoxic concentrations of PAMAM G2.5 also for the spring season, it was hard or even impossible to observe any effect of dendrimer tested at 5 μM on mitochondria, regardless of 10 min pre-incubation of G2.5 with mitochondria before the measurements. In support of this notion, no impact of G2.5 was observed on calcium transport, mitochondrial potential and membrane fluidity, thus suggesting that the used G2.5 concentration probably was too low for spring mitochondria under the tested conditions. Additionally, the experiment with JC-1 using PAMAM G2.5 in the range of concentrations: 1 - 50 μM and rat liver mitochondria carried out in summer (August) showed that none of the tested concentrations had the impact of mitochondrial potential (data not shown).

In addition to assessing of the main mitochondrial parameters, the effect of mitochondrial "aging", based on the results obtained from membrane potential evaluations, has been studied. It was proven that the same mitochondrial samples revealed statistically significant membrane depolarization measured after 30 min post-incubation (in dark and room temperature). It was clearly evidenced that the passage of time plays an important role during mitochondrial measurements. Nevertheless, it is not a new discovery and mitochondrial functional impairment with aging was evidenced earlier [28,29].

Summarizing, the present data revealed that PAMAM dendrimer G2.5, used at the concentration of 5 μM (with and without pre-incuabtion), had no impact on rat liver mitochondrial functions. On the other hand, however, it was proven that it might be too low concentration in order to use it as a regulator of calcium ions in the mitochondrial therapy. These results indicate that in the future the dendrimer G2.5 could be used successfully in further mitochondrial examinations not only in a combination with calcium ions. Nevertheless, all experiments should be finished during the short interval of time in order to limit the influence of seasonal changes on the acquired data.

5. Conclusion

The primary objective of this study was the using of PAMAM G2.5 as a regulator of detrimental effect of calcium ions on the main mitochondrial functions. The goal was to find the destructive concentration of calcium ions and the neutral concentration of G2.5 achieved with the use of liver mitochondria isolated from Wistar rats in different seasonal periods (autumn and spring). Based on the obtained results it is worth of noting that seasonal fluctuations have an important effect on measured mitochondrial parameters. Furthermore, the unsuitable experimental design and their inappropriate statistical analysis (e.g. not including block designed ANOVA) often leads to the misleading conclusions. Therefore, to determine the targeting experimental tools and therapeutic agents to mitochondria with using animals in different age or derived in different seasons requires to demonstrate the great care and knowledge about the entire research model. With greater understanding of physiological differences between tested samples/animals and advances in biomaterial design, there is an opportunity to develop systems that respond to physiological stimuli, seasonal fluctuations as well as the potential to use compounds/ chemicals as promising drug(s). Use of safe and effective multi-functional nano-platforms promises to alleviate many of the mitochondrial disturbances what may be benefit for treatment of diseases involving mitochondrial disorders in the future.

6. Acknowledgements

This work has been funded by grant from Ministry of Science and Higher Education, N N405 261037. Authors thank Anna Jarosz for her excellent technical support. Authors are also grateful to Prof. Cezary Watala for the opportunity to use the laboratory of the Department of Haemostasis and Haemostatic Disorders (Medical University of Lodz) during the collecting of data for this study.

REFERENCES

[1] H. Kawamata, G. Manfredi, "Mitochondrial Dysfunction and Intracellular Calcium Dysregulation in ALS," *Mechanism of Ageing and Development*, Vol. 131, No. 7-8, 2010, pp. 517-526. doi:10.1016/j.mad.2010.05.003

[2] M. Labieniec and T. Gabryelak, "Preliminary Biological Evaluation of Poli(Amidoamine) (PAMAM) Dendrimer G3.5 on Selected Parameters of Rat Liver Mitochondria," *Mitochondrion*, Vol. 8, No. 4, 2008, pp. 305-312. doi:10.1016/j.mito.2008.07.001

[3] S. M. Cohen, S. Petoud and K. N. Raymond, "Synthesis and Metal Binding Properties of Salicylate-, Catecholate-, and Hydroxypyridinonate-Functionalized Dendrimers," *Chemistry—A European Journal*, Vol. 7, No. 1, 2001, pp. 272-279. doi:10.1002/1521-3765(20010105)7:1<272::AID-CHEM 272>3.0.CO;2-Y

[4] M. S. Diallo, S. Christie, P. Swaminathan, L. Balogh, X. Shi, W. Um, C. Papelis, W. A. Goddard III and J. H. Johnson Jr., "Dendritic Chelating Agents. 1. Cu(II) Binding to Ethylene Diamine Core Poly(Amidoamine) Dendrimers in Aqueous Solutions," *Lanqmuir*, Vol. 20, No. 7, 2004, pp. 2640-2651. doi:10.1021/la036108k

[5] S. Sekowski, A. Kazmierczak, J. Mazur, M. Przybyszewska and T. Gabryelak, "The Interaction between PAMAM G3.5 Dendrimer, Cd^{2+}, Dendrimer-Cd^{2+} Complexes and Hu-

man Serum Albumin," *Colloids and Surfaces B: Biointerfaces*, Vol. 69, No. 1, 2009, pp. 95-98. doi:10.1016/j.colsurfb.2008.11.006

[6] M. Labieniec, T. Przygodzki, J. Carsky, D. Malinska, J. Rysz and C. Watala, "Effects of Resorcylidene Aminoguanidine (RAG) on Selected Parameters of Isolated Rat Liver Mitochondria," *Chemico-Biological Interactions*, Vol. 179, No. 2-3, 2009, pp. 280-287. doi:10.1016/j.cbi.2008.11.005

[7] Ch. J. Feeney, P. S. Pennefather and A. V. Gyulkhandanyan, "A Cuvette-Based Fluorometric Analysis of Mitochondrial Membrane Potential Measured in Cultured Astrocyte Monolayers," *Journal of Neuroscience Methods*, Vol. 125, No. 1-2, 2003, pp. 13-25. doi:10.1016/S0165-0270(03)00027-X

[8] M. Shinitzky, "Membrane Fluidity and Cellular Functions," In: M. Shinitzky, Ed., *Physiology of Membrane Fluidity*, Vol. 12, Plenum Press, New York, 1984, pp. 1-51. doi:10.1007/978-1-4684-4667-8_20

[9] P. S. Brookes, Y. Yoon, J. L. Robotham, M. W. Anders and S.-S. Sheu, "Calcium, ATP, and ROS: A Mitochondrial Love-Hate," *American Journal of Physiology: Cell Physiology*, Vol. 287, No. 4, 2004, pp. C817-C833.

[10] I. J. Majoros, Ch. R. Williams, D. A. Tomalia and J. R. Baker Jr., "New Dendrimers: Synthesis and Characterization of Popam-Pamam Hybrid Dendrimers," *Macromolecules*, Vol. 41, No. 22, 2008, pp. 8372-8379. doi:10.1021/ma801843a

[11] T. P. Thomas, I. Majoros, A. Kotlyar, D. Mullen, M. M. Holl and J. R. Baker Jr., "Cationic Poly(Amidoamine) Dendrimer Induces Lysosomal Apoptotic Pathway at Therapeutically Relevant Concentrations," *Biomacromolecules*, Vol. 10, No. 12, 2009, pp. 3207-3214. doi:10.1021/bm900683r

[12] J. H. Lee, K. E. Cha, M. S. Kim, H. W. Hong, D. J. Chung, G. Ryu and H. Myung, "Nanosized Polyamidoamine (PAMAM) Dendrimer-Induced Apoptosis Mediated by Mitochondrial Dysfunction," *Toxicology Letters*, Vol. 190, No. 2, 2009, pp. 202-207. doi:10.1016/j.toxlet.2009.07.018

[13] M. Labieniec, O. Ulicna, O. Vancova, J. Kucharska, T. Gabryelak and C. Watała, "Effect of Poly(Amido)Amine (PAMAM) G4 Dendrimer on Heart and Liver Mitochondria in an Animal Model of Diabetes," *Cell Biology International*, Vol. 34, No. 1, 2010, pp. 89-97.

[14] S. P. Mukherjee, F. M. Lyng, A. Garcia, M. Davoren and H. J. Byrne, "Mechanistic Studies of *in Vitro* Cytotoxicity of Poly(Amidoamine) Dendrimers in Mammalian Cells," *Toxicology and Applied Pharmacology*, Vol. 248, No. 3, 2010, pp. 259-268. doi:10.1016/j.taap.2010.08.016

[15] M. A. Kaczorowska and H. J. Cooper, "Electron Capture Dissociation and Collision-Induced Dissociation of Metal Ion (Ag^+, Cu^{2+}, Zn^{2+}, Fe^{2+}, and Fe^{3+}) Complexes of Polyamidoamine (PAMAM) Dendrimers," *Journal of the American Society for Mass Spectrometry*, Vol. 20, No. 4, 2009, pp. 674-681. doi:10.1016/j.jasms.2008.12.013

[16] M. Labieniec-Watala, K. Siewiera and Z. Jozwiak, "Resorcylidene Aminoguanidine (RAG) Improve Cardiac Mitochondrial Bioenergetics Impaired by Hyperglycaemia in a Model of Experimental Diabetes," *International Journal of Molecular Sciences*, Vol. 12, No. 11, 2011, pp. 8013-8026. doi:10.3390/ijms12118013

[17] K. Siewiera and M. Labieniec-Watala, "Ambiguous Effect of Dendrimer PAMAM G3 on Rat Heart Respiration in a Model of an Experimental Diabetes—Objective Causes of Laboratory Misfortune or Unpredictable G3 Activity?" *International Journal of Pharmaceutics*, Vol. 430, No. 1-2, 2012, pp. 258-265.

[18] M. L. Carras, E. Brenowitz and E. W. Rubel, "Peripheral Auditory Processing Changes Seasonally in Gambel's White-Crowned Sparrow," *Journal of Comparative Physiology A: Neuroethology Sensory, Neural and Behavioral Physiology*, Vol. 196, No. 8, 2010, pp. 581-599. doi:10.1007/s00359-010-0545-1

[19] L. B. Martin, Z. M. Weil and R. J. Nelson, "Seasonal Changes in Vertebrate Immune Activity: Mediation by Physiological Trade-Offs," *Philosophical Transactions of the Royal Society B: Biological Sciences*, Vol. 363, No. 1490, 2008, pp. 321-339. doi:10.1098/rstb.2007.2142

[20] M. T. Avey, A. Rodriguez and Ch. B. Sturdy, "Seasonal Variation of Vocal Behaviour in a Temperate Songbird: Assessing the Effects of Laboratory Housing on Wild-Caught, Seasonally Breeding Birds," *Behavioural Processes*, Vol. 88, No. 3, 2011, pp. 177-183. doi:10.1016/j.beproc.2011.09.005

[21] B. J. Prendergast, A. Kampf-Lassin, J. R. Yess, J. Galang, N. McMaster and L. M. Kay, "Winter Lengths Enhance T Lymphocyte Phenotypes, Inhibit Cytokine Responses, and Attenuate Behavioral Symptoms of Infection in Laboratory Rats," *Brain Behavior, and Immunity*, Vol. 21, No. 8, 2007, pp. 1096-1108. doi:10.1016/j.bbi.2007.05.004

[22] A. Konior, E. Klemenska, M. Brudek, E. Podolecka, E. Czarnowska and A. Beręsewicz, "Seasonal Superoxide Overproduction and Endothelial Activation in Guinea-Pig Heart; Seasonal Oxidative Stress in Rats and Humans," *Journal of Molecular and Cellular Cardiology*, Vol. 50, No. 4, 2011, pp. 686-694. doi:10.1016/j.yjmcc.2010.11.010

[23] S. Masumura, F. Furui, M. Hashimoto and Y. Watanabe, "The Effects of Season and Exercise on the Levels of Plasma Polyunsaturated Fatty Acids and Lipoprotein Cholesterol in Young Rats," *Biochimica and Biophysica Acta: Lipids and Lipid Metabolism*, Vol. 1125, No. 3, 1992, pp. 292-296.

[24] E. J. Mock and A. I. Frankel, "A Seasonal Influence on Tests Weight and Serum Gonadotropin Levels of the Mature Male Laboratory Rat," *Biology Reproduction*, Vol. 18, No. 5, 1978, pp. 772-778. doi:10.1095/biolreprod18.5.772

[25] M. J. Devinney, L. M. Malaiyandi, O. Vergun, D. B. DeFranco, T. G. Hastings and K. E. Dineley, "A Comparison of Zn^{2+}- and Ca^{2+}-Triggered Depolarization of Liver Mitochondria Reveals No Evidence of Zn^{2+}-Induced Permeability Transition," *Cell Calcium*, Vol. 45, No. 5, 2001, pp. 447-455. doi:10.1016/j.ceca.2009.03.002

[26] O. Vergun and I. Reynolds, "Distinct Characteristics of Ca^{2+}-Induced Depolarization of Isolated Brain and Liver Mitochondria," *Biochimica and Biophysica Acta: Bio-

energetics, Vol. 1709. No. 2, 2005, pp. 127-137.

[27] P. M. Silva, E. Tanabe, A. P. Hermoso, C. A. Bersani-Amado, A. Bracht, E. L. Ishii-Iwamoto and C. L. Salgueiro-Pagadigorria, "Changes in Calcium-Dependent Membrane Permeability Properties in Mitochondria of Livers from Arthritic Rats," *Cell Biochemistry and Function*, Vol. 26, No. 4, 2008, pp. 443-450. doi:10.1002/cbf.1461

[28] M. Picard, D. Ritchie and K. J. Wright, "Mitochondrial Functional Impairment with Aging Is Exaggerated in Isolated Mitochondria Compared to Permeabilized Myofibers," *Aging Cell*, Vol. 9, No. 6, 2010, pp. 1032-1046. doi:10.1111/j.1474-9726.2010.00628.x

[29] M. Picard, D. Ritchie, M. M. Thomas, K. J. Wright and R. T. Hepple, "Alterations in Intrinsic Mitochondrial Function with Aging Are Fiber Type-Specific and Do Not Explain Differential Atrophy between Muscles," *Aging Cell*, Vol. 10, No. 6, 2011, pp. 1047-1055. doi:10.1111/j.1474-9726.2011.00745.x

Abbreviations

PAMAM dendrimer: poly(amido)amine dendrimer;
mPT pore: mitochondrial permeability transition pore.

5

An Analysis of the Correlation between the Changes in Satellite DNA Methylation Patterns and Plant Cell Responses to the Stress

Darina A. Sokolova, Galina S. Vengzhen, Alexandra P. Kravets
Department of Plant Biophysics and Radiobiology, Institute of Cell Biology and Genetic Engineering,
National Academy of Science of Ukraine, Kiev, Ukraine
Email: kaplibra@gmail.com

ABSTRACT

The differences in satellite DNA methylation pattern of corn seedlings with various spontaneous chromosome aberration yields and changes in methylation pattern of these DNA sequences under different exposure modes of acute UV-C and chronic gamma-irradiations have been investigated. The obtained experimental data and the conducted correlation analysis demonstrated the significant correlation between the satellite DNA methylation pattern varieties and chromosome aberration yields under various stress exposure modes. The role of satellite DNA methylation pattern variability and its changing in key responses to stress such as mobile elements' activation, cell's passage of checkpoints, and homological repair was discussed.

Keywords: Stress Response; Plant Resistance; Satellite DNA Methylation Pattern; Brave-Pirson Linear Correlation

1. Introduction

Changing organism's resistance to stress factors, various reactions, which role in this process depends on factor's acting rate, duration and/or periodicity. Complexity of interactions in stress reactions can also be attributed to hierarchical-structural and functional, organism organization, where different processes have various sensitivities and times of development.

DNA methylation is one of the most important and polyfunctional mechanisms of biological regulation, which has a great significance in such epigenetic processes as genomic imprinting, differentiation, apoptosis and morphogenesis, aging of an organism, regulation of mobile elements' activity [1-3].

It is also known that methylation of cytosine is the natural factor of mutagenesis [2] and at the same time it is a factor affecting regional DNA structure's organization that is necessary for successful passage of enzymatic reactions, related to reading-out of information and reparation.

Plants contain most of methylated cytosine (up to 30%); the DNA methylation of these organisms is the result of functioning four groups of methyltransferases [2] that provide a great methylation sites' variety.

It may be claimed that various methylation pathways can play important roles in stress response reactions and rearrangements of their resistance, whereas in alternative "to run or to fight" these organisms choose "fighting" at all levels of organization. A number of last investigations show some changes in level and pattern of DNA methylation under biotic [3,4] and different forms of abiotic stress—dryness [3], salinization [3-6], radiation exposures with various dose rates [7-9] and duration [8,10].

Polyfunctional of DNA methylation process also allows different ways of its participation both in failure (e.g. activization of mobile elements, initialization of genome instability) and/or formation of active protective reactions, associated with metabolism reorganization. Thus changes in DNA methylation level and/or pattern under different stress exposure [3-10] still require specification of their biological significance.

The appearance of DNA micro array technology made a revolution in studying changes in gene expression under stress exposures. Obtained data have confirmed the connection between changes in methylation pattern of transcribed DNA with changes in expression of major gene groups, metabolism rearrangements and resistance changes under stress exposure [11-13].

In parallel with studying the majority of changes in transcribed DNA methylation pattern in their responsiveness under stress exposures, great changes in satellite DNA methylation pattern have been detected [14].

It is known that satellite DNA is true to type component of eukaryotic genome. It consists of tandem organized repeats, and it is never transcribed or encoded proteins and is located in heterochromatin part of chromosome [15]. A high methylation level of satellite DNA' cytosine has been shown but satellite DNA's biological importance still hasn't been understood. A question about biological role of changes in DNA methylation pattern under stress exposures and subsequent changes of cell resistance is also unexplored now.

The paper is dedicated to investigate the connection between variability of satellite DNA methylation pattern and spontaneous chromosome aberration' rate as well as changes in methylation pattern of satellite DNA under different modes of acute UV-C and chronic gamma-exposure of seedlings. The study of DNA methylation pattern is performed by comparing the chromosomal aberrations yielded in meristematic tissues as the independent index that allowed to estimate the plant cell resistance.

2. Material and Method

The investigation of connection between satellite DNA methylation statuses with plant cell resistance to stress exposure was carried out in three series of experiments:

1) Acute UV-C exposure of epigenetically different corn seedlings (EDS). Preliminary three groups of corn' seedlings with different germination rates were empiric selected: fast germinating (F-G), middle germinating (M-G), and slowly germinating (S-G). A great connection between germination rates and differences in transcribed DNA methylation pattern has been preinstalled;

2) Acute UV-C exposure in the mode of "adaptive exposure-challenge exposure" with different ranges between the adaptive UV-C irradiation and challenge one (different mode UV-C exposure). The adaptive dose was 1 kJ/m^2 and the challenge one—6.2 kJ/m^2;

Combined exposure: preliminary chronic gamma-exposure of dry seeds with various accumulative dose and subsequent seedlings acute UV-C exposure.

Two intervals between the adaptive UV-C irradiation and challenge one were investigated: 4 hours and 24 hours. The necessity to expose seedlings in challenge dose (6.2 kJ/m^2) and whole dose (7.2 kJ/m^2) in the same physiological state was taken into account. Thus such variants of irradiation were used:

1) Non UV-C irradiated seedlings;

2) Adaptive exposure (1 kJ/m^2);

3) Adaptive exposure, in 4 hours-challenging one (6.2 kJ/m^2);

4) Whole dose exposure (7.2 kJ/m^2); exposure simultaneously with the challenging irradiation of variant 3;

5) Adaptive exposure, in 24 hours—challenging one (6.2 kJ/m^2);

6) Whole dose exposure (7.2 kJ/m^2); irradiation simultaneously with the challenging irradiation of variant 5.

Such ways of irradiation were conducted both with seedlings from non preliminary gamma-irradiated seeds (NPI) and with seedlings from preliminary gamma-irradiated seeds (PI).

The study was performed using 3 - 7-days maize seedlings, sort Titan. Seeds' sprouting was conducted on bottom plates with wet filter paper, in thermostat under the temperature +23˚C - +24˚C. Bactericidal irradiator of the open type OBN-150M (Ukraine) with Philips Special TUV 30 W lamps was used. Three-day seedlings were exposed by UV-C in whole doses of 7.2 kJ/m^2 (dose rate was 6.2 W/m^2) in the range 4 hours and 24 hours between adaptive and challenging irradiation as described above.

A glass container with $^{137}CsCl_2$ was used for investigation of chronic exposure effects; dry seeds were exposed with dose rate 30 mR/h, accumulated dose reached 3.5 Gy.

The apical root meristems were used as an object for cytogenetic analysis. Sampling was carried out on the 4th day after irradiation. Detached apexes have been put to the Brodsky' fixative (acetic acid: ethanol: formalin = 0.3:1:3) for two hours with following washing by 70% ethanol (3 - 4 times). Maceration has been performed by alkaline hydrolysis with 20% NaOH over two hours. Then preparations have been washed in distilled water for 15 minutes. Staining was carried out by acetoarsein and hydrochloric acid mixture (acetoarsein: 1M HCl = 1:1) over 16 - 18 hours. Stained samples have been washed in 45% CH_3COOH with following preparation the crushed specimens. Ten alternative apexes were used and 5 - 10 thousands of cells were analyzed for every variant. The unstable chromosomal aberrations were detected using anaphase-telophase technique due to plant tissue specificity. In spite of this cells' sampling has averaged over 300 - 350 chromosomal aberrations during the anaphase in each preparation. A cytogenetic analysis was conducted on the light microscope "Jenaval" (Germany). Independent cytogenetic analisis was performed 8 times. Significance level (α) of assessment is 0.05.

Isolation of DNA was performed from the 6-day-old corn seedlings with the set of reagents Diatom™ DNA Prep100 based on NucleoS-sorbent. The standard protocol for DNA extraction provided by the manufacturer was used. The concentration of DNA solution was measured by BioPhotometer Plus Eppendorf v.1.35 using standard technique [16,17].

The PCR was carried out in the four-channel DNA-amplifier "Tercik" ("DNA-Technology", Moscow). One primer has been used: inter simple sequence repeat-ISSR (15-soro, sequence-5'-AC-AC-AC-AC-AC-AC-AC-AC-<C>-3'), were synthesized by company "Metabion" (Germany) [18].

The restriction analysis as well as the PCR was carried out in the four-channel DNA-amplifier "Tercik" ("DNA-Technology", Moscow). Two types of restriction enzymes-isoschizomers were used: HpaII (5'.C CGG.3'), MspI (5'.C CGG.3') and restrictase MboI ("Fermentas", Germany). Reactions were performed according to the conventional manual by the supplier (**Table 1**).

The reaction mixture for the HpaII-analysis (total volume 25 µl) contained: 0.2 µl HpaII, 2.0 µl 10xBuffer Tango, 1.5 µg total DNA and 17.7 µl deionized water. The mixture has been covered with the 20 µl of mineral oil.

The reaction mixture for the MspI-analysis (total volume 25 µl) contained: 0.6 µl MspI, 2.0 µl 10xBuffer Tango, 1.5 µg total DNA and 17.1 µl deionized water. The mixture has been covered with the 20 µl of mineral oil.

The reaction mixture for the MboI-analysis (total volume 25 µl) contained: 0.2µl MboI, 2.0 µl 10xBuffer Tango, 1.5 µg total DNA and 17.7 µl deionized water. The mixture has been covered with the 20 µl of mineral oil.

The conditions for restriction reactions were: 16 hours under 37˚C, then 20 min under 65˚C (for HpaII and MboI) and 20 min under 80˚C (for MspI) to stop the reactions.

Products of PCR and restriction analysis were separated in 1.0% agarose gel with TBE-buffer at the presence of ethidium bromide, and visualized in UV-transilluminator. The same volume of PCR and restriction products (10 µl) was brought into the gel pockets. The FastRuler High Range DNA Ladder ("Fermentas", Germany) with fragments' length 10,000, 4000, 2000, 1000 and 500 base pairs and the FastRuler Low Range DNA Ladder ("Fermentas", Germany) with fragment length 1500, 850, 400, 200 and 50 base pairs were used as a molecular weight markers. Independent ISSR-PCR was performed 8 times also.

Experimental findings statistical analysis–the variance value and the Brave-Pirson's correlation coefficient-were calculated with traditional method [19].

Table 1. Restriction enzymes and their sites of recognition/restriction.

Restriction enzyme	Sites of recognition/restriction
MspI	5'C...C*CG, C...5'
HpaII	5'...C*CGG...3'
	3'G...G C*C...5'
MboI	5'...C*CGC...3'
	3'...CT...AG *C...5'

3. Results and Discussion

The obtained cytogenetic data pointed out major varieties in chromosome aberrations' yield (Ab, %) appeared among groups F-G, M-G and S-G seedling (**Figure 1**).

The electrophoregram of isolated DNA nativity is shown in **Figure 2**.

The electrophoregram of native DNA amplification with ISSR primers (**Figure 3**) shows specific differences

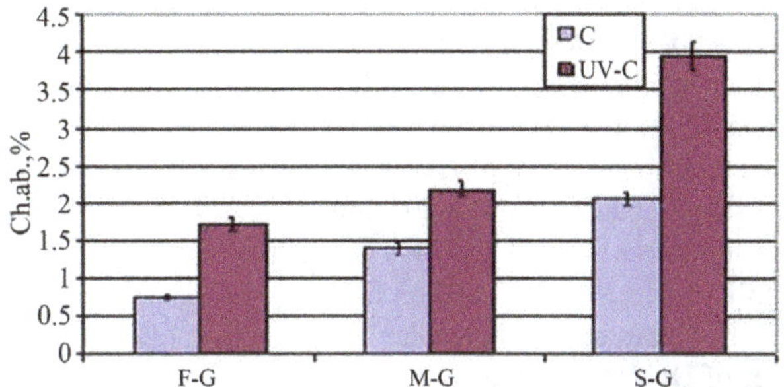

Figure 1. The chromosome aberration yield (α = 0.05) in root meristem of corn seedlings with various germination rates; C-non-irradiated seedlings; UV-C-seedlings irradiated with UV-C.

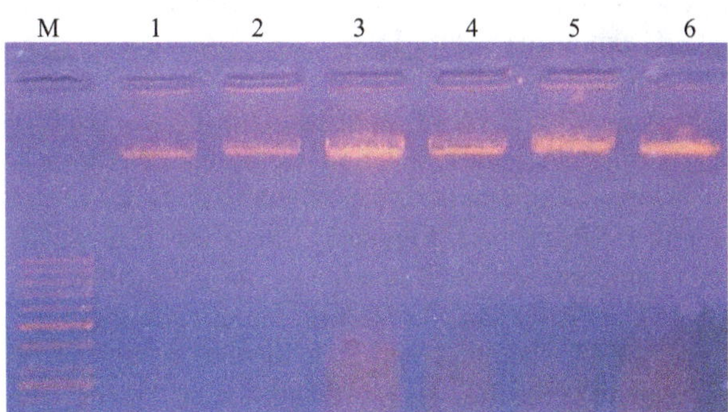

Figure 2. The electrophoregram of isolated DNA quality. M—high-molecular-weight marker; 1—"FG" sample; 2—"FG + UV-C" sample; 3—"MG" sample; 4—"MG + UV-C" sample; 5—"SG" sample; 6—"SG + UV-C" sample.

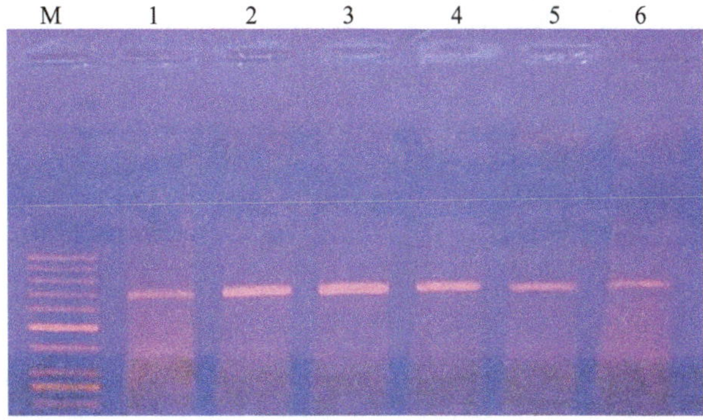

Figure 3. The electrophoregram of native DNA amplification products with ISSR primers. M—high-molecular-weight marker; 1—"FG" sample; 2—"FG+UV-C" sample; 3—"M-G" sample; 4—"M-G + UV-C" sample; 5—"SG" sample; 6—"S-G + UV-C" sample.

in amplicons range of irradiated and unirradiated fast-growing seedlings (positions 1 and 2).

These data do not contradict with data about good nativity of isolated DNA. The most appropriate explanation is connected with appearance of damage during PCR that might indirectly evidence about low methylation level of this DNA part in fast-growing seedlings resulting to greater vulnerability of these DNA samplers 1.

An electrophoregram of the amplification products obtained by ISSR-PCR of the MspI restriction products (**Figure 4**) illustrated the differences in DNA methylation pattern among seedlings with various germination rates (positions 1, 3, 5).

The electrophoregram of fast-germinated seedlings (F-G, position 1) had four distinct groups of amplicons with almost the same number of DNA fragments. The groups of amplicons (positions 3 and 5) for variants "M-G" and "S-G" had the same molecular weight, but different number of DNA fragments.

The comparison of positions 1 and 2, 3 and 4, 5 and 6 of this electrophoregram (**Figure 4**) shows great changes of satellite DNA methylation pattern after irradiation. Positions 2, 4, 6 are also differing from each others that correspond to increased chromosome aberration' yield after UV-C exposure (**Figure 1**).

Also considerable differences between methylation patterns of satellite DNA of seedlings that initially had various germination rates (positions 1, 3, 5) were observed in separating amplification products of MboI restricts with ISSR–primers (**Figure 5**). There was just one type of amplicons for "F-G" seedlings and great differences between "M-G" and "S-G" variants. Electrophoregram for "M-G" seedlings had four distinct groups of amplicons with comparatively more high-molecular fragments.

The comparison of positions 1 and 2, 3 and 4, 5 and 6 of this electrophoregram (**Figure 5**) shows great changes of satellite DNA methylation pattern after irradiation.

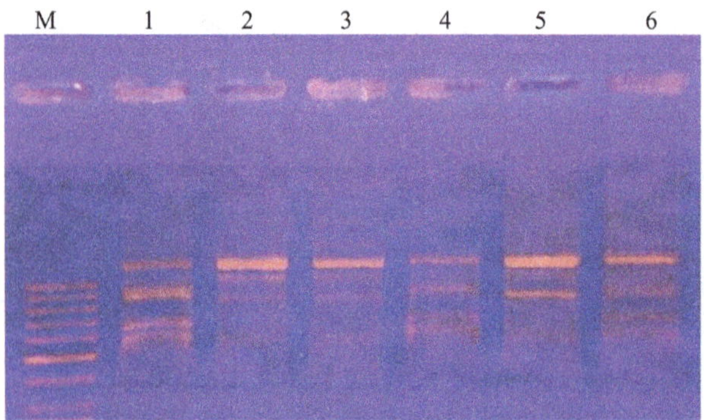

Figure 4. The electrophoregram of the amplification products obtained by ISSR-PCR of the MspI restriction products. M—high-molecular-weight marker; 1—"FG" sample; 2—"FG + UV-C" sample; 3—"MG" sample; 4—"MG + UV-C" sample; 5—"SG" sample; 6—"SG + UV-C" sample.

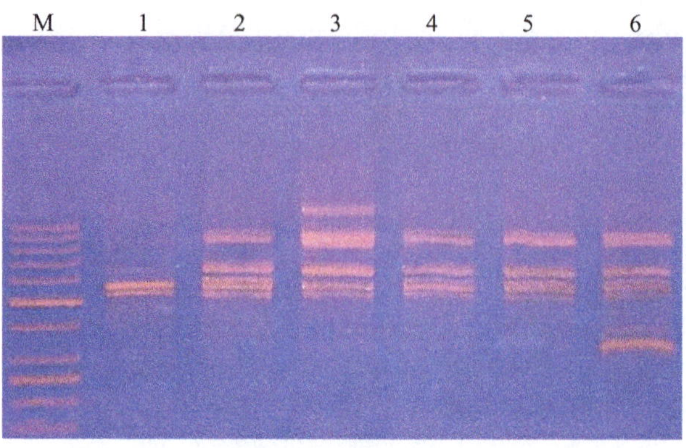

Figure 5. The electrophoregram of the amplification products obtained by ISSR-PCR of the MboI restriction products. M—high-molecular-weight marker; 1—"FG" sample; 2—"FG + UV-C" sample; 3—"MG" sample; 4—"MG + UV-C" sample; 5—"SG" sample; 6—"SG + UV-C" sample.

Positions 2, 4 do not have major differences from each other. The greatest difference is observed between positions 4 and 6. Such differences correspond to various increasings in the chromosome aberration' yield after UV-C exposure (**Figure 1**).

Thus original difference in satellite DNA methylation pattern is connected to differences in pattern changes under irradiation exposure and chromosome aberration' yield. This indicates both different effectiveness of repair processes or various original sensitiveness to damage.

Consider the data about acute UV-C exposure mode "adaptive-challenging irradiation" as well as combined exposure whereby seedlings growing from preliminary gamma-irradiated seeds have been exposed.

Chromosome aberration yield in root meristematic tissue (**Figure 6**) indicates to major differences in appearance of seedlings' adaptive reactions that have grown from unexposed and gamma-exposed seeds. Chronic radiation exposure of seeds causes increase of chromosome aberration rate in seedlings' root meristematic tissues. UV-C exposure of seedlings from preliminary unirradiated seeds with adaptive dose leads to increasing chromosome aberration yield whereas exposure of seedlings from preliminary irradiated seeds causes the hormetic effect. Exposure mode "adaptive, in 4 hours-challenging" causes the appearance of adaptive response for seedlings without preliminary irradiation exposure; with interval in 24 hours between adaptive and challenging exposure the adaptive response haven't been observed. Seedlings from preliminary irradiated seeds didn't show the adaptive response with both intervals between adaptive and challenging irradiation.

An explanation of such phenomena from the standpoint about meristematic tissue' heterogeneity and possibility of two forms of repopulation renewal is given in paper [14]. The object of this study is to compare stability changes to stress factor affecting and changes in sat-

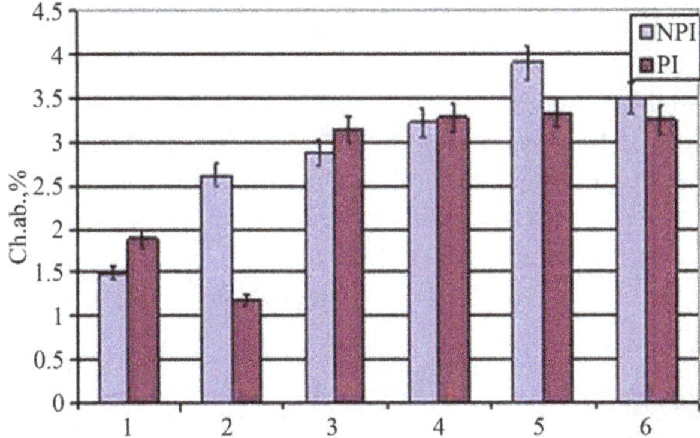

Figure 6. The chromosome aberration yield (α = 0.05) in root meristem of corn seedlings from non preliminary gamma-irradiated seeds (NPI), and preliminary gamma-irradiated seeds, (PI) with UV-C irradiation mode as described in " Material and Methods".

ellite DNA methylation pattern.

The electrophoregrams of DNA quality checking are shown in **Figure 7**. These results demonstrate the absence of meaningful DNA fragmentation that would take a place in apoptosis inducing by UV-C irradiation.

The electrophoregram of native DNA amplification shows specific differences in comparison with all other variants of amplicon's range for variant "adaptive-challenge exposure in 24 hours" (Position 6). The most appropriate explanation is connected to appearance of ulterior (single-stranded) injuries during PCR. It's essential that the features of DNA fragmentation are observed in a variant exposed with full dose at a time.

For **Figures 7-11**: 1. Total control (NPI and non UV-C irradiated seedlings); 2. NPI + adaptive exposure; 3. NPI + adaptive exposure, in 4 hours-challenging one; 4. NPI + whole dose exposure (7.2 kJ/m^2); exposure simultaneously with the challenging irradiation of variant 3); 5. NPI + adaptive exposure, in 24 hours-challenging one; 6. NPI+ whole dose exposure; irradiation simultaneously with the challenging irradiation of variant 5; 7. PI + non UV-C irradiation; 8. PI + adaptive exposure; 9. PI + adaptive exposure, in 4 hours–challenging one;10.PI + whole dose exposure; exposure simultaneously with the challenging irradiation of variant 3 and 9); 11. PI + adaptive exposure, in 24 hours-challenging one; 12. PI + whole dose exposure; irradiation simultaneously with the challenging irradiation of variants 5 and 11.

The electrophoregram of MboI restricts' ISSR amplification shows various differences in DNA methylation patterns according to exposure mode. Comparison of positions 1 (seedlings from seeds without preliminary irradiation) and 7 (seedlings from preliminary gamma-irradiated seeds) indicates to major differences in range of amplicons: as a result of dry seeds chronic exposure the satellite DNA methylation pattern of seedlings shows some complication on electrophoregram because of ap-

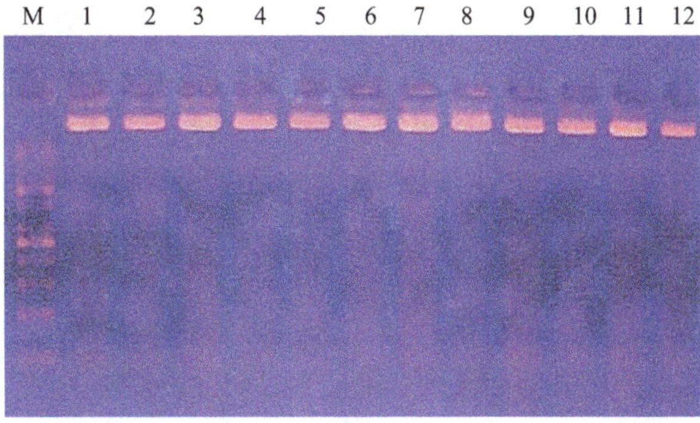

Figure 7. The electrophoregram of isolated DNA quality.

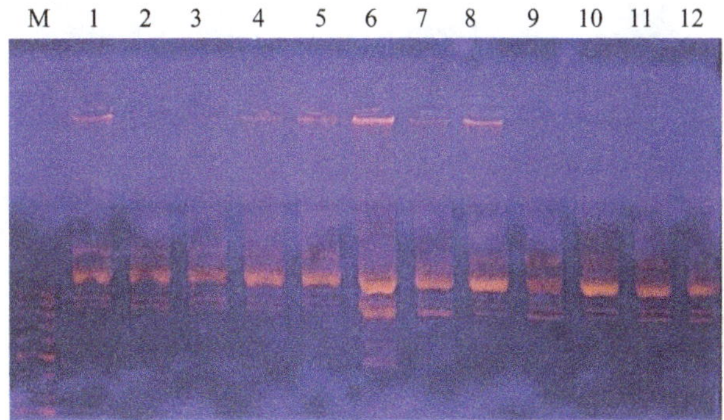

Figure 8. The electrophoregram of native DNA ISSR-amplification.

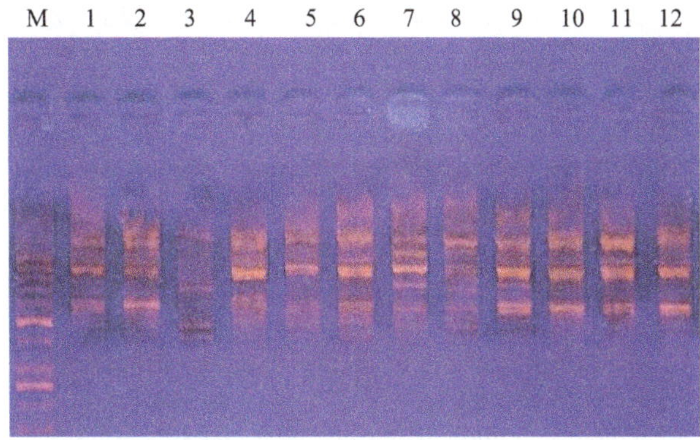

Figure 9. The electrophoregram of MboI restricts' ISSR amplification.

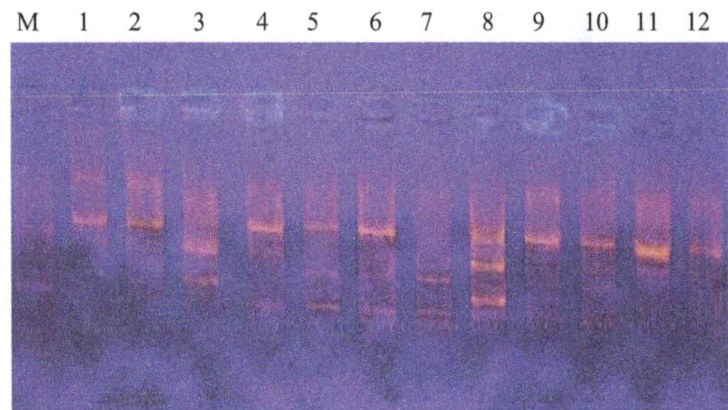

Figure 10. The electrophoregram of ISSR-amplification of MspI restricts.

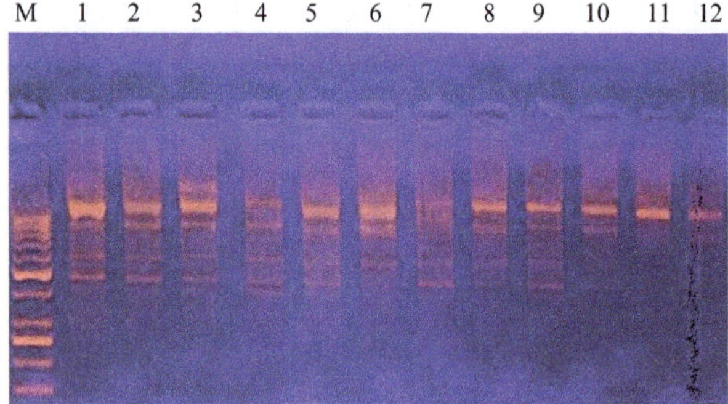

Figure 11. The electrophoregram of HpaII-restricts ISSR-amplification.

pearance of amplicons with low and middle weight. It's essential that variants 9 - 12 show identical ranges and parallels with the same chromosome aberration yield in the variants are also possible (**Figure 6**).

The electrophoregram of ISSR-amplification of MspI restricts shows various changes in DNA methylation pattern according to exposure mode. Comparison of positions 1 (seedlings from seeds without preliminary irradiation) and 7 (seedlings from preliminary gamma-exposed seeds) indicates to major differences in amplicons range: after dry seeds' chronic irradiation satellite DNA methylation pattern leads to great complication of electrophoregram because of appearing low-weight amplicons indicated to increase of restriction MspI sites. Major difference of 8th variant is observed, which demonstrates hormesis effect in terms of chromosome aberration yield. It's essential that variants 9 - 12 show identical ranges as well as MboI enzyme.

Electrophoregram of HpaII-restricts' ISSR-amplification indicates to less dependence of amplicon range from exposure mode. The differences between positions 1 (seedlings from seeds without preliminary irradiation) and 7 (seedlings from preliminary gamma-irradiated seeds) are also visible because of less content of high-molecular weight fragments. Great difference of variants 9 and 10-12 that corresponds almost identical chromosome aberration yield is observed (**Figure 6**).

Quantify connection between changes in satellite DNA methylation patterns and chromosome aberrations' yield under various affects using Brave-Pirson's linear correlation. To perform such approach we have to suggest some principals of quantifying various changes on electrophoregrams and their degrees.

There are several significant quantitative characteristics of DNA methylation pattern changes that could be registered on electrophoregrams:

1) Change of general amplicons' number; wherein following versions are possible:

a) Changes in molecular mass of amplicons, *i.e.* position related to ladder bands on electrophoregram but within the amplicons' mass of control variant; nevertheless the number of new control bands or their disappearance could be various;

b) The appearance of amplicons with mass that greatly exceed the limits of control bands' mass both in the range of more high molecular mass and;

2) Change of bands' brightness–that indicates to changing number of amplicone' fragments of the same mass;

3) The combination of the listed above quantitative indicators.

Interactions between these various indicators greatly exceed the classification possibilities of changes in DNA methylation pattern and accordingly the correlation estimation between their changing rates.

Consider the simplest connection type-linear correlation between the number of amplicons and the chromosome aberrations' yield for various series of experiments. The statistical analysis for each experimental series was conducted separately.

Correlation indexes shown in **Table 2** indicate to existence of significant (α = 0.05) positive correlation between amplicons' number and chromosome aberration yield for experiments with acute UV-C exposure and MboI enzyme and significant negative correlation (α= 0.01) just for experiments with composed radiation exposure and MspI ans MboI enzymes.

To continue the correlation analysis using more detail approach via determination of 5 grades (from 0 to 4) of methylation patterns' varieties. It will be used following indexes:

0—The absence of differences according to control variant;

1—The differences in amplicons' number, which mass is in the range of control amplicons' mass;

2—The differences in amplicons' number, which mass is in the range of control amplicons' mass + differences in brightness of bands that indicates to various number of fragments in one amplicon;

3—The differences in amplicons' number, which mass is not in the range of control amplicons' mass;

4—The differences in amplicons' number, which mass is not in the range of control amplicons' mass + differ-

Table 2. The coefficient of correlation between amplicons' number and chromosome aberration yield.

Experimental series	Correlation Coefficient, R		
	MspI	HpaII	MboI
1.	0.29	-	0.72
2	0.69	0.105	0.82*
3	–0.89**	–0.73	–0.89**

Significance of a correlation coefficient, $^{*}\alpha$ = 0.05, $^{**}\alpha$ = 0.01.

ences in brightness of bands.

Results of this way of correlation assessment are shown in **Table 3**.

Thus such approach for determination the degree of methylation pattern changes increased the correlation index for some variants and decreased it for another one.

Continue the specification of approach to correlation assessment via determination of 9 grades (from 0 to 8) of methylation patterns' varieties. It will be used following indexes:

0—The absence of differences according to control variant;

1—The differences in amplicons' number (n), which mass is in the range of control amplicons' mass, $n \leq 3$;

2—The differences in amplicons' number (n), which mass is in the range of control amplicons' mass, $n \leq 3$ + differences in their brightness;

3—The differences in amplicons' number ($n > 3$), which mass is in the range of control amplicons' mass;

4—The differences in amplicons' number ($n > 3$), which mass is in the range of control amplicons' mass + differences in their brightness;

5—The differences in amplicons' number ($n \leq 3$), which mass is not in the range of control amplicons' mass;

6—The differences in amplicons' number ($n \leq 3$), which mass is not in the range of control amplicons' mass + differences in their brightness;

7—The differences in amplicons' number ($n > 3$), which mass is not in the range of control amplicons' mass;

8—The differences in amplicons' number ($n > 3$), which mass is not in the range of control amplicons' mass + differences in their brightness.

Results of this way of correlation assessment are shown in **Table 4**.

Specification of differences between electrophoregrams and number of their grades could be continued using additional characteristics of electrophoregrams and their combination. However performed correlation analysis using three approaches allows to make general

Table 3. The coefficient of correlation between 5 grades of electrophoregram varieties and chromosome aberration yield.

Experimental series	Correlation Coefficient, R		
	MspI	HpaII	MboI
1	0.27	-	0.57
2	0.81*	0.77	0.91**
3	0.43	0.43	0.64

*$\alpha = 0.05$; **$\alpha = 0.01$.

Table 4. The coefficient of correlation between 9 grades of electrophoregram varieties and chromosome aberration yield.

Experimental series	Correlation Coefficient, R		
	MspI	HpaII	MboI
1	0.57	-	0.64
2	0.87*	0.84*	0.89**
3	0.21	0.71	0.7

*$\alpha = 0.05$; **$\alpha = 0.01$.

conclusion about existence of quantitative connection between chromosome aberration yield like both integral cell stress response and changes in satellite DNA methylation pattern.

Performed analysis also show that unique approach to quantify connection between chromosome aberration yield and their rates on electrophoregrams doesn't exist. Such suggestion points to possible difference in mechanisms of cell response to exposure type (physical exposure, exposure rate and duration and so on).

The investigation of differential gene activity using micro array methods and changes of DNA methylation pattern indicated that according to exposure type and intensity the activity of various gene groups had changed. That's for satellite DNA–its direct or indirect participation in cell stress response could be related to different mechanisms according to exposure type.

Despite of ways of satellite DNA participation in stress reaction such mechanisms are different as well as for transcribed DNA.

It should have been emphasized that for experimental series 1 with 3 polymorphic groups of plants significant correlations weren't obtained with any criteria. Reason for such phenomena is connected to original epigenetical polymorphism of biological material and deficient sampling from 6 variants for correlation assay.

4. Conclusions

Comparison of the results of cytogenetic analysis with changes in methylation patterns of satellite DNA after irradiation pointed out to their connections with different stress tolerance.

Change of the satellite DNA methylation profile may reflect the mobile elements activization, mostly associated with satellite DNA [13], and indicate the damage's progress. Such ability is especially essential for corn; it's known that nearly 50% satellite DNA of the plant are represented with mobile elements [2,3,15].

At the same time, it can result in DNA configuration changes and has the protective effect. Since functional importance of satellite DNA was explained in part by conceptions, it was assumed to have a structural role in spatial organization of genome, and take part in homologous chromosomes' conjugation during meiosis and replication of chromosomes' telomeric sites [15]. Probably in this case different methylation patterns of satellite DNA, which meant various chromatin conformations, could have interactive character: specific methylation patterns of transcribed DNA may play role in transcription processes only under definite conformation of all the chromatin.

Interaction between satellite DNA methylation pattern and resistance to external exposures might have another explanation. It could result not only from efficient functioning of repair systems of spontaneous и inducible DNA injuries, but also from systems responsible for passing cell cycle checkpoints and complete repair of double-stranded DNA breaks. It was known, that effective repair of double-stranded DNA breaks with the mechanism of homologous recombination was possible only under conditions of certain level of chromatin relaxation [20], so it was also associated directly to the conformation of satellite DNA.

Thus conducted research provides grounds to suggest that satellite DNA methylation patterns and their changes might have various roles in cell response to stress factor. All the functions are mediated by conformation changes of these DNA sequences.

5. Acknowledgements

Funding for the study was provided by the Academy of Science of Ukraine, Grant No. III-3-08 "Epigenetic components of plant adaptation".

We thank PhD, Head of Laboratory of Molecular Genetics Morgun B.V., Institute of Cell Biology and Genetic Engineering, National Academy of Science of Ukraine for help in method mastering.

REFERENCES

[1] R. L. P. Adams, "DNA Methylation," *Principles of Medical Biology*, Vol. 5, 1996, pp. 33-66. doi:10.1016/S1569-2582(96)80107-3

[2] E. N. Tishchenko and O. V. Dubrovnaya, "Epigenetic Regulation. DNA Methylation of Genes and Transgenes in Plants," Logos, Kiev, 2004, 384 Pages.

[3] M.-T. Hauser, W. Aufsatz, C. Jonak and Ch. Luschnig, "Transgenerational Epigenetic Inheritance in Plants," *Biochimica at Biophysica Acta*, Vol. 1809, No. 8, 2011, pp. 459-468. doi:10.1016/j.bbagrm.2011.03.007

[4] A. Agorio and P. Vera, "ARGONAUTE4 Is Required for Resistance to *Pseudomonas syringae* in Arabidopsis," *The Plant Cell*, Vol. 19, No. 11, 2007, pp. 3778-3790. doi:10.1105/tpc.107.054494

[5] A. Bilichak, Y. Ilnystkyy, Y. Hollunder and I. Kovalchuk, "The Progeny of Arabidopsis Thaliana Plants Exposed to Salt Exhibit Changes in DNA Methylation, Histone Modifications and Gene Expression," *PLoS One*, Vol. 7, No. 1, 2012, pp. 1-15.

[6] L. Zhong, Y. Xu and J. Wang, "DNA-Methylation Changes Induced by Salt Stress in Wheat *Triticum aestivum*," *African Journal of Biotechnology*, Vol. 8, No. 22, 2009, pp. 6201-6207.

[7] O. Kovalchuk, P. Burke, A. Arkhipov, N. Kuchma, S. J. James, I. Kovalchuk and I. Pogribny, "Genome Hypermethylation in *Pinus silvestris* of Chernobyl—A Mechanism for Radiation Adaptation?" *Mutation Research*, Vol. 529, No. 1-2, 2003, pp. 13-20. doi:10.1016/S0027-5107(03)00103-9

[8] I. Kovalchuk, V. Abramov, I. Pogribny and O. Kovalchuk, "Molecular Aspects of Plant Adaptation to Life in the Chernobyl Zone," *Plant Physiology*, Vol. 135, No. 1, 2004, pp. 357-363. doi:10.1104/pp.104.040477

[9] I. Pogribny, I. Koturbash, V. Tryndyak, D. Hudson, S. M. L. Stevenson, O. Sedelnikova, W. Bonner and O. Kovalchuk, "Fractionated Low-Dose Radiation Exposure Leads to Accumulation of DNA Damage and Profound Alterations in DNA and Histone Methylation in the Murine Thymus," *Molecular Cancer Research*, Vol. 3, No. 10, 2005, pp. 553-561. doi:10.1158/1541-7786.MCR-05-0074

[10] A. P. Kravets, T. A. Mousseau, A. V. Litvinchuk, Sh. Ostermiller and G. Vengzhen, "Changes in DNA Methylation Pattern in Weat Plants under Chronical γ—Exposure of Seeds," *Cytology and Genetics*, Vol. 44, No. 5, 2010, pp. 18-22. doi:10.3103/S0095452710050038

[11] M. A. Coleman, E. Yin and L. Peterson, "Low-Dose Irradiation Alters the Transcript Profiles of Human Lymphoblastoid Cells Including Genes Associated with Cytogenetic Radioadaptive Response," *Radiation Research*, Vol. 164, No. 4, 2005, pp. 369-382. doi:10.1667/RR3356.1

[12] M. Banda, A. Bommineni, R. A. Thomas, L. S. Luckinbill and J. D. Tucker, "Evaluation and Validation of Housekeeping Genes in Response to Ionizing Radiation and Chemical Exposure for Normalizing RNA Expression in Real-Time PCR," *Mutation Research*, Vol. 8, No. 649, 2008, pp. 126-134.

[13] A. M. Serebryanyi, "Radiation Adaptive Response as a Stress Reaction of a Cell," *Radiatsionnaya Biologiya. Radioekologiya*, Vol. 51, No. 4, 2011, pp. 399-405.

[14] A. Kravets, D. Sokolova, G. Vengzhen and D. Grodzinsky "Corn Plant DNA Methylation Pattern Changes at UV-C Irradiation Fractionating," *Cytology and Genetics*, Vol. 47, No. 1, 2013, pp. 29-35. doi:10.3103/S0095452713010052

[15] V. Hemleben, T. G. Beridze, L. Bakhman, Y. Kovarik and R. Torres, "Satellite DNA," *Uspehi Biologicheskoy Khimii*, Vol. 43, 2003, pp. 267-306.

[16] F. M. Ausubel, "Current Protocols in Molecular Biology," Biophotometer Operating Manual, 2004. http://www.eppendorf.com

[17] J. M. S. Bartlett and D. Stirling, "PCR Protocols," Hu-

mana Press Incorporate, Humana, 2003. http://www.dartmouth.edu/~eprctr/biodose2008/pdf/B10.pdf

[18] Y. M. Tikunov and L. I. Khrystaleva, "Application of ISSR Markers in the Genus Lycopersicon," *Euphitica*, Vol. 131, No. 1, 2003, pp. 71-80. doi:10.1023/A:1023090318492

[19] A. V. Lakin, "Biometry," High School, Moskva, 1990.

[20] A. I. Gaziev, "Deterioration of Critical DNA Damage Repair Efficiency under Low Dose Irradiation," *Radiatsionnaya Biologiya Radioekologiya*, Vol. 51, No. 5, 2011, pp. 512-529.

6

Antibody to MyoD or Myogenin Decreases Acetylcholine Receptor Clustering in C2C12 Myotube Culture

Matthew K. Ball[1], David H. Campbell[2], Kelly Ezell[1], Jessica B. Henley[1], Paul R. Standley[2], Wade A. Grow[1*]

[1]Department of Anatomy, Arizona College of Osteopathic Medicine, Midwestern University, Glendale, USA

[2]Department of Basic Medical Sciences, University of Arizona College of Medicine, Phoenix, USA

Email: *wgrowx@midwestern.edu

ABSTRACT

Skeletal muscle development is influenced by myogenic regulatory factors, including the expression of MyoD and myogenin. Our objective was to use the C2C12 cell culture model to test the hypothesis that both MyoD and myogenin were required for agrin-induced acetylcholine receptor (AChR) clustering and the fusion of myoblasts into myotubes. We induced fusion of myoblasts into myotubes by switching from growth medium (GM) to differentiation medium (DM). During myotube formation AChRs cluster spontaneously, but treatment with motor neuron derived agrin increases clustering of AChRs and other postsynaptic components of the neuromuscular synapse. We examined the normal expression pattern of MyoD and myogenin in C2C12 cell culture using immunofluorescence. MyoD was highly expressed while myoblasts were in GM, but expression declined within 72 hours after cell cultures were switched to DM. Myogenin expression was low in GM, but increased when cell cultures were switched to DM. Next we used antibodies to decrease MyoD and/or myogenin function. Fluorescence microscopy images were captured and then analyzed to assess agrin-induced AChR clustering with or without antibody treatment. Finally we calculated the proportion of nuclei in myotubes and myoblasts by creating digital overlays of phase contrast and DAPI stained microscopy images. This allowed the comparison of myotube formation with or without antibody treatment. We report that antibody to either MyoD or myogenin decreases the frequency of agrin-induced AChR clustering without affecting myotube formation. We conclude that agrin-induced AChR clustering requires both MyoD and myogenin.

Keywords: Agrin; Acetylcholine Receptor; MyoD; Myogenin; C2C12

1. Introduction

Skeletal muscle development is guided by myogenic regulatory factors including MyoD (myf3), myogenin (myf4), myf5, and MRF4 (myf6). This family of basic helix-loop-helix transcription factors binds to the E-box found in the promoters or enhancers of many muscle-specific genes which results in high levels of transcription [1-5]. Experiments with knockout mice helped establish the temporal expression pattern of the myogenic regulatory factors. MyoD and myf5 are essential for myoblast identity [6-8], while myogenin is essential for myoblast differentiation [9,10]. MRF4 does not appear to be essential for myogenesis, but instead is highly expressed in adult skeletal muscle fibers [11-15]. While single null mutations of MyoD, myf5, and MRF4 are not lethal [6,7,16], the null mutation for myogenin results in severe muscle deficiency due to inadequate secondary muscle fiber development, and subsequent neonatal death [9,10,17].

Using C2C12 cell culture, western blots revealed that MyoD was expressed in proliferating myoblasts and myotubes, while myogenin was expressed in myotubes only, and MyoD increased myogenin gene expression [18]. Skeletal muscle cell cultures, such as the C2C12 cell line derived from mouse hindlimb, provide simplified systems for studying myogenesis as well as the development of the postsynaptic component of the neuromuscular synapse [19,20].

During development of the neuromuscular synapse both the concentration and location of acetylcholine receptors (AChRs) on the skeletal muscle cell surface is regulated [21], resulting in a mature neuromuscular synapse with a concentration of AChRs that is 1000 times as great as that found extrasynaptically [22]. AChRs aggre-

*Corresponding author.

gate and co-localize with a large number of other molecules, including a muscle specific kinase (MuSK) [23] and rapsyn [24,25]. Indeed, MuSK is essential for the signaling events that precede neuromuscular synapse formation [26-29] and rapsyn is essential for the formation of AChR clusters during neuromuscular synapse formation [30]. In addition to increasing myogenin gene expression, MyoD also targets MuSK and rapsyn gene expression, while myogenin targets rapsyn but not MuSK gene expression [18]. Moreover, myogenin activates genes for AChR subunits [31,32]. This suggests that myogenic regulatory factors like MyoD and myogenin may be intricately linked to the development of the postsynaptic component of the neuromuscular synapse.

Cultured myotubes cluster AChRs spontaneously and respond to application of motor neuron derived agrin with an increase in the frequency of AChR clusters [33,34]. Agrin was first isolated because of its ability to cluster AChRs in cell culture [35], and plays a major role in assembly of the postsynaptic component of the neuromuscular synapse [36]. Agrin binds to low-density lipoprotein receptor-related protein 4 (Lrp4) [37,38] to stimulate tyrosine phosphorylation of MuSK [28] and the consequent signaling pathway that includes the AChR β subunit and leads to increased AChR clustering [39,40]. In addition, AChRs are required for the agrin-induced aggregation of MuSK at the neuromuscular synapse [41].

Using an immortalized rat muscle cell line, RNA interference experiments revealed that myogenin expression was necessary for robust spontaneous AChR clustering [42]. Our objective was to use the C2C12 cell culture model to test the hypothesis that both MyoD and myogenin were required for agrin-induced AChR clustering and the fusion of myoblasts into myotubes. We report that antibody to either MyoD or myogenin decreases the frequency of agrin-induced AChR clustering without affecting myotube formation. We conclude that agrin-induced AChR clustering requires both MyoD and myogenin.

2. Materials and Methods

2.1. Cell Culture Maintenance

C2C12 myoblasts were derived from mouse hind limb [19,20], and are commonly used for skeletal muscle cell culture experiments. They are ideal for studying acetylcholine receptor (AChR) clustering and myoblast fusion to form myotubes. For normal maintenance of C2C12 cell culture, myoblasts were first plated in growth medium (GM) on 10 cm plates at approximately 20% confluence. GM consists of Dulbecco's modified Eagle's medium (DMEM) plus 20% fetal bovine serum, 0.5% chick embryo extract and 100 U/ml penicillin. GM was replaced daily, and myoblast cultures were split into new plates at approximately 60% confluence. For formation of myotubes, myoblasts were plated in GM on 22 × 22 mm cover slips that had been flamed in 200-proof ethanol and placed in 6-well plates. GM was replaced daily. After 48 hours in GM, myoblast cultures typically reached 80% confluence, and cultures were then switched to differentiation medium (DM). DM consists of DMEM plus 2% horse serum and 100 U/ml penicillin. DM was replaced daily as myoblasts fused to form myotubes, and cultures were maintained for 72 hours in DM. The incubator was maintained at 37°C under 8% carbon dioxide and 100% humidity.

2.2. MyoD and Myogenin Expression in Cell Culture

Some C2C12 cell cultures were fixed each day cells grew on 22 × 22 mm cover slips in GM and DM. Cover slips were rinsed three times with phosphate buffered saline (PBS), fixed for 10 minutes with 2% paraformaldehyde in PBS, rinsed three times with PBS, incubated for 10 minutes with 0.2% Triton X-100 in PBS, rinsed three times with PBS, and then incubated for 60 minutes with 5% bovine serum albumin (BSA) in PBS as a blocking agent. Cover slips were then incubated for 60 minutes with a mouse monoclonal primary antibody to MyoD (Santa Cruz Biotechnology sc-71629) or myogenin (Santa Cruz Biotechnology sc-12732) at 1:10 in the BSA blocking agent, rinsed three times in PBS, and then incubated with a TRITC anti-mouse fluorescent secondary antibody (Molecular Probes) at 1:200 in the BSA blocking agent. After rinsing three times in PBS, cover slips were dehydrated in cold methanol for 5 minutes at −20°C, and mounted on microscope slides in Vectashield Mounting Medium for Fluorescence (Vector Laboratories). Fluorescent staining was visualized with an IX70 Olympus inverted microscope under the 20× objective (yielding a total magnification of 200×), and fluorescent images were captured as high-resolution JPG files with an Olympus camera with Magnafire digital imaging software. MyoD and myogenin expression was determined by counting fluorescent and non-fluorescent nuclei and calculating what percentage of nuclei was fluorescent each day in GM and DM. Nuclei were counted from images captured as JPG files for each day in GM and DM.

2.3. Endo-Porter Use in Cell Culture

To verify the ability of Endo-Porter (Gene Tools) to increase intracellular antibody, some cell cultures were treated with 1.00 μg/ml antibody for MyoD and 6 μM Endo-Porter at 24 hours in GM and fixed at 48 hours in GM, while other cell cultures were treated with 1.00 μg/ml antibody for myogenin and 6 μM Endo-Porter at

48 hours in DM and fixed at 72 hours in DM. Antibody and Endo-Porter were added 24 hours prior to fixation to maximize the possibility of visualizing increased intracellular antibody. These cell cultures were compared to others that were treated with antibody alone for the same time period. The Endo-Porter concentration used was recommended by the manufacturer as allowing optimal cell access with minimal cell damage. Cover slips were rinsed three times with PBS, fixed for 10 minutes with 2% paraformaldehyde in PBS, rinsed three times with PBS, incubated for 10 minutes with 0.2% Triton X-100 in PBS, rinsed three times with PBS, and then incubated for 60 minutes with 5% BSA in PBS as a blocking agent. Cover slips were then incubated with a TRITC anti-mouse fluorescent secondary antibody at 1:200 in the BSA blocking agent. After rinsing three times in PBS, the cover slips were dehydrated in cold methanol for 5 minutes at −20°C, and mounted on microscope slides in Vectashield Mounting Medium for Fluorescence. Fluorescent staining was visualized with an IX70 Olympus inverted microscope under the 20× objective (yielding a total magnification of 200×), and fluorescent images were captured as high-resolution JPG files with an Olympus camera with Magnafire digital imaging software.

2.4. Experimental Manipulations with Antibody for MyoD or Myogenin

C2C12 cell cultures were either maintained as controls or were exposed to antibody for MyoD or myogenin or both, with new antibody added each time the media was changed beginning when cells were first plated on 22 × 22 mm cover slips. Antibody concentrations 0.001 μg/ml, 0.01 μg/ml, 0.10 μg/ml, 1.00 μg/ml, and 2.00 μg/ml were tested for the ability to decrease agrin-induced AChR clustering. To optimize intracellular antibody, media with 6 μM Endo-Porter was used for the first 24 hours in GM. This is much earlier than the experiments performed to verify that Endo-Porter could increase intracellular antibody, and was intended to minimize any potential effect of Endo-Porter on later myotube formation or AChR clustering. A consequence could be that this early treatment led to less Endo-Porter available each subsequent day when antibody was added to cell culture. Cultures were exposed to 10 ng/ml agrin (R&D Systems) for the last 16 hours of 72 hours in DM to induce AChR clustering.

2.5. Acetylcholine Receptor Clustering Assay

AChRs were labeled by the binding of α-bungarotoxin conjugated to tetramethyl rhodamine (Molecular Probes) [43]. Cultures were incubated in the toxin-containing medium for 30 minutes at 37°C to label AChRs after 72 hours in DM. Cover slips were rinsed three times with PBS, fixed for 10 minutes with 2% paraformaldehyde in PBS, rinsed three times with PBS, dehydrated in cold methanol for 5 minutes at −20°C, and mounted on microscope slides in Vectashield Mounting Medium for Fluorescence. For some experiments the mounting medium contained 4'6-diamidino-2-phenylindole (DAPI) to visualize nuclei. Nuclei were counted from images captured as JPG files for each day in GM and DM. Fluorescent staining was visualized with an IX70 Olympus inverted microscope under the 20× objective (yielding a total magnification of 200×), and fluorescent images were captured as high-resolution JPG files with an Olympus camera with Magnafire digital imaging software. Bright clusters of AChRs were observed on all aspects of myotubes in fluorescent images. The frequency of AChR clustering was determined by using an algorithm developed for Cell Profiler [44]. AChR clusters were counted from images captured as JPG files from cover slips. These data were utilized to assay agrin-induced AChR clustering after exposure to antibody for MyoD or myogenin or both. Comparisons of control cultures with cultures exposed to antibody were analyzed by Student's t-test to determine statistically different results at $p < 0.01$.

2.6. Cell Profiler Algorithm

Each grayscale image was analyzed using Cell Profiler's object identification algorithm. The threshold used in the algorithm was defined as the minimum fluorescent intensity that a pixel must display to be counted as part of a cluster. Experimentation with the threshold revealed that a minimum brightness of 30% best defined a pixel with enough fluorescence to be counted as part of a cluster, and this threshold was used for all analyses. To ensure objectivity and consistent quantification, the threshold and all other settings were kept constant across all groups and images. The diameter range for identifying an AChR cluster was set at 4 - 150 pixels (2.93 μm - 109.95 μm). Contiguous pixels meeting both the intensity and size requirements were counted as parts of AChR clusters. The total clustered pixels per image were converted to AChR clusters as percentage of field, to reflect what percentage of the pixels in an image were counted as containing clustered AChRs. Analysis of data trends via unpaired t-tests were completed using GraphPad Prism.

2.7. Myotube Formation Index

Cell cultures were visualized with an IX70 Olympus inverted microscope under the 20× objective (yielding a total magnification of 200×), and representative phase contrast and DAPI images were captured as high-resolution JPG files with an Olympus camera with Magnafire digital imaging software. These JPG files were utilized to quantify myotube formation by modifying a myoblast fusion index paradigm [45-48]. In brief, the number of nuclei in myoblasts (defined as cells with one or two nuclei) and myotubes (defined as cells with three or more nuclei) were counted after 72 hours in DM. Only nuclei obviously in myotubes were counted as such. All nuclei for which a designation was difficult were grouped with the nuclei in myoblasts. This method biased the data toward fewer nuclei in myotubes for both control and experimental groups. A total of five pairs of images (phase contrast and DAPI) were analyzed for control cultures, cultures exposed to 1.00 μg/ml antibody for MyoD, cultures exposed to 1.00 μg/ml antibody for myogenin, and cultures exposed to 1.00 μg/ml of both antibodies. For each image, nuclei were determined to be either in a myotube or not prior to being counted. The myotube formation index was then calculated as nuclei in myotubes divided by total nuclei in the image and reported in **Table 1**.

3. Results

The results reported here demonstrate that antibody to either MyoD or myogenin during the specific time period when myoblasts fuse to form myotubes decreases the frequency of agrin-induced acetylcholine receptor (AChR) clustering without affecting myotube formation.

3.1. MyoD Expression Peaks during Myoblast Proliferation while Myogenin Expression Peaks during Myotube Formation

MyoD and myogenin expression were determined by-counting fluorescent and non-fluorescent nuclei and calculating what percentage of nuclei were fluorescent each day in GM and DM. MyoD expression was highest in GM and myogenin expression was highest in DM (**Figure 1**). Nuclei were counted from 50 images captured as JPG files for each day in GM and DM. Representative images were chosen that had MyoD or myogenin expression consistent with the percentages calculated each day in GM and DM, and then assembled into **Figure 2**. Percentages of fluorescent nuclei were as follows: 38% MyoD and 8% myogenin at 48 hours in GM, 48% MyoD and 36% myogenin at 24 hours in DM, and 5% MyoD and 47% myogenin at 72 hours in DM (data not shown).

3.2. Intracellular Antibody to MyoD or Myogenin Increases When Applied with Endo-Porter

To verify the ability of Endo-Porter to increase intracellular antibody, cell cultures were treated with antibody to MyoD and Endo-Porter at 24 hours in GM and fixed at 48 hours in GM. Other cell cultures were treated with antibody to myogenin and Endo-Porter at 48 hours in DM and fixed at 72 hours in DM. In both cases a TRITC fluorescent secondary antibody was used to localize the primary antibody. These cell cultures were compared to others treated with antibody but without Endo-Porter for the same time period. Antibody and Endo-Porter were added 24 hours prior to fixation to maximize the possibility of visualizing increased intracellular antibody. Fluorescent images were captured as high-resolution JPG files with an Olympus camera with Magnafire digital imaging software. Intracellular antibody to MyoD or myogenin was greatly increased when applied with Endo-Porter when compared with antibody application alone (**Figure 3**). Virtually all intracellular antibodies were observed in cytoplasm, with little or no antibody in nuclei. In fact, nuclei were so devoid of antibody that the location of nuclei was readily apparent as dark ovals amidst the cytoplasmic staining.

3.3. Antibody to Either MyoD or Myogenin Decreases Agrin-Induced AChR Clustering

AChRs cluster spontaneously with a baseline frequency on C2C12 myotubes and this clustering is increased with

Table 1. Myotube formation index. Each pairing of control or antibody treatment was a separate experiment. Percentage of nuclei in myotubes was calculated from counts of nuclei in myotubes and myoblasts. Treatment with anti-MyoD, anti-myogenin, or both did not alter the percentage of nuclei in myotubes, but antibody to myogenin may decrease myoblast viability.

	Nuclei in Myotubes	Nuclei NOT in Myotubes	Total Nuclei	Percentage of Nuclei in Myotubes
Control	145	1101	1246	12%
Anti-MyoD (1 μg/ml)	190	1094	1284	15%
Control	499	1405	1904	26%
Anti-Myogenin (1 μg/ml)	114	231	345	33%
Control	752	1442	2194	34%
Anti-MyoD and Anti-Myogenin (1 μg/ml)	344	737	1081	32%

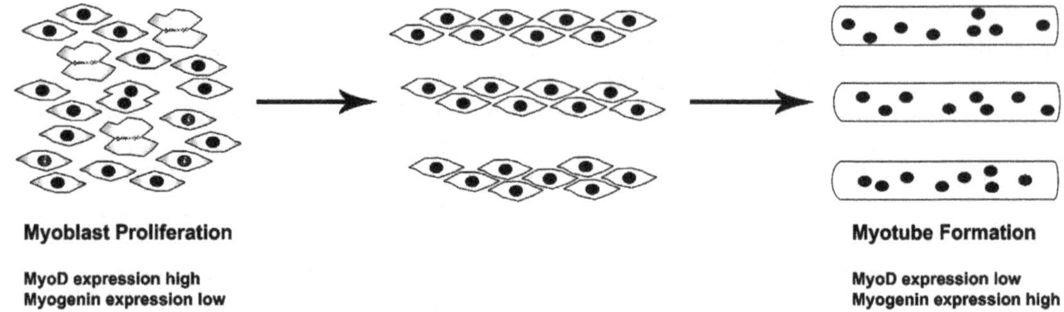

Figure 1. During skeletal muscle development, proliferating myoblasts express a high level of MyoD and a low level of myogenin. As myoblasts fuse into myotubes they express a high level of myogenin and a low level of MyoD.

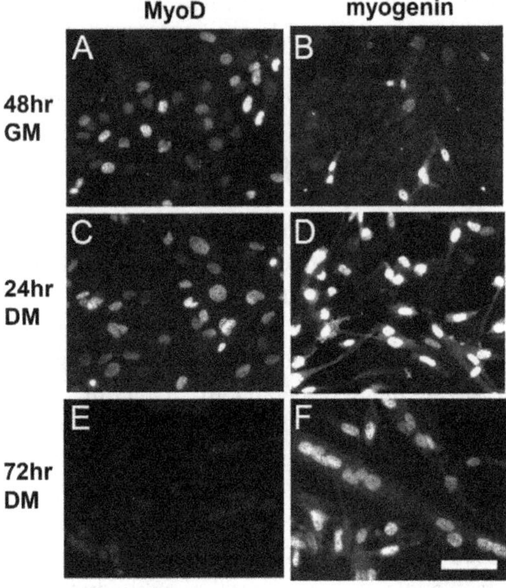

Figure 2. Fluorescent images were captured showing MyoD ((A), (C), (E)) or myogenin ((B), (D), (F)) expression in C2C12 skeletal muscle cell culture at the following time points: 48 hours in GM ((A), (B)), 24 hours in DM ((C), (D)), and 72 hours in DM ((E), (F)). MyoD is observed in more nuclei at 48 hours in GM. Myogenin is observed in more nuclei at 72 hours in DM. GM = growth medium, DM = differentiation medium. Scale bar = 100 μm.

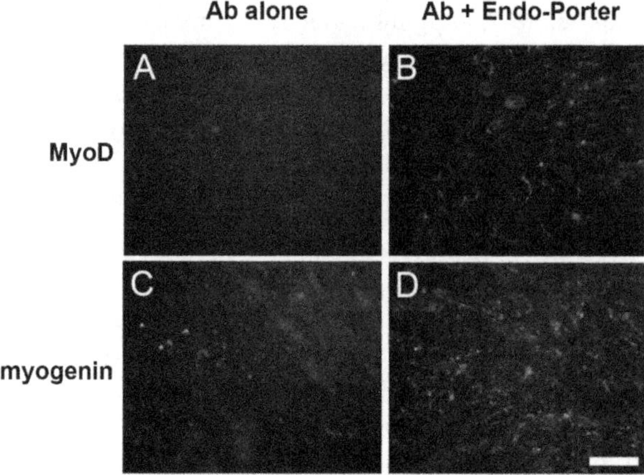

Figure 3. Fluorescent images were captured showing Anti-MyoD ((A), (B)) or Anti-myogenin ((C), (D)) localization in C2C12 myoblasts or myotubes when treated with 1.00 μg/ml antibody alone ((A), (C)) or 1.00 μg/ml antibody with 6 μM Endo-Porter (B,D). Anti-MyoD was added at 24 hours in GM and myoblasts were fixed at 48 hours in GM. Anti-myogenin was added at 48 hours in DM and myotubes were fixed at 72 hours in DM. In both cases a TRITC secondary antibody was used to localize the primary antibody, with fluorescence indicating presence of MyoD or myogenin. Anti-MyoD and Anti-myogenin enter cells in much greater concentration when applied with Endo-Porter, and are observed primarily in cytoplasm. Ab = antibody. Scale bar = 100 μm.

agrin treatment [33,34]. C2C12 cell cultures were switched from GM to DM at 80% confluence, 10 ng/ml agrin was added for the last 16 hours in DM, and myotubes were examined for AChR clustering after 72 hours in DM. AChRs in clusters were quantified by analyzing fluorescent images captured at 72 hours in DM with Cell Profiler software, using a method previously optimized and reported [44]. Cell Profiler counted the total clustered pixels per image. We converted that data to AChR clusters as percentage of field. We had previously used human counts of the number of AChR clusters per image which did not account for the size of the clusters. Images were captured from agrin-induced cultures that were untreated by antibody or treated daily with antibodies to MyoD, myogenin, or both. All control and experimental cultures were treated with Endo-Porter when the cells were first plated in GM. This is much earlier than the experiments performed to verify that Endo-Porter could increase intracellular antibody, and was intended to minimize any potential effect of Endo-Porter on later myotube formation or AChR clustering. A consequence could be that this early treatment led to less Endo-Porter available each subsequent day when antibody was added to cell culture. For antibody to MyoD, concentrations as low as 0.01 g/ml decreased agrin-induced AChR clustering relative to untreated cultures, using Student's t-test at $p < 0.01$ (**Figures 4(A)** and **5**). For antibody to myogenin or the combination of antibodies to MyoD and myogenin, a concentration of 1.00 μg/ml was necessary

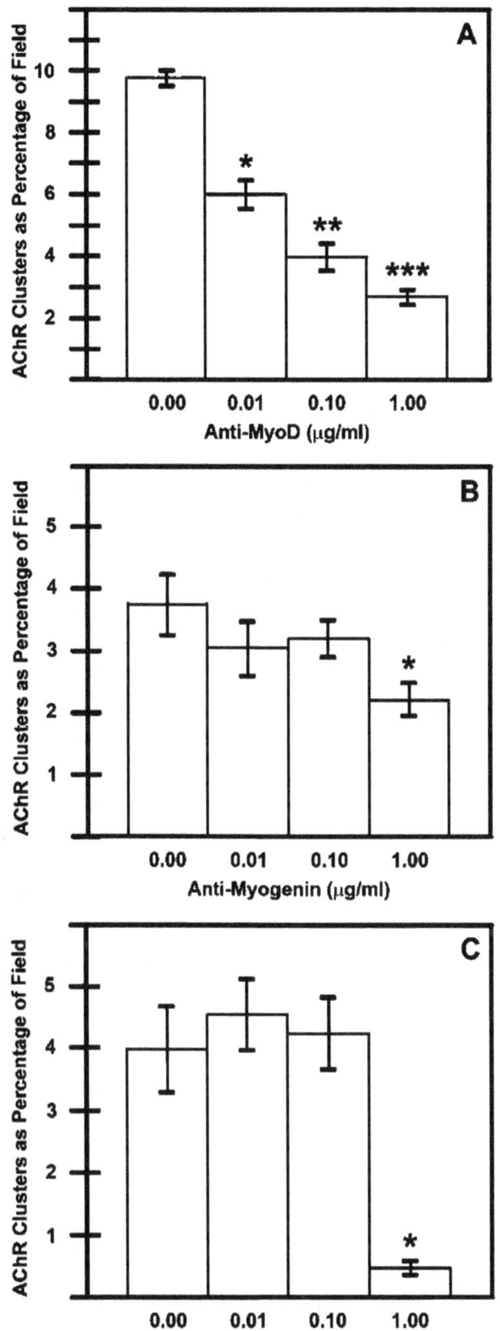

Figure 4. AChRs in clusters were quantified by analyzing fluorescent images captured at 72 hours in DM with Cell Profiler software. Images were captured from agrin-induced cultures that were untreated by antibody (control cultures) or treated daily with Anti-MyoD (A); Anti-myogenin (B); or both Anti-MyoD and Anti-myogenin (C). All cultures were treated with Endo-Porter when the cells were first plated in GM. The histograms reveal that daily treatment of 1.00 μg/ml of either antibody is sufficient to decrease agrin-induced AChR clustering relative to control cultures, using Student's t-test at $p < 0.01$. (*) statistically decreased from 0.00 μg/ml. (**) statistically decreased from 0.01 μg/ml. (***) statistically decreased from 0.10 μg/ml.

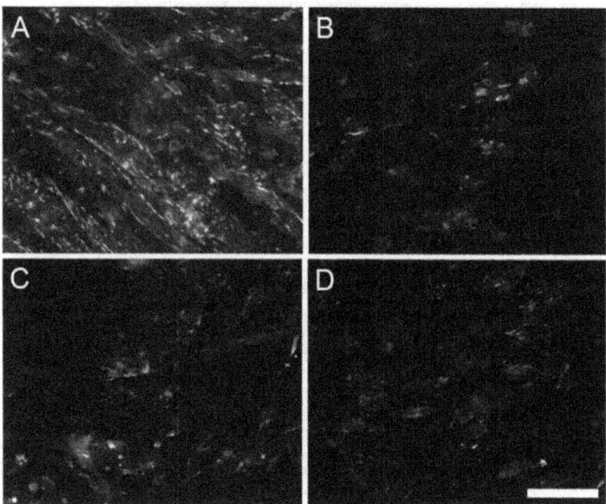

Figure 5. Examples of fluorescent images captured from agrin-induced cultures that were untreated by antibody (control cultures) (A), or treated daily with 1.00 μg/ml Anti-MyoD (B), 1.00 μg/ml Anti-myogenin (C), or 1.00 μg/ml Anti-MyoD and Anti-myogenin (D). Images were captured at 72 hours in DM. The fluorescent areas are clusters of AChRs. Scale bar = 100 μm.

to decrease agrin-induced AChR clustering relative to untreated cultures, using Student's t-test at $p < 0.01$ (**Figures 4(B)** and **(C)**, **Figure 5**). Further experiments comparing 1.00 μg/ml with 2.00 μg/ml revealed that the higher concentration further decreased agrin-induced AChR clustering with antibody to MyoD but not with antibody to myogenin (**Figure 6**). Each graph in a figure presents the data from a single experiment, and only data within a single experiment is compared. Human counts of the number of AChR clusters per image were also conducted for the images used for data collection for **Figures 4** and **6**. Similar differences in the effect of antibodies were calculated, with 1.00 μg/ml of either antibody sufficient to decrease agrin-induced AChR clustering relative to untreated cultures (data not shown). Experiments were also performed using antibodies to MyoD and/or myogenin without using Endo-Porter. In those experiments results were inconsistent (data not shown).

3.4. Antibody to Either MyoD or Myogenin Has No Effect on Myotube Formation

Myotube formation was quantified by modifying a myoblast fusion index paradigm [45-48]. In brief, the number of nuclei in myoblasts (defined as cells with one or two nuclei) and myotubes (defined as cells with three or more nuclei) were counted after 72 hours in DM. Only nuclei obviously in myotubes were counted as such. All nuclei for which a designation was difficult were grouped with the nuclei in myoblasts. This method biased the data toward fewer nuclei in myotubes for both control and experimental groups. Control cultures were

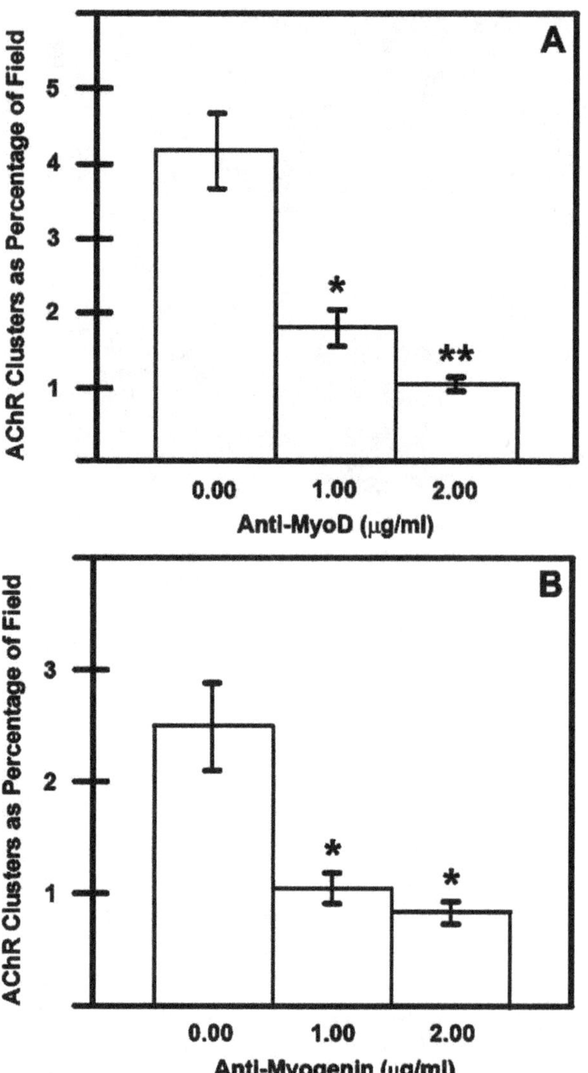

Figure 6. AChRs in clusters were quantified by analyzing fluorescent images captured at 72 hours in DM with Cell Profiler software. Images were captured from agrin-induced cultures that were untreated by antibody (control cultures) or treated daily with Anti-MyoD (A) or Anti-myogenin (B). All cultures were treated with Endo-Porter when the cells were first plated in GM. The histograms reveal that daily treatment with 2.00 µg/ml Anti-MyoD decreased agrin-induced AChR clustering greater than 1.00 µg/ml Anti-MyoD, while 2.00 µg/ml Anti-myogenin decreased agrin-induced AChR clustering similar to 1.00 µg/ml Anti-myogenin, using Student's t-test at $p < 0.01$. (*) statistically decreased from 0.00 µg/ml. (**) statistically decreased from 1.00 µg/ml.

compared with cultures exposed to 1.00 µg/ml antibody for MyoD, or 1.00 µg/ml antibody for myogenin, or both, in separate experiments. Each pairing of control or antibody treatment was a separate experiment, combining data from multiple culture plates. With antibody to MyoD the number of cells and the percentage of nuclei in myotubes were similar to control. With antibody to myogenin or both MyoD and myogenin, the percentage of nuclei in myotubes was similar to control, suggesting that myotube formation was unaffected. However, the number of cells (and thus nuclei) was decreased by antibody to myogenin or both MyoD and myogenin, suggesting that myoblast viability may be decreased (**Table 1**). If so, this could contribute to the decrease in agrin-induced AChR clustering resulting from antibody to myogenin.

4. Discussion

The results reported here demonstrate that treatment with antibody to MyoD or myogenin during the specific time period when myoblasts fuse to form myotubes decreases the frequency of agrin-induced acetylcholine receptor (AChR) clustering without affecting myotube formation. We conclude that agrin-induced AChR clustering requires both MyoD and myogenin.

Previously it was reported that RNA interference of myogenin expression in an immortalized rat muscle cell line decreased the number of spontaneous AChR clusters per nuclei by approximately three-fold, and decreased the percentage of muscle fibers with agrin-induced AChR clusters by approximately four-fold [42]. However, the percentage of muscle fibers with AChR clusters was identical with or without agrin treatment. Moreover, the data regarding agrin-induced AChR clustering did not include AChR clusters per nuclei or other more standard measures. We used the C2C12 cell culture model to test the hypothesis that both MyoD and myogenin are required for agrin-induced AChR clustering. We analyzed the images captured from control and treatment groups with two different methods. First, we used a standard method that counted the number of AChR clusters per image. Second, we used the software program Cell Profiler to determine the total clustered pixels per image, and then converted that data to AChR clusters as a percentage of field. This second method allowed us to calculate how many AChRs were clustered, rather than just counting how many clusters of AChRs were present. Regardless, with both methods we saw a consistent decrease in agrin-induced AChR clustering by antibody to MyoD or myogenin, with 1.00 µg/ml of either antibody being sufficient. By using antibodies rather than RNA interference we could directly test the effect of decreased protein function without the potential confounders that may arise with disruption of normal RNA function.

Skeletal muscle development is guided by myogenic regulatory factors including MyoD and myogenin. These transcription factors are produced in the cytoplasm but bind to DNA in the nuclei of cells to activate or inhibit transcription of specific genes. In the experiments reported here, some cell cultures were treated with antibody to MyoD or myogenin after cell fixation, and then treated with a fluorescent secondary antibody. In those cultures, MyoD and myogenin staining was concentrated in the nuclei of myoblasts or myotubes. Other cell cultures were treated with antibody to MyoD or myogenin while still in differentiation medium, fixed, and then treated with a fluorescent secondary antibody. In those cultures, MyoD and myogenin staining was limited to the cytoplasm of myoblasts or myotubes and was not observed in nuclei. This may reflect the difficulty of antibody or antibody-bound MyoD or myogenin to reach the nuclei in cell culture, possibly due to size or charge. Thus, any antibody-bound MyoD or myogenin visualized with fluorescent staining was limited to the cytoplasm. For the experiments investigating agrin-induced AChR clustering or myotube formation, with antibody to MyoD or myogenin added to cell cultures over multiple days, the antibodies most likely bound to targets in the cytoplasm and prevented them from reaching cell nuclei and affecting gene transcription.

MyoD activates genes involved in synaptic function, including the muscle specific kinase (MuSK) and rapsyn. Interestingly, myogenin also activates the rapsyn gene but not the MuSK gene [18]. Furthermore, myogenin activates genes for AChR subunits [31,32]. This may reflect a role for MyoD earlier in differentiation involving MuSK and rapsyn, and a role for myogenin later in differentiation involving rapsyn and AChR but not MuSK. In the data reported here, 1.00 μg/ml of antibody to myogenin was required to decrease agrin-induced AChR clustering, with no further decrease at 2.00 μg/ml. In contrast, there was a dose response to antibody to MyoD that began at 0.01 μg/ml and continued through 2.00 μg/ml, with increasing concentrations of antibody further decreasing agrin-induced AChR clustering. This may reflect a MyoD-specific role in MuSK expression, where decreased MyoD decreases MuSK expression which then decreases agrin-induced AChR clustering. Low-density lipoprotein receptor-related protein 4 (Lrp4) serves as an agrin receptor that forms a complex with MuSK and mediates MuSK activation by agrin [37,38]. Specifically agrin binds to the N-terminal region of Lrp4 and stimulates the association between Lrp4 and the first immunoglobulin-like domain in MuSK [49]. In response to agrin binding Lrp4, MuSK is tyrosine phosphorylated [28], beginning a signaling cascade that includes the tyrosine phosphorylation of the AChR β subunit [39,40]. Suppression of Lrp4 expression attenuates agrin binding activity, agrin-induced MuSK tyrosine phosphorylation, and agrin-induced AChR clustering [38]. The agrin/MuSK interaction to increase AChR clustering may be decreased when antibody to MyoD decreases MuSK gene activation. Alternatively, using antibody to decrease the amount of MyoD may decrease agrin-induced AChR clustering by decreasing myogenin expression, with myogenin more directly driving AChR clustering and presumably synapse formation. The myogenin gene is activated by MyoD during myogenesis [18].

Myogenin appears to be required for late but not early events in myogenesis. First, myogenin is not required for myotome formation or the appearance of myoblasts [17]. Second, MyoD initiates expression of early genes but requires combined activity with myogenin to initiate expression of late genes. In contrast, myogenin inefficiently activates genes in the absence of MyoD but acts synergistically with MyoD on a set of genes normally expressed late in the program of myogenic differentiation [50]. Third, overexpression of myogenin in transgenic mice elevates mRNA and protein levels of all five AChR subunits, while reducing MyoD protein levels [51]. Fourth, mice with a mutation in the myogenin gene have severe muscle defects resulting in perinatal death [17]. Finally, myogenin null mice die at birth due to a lack of secondary muscle fiber development [9]. The amount of myogenin antibody used in our experiments was insufficient to affect myotube formation, but was sufficient to decrease agrin-induced AChR clustering and may decrease myoblast viability. This may reflect a difference in myogenin sensitivity between AChR clustering and myotube formation. More importantly, since myogenin activates genes for AChR subunits [31,32], antibody to myogenin decreases the number of AChRs available for clustering in response to agrin.

5. Acknowledgements

This work was supported in part by the Office of Research and Sponsored Programs at Midwestern University, which provided intramural funding, as well as providing a Summer Fellowship for Matthew K. Ball. In addition, David H. Campbell and Paul R. Standley were supported by funding provided by the A.T. Still University Strategic Research Initiative.

REFERENCES

[1] J. Piette, J. L. Bessereau, M. Huchet and J. P. Changeux, "Two Adjacent MyoD1-Binding Sites Regulate Expression of the AChR Alpha-Subunit Gene," *Nature*, Vol. 345, No. 6273, 1990, pp. 353-355. doi:10.1038/345353a0

[2] B. P. Gilmour, G. R. Fanger, C. Newton, E. M. Evans and P. D. Gardner, "Multiple Binding Sites for Myogenic Regulatory Factors Are Required for Expression of the AChR Gamma-Subunit Gene," *The Journal of Biological Chemistry*, Vol. 266, No. 30, 1991, pp. 19871-19874.

[3] M. Numberger, I. Durr, W. Kues, M. Koenen and V. Witzcmann, "Different Mechanisms Regulate Muscle-Specific AChR Gamma- and Epsilon-Subunit Gene Expression," *The EMBO Journal*, Vol. 10, No. 10, 1991, pp. 2957-2964.

[4] C. A. Prody and J. P. Merlie, "A Developmental and Tis-

sue-Specific Enhancer in the Mouse Skeletal Muscle AChR Alpha-Subunit Gene Regulated by Myogenic Factors," *The Journal of Biological Chemistry*, Vol. 266, No. 23, 1991, pp. 22588-22596.

[5] A. M. Simon and S. J. Burden, "An E Box Mediates Activation and Repression of the AChR Delta-Subunit Gene during Myogenesis," *Molecular and Cellular Biology*, Vol. 13, No. 9, 1993, pp. 5133-5140.

[6] T. Braun, M. A. Rudnicki, H. H. Arnold and R. Jaenisch, "Targeted Inactivation of the Muscle Regulatory Gene Myf-5 Results in Abnormal Rib Development and Perinatal Death," *Cell*, Vol. 71, No. 3, 1992, pp. 369-382. doi:10.1016/0092-8674(92)90507-9

[7] M. A. Rudnicki, T. Braun, S. Hinuma and R. Jaenisch, "Inactivation of MyoD in Mice Leads to Up-Regulation of the Myogenic HLH Gene Myf-5 and Results in Apparently Normal Muscle Development," *Cell*, Vol. 71, No., 1992, pp. 383-390. doi:10.1016/0092-8674(92)90508-A

[8] M. A. Rudnicki, P. N. Schnegelsberg, R. H. Stead, T. Braun, H. H. Arnold and R. Jaenisch, "MyoD or Myf-5 Is Required for the Formation of Skeletal Muscle," *Cell*, Vol. 75, No. 7, 1993, pp. 1351-1359. doi:10.1016/0092-8674(93)90621-V

[9] P. Hasty, A. Bradley, J. H. Morris, D. G. Edmondson, J. M. Venuti, E. N. Olson and W. H. Klein, "Muscle Deficiency and Neonatal Death in Mice with a Targeted Mutation in the Myogenin Gene," *Nature*, Vol. 364, No. 6437, 1993, pp. 501-506. doi:10.1038/364501a0

[10] Y. Nabeshima, K. Hanaoka, M. Hayasaka, E. Esumi, S. Li, I. Nonaka and Y. Nabeshima, "Myogenin Gene Disruption Results in Perinatal Lethality Because of Severe Muscle Defect," *Nature*, Vol. 364, No. 6437, 1993, pp. 532-535. doi:10.1038/364532a0

[11] S. J. Rhodes and S. F. Konieczny, "Identification of MRF4: A New Member of the Muscle Regulatory Factor Gene Family," *Genes & Development*, Vol. 3, No. 12b, 1989, pp. 2050-2061. doi:10.1101/gad.3.12b.2050

[12] E. Bober, G. E. Lyons, T. Braun, G. Cossu, M. Buckingham and H. H. Arnold, "The Muscle Regulatory Gene, Myf-6, Has a Biphasic Pattern of Expression during Early Mouse Development," *The Journal of Cell Biology*, Vol. 113, No. 6, 1991, pp. 1255-1265. doi:10.1083/jcb.113.6.1255

[13] T. J. Hinterberger, D. A. Sassoon, S. J. Rhodes and S. F. Konieczny, "Expression of the Muscle Regulatory Factor MRF4 during Somite and Skeletal Myofiber Development," *Developmental Biology*, Vol. 147, No. 1, 1991, pp. 144-156. doi:10.1016/S0012-1606(05)80014-4

[14] A. Buonanno, L. Apone, M. I. Morasso, R. Beers, H. R. Brenner and R. Eftimie, "The MyoD Family of Myogenic Factors Is Regulated by Electrical Activity: Isolation and Characterization of a Mouse Myf-5 cDNA," *Nucleic Acids Research*, Vol. 20, No. 3, 1992, pp. 539-544. doi:10.1093/nar/20.3.539

[15] K. Hannon, C. K. Smith, K. R. Bales and R. F. Santerre, "Temporal and Quantitative Analysis of Myogenic Regulatory and Growth Factor Gene Expression in the Developing Mouse Embryo," *Developmental Biology*, Vol. 151, No. 1, 1992, pp. 137-144. doi:10.1016/0012-1606(92)90221-2

[16] J. K. Yoon, E. N. Olson, H. H. Arnold and B. J. Wold, "Different MRF4 Knockout Alleles Differentially Disrupt Myf-5 Expression: Cis-Regulatory Interactions at the MRF4/Myf-5 Locus," *Developmental Biology*, Vol. 188, No. 2, 1997, pp. 349-362. doi:10.1006/dbio.1997.8670

[17] J. M. Venuti, J. H. Morris, J. L. Vivian, E. N. Olson and W. H. Klein, "Myogenin Is Required for Late but Not early Aspects Of Myogenesis during Mouse Development," *Journal of Cell Biology*, Vol. 128, No. 4, 1995, pp. 563-576. doi:10.1083/jcb.128.4.563

[18] A. Blais, M. Tsikitis, D. Acosta-Alvear, R. Sharan, T. Kluger and B. D. Dynlacht, "An Initial Blueprint for Myogenic Differentiation," *Genes & Development*, Vol. 19, No. 5, 2005, pp. 553-569. doi:10.1101/gad.1281105

[19] D. Yaffe and O. Saxel, "Serial Passaging and Differentiation of Myogenic Cells Isolated from Dystrophic Mouse Muscle," *Nature*, Vol. 270, No. 5639, 1977, pp. 725-727. doi:10.1038/270725a0

[20] H. M. Blau, G. K. Pavlath, E. C. Hardeman, C. P. Chiu, L. Silberstein, S. G. Webster, S. C. Miller and C. Webster, "Plasticity of the Differentiated State," *Science*, Vol. 230, No. 4727, 1985, pp. 758-766. doi:10.1126/science.2414846

[21] D. M. Fambrough, "Control of Acetylcholine Receptors in Skeletal Muscle," *Physiological Reviews*, Vol. 59, No. 1, 1979, pp. 165-227.

[22] H. C. Fertuck and M. M. Salpeter, "Quantitation of Junctional and Extrajunctional Acetylcholine Receptors by Electron Microscope Autoradiography after 125I-Alpha-Bungarotoxin Binding at Mouse Neuromuscular Junctions," *Journal of Cell Biology*, Vol. 69, No. 1, 1976, pp. 144-158. doi:10.1083/jcb.69.1.144

[23] C. Fuhrer, J. E. Sugiyama, R. G. Taylor and Z. W. Hall, "Association of Muscle-Specific Kinase MuSK with the Acetylcholine Receptor in Mammalian Muscle," *The EMBO Journal*, Vol. 16, No. 16, 1997, pp. 4951-4960. doi:10.1093/emboj/16.16.4951

[24] W. J. La Rochelle and S. C. Froehner, "Determination of the Tissue Distributions and Relative Concentrations of the Postsynaptic 43 kDa Protein and the Acetylcholine Receptor in Torpedo," *Journal of Cell Biology*, Vol. 261, No. 12, 1986, pp. 5270-5274.

[25] P. G. Noakes, W. D. Phillips, T. A. Hanley, J. R. Sanes and J. P. Merlie, "43 k Protein and Acetylcholine Receptors Colocalize during the Initial Stages of Neuromuscular Synapse Formation *in Vivo*," *Developmental Biology*, Vol. 155, No. 1, 1993, pp. 275-280. doi:10.1006/dbio.1993.1025

[26] D. C. Bowen, J. S. Park, S. Bodine, J. L. Stark, D. M. Valenzuela, T. N. Stitt, G. D. Yancopoulos, R. M. Lindsay, D. J. Glass and P. S. DiStefano, "Localization and Regulation of MuSK at the Neuromuscular Junction," *Developmental Biology*, Vol. 199, No. 2, 1998, pp. 309-319. doi:10.1006/dbio.1998.8936

[27] M. Gautam, T. M. DeChiara, D. J. Glass, G. D. Yancopoulos and J. R. Sanes, "Distinct Phenotypes of Mutant Mice Lacking agrin, MuSK, or rapsyn," *Brain Research*.

Developmental Brain Research, Vol. 114, No. 2, 1999, pp. 171-178. doi:10.1016/S0165-3806(99)00013-9

[28] D. J. Glass, D. C. Bowen, T. N. Stitt, C. Radziejewski, J. Bruno, T. E. Ryan, D. R. Gies, S. Shah, K. Mattsson, S. J. Burden, P. S. DeStefano, D. M. Valenzuela, T. M. DeChiara and G. D. Yancopolous, "Agrin Acts via a MuSK Receptor Complex," *Cell*, Vol. 85, No. 4, 1996, pp. 513-523. doi:10.1016/S0092-8674(00)81252-0

[29] J. E. Sugiyama, D. J. Glass, G. D. Yancopoulos and Z. W. Hall, "Laminin-Induced Acetylcholine Receptor Clustering: An Alternative Pathway," *Journal of Cell Biology*, Vol. 139, No. 1, 1997, pp. 181-191. doi:10.1083/jcb.139.1.181

[30] M. Gautam, P. G. Noakes, J. Mudd, M. Nichol, G. C. Chu, J. R. Sanes and J. P. Merlie, "Failure of Postsynaptic Specialization to Develop at Neuromuscular Junctions of Rapsyn-Deficient Mice," *Nature*, Vol. 377, No. 6546, 1995, pp. 232-236. doi:10.1038/377232a0

[31] H. Tang, Z. Sun and D. Goldman, "CaM Kinase II-Dependent Suppression of Nicotinic Acetylcholine Receptor Delta-Subunit Promoter Activity," *Journal of Biological Chemistry*, Vol. 276, No. 28, 2001, pp. 26057-26065. doi:10.1074/jbc.M101670200

[32] H. Tang, P. Macpherson, L. S. Argetsinger, D. Cieslak, S. T. Suhr, C. Carter-Su and D. Goldman, "CaM Kinase II-Dependent Phosphorylation of Myogenin Contributes to Activity-Dependent Suppression of nAChR Gene Expression in Developing Rat Myotubes," *Cell Signal*, Vol. 16, No. 5, 2004, pp. 551-563. doi:10.1016/j.cellsig.2003.09.006

[33] E. W. Godfrey, R. M. Nitkin, B. G. Wallace, L. L. Rubin and U. J. McMahan, "Components of Torpedo Electric Organ and Muscle That Cause Aggregation of Acetylcholine Receptors on Cultured Muscle Cells," *Journal of Cell Biology*, Vol. 99, No. 2, 1984, pp. 615-627. doi:10.1083/jcb.99.2.615

[34] R. M. Nitkin, M. A. Smith, C. Magill, J. R. Fallon, Y. M. Yao, B. G. Wallace and U. J. McMahan, "Identification of Agrin, a Synaptic Organizing Protein from Torpedo Electric Organ," *Journal of Cell Biology*, Vol. 105, No. 6, 1987, pp. 2471-2478. doi:10.1083/jcb.105.6.2471

[35] U. J. McMahan, "The Agrin Hypothesis," *Cold Spring Harbor Symposia on Quantitative Biology*, Vol. 55, 1990, pp. 407-418. doi:10.1101/SQB.1990.055.01.041

[36] M. A. Bowe and J. R. Fallon, "The Role of Agrin in Synapse Formation," *Annual Review of Neuroscience*, Vol. 18, 1995, pp. 443-462. doi:10.1146/annurev.ne.18.030195.002303

[37] N. Kim, A. L. Stiegler, T. O. Cameron, P. T. Hallock, A. M. Gomez, J. H. Huang, S. R. Hubbard, M. L. Dustin and S. J. Burden, "LRP4 Is a Receptor for Agrin and Forms a Complex with MuSK," *Cell*, Vol. 135, No. 2, 2008, pp. 334-342. doi:10.1016/j.cell.2008.10.002

[38] B. Zhang, S. Luo, Q. Wang, T. Suzuki, W. C. Xiong and L. Mei, "LRP4 Serves as a Co-Receptor of Agrin," *Neuron*, Vol. 60, No. 2, 2008, pp. 285-297. doi:10.1016/j.neuron.2008.10.006

[39] M. Ferns, M. Deiner and Z. Hall, "Agrin-Induced Acetylcholine Receptor Clustering in Mammalian Muscle Requires Tyrosine Phosphorylation," *The Journal of Cell Biology*, Vol. 132, No. 5, 1996, pp. 937-944. doi:10.1083/jcb.132.5.937

[40] B. G. Wallace, Z. Qu and R. L. Huganir, "Agrin Induces Phosphorylation of the Nicotinic Acetylcholine Receptor," *Neuron*, Vol. 6, No. 6, 1991, pp. 869-878. doi:10.1016/0896-6273(91)90227-Q

[41] W. A. Grow and H. Gordon, "Acetylcholine Receptors Are Required for Postsynaptic Aggregation Driven by the Agrin Signalling Pathway," *European Journal of Neuroscience*, Vol. 12, No. 2, 2000, pp. 467-472. doi:10.1046/j.1460-9568.2000.00923.x

[42] P. C. D. Macpherson, D. Cieslak and D. Goldman, "Myogenin-Dependent nAChR Clustering in Aneural Myotubes," *Molecular and Cellular Neuroscience*, Vol. 31, No. 4, 2006, pp. 649-660. doi:10.1016/j.mcn.2005.12.005

[43] P. Ravdin and D. Axelrod, "Fluorescent Tetramethyl Rhodamine Derivatives of Alpha-Bungarotoxin: Preparation, Separation, and Characterization," *Analytical Biochemistry*, Vol. 80, No. 2, 1977, pp. 585-592. doi:10.1016/0003-2697(77)90682-0

[44] D. H. Campbell, M. Hicks, W. Grow and P. Standley, "Improved Quantification of Acetylcholine Receptor (AChR) Clustering in C2C12 Myotubes Using an Objective Computational Algorithm," *Arizona Imaging & Microscopy Society Conferenc* 2011. .

[45] T. A. Rando and H. M. Blau, "Primary Mouse Myoblast Purification, Characterization, and Transplantation for Cell-Mediated Gene Therapy," *The Journal of Cell Biology*, Vol. 125, No. 6, 1994, pp. 1275-1287. doi:10.1083/jcb.125.6.1275

[46] T. J. Miller and W. A. Grow, "Mercury Decreases the Frequency of Induced But Not Spontaneous Clustering of Acetylcholine Receptors," *Cell and Tissue Research*, Vol. 316, No. 2, 2004, pp. 211-219. doi:10.1007/s00441-004-0878-6

[47] B. W. Steffens, L. M. Batia, C. J. Baarson, C.-K. C. Choi and W. A. Grow, "The Pesticide Methoxychlor Decreases Myotube Formation in Cell Culture by Slowing Myoblast Proliferation," *Toxicology in Vitro*, Vol. 21, No. 5, 2007, pp. 770-781. doi:10.1016/j.tiv.2007.01.007

[48] D. B. Owen, K. T. Chamberlain, S. Shishido and W. A. Grow, "Ethanol Decreases Agrin-Induced Acetylcholine Receptor Clustering in C2C12 Myotube Culture," *Toxicology in Vitro*, Vol. 24, No. 2, 2010, pp. 645-651. doi:10.1016/j.tiv.2009.09.020

[49] W. Zhang, A. S. Coldefy, S. R. Hubbard and S. J. Burden, "Agrin Binds to the N-Terminal Region of LRP4 Protein and Stimulates Association between LRP4 and the First Immunoglobulin-Like Domain in Muscle-Specific Kinase (MuSK)," *The Journal of Biological Chemistry*, Vol. 286, 2011, pp. 40624-40630. doi:10.1074/jbc.M111.279307

[50] Y. Cao, R. M. Kumar, B. H. Penn, C. A. Berkes, C. Kooperberg, L. A. Boyer, R. A. Young and S. J. Tapscott, "Global and Gene-Specific Analyses Show Distinct Roles for Myod and Myog at a Common Set of Promoters," *The EMBO Journal*, Vol. 25, 2006, pp. 502-511. doi:10.1038/sj.emboj.7600958

[51] K. Gundersen, I. Rabben, B. J. Klocke and J. P. Merlie, "Overexpression of Myogenin in Muscles of Transgenic Mice: Interaction with Id-1, Negative Crossregulation of Myogenic Factors, and Induction of Extrasynaptic Acetylcholine Receptor Expression," *Molecular and Cellular Biology*, Vol. 15, No. 12, 1995, pp. 7127-7134.

7

Role of Active Principles of *Podophyllum hexandrum* in Amelioration of Radiation Mediated Lung Injuries by Reactive Oxygen/Nitrogen Species Reduction

Rashmi Saini, Savita Verma, Abhinav Singh, Manju Lata Gupta*
Institute of Nuclear Medicine and Allied Sciences (INMAS), Defence Research and Development Organization (DRDO), New Delhi, India
Email: *drmanjugupta2003@yahoo.com

ABSTRACT

Radiation induced reactive oxygen/nitrogen species (ROS/RNS) are reported to cause lung injuries such as pneumonitis and fibrosis which may be fatal at times. Current study is designed to analyse the radioprotective efficacy of *P. hexandrum* active principles (G-002M) on lungs of mice exposed to high dose of gamma irradiation (7 Gy). Cellular profiles and inflammatory cell infiltrates of irradiated bronchoalveolar lavage fluid (BALF) have shown correlations with lung pathology. Cell counts were determined in BALF of control, 7 Gy radiation exposed and radiation with G-002M pretreated mice. ROS/Nitric Oxide (NO) production was measured by 2,7 dichlorodihydrofluorescein diacetate (DCF-DA) and diaminofluorescein diacetate (DAF-2DA) through microscopy and flow cytometry respectively. Immunostaining of inducible nitric oxide synthase (iNOS) in BALF cells and lung sections was also observed microscopically. iNOS expression was observed in lungs by western blotting. BALF was also processed to estimate total protein, LDH, and phospholipids content. Catalase, reduced Glutathione (GSH), Glutathione reductase (GR) and lipid peroxidation were estimated in lung tissues. Pre-administration of G-002M significantly decreased radiation mediated neutrophils count in BALF of irradiated mice. ROS generation, iNOS expression, total protein, LDH and phospholipids were found less affected in G-002M pretreated group in comparison to radiation alone group. Radiation exposure to mice was found apparently leading to parenchymal fibrosis, an architectural distortion of the lung tissue with edema, infiltration of inflammatory blood cells with increased immunolabeling of iNOS. G-002M pretreatment significantly countered radiation mediated increased lipid peroxidation and decreased GR, catalase and GSH in mice. Current study demonstrates possible role of *P. hexandrum* (G-002M) in minimizing lung damage induced by radiation mediated ROS/RNS generation.

Keywords: Bronchoalveolar Lavage Fluid; Lung Injury; Reactive Nitrogen Species; *Podophyllum hexandrum*; Radiation

1. Introduction

Radiation inflicts various histopathological changes in the lung that can be diversed with overlapping features, characterized by varying degrees of inflammation and fibrosis [1,2]. Architectural distortion of the lung tissue along with thickening of alveolar wall (edema), increased degree of vascular congestion, hyaline membrane formation in distal airway and inflammatory cell accumulation are found to be common in pulmonary disease cases. Under pathological conditions, including pulmonary fibrosis, the production of reactive oxygen species (ROS) is augmented by a variety of mechanisms, resulting into hydroxyl radicals and reactive nitrogen species (RNS) generation [3]. Lungs like other tissues have highly specialized and compartmentalized antioxidant defence mechanisms to protect against ROS and RNS. Several studies have demonstrated that nitric oxide (NO) is a significant mediator of radiation-induced acute lung injury [3-6]. Inhibition in inducible nitric oxide synthase (iNOS) isoform expression, enhanced in the irradiated lungs, can decrease NO production mediated lung damage [7]. Production of NO, being a small and diffusible signalling molecule, is tightly regulated and is produced at the right time, in right amount and at right place [8].

*Corresponding author.

Thus, the localization of the ROS/RNS generation and the concentration of protective antioxidants play the lead role in expression of radiation inflicted lung injuries and mediating in pathologic response. Lung inflammation, as manifested by infiltration of large numbers of neutrophils, eosinophils, macrophages, and lymphocytes into the lung, can be monitored by examining presence of inflammatory cells in bronchoalveolar lavage fluid (BALF) and/or histologically prepared lung tissue sections. Measurement of inflammatory cell infiltrates of BALF has been shown to correlate with disease severity [9-12]. Cellular differential profiles in BALF are used to determine the nature of the disease and their appearance in the interstitium of lungs may have diagnostic or therapeutic impact [13-17]. Several interstitial lung disorders can be characterized by evaluating the distribution of cell types and their subpopulations in BALF [10,18,19]. BALF study is well reported tool in the sampling of the lower respiratory tract both for clinical indications and for research related investigations [9]. Quantitation of the cell types recovered from BALF is also among vital parameters for evaluation of the biological effect of ionizing radiation in lungs.

Increased threat of nuclear or radiological exposures by the multifaceted use of radiation has expressed the need for development of safe and effective radioprotector. Undesired toxicity of synthetic radioprotectors and their analogues have precluded them from clinical use and necessitated the search for safe alternative agents. Among the high altitude plants screened for radioprotective efficacy, *Podophyllum hexandrum* (Division: Magnoliophyta; Family: Berberidaceae) has been found to be a potent scavenger of free radicals induced by ionizing radiation [20-24]. The present study has focused on the radioprotective potential of G-002M (formulation prepared by combination of three active compounds isolated from *P. hexandrum*) on ROS/RNS mediated lung tissue damages.

The current study has been specifically designed to evaluate the modulatory effect of formulation of P. hexandrum (G-002M) on 1) cellularity and differential cell counts in BALF; 2) ROS/NO production; 3) expression of iNOS and 4) alterations in antioxidant enzymes(GSH, GR catalase) and lipid peroxidation following radiation inflicted lung injury.

2. Materials & Methods

2.1. Reagents and Antibodies

Goat anti-rabbit HRP (Cat No. Sc-2030) was procured from Santa Cruz (CA). Polyclonal antibodies to iNOS (Cat No.-N7782), Goat anti rabbit FITC 488 (Cat No. F4018), 2,7 dichlorodihydrofluorescein diacetate (DCF-DA), diaminofluorescein diacetate (DAF-2DA) and all other required chemicals were obtained from Sigma Aldrich Co. (St Louis, MO).

2.2. Preparation of G-002M

The drug (G-002M) was prepared by combining three compounds isolated from dried rhizomes of P. hexandrum plant, collected from high altitude region of Leh and Ladakh, India. The dried rhizomes were crushed to get fine powder which was processed further using standard protocols. The three active principles isolated after elaborate processing, were analysed on HPLC for their chemical identification and purity. The active principles were identified as podophyllotoxin, podophyllotoxin-β-D-glucoside and rutin. All three molecules were in their > 97% purity. G-002M was prepared freshly at the time of administration by dissolving in DMSO which was diluted further in distilled water to the final concentration of 1:9 (DMSO: distilled water). The injectable volume was 200 μl/mice administered intramuscularly one hour before irradiation of animals.

2.3. Animals and γ-Ray Irradiation

Strain "A" female mice (25 - 28 g), 8 - 10 weeks old were obtained from the institute animal house. All the experiments were based on institutional ethical guidelines. Mice were exposed to a single dose of 7.0 Gy in ^{60}CO gamma chamber (Cobalt Teletherapy Bhabhatron-II) at the dose rate of 0.925 - 0.828 cGy/sec. Radiation dose calibration was done by Fricke's dosimetry method by radiation physicists. All the experiments were repeated thrice.

2.4. Experimental Design

As per previous reports [3,25,26], the detailed studies in mice BALF were performed after 2 weeks of irradiation as acute lung damage peaked at week 2. Animals were divided into four experimental groups: 1) control; 2) irradiated; 3) G-002M plus radiation; 4) G-002M only. In irradiated group, the mice were exposed to 7 Gy (whole body). G-002M plus radiation group, was administered with G-002M intramuscularly prior to irradiation (−1 hr). G-002M was dissolved freshly in DMSO and diluted with water maintaining the final concentration of DMSO to 10% of total injectable (200 μl/mice).

2.5. Collection of BALF and Its Cells Count

Mice were sacrificed and lungs were lavaged 5 times with 200 μl of buffered saline at 37°C [27]. Cells were collected individually from three animals, after centrifugation (2000 rpm for 10 min at 4°C) were processed for cytological evaluation. BALF cells were quantified with a haemocytometer by the conventional method and also by fully automated 5 part Hematology Analyzer (ADVIA 2120 Seimens Diagnostics, USA). Data from the Perox and Baso channels were used to determine BALF total

WBC count and differential analysis [28]. Differential cell counts (lymphocytes, neutrophils) were calculated from the percentage of each cell type taken from a count of at least 200 cells on a stained preparation of BALF. Fluid from the lavage was used for evaluation of biochemical changes after cell removal.

2.6. BALF Smear

BALF smears were prepared for microscopic examination [29]. Briefly, the aspirated cells from lavage fluid were collected and centrifuged at 2000 rpm for 10 min. The supernatant was discarded, cells mixed with fetal bovine serum and smeared on clean glass slide. Slides were air-dried, stained with May-Grünwald Giemsa stain for 10 min and total 200 different cell types (neutrophils and lymphocytes) were counted to determine percentage of differential cell counts in BALF.

2.7. ROS/NO Measurement in BALF Cells by Flow Cytometry

BALF cells (2×10^4/ml) were washed with PBS (137 mM NaCl, 27 mM KCl, 32 mM Na_2HPO_4, 15 mM K_2HPO_4, 15 mM KH_2PO_4, pH 7.4) and incubated for 30 min with DCF-DA (10 μm) and DAF-2DA (10 μm) dye for estimating ROS generation and NO production respectively [8,30]. Real time ROS generation and NO production was measured by acquiring 10,000 cells from each sample in flow cytometer and analyzed by the Cell Quest program.

2.8. DCF-DA and DAF-2DA Fluorescence Intensity Measurement in BALF Cells

BALF cells (2×10^4/ml) were washed with PBS, and incubated for 15 min with DCF-DA (10 μm) dye. DCF-DA loaded cells were dispensed on 0.01% (w/v) poly-L-lysine coated coverslips and monitored for real time ROS generation under fluorescence microscope (Olympus, Model DP 72). For estimation of NO production, BALF cells (2×10^4/ml) were incubated for 15 min with DAF-2DA (10 μm) dye, dispensed on 0.01% (w/v) poly-L-lysine coated coverslips and monitored for real time NO production under fluorescence microscope. Data was captured with Image-Pro Plus Software 7.0, and photographs were processed using Adobe Photoshop software San Jose, CA. Cells without dyes or fixed cells (1 hr with 4% paraformaldehyde), incubated with dyes were used as negative controls.

2.9. BALF Cells Studies Using Immunocytochemistry

BALF cells (2×10^4 cells/ml) were fixed overnight at 25°C in 4% (w/v) paraformaldehyde in PBS (pH 7.4) and washed two times for 5 min each with PBS containing 0.5% (w/v) glycine. The washed cells were allowed to adhere on 0.01% (w/v) poly-L-lysine coated coverslips, permeabilized with 0.2 % (v/v) Triton X-100 (5 - 10 min) and blocked with 1% BSA for 2 hrs. Cells were incubated overnight at 4°C with antibodies against iNOS at a dilution of 1:100 and were subsequently stained at 4°C for 4 h with FITC tagged secondary antibody (1:500). Nuclei were stained with DAPI (0.5 μg/ml) at 25°C for 15 min. Control samples were separately processed by omitting primary antibodies. Mounted coverslip images were acquired by fluorescence microscope and were presented using Adobe Photoshop software.

2.10. Histological Studies in Lung Tissues

Anesthesized mice were dissected and lungs tissues were excised. After removing the blood clots, lung tissues were fixed in 10% buffered formalin. Samples were processed using standard histological procedures. 3 - 5 μm sections were cut from lung tissues embedded in paraffin by using semi automated microtome. Sections were stained with hematoxylin-eosin (H-E) and evaluated under the light microscope.

2.11. Immunohistochemistry of Lungs Sections

Excised tissues were fixed in 10% buffered formalin and embedded in paraffin. 5 μm sections were deparaffinised, rehydrated, rinsed in distilled water and washed in Tris buffer saline. The slides were immersed in citrate buffer (pH 6.0) and antigen retrieval was done using a domestic microwave oven at 600 wattage. Immunohistochemistry was then performed as described previously [31]. In brief, lung tissue sections were adhered on coverslips, deparaffinised and incubated with primary antibody (iNOS, 1:200 dilution) overnight at 4°C and secondary antibody tagged with FITC (1:500 dilution) for 4 hrs at 4°C. Slides were stained with DAPI (0.5 mg/ml) and sections were visualized under fluorescence microscope.

2.12. Assessment of iNOS Expression by Western Blotting

Frozen lung tissue was homogenized in RIPA buffer (50 mM Tris-HCl, pH 7.4, 150 mM NaCl, 1 mM EDTA, 0.1% SDS, 0.5% Sodium deoxycholate, 1% Nonidet-P40, and a mixture of protease inhibitors), and protein concentration was determined using the Bradford Assay. Protein (50 μg) was solubilized in SDS sample buffer, separated on 10% SDS-PAGE and transferred onto a nitrocellulose membrane at 120 V for 1 h, by the method of Saini *et al*, with modifications [8]. The membrane was stained with Ponceau S stain to check the efficiency of transfer of proteins and blocked with 5% skimmed milk in PBST, 10 mM Tris-HCl, pH 7.5, 150 mM NaCl containing 0.1% Tween 20. The membrane was then incubated with anti-iNOS antibody (1:1000) overnight at 4°C. The membrane was washed three times (5 min each) with blocking buffer and then incubated with HRP-conjugated anti-rabbit IgG (1:5000) for 2 hrs. After being washed three times (5 min each), the blot was developed with ECL reagents and exposed to X-ray film to visualize the protein bands.

2.13. Measurement of Total Protein Concentration and LDH Activity in BALF

Determination of LDH activity in BALF was done according to the method of Seth *et al.*, 1994 [32]. Briefly, the samples were incubated with sodium pyruvate and the reaction was initiated by nicotinamide adenine dinucleotide (NADH). Absorbance was recorded at 340 nm at the difference of 15 secs for 1 min. Protein contents were measured by Bradford, 1976 method [33]. Standard curve was plotted by using different concentrations of bovine serum albumin (BSA).

2.14. Phospholipids Content in BALF

Phospholipids content was measured with Dithiocynatoiron method [34]. Briefly, 0.2 ml BALF was heated for 5 min at 60°C. After addition of 0.5 ml thiocynatoiron reagent and 0.3 ml of 0.17 N hydrochloric acid, the mixture was incubated for 5 min at 35°C. The thiocynatoiron-phospholipid complex formed was extracted with 1.5 ml 1, 2-dichloroethane by vigorous shaking for 2 min. After centrifugation, the absorbance of the lower layer containing thiocynatoiron-phospholipid complex was measured at 470 nm against a blank without BALF. Absorbance of thiocynatoiron-phospholipid complex is directly proportional to phopholipids content in the sample.

2.15. Biochemical Estimations in Lung Homogenate

2.15.1. Lipid Peroxidation Estimation

Lipid peroxidation (LPX) level was estimated as nmoles of malondialdehyde (MDA) formed, by the method of Buege and Aust, 1978 [35]. In brief, reaction mixture containing lung tissue homogenate, 15% w/v trichloroacetic acid (TCA) and 0.37% w/v tribarbituric acid (TBA) was boiled at 95°C (30 min), centrifuged at 8000 g for 10 min and absorbance of the supernatants was recorded at 535 nm. The concentration of thiobarbituric acid reactive species (TBARS) was determined by using an extinction coefficient of $1.56 \times 10^5\ M^{-1} \cdot cm^{-1}$. Results were expressed as nmoles of MDA formed per mg of total protein.

2.15.2. Reduced Glutathione (GSH) Determination

GSH concentration was estimated in lungs tissues according to the method described by Beutler *et al*, 1975 [36]. Briefly, the proteins of the samples were precipitated by adding precipitating solution (1.67% metaphosphoric acid, 0.2% EDTA, 30% NaCl). After centrifugation for 10 min, 8ml phosphate solution (0.3M) was added to 2 ml of supernatant. After addition of DTNB to the reaction mixture absorbance was recorded at 412 nm. Glutathione concentration was expressed as µM GSH/mg protein.

2.15.3. Glutathione Reductase (GR) Estimation

GR activity was assayed by the method described previously [37]. The assay mixture containing 0.2 M potassium phosphate buffer (pH 7.0), 2 mM nicotinamide adenine dinucleotide phosphate (NADPH) and 20 mM oxidized glutathione was mixed with distilled water to make final volume of 3 ml. Enzyme preparation/sample (0.1 ml) was added to initiate the reaction and activity absorbance was measured at 340 nm for one minute at 15 sec interval. GR activity was expressed as mU/min/mg protein.

2.15.4. Catalase Estimation

Catalase activity was assayed by the method of Sinha *et al.* [38]. Briefly, the assay mixture consisted of 1.96 ml phosphate buffer (0.01 M, pH 7.0), 1.0 ml hydrogen peroxide (0.2 M) and 0.04 ml phenazine methosulfate (PMS) (10%) in a final volume of 3.0 ml. About 2 ml dichromate acetic acid reagent was added in 1 ml of reaction mixture, boiled for 10 min, and was cooled. Absorbance of the supernatants was recorded at 570 nm.

2.16. Statistical Analysis

The experiment results were expressed as mean ± SD of three replicates. Student's paired t-test was used for analyzing the data. A value of $p < 0.05$ was considered as statistically significant.

3. Results

3.1. Total and Differential Cell Counts in BALF

Preparation of the microscope slide smear technique yielded well-preserved cell morphology. The total numbers of BALF cells were found increased in radiation exposed mice to controls. Differential cell counts in control group revealed an average proportion of 72.3% ± 2.7% macrophages, 18.0% ± 2.2% lymphocytes, 8.0% ± 2.3% neutrophils. However, BALF collected from irradiated mice expressed increased percentage of neutrophils (**Table 1**). The increased number of neutrophils in BALF (16%) was found to be decreased (9%) in G-002M pretreated irradiated group. The radiation mediated increased number of total cell counts, macrophages and lymphocytes in BALF were also significantly countered

Table 1. Total number of cells in BALF of experimental mice.

S. No	Experimental Groups (n = 6)	Total cells in BALF/(mm^3)	Macrophages/(mm^3)	Lyphocytes/(mm^3)	Neutrophils/(mm^3)
1.	Control	500 ± 35	360 ± 15	90 ± 10	40 ± 8
2.	Radiation (7 Gy)	1440 ± 62^{a**}	918 ± 32^{a**}	320 ± 25^{a**}	230 ± 15^{a**}
3.	G-002M + Radiation	710 ± 46^{b**}	524 ± 37^{b**}	100 ± 12^{b**}	70 ± 12^{b**}

Values are mean ± SD of the cells obtained from BALF collected individually from all the groups of experimental animals. Levels of significance of the differences are indicated as p values. Study was repeated in three different sets of mice having three animals in each set. [a]Radiation only group compared with controls; [b]G-002M+Radiation group compared with radiation only group. $^{**}p < 0.001$.

by our formulation (**Table 1**). The percentage of differential cell counts obtained from ADVIA 120 Hematology system had been found in consonance with the percentage of cells counted from smear slide preparations of BALF (**Figure 1**).

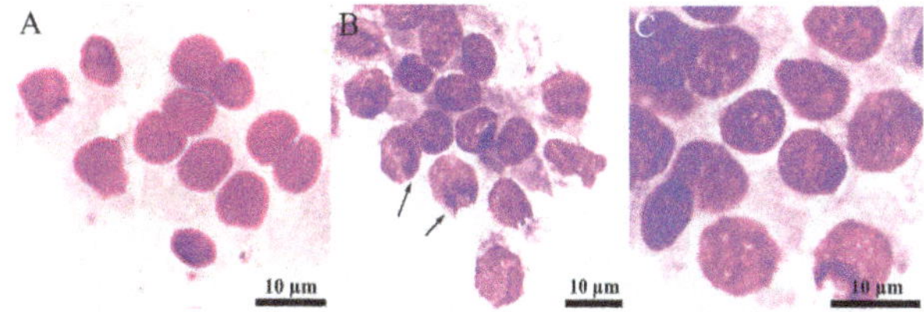

Figure 1. Images of BALF smears stained with May-Grünwald Giemsa stain and viewed at 1000× under trinocular microscope. (A) Control, showing macrophages; (B) BALF collected from irradiated mice showing increased number of neutrophils as evident by multilobulated nuclei (arrows) in most of the cells; (C) BALF collected from G-002M pretreated irradiated mice showing decreased neutrophils. Mice were dissected after 2 week of experimentation.

3.2. Reactive Oxygen Species (ROS)/NO Generation in BALF Cells

ROS generation was measured in the BALF cells by using flow cytometry. Histograms (**Figures 2(G)** and **(H)**) represent differential ROS and NO generation respectively in all the three studied groups (control, irradiated and G-002M pretreated). In radiation only group ROS generation was found significantly increased when compared with controls. G-002M pretreatment apparently countered ROS upregulation. DCF-DA and DAF-2DA fluorescence observed under fluorescence microscope represented radiation mediated increase in ROS (**Figure 2(B)**) and NO production (**Figure 2(E)**) in comparison to corresponding controls (**Figures 2(A)** and **(D)**). G-002M significantly countered formation of both ROS (**Figure 2(C)**) and NO (**Figure 2(F)**).

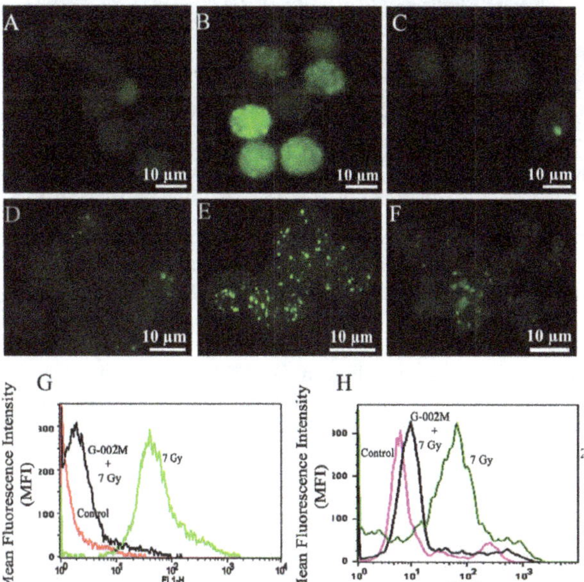

Figure 2. ROS/NO generation in the BALF cells. Images were viewed at 1000X under fluorescence microscope. ROS generation in (A) BALF cells from control animals; (B) 7Gy irradiated mice (C) Mice pretreated with G-002M formulation prior to radiation. Fluorescence intensity expressed the real time ROS generation in BALF cells. NO production observed in (D) BALF cells from control animals; (E) 7Gy irradiated mice; (F) Mice pretreated with G-002M formulation prior to radiation; (G) Histogram represents the ROS generation and (H) NO production studied by flow cytometry.

3.3. Expression of Inducible Nitric Oxide Synthase (iNOS) in BALF Cells

Immunolabeling of iNOS in BALF cells was conducted in controls, irradiated and G-002M pretreated irradiated animals. Radiation (7 Gy, whole body) increased the expression of iNOS in BALF cells as evident from the **Figure 3(D)**. BALF cells collected from mice pretreated with formulation prior to radiation demonstrated low intensity of iNOS immunostaining (**Figure 3(F)**). Cells were counterstained with DAPI (**Figures 3(A)**, **(C)** and **(E)**) as a confirmatory indication for presence of nucleus.

3.4. Histology of Lung Tissues

Histopathological examination performed on 15th day after 7 Gy radiation exposure led to significant changes in the lung architecture like alveolar edema, damage of the alveolar cell lining, destruction of the interalveolar septae, inflammation and infiltration of lymphocytes.

Acute pneumonitis was also visualized (**Figure 4(B)**). Administration of G-002M, 1 hr prior to irradiation showed nonsignificant infiltration of lymphocytes in lung alveoli, minimal edematous changes and nil expression of acute pneumonitis (**Figure 4(C)**).

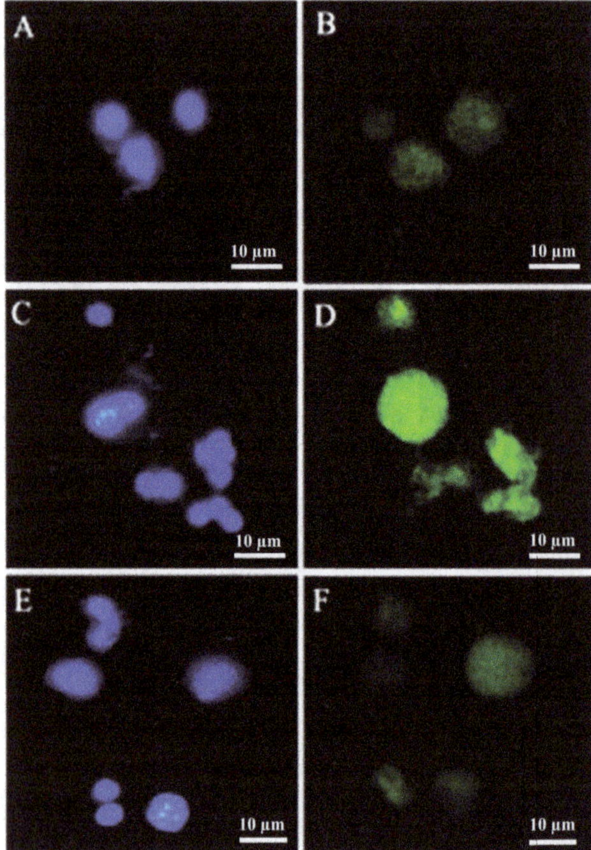

Figure 3. iNOS immunostaining in the BALF cells (A) Nucleus staining in controls (B) iNOS immunolabeling in the controls; (C) Nucleus stained with DAPI in irradiated groups of mice (D) iNOS immunolabeling in irradiated mice. (E) Nucleus staining in G-002M pretreated irradiated mice and (F) iNOS immunolabeling in G-002M pretreated mice. BALF cells were studied under fluorescence microscope at 1000×.

3.5. Immunohistochemical Detection of iNOS

Effect of ionizing radiation on expression of iNOS was explored in lung sections by immunohistochemistry. Advanced parenchymal fibrosis imparting a honeycomb architectural distortion of the lung tissue was observed in irradiated mice along with increased immunolabeling of iNOS. Lung architecture was found to be restored in the mice pretreated with G-002M before radiation exposure with less intense iNOS immunostaining in lung sections (**Figure 5**). Infilterated cells were seen in the air spaces, (**Figure 5(D)**) which were not observed in the G-002M pretreated groups (**Figures 5(F)**). **Figure 5(G)** shows the increased expression of iNOS in the irradiated groups which was found less in the G-002M pretreated groups. The expression of iNOS was not found to be detectable in the control group.

3.6. Lung Permeability and Lung Cytotoxicity

Lung permeability and damage was assessed by estimateing total protein and phospholipids content in BALF. For estimation of lung cytotoxicity, LDH activity was measured in BALF. Radiation exposed animals showed about 2 and 4 folds increase in phospholipids ($p < 0.001$) and protein contents ($p < 0.001$) respectively as compared to controls (**Figures 6(A) and (B)**). Lung cytotoxicity in the same group of mice, as evident by LDH activity (**Figure 6(A)**) was also found extensively increased (6 folds, $p < 0.001$). Pretreatment of G-002M formulation resulted in decreased lung damage expressed by very low amount of protein, LDH and phospholipids content in BALF (**Figure 6**).

3.7. Biochemical Studies in Lung Homogenates

Glutathione reductase (GR), GSH and catalase along with lipid peroxidation (LPX) were measured in lung

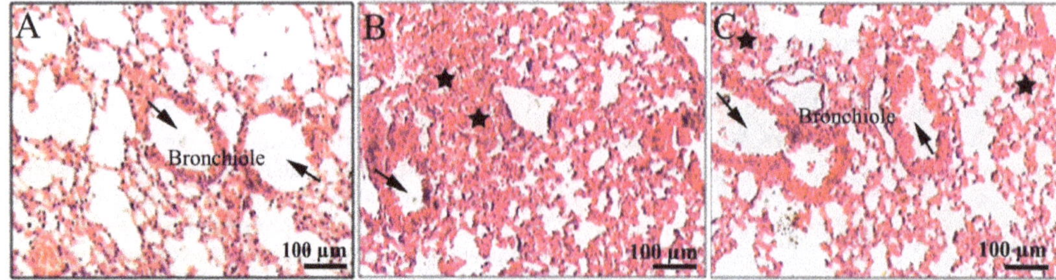

Figure 4. Histological examination of the lungs of mice of different treatment groups: (A) Lungs of controls had not expressed any pathological changes. Bronchiole with large air spaces (arrows) shown (B) lung from irradiated mice showing edema (black solid stars), infiltration of inflammatory cells, and reduced air spaces (arrow) (C) lung from G003M pretreated irradiated mice showing less edematous changes (black solid stars), nil infiltration of cells and bronchiole with large air spaces (arrows) as compared to irradiated group.

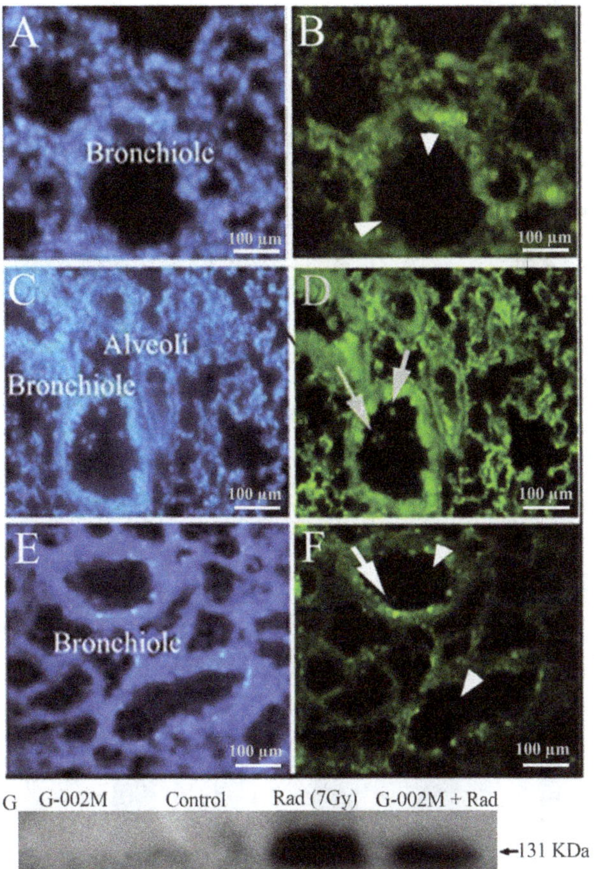

Figure 5. Images showing iNOS immunostaining in lung tissue sections (A) Nucleus (DAPI) staining in controls, (B) iNOS immunostaining in the controls showing bronchiole having large air spaces (arrowheads); (C) Nucleus stained with DAPI in irradiated groups of mice (D) iNOS immunostaining in the irradiated groups of mice showing intense iNOS expression (fluorescence) and infiltration of cells (arrows); (E) DAPI staining in the lung section of G-002M pretreated irradiated group (F) iNOS immunostaining in the G-002M pretreated irradiated group showing decreased iNOS expression (fluorescence), bronchioles with large air spaces (arrowheads) showing negligible infiltration of cells (arrow). Paraffin sections of lung tissue were stained with iNOS antibody and FITC coupled secondary antibody and observed under fluorescence microscope; (G) Western Blotting of iNOS in control, irradiated and G-002M pretreated group.

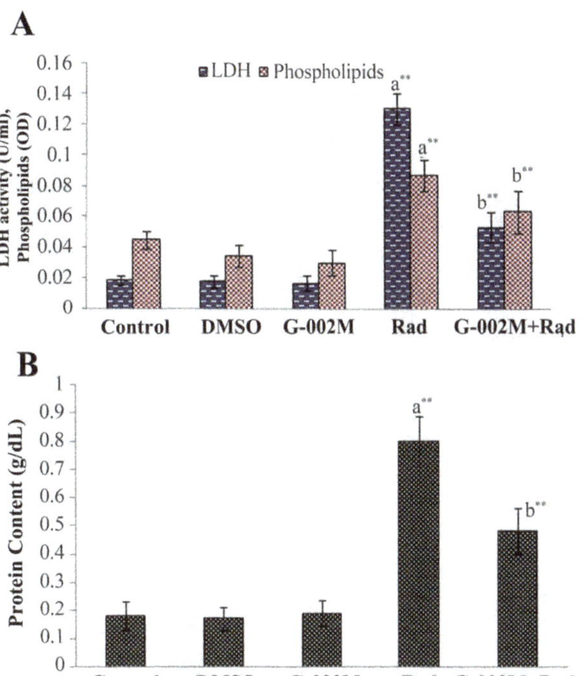

Figure 6. (A) Effect of G-002M on radiation induced alteration in LDH and phospholipids content in BALF of mice exposed to 7Gy whole body irradiation. Mice were sacrificed and lungs were lavaged 5 times with buffered saline. For LDH estimation samples were incubated with sodium pyruvate and the reaction was initiated by addition of NADH. Samples were measured at 340 nm for 15 sec intervals. Phospholipids content was measured by recording absorbance at 470 nm against a blank without BALF. Study was repeated in three different sets of mice having three animals in each group; (B) Effect of G-002M on radiation induced alteration in total protein content in BALF of mice exposed to 7 Gy whole body irradiation. Mice were sacrificed after 2 weeks of experimentation and and lungs were lavaged 5 times with buffered saline. After centrifugation, BALF was collected for total protein estimation. Samples were incubated with Bradford reagent for 5 min at room temperature and absorbance was recorded at 595 nm. Protein concentration was calculated by standard curve prepared by using different concentrations of BSA. [a]Radiation only group compared with controls, [b]G-002M + Radiation group compared with radiation only group. $^{**}p < 0.001$.

homogenates of irradiated mice, mice pretreated with G-002M and exposed to radiation, G-002M only, vehicle and untreated control group. GSH and catalase enzyme activities (**Figures 7(A)** and **(B)**) found decreased in the lung homogenates of irradiated mice (17.45 µM/mg protein and 2.66 U/mg protein respectively), were restored in the homogenates of G-002M pretreated mice (27.89 µM/mg protein and 3.32 U/mg protein respectively). Radiation enhanced peroxidation of lipids, though marginally (0.231 nM MDA × 1000/mg protein), was found decreased by G-002M (0.209 nM MDA ×1000/mg protein) and values were corresponding to controls (**Figure 7(C)**). GR activity, which got declined in irradiated group (15.03 mU/min/mg protein), was found enhanced by G-002M pretreatment (17.04 mU/min/mg protein), but the difference was marginal when compared to control group of mice (25.60 mU/mins/mg protein). Values retrieved from the vehicle and drug only groups were almost parallel to controls (**Figure 7**).

4. Discussion

Radiation exposure may deregulate biological system

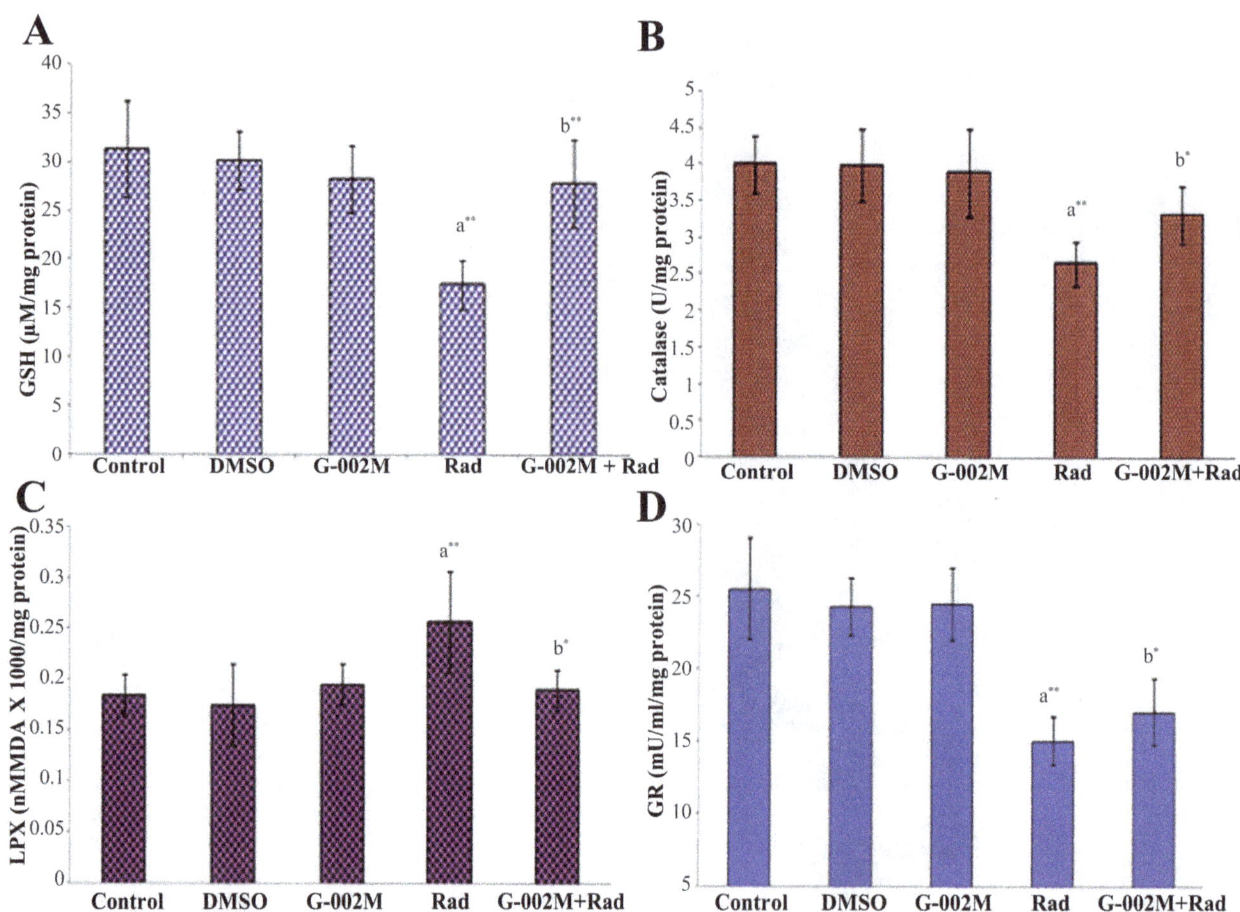

Figure 7. (A) Effect of G-002M on radiation induced alteration in GSH of lungs of mice exposed to 7 Gy whole body irradiation. Samples absorbance was recorded at 412 nm. Glutathione concentration is expressed as μM GSH/mg protein; (B) Effect of G-002M on radiation induced alteration in catalase activity of lungs of mice exposed to 7 Gy whole body irradiation. Supernatants of the lung homogenates were added to the assay mixture consisted of phosphate buffer and hydrogen peroxide, mixed with dichromate acetic acid reagent. Absorbance of the supernatants was recorded at 570 nm; (C) Effect of G-002M on radiation induced alteration in lipid peroxidation of lungs of mice exposed to 7 Gy whole body irradiation. Lung homogenates from different treated groups of mice were mixed with TBA-TCA and boiled at 95°C. Absorbance of the supernatants was recorded at 535 nm. Results are expressed as nmoles of MDA formed per mg of protein; (D) Effect of G-002M on radiation induced alteration in GR activity of lungs of mice exposed to 7 Gy whole body irradiation. The assay mixture containing 0.2 M potassium phosphate buffer, 2 mM nicotinamide adenine dinucleotide phosphate (NADPH) and 20 mM oxidized glutathione was mixed with distilled water and the reaction was initiated by addition of samples. Absorbance was measured at 340 nm for one minute at 15 sec interval. GR activity is expressed as mU/min/mg protein. Study was repeated in three different sets of mice having three animals in each group. [a]Radiation only group compared with controls, [b]G-002M + Radiation group compared with radiation only group. $^{**}p < 0.001$, $^{*}p < 0.01$.

also by hydrolysis of water in the cellular milieu thereby, generating reactive oxygen species, which initiate chemical peroxidative processes and affect the functionality of biomolecules [39]. Radiation induced free radical generation leads to oxidative stress which further results into nitrosative stress mediated by a chain of reactions involving reactive nitrogen species (RNS) mainly nitric oxide (NO) and peroxynitrite (ONOO-) [40]. Radiation induced damage to normal tissues can be averted by the use of safe prophylactic agents [41]. Development of radioprotector is an area of great significance due to its possible applications against unplanned/planned radiation exposures [41]. Enormous research and developmental efforts have been witnessed worldwide to develop safe radioprotectors to guard normal cells against radiotherapy, radiation explosions related rescue missions and space explorations. However, only a single compound, WR-2721, has so far received US FDA approval. Number of the investigations into radioprotector development have shown that the potential use of plants and their products can minimize the radiation mediated tissue damages [24,42,43] with minimal toxic effects. *P. hexandrum*, a high altitude plant, has been in use under traditional medicine practice against rheumatism, plague, allergic and inflammatory conditions of the skin and various other types of cancers. Our studies with this plant have documented its use as a potent radioprotector against lethal doses of radiation in *in-vitro*, *in-vivo*, *ex-vivo* model systems [22,44]. Studies on *P. hexandrum* utilizeing semipurified fractions have proved its potent radioprotective effect against lethal and supralethal doses of radiation in rodents [20,22,44]. Recently, our group has reported the radioprotective properties of purified fractions of *P. hexandrum* [23]. The present study has been aimed to explore the mechanism by which our formulation leads to repair the radiation inflicted damage in the pulmonary system.

In the current study, degree of radiation induced pulmonary damage and its modification by G-002M was performed by the analysis of BALF using well reported parameters [45]. Scoring of neutrophils, important for marking inflammatory response in bronchoalveolar region, was among other markers evaluated currently. Studies on cellular enzyme lactate dehydrogenase (LDH), protein and phospholipids content and smear preparation for quantitation of total and differential cell counts in BALF (**Table 1**) have supported the evaluation on lung tissue damage and its recovery by G-002M.

Nitric oxide readily diffuses across the cells and is considered an important molecule in cellular signaling. Reactive nitrite species (nitric oxide) quenching potential of *P. hexandrum* has been described by Sagar *et al.* [46]. There are several reports suggesting increased nitric oxide synthase during radiation-induced oxidative stress [3], and some researchers have reported the role of nitric oxide in the pathogenesis of the inflammatory response in radiation enteritis [47]. Giannopoulou *et al.* [48] reported that amifostine exhibits antiangiogenesis activity by reducing the levels of nitric oxide; authors expressed that reduction in the levels of nitric oxide reduces the radiation mediated inflammatory responses. Earlier studies have also reported radiation mediated iNOS induction, as a major factor in inflicting lung injuries and nitrotyrosine formation may participate in the NO-induced pathogenesis [3]. Since iNOS is the major isoform related to NO production and RNS generation after radiation exposure, its expression was explored in mice BALF cells and in the lungs during current study. G-002M pretreatment exhibited considerable reduction in the iNOS expression. An increase of LDH levels in BALF after radiation exposure appeared to be associated with increased number of macrophages and neutrophils in lavage fluid (**Figure 1**). Pretreatment of G-002M formulation prior to radiation exposure resulted in decreased lung permeability and tissue damage as evident by low amount of protein and phospholipids in BALF and decreased LDH activity (**Figure 6**).

Oxidative stress is also known to be mitigated by the activation of several functionally interrelated antioxidant enzymes such as catalase, peroxidise and glutathione reductase [49,50]. These enzymes enhance acclimatisation mechanisms to scavenge the toxic free radicals of oxygen produced under stress condition. Oxidative stress accompanied by the synthesis of hydrogen peroxide is normally detoxified by catalase activity [51]. Radiation is known to downregulate the endogenous antioxidant enzymes (GSH, GR, catalase). These enzymes were found significantly upregulated by prophylactic administration of our *P. hexandrum* formulation (G-002M) in lethally irradiated mice confirming the scavenging of ROS/NO formation. Though G-002M pretreatment exhibited considerable restoration of GSH and catalase enzyme activeties (**Figures 5(A)** and **(B)** respectively) however, GR enzyme activity was found only marginally increased.

As well documented in other studies and confirmed by our studies too, support to immune system [22], hematopoietic system [21,46] and cellular DNA [21,22] by semipurified and purified formulations of *P. hexandrum,* is predominantly attributed to their free radical scavenging property, reduced lipid peroxidation [44], transient metal-chelation [21] and protection to endogenous defense enzymes and to cellular macromolecules [44]. The current study also demonstrates the possible involvement of *P. hexandrum* formulation (G-002M) in modulation of radiation associated iNOS mediated RNS formation which ultimately resulted into the reduced damage of the lungs of irradiated mice. This drug with numerable radioprotective properties may be studied further to reveal its potential against radiation mediated respiratory syndromes in humans.

5. Acknowledgements

Authors are grateful to Defence Research Development and Organization (DRDO) for grant assistance. The assistance rendered by Mrs. Namita Kalra for flow cytometry measurement and Ms. Neetu Singh for support in histological studies is highly appreciated.

REFERENCES

[1] J. C. Lee, P. A. Kinniry, E. Arguiri, M. Serota, S. Kanterakis, S. Chatterjee, *et al.*, "Dietary Curcumin Increases Antioxidant Defenses in Lung, Ameliorates Radiation-Induced Pulmonary Fibrosis, and Improves Survival in Mice," *Radiation Research*, Vol. 173, No. 5, 2010, pp. 590-601. doi:10.1667/RR1522.1

[2] V. Mehta, "Radiation Pneumonitis and Pulmonary Fibrosis in Non-Small-Cell Lung Cancer: Pulmonary Function, Prediction, and Prevention," *International Journal of Radiation Oncology Biology Physics*, Vol. 63, No. 1, 2005, pp. 5-24. doi:10.1016/j.ijrobp.2005.03.047

[3] C. Tsuji, Shioya, Y. Hirota, N. Fukuyama, D. Kurita, T. Tanigaki, *et al.*, "Increased Production of Nitrotyrosine in

Lung Tissue of Rats with Radiation-induced Acute Lung Injury," *American Journal of Physiology Lung Cellular and Molecular Physiology*, Vol. 278, No. 4, 2000, pp. 719-725.

[4] P. G. Tsoutsou and M. I. Koukourakis, "Radiation Pneumonitis and Fibrosis: Mechanisms Underlying its Pathogenesis and Implications for Future Research," *International Journal of Radiation Oncology Biology Physics*, Vol. 66, No. 5, 2006, pp. 1281-1293. doi:10.1016/j.ijrobp.2006.08.058

[5] A. Giaid, S. M. Lehnert, B. Chehayeb, D. Chehayeb, I. Kaplan and G. Shenouda, "Inducible Nitric Oxide Synthase and Nitrotyrosine in Mice with Radiation-Induced Lung Damage," *American Journal of Clinical Oncology*, Vol. 26, No. 4, 2003, pp. 67-72. doi:10.1097/01.COC.0000077940.05196.86

[6] J. A. Royall, N. W. Kooy and J. S. Beckman, "Nitric Oxide-Related Oxidants in Acute Lung Injury," *New Horizons*, Vol. 3, No. 1, 1995, pp. 113-122.

[7] Y. Nozaki, Y. Hasegawa, A. Takeuchi, Z. H. Fan, K. I. Isobe, I. Nakashima, *et al.*, "Nitric Oxide as an Inflammatory Mediator of Radiation Pneumonitis in Rats," *American Journal of Physiology Lung Cellular and Molecular Physiology*, Vol. 272, No. 4, 1997, pp. 651-658.

[8] R. Saini, S. Patel, R. Saluja, A. A Sahasrabuddhe, M. P. Singh, S. Habib, V. K. Bajpai, *et al.*, "Nitric Oxide Synthase Localization in the Rat Neutrophils: Immunocytochemical, Molecular and Biochemical Studies," *Journal of Leukocyte Biology*, Vol. 79, No. 3, 2006, pp. 519-528. doi:10.1189/jlb.0605320

[9] K. J. Park, Y. T. Oh, W. J. Kil, W. Park, S. H. Kang and M. Chun, "Bronchoalveolar Lavage Findings of Radiation Induced Lung Damage in Rats," *Journal of Radiation Research*, Vol. 50, No. 3, 2009, pp. 177-182. doi:10.1269/jrr.08089

[10] U. Costabel and J. Guzman, "Bronchoalveolar Lavage in Intestinal Lung Disease," *Current Opinion in Pulmonary Medicine*. Vol. 7, No. 5, 2001, pp. 255-261.

[11] J. Bousquet, P. Chanez, J. Y. Lacoste, G. Barnéon, N. Ghavanian, I. Enander, *et al.*, "Eosinophilic Inflammation in Asthma," *New England Journal of Medicine*, Vol. 323, No. 15, 1990, pp. 1033-1039. doi:10.1056/NEJM199010113231505

[12] J. Y. Lacoste, J. Bousquet, P. Chanez, T. Van Vyve, J. Simony-Lafontaine, N. Lequeu, *et al.*, "Eosinophilic and Neutrophilic Inflammation in Asthma, Chronic Bronchitis, and Chronic Obstructive Pulmonary Disease," *Journal of Allergy and Clinical Immunology*, Vol. 92, No. 4, 1993, pp. 537-548. doi:10.1016/0091-6749(93)90078-T

[13] D. A. Campbell, L. W. Poulter and R. M. du Bois, "Immunocompetent Cells in Bronchoalveolar Lavage Reflect the Cell Populations Intransbronchial Biopsies in Pulmonary Sarcoidosis," *American Review of Respiratory Disease*, Vol. 132, No. 6, 1985, pp. 1300-1306.

[14] P. J. Haslam, C. W. G. Turton, B. Heard, A. Lukoszek, J. V. Collins, A. J. Salsbury, *et al.*, "Bronchoalveolar Lavage in Pulmonary Fibrosis: Comparison of Cells Obtained with Lung Biopsy and Clinical Features," *Thorax*, Vol. 35, No. 1, 1980, pp. 9-18. doi:10.1136/thx.35.1.9

[15] G. W. Hunninghake, O. Kawanami, V. J. Ferrans, R. C. Jr Young, W. C. Roberts and R. G. Crystal, "Characterization of Inflammatory and Immune Effector Cells in the Lung Parenchyma of Patients with Interstitial Lung Disease," *American Review of Respiratory Disease*, Vol. 123, No. 4, 1981, pp. 407-412.

[16] I. L. Paradis, J. H. Dauber and B. S. Rabin, "Lymphocyte Phenotypes in bronchoalveolar Lavage and Lung Tissue in Sarcoidosis and Idiopathic Pulmonary Fibrosis," *American Review of Respiratory Disease*, Vol. 133, No. 5, 1986, pp. 855-860.

[17] G. Semenzato, M. Chilosi, E. Ossi, L. Trentin, G. Pizzolo, A. Cipriani, *et al.*, "Bronchoalveolar Lavage and Lung histology: Comparative Analysis of Inflammatory and Immunocompetent Cells in Patients with Sarcoidosis and Hypersensitivity Pneumonitis," *American Review of Respiratory Disease*, Vol. 132, No. 2, 1985, pp. 400-404.

[18] M. B. Gotway, "Interstitial lung Diseases: Imaging Evaluation," *Applied radiology*, Vol. 29, No. 9, 2000, pp. 31-46. doi:10.1016/S0160-9963(00)80215-0

[19] M. D. Rossman, J. A. Kern, J. A. Elias, M. R. Cullen, P. E. Epstein and O. P. Preuss, "Proliferative Response of Broncho Alveolar Lymphocytes to Beryllium," *Annals of Internal Medicine*, Vol. 108, No. 5, 1988, pp. 687-693. doi:10.7326/0003-4819-108-5-687

[20] M. Lata, J. Prasad, S. Singh, R. Kumar, L. Singh, P. Chaudhary, *et al.*, "Whole Body Protection against Lethal Ionizing Radiation in Mice by REC-2001: A Semi-Purified Fraction of *Podophyllum hexandrum*," *Phytomedicine*, Vol. 16, No. 1, 2009, pp. 47-55. doi:10.1016/j.phymed.2007.04.010

[21] M. L. Gupta, V. Gupta, S. K. Shukla, S. Verma, S. Sankhwar, A. Dutta, *et al.*, "Inhibition in Radiation Mediated Cellular Toxicity by Minimizing Free Radical Flux: One of the Possible Mechanisms of Biological Protection against Lethal Ionizing Radiation by a Subfraction of *Podophyllum hexandrum*," *Cellular and Molecular Biology*, Vol. 56, Suppl. OL1341-9, 2010. doi:10.1170/153

[22] S. Sankhwar, M. L. Gupta, V. Gupta, S. Verma, K. A. Suri, M. Devi, *et al.*, "*Podophyllum hexandrum* Mediated Survival Protection and Restoration of Other Cellular Injuries in Lethally Irradiated Mice," *Evidence Based Complementary and Alternative Medicine*, 2011, Article ID: 175140. doi:10.1093/ecam/nep061

[23] A. Dutta, S. Verma, S. Sankhwar, S.J. Flora and M. L. Gupta, "Bioavailability, Antioxidant and Non Toxic Properties of a Radioprotective Formulation Prepared from Isolated Compounds of *Podophyllum hexandrum*: A Study in Mouse Model," *Cellular and Molecular Biology*, Vol. 58, 2012.

[24] S. K. Shukla and M. L. Gupta, "Approach Towards Development of a Radioprotector Using Herbal Source Against Lethal Irradiation," *International Research Journal of Plant Science*, Vol. 1, No. 1, 2010, pp. 118-125.

[25] A. Kawana, S. Shioya, H. Katoh, C. Tsuji, M. Tsuda and Y. Ohta, "Expression of Intercellular Adhesion Molecule-1 and Lymphocyte Function-Associated Antigen-1 on Alveolar Macrophages in the Acute Stage of Radia-

tion-Induced Lung Injury in Rats," *Radiation Research*, Vol. 147, No. 4, 1997, pp. 431-436. doi:10.2307/3579499

[26] S. Shioya, C. Tsuji, D. Kurita, H. Katoh, M. Tsuda, M. Haida, *et al.*, "Early Damage to Lung Tissue after Irradiation Detected by the Magnetic Resonance T2 Relaxation Time," *Radiation Research*, Vol. 148, No. 4, 1997, pp. 359-364. doi:10.2307/3579521

[27] Q. Zhang, Y. Kusaka, Q. Zhang, L. He, Z. Zhang and K. Sato, "Dynamic Changes of Constituents in Bronchoalveolar Lavage Fluid in Experimental Silicotic Rats," *Industrial Health*, Vol. 34, No. 4, 1996, pp. 379-388. doi:10.2486/indhealth.34.379

[28] M. Natiello, G. Kelley, J. Lamca, D. Zelmanovic, R. W. Chapman and J. E. Phillips, "Manual and Automated Leukocyte Differentiation in Bronchoalveolar Lavage Fluids from Rodent Models of Pulmonary Inflammation," *Comparative Clinical Pathology*, Vol. 18, No. 2, 2009, pp. 101-111. doi:10.1007/s00580-008-0772-9

[29] A.B. Thompson, H. Teschler, Y. M. Wang, N. Konietzko and U. Costabel, "Preparation of Bronchoalveolar Lavage Fluid with Microscope Slide Smears". *European Respiratory Journal*, Vol. 9, No. 3, 1996, pp. 603-608. doi:10.1183/09031936.96.09030603

[30] P. Sharma, S. A. V. Raghavan, R. Saini, and M. Dikshit, "Functional Role of Ascorbic Acid in the Regulation of Free Radical Generation and Phagocytosis by Polymorphonuclear Leukocytes: A No-Mediated Effect," *Journal of Leukocyte Biology*, Vol. 75, No. 6, 2004, pp. 1070-1078. doi:10.1189/jlb.0903415

[31] B. Vojtesek, J. Bartek, C.A. Midgley and D. P. Lane, "An Immunochemical Analysis of the Human Nuclear Phosphoprotein P53. New Monoclonal Antibodies and Epitope Mapping Using Recombinant P53," *Journal of Immunological Methods*, Vol. 151, No. 1-2, 1992, pp. 237-244. doi:10.1016/0022-1759(92)90122-A

[32] P. Seth, R. Kumari, M. Dikshit and R. C. Srimal, "Modulation of Rat Peripheral Polymorphonuclear Leukocyte Response by Nitric Oxide and Arginine," *Blood*, Vol. 84, No. 8, 1994, pp. 2741-2748.

[33] M. Bradford, "A Rapid And Sensitive Method for the Quantitation of Microgram Quatities of Protein Utilizing the Principle of Protein-Dye Binding," *Analytical Biochemistry*, Vol. 72, No. 1-2, 1976, pp. 248-254. doi:10.1016/0003-2697(76)90527-3

[34] Y. Yoshida, E. Furuya and K. Tagawa, "A Direct Colorimetric Method for the Determination of Phospholipids with Dithiocyanatoiron Reagent," *Journal of Biochemistry*, Vol. 88, No. 2, 1980, pp. 463-468.

[35] J. A. Buege and S. D Aust, "Microsomal Lipid Peroxidation," *Methods Enzymology*, Vol. 52, 1978, pp. 302-310. doi:10.1016/S0076-6879(78)52032-6

[36] E. Beutler, "Reduced Glutathione-GSH, U: Beutler E. (Ur.) Red Cell Metabolism: A Manual of Biochemical Methods," Grane and Straton, New York, 1975.

[37] I. Carlberg and B. Mannervik, "Glutathione Reductase," *Methods Enzymology*, Vol. 113, 1985, pp. 484-490. doi:10.1016/S0076-6879(85)13062-4

[38] A. K. Sinha, "Colorimetric Assay of Catalase," *Analytical Biochemistry*, Vol. 47, No. 2, 1972, pp. 389-394.

[39] Y. Z. Fang, S. Yang and G. Wu, "Free Radicals, Antioxidants, and Nutrition," *Nutrition Journal*, Vol. 18, No. 10, 2002, pp. 872-879. doi:10.1016/S0899-9007(02)00916-4

[40] M. Kirsch and H. de Groot, "Formation of Peroxynitrite from Reaction of Nitroxyl Anion with Molecular Oxygen," *The Journal of Biological Chemistry*, Vol. 277, No. 16, 2002, pp. 13379-13388. doi:10.1074/jbc.M108079200

[41] S. J. Hosseinimehr, "Trends in the Development of Radioprotective Agents," *Drug Discovery Today*, Vol. 12, No. 19-20, 2007, pp. 794-805. doi:10.1016/j.drudis.2007.07.017

[42] S. Verma, M. L. Gupta, A. Dutta, S. Sankhwar, S. K. Shukla and S. J. Flora, "Modulation of Ionizing Radiation Induced Oxidative Imbalance by Semi-Fractionated Extract of Piper Betle: An *in Vitro* and *in Vivo* Assessment," *Oxidative Medicine and Cellular Longevity*, Vol. 3, No. 1, 2010, pp. 44-52. doi:10.4161/oxim.3.1.10349

[43] D. P. Uma Devi, A. Ganasoundari, B. S. Rao and K. K. Srinivasan, "*In Vivo* Radioprotection by Ocimum Flavonoids: Survival of Mice," *Radiation Research*, Vol. 151, No. 1, 1999, pp. 74-78. doi:10.2307/3579750

[44] M. L. Gupta, S. Tyagi, S. J. Flora, P. K. Agrawala, P. Choudhary, S. C. Puri, *et al.*, "Protective Efficacy of Semi Purified Fraction of High Altitude *Podophyllum hexandrum* Rhizomes in Lethally Irradiated Swiss Albino Mice," *Cellular and Molecular Biology*, Vol. 53, No. 5, 2007, pp. 29-41.

[45] H. Y. Reynolds, "Bronchoalveolar Lavage," *American Review of Respiratory Disease*, Vol. 135, No. 1, 1987, pp. 250-263.

[46] R. K. Sagar, R. Chawla, R. Arora, S. Singh, B. Krishna, R. K. Sharma, *et al.*, "Protection of Hematopoietic System by *Podophyllum hexandrum* against Gamma Radiation-Induced Damage," *Planta Medica*, Vol. 72, No. 2, 2006, pp. 114-120. doi:10.1055/s-2005-873148

[47] Y. Erbil, C. Dibekoglu, U. Turkoglu, E. Ademoglu, E. Berber, A. Kizir, *et al.*, "Nitric Oxide and Radiation Enteritis," *European Journal of Surgery*, Vol. 164, No. 11 1998, pp. 863-868. doi:10.1080/110241598750005291

[48] E. Giannopoulou, P. Katsoris, D. Kardamakis and E. Papadimitriou, "Amifostine Inhibits Angiogenesis *in Vivo*," *Journal of Pharmacology and Experimental Therapy*, Vol. 304, No. 2, 2003, pp. 729-737. doi:10.1124/jpet.102.042838

[49] P. V. Limaye, N. Raghuram and S. Sivakami, "Oxidative Stress and Gene Expression of Antioxidant Enzymes in the Renal Cortex of Streptozotocin Induced Diabetic Rats," *Molecular Cell Biochemistry*, Vol. 243, No. 1-2, 2003, pp. 147-152.

[50] I. Zelen, P. Djurdjevic, S. Popovic, M. Stojanovic, V. Jakovljevic, S. Radivojevic, *et al.*, "Antioxidant Enzymes Activities and Plasma Levels of Oxidative Stress Markers in B-Chronic Lymphocytic Leukemia Patients," *Journal of Balkan Union of Oncology*, Vol. 15, No. 2, 2010, pp. 330-336.

[51] C. H. Foyer, H. Lopez-Delgado, J. F. Dat and I. M. Scott, "Hydrogen Peroxide- and Glutathione-Associated Mechanisms of Acclamatory Stress Tolerance and Signalling," *Physiologia Plantarum*, Vol. 100, No. 2, 1997, pp. 241-254. doi:10.1111/j.1399-3054.1997.tb04780.x

Therapeutic Efficacy of Genistein-Topotecan Combination Compared to Vitamin D_3-Topotecan Combination in LNCaP Prostate Cancer Cells

Shreyasee Chakraborty, Bibiana Sandoval-Bernal, James Kumi-Diaka
Department of Biological Sciences, College of Sciences, Florida Atlantic University at Davie, Davie, USA
Email: bsandov1@fau.edu

ABSTRACT

Background: Prostate cancer is the most common cancer in men over the age of 60 in Western countries. An estimated 241,740 new cases of prostate cancer have been diagnosed in the United States in 2012 with a death toll of 28,170. Varieties of natural phytochemicals such as genistein and topotecan have shown potential chemotherapeutic capacities and are being used to inhibit the growth and proliferation of cell in prostate cancer. **Purpose of Study:** In this study, we aim to determine the efficacy of Vitamin D_3-Topotecan combination compared to Genistein-Topotecan in apoptosis induction in LNCaP prostate cancer cells. **Methods:** LNCaP cells were grown in complete RPMI medium and cultured at 37˚C, 5% CO_2 for 23 - 48 hrs to achieve 70% - 80% confluence. The cells were then treated with Genistein-Topotecan, Vitamin D_3-Topotecan combination and TPT alone for 24 - 48 hours. In addition, post-treatment assayed using: Trypan Blue exclusion and MTT for cell viability, Ethidium bromide/Acridine orange to determine apoptosis induction, Rhodamine 123/Ethidium bromide to differentiate between viable, apoptotic, and necrotic cells, as well as to assess possible apoptotic mechanism, and DNA fragmentation to discriminate between apoptotic and necrotic cell death. **Results:** The overall data indicated the dose-and time-dependent cell death in the LNCaP cells and apoptosis as the major mechanism of treatment-induced cell growth arrest. **Conclusion:** The Genistein-Topotecan combination treatment was significantly more efficacious in growth inhibition of LNCaP cells compared to Vitamin D_3-Topotecan or Topotecan alone.

Keywords: Topotecan; Genistein; Vitamin D; Prostate Cancer

1. Introduction

In 2012 about 241,740 new cases of prostate cancer were diagnosed making this type of cancer the second most common type of male cancer in America, preceded only by skin cancer. Furthermore, nearly 28,170 died this year from this condition [1]. American men have a 1 in 6 chance of developing prostate cancer during their lifetime. Prostate cancer occurs in older men usually over the age of 60. The risk factors for prostate cancer include: age, ethnicity, family history, diet, and obesity. A variety of natural dietary phytochemicals such as Genistein in soybean, have been proven to have chemotherapeutic capabilities. Genistein isoflavone (4'5'7-trihydroxy-isoflavone) is the major dietary flavonoid found in soy (derived from soybean). Soy naturally contains genistin: beta-glucoside, which is broken down in the gastrointestinal tract, into genistein through fermentation by microbes [2,3]. Exposure of malignant cells to genistein has shown genistein's ability to inhibit cell growth and proliferation. Potential mechanisms of action have been identified in various studies and include: alteration of signal transduction pathways, caspase protease activation and regulation of the cell cycle [4-7]. In addition genistein inhibits topoisomerase II enzyme, and angiogenesis through the blockage of VEGF signaling. Genistein has structural similarity to estrogen [7-9]. Genistein-induced apoptosis in carcinoma cells has been shown to be due to genistein's ability to control expression of apoptosis-related genes, such as up-regulation of Bax, and utilization of an independent p53 pathway. Genistein also induces apoptosis via other signal pathways including: increase in caspase-3 protease activity, initiation of DNA damage and halting of the cell cycle at the G2/M phase [5-10]. Studies have proven its low cytotoxicity against other chemo and radiation therapy.

Another micronutrient of potential prophylactic/therapeutic significance is Vitamin D_3. Vitamin D_3 also known as the "Sunshine Vitamin" is a group of fat-soluble secosteroids. It can be ingested as cholecalciferol (Vitamin D_3) or ergocalciferol (Vitamin D_2) and the human body can also synthesize it when sun exposure is adequate. The active form of vitamin D in the body is 1,25-dihydroxyvitamin D, or calcitriol, which can be made from either vitamin D_2 or vitamin D_3. To make the active form, vitamin D_2 and vitamin D_3 are modified in the liver to produce 25-hydroxyvitamin D, which travels through the blood to the kidneys, where it is modified further to make 1,25-dihydroxyvitamin D.

Previous research indicates that the human prostate cancer cell line possesses Vitamin D receptors (VDR) [11-13]. Calcitriol [1alpha,25-dihydroxyvitamin D3] (VD_3) is the natural ligand of the vitamin D receptor (VDR). Studies show that VD_3 could down regulate the antiapoptotic proteins Bcl-2, Bcl-X(L), and Mcl-1, BAG1L, XIAP, cIAP1, and cIAP2 (without altering the proapoptotic Bax and Bak) in association with increase in apoptosis [14]. In VDR sensitive LNCaP cell lines, VD_3 activates downstream effector protease, caspase-3, and upstream initiator protease caspase-9, the apical protease in the mitochondrial ("intrinsic") pathway for apoptosis. VD_3 induced declines in antiapoptotic proteins and also stimulated cytochrome c release from mitochondria by a caspase-independent mechanism [15]. Moreover, apoptosis induction by VD_3 was suppressed by overexpressing Bcl-2, a known blocker of cytochrome c release, whereas the caspase-8 suppressor CrmA afforded little protection [15]. Thus, VD_3 is capable of inhibiting expression of multiple antiapoptotic proteins in VDR-expressing prostate cancer cells, leading to activation of the mitochondrial pathway for apoptosis. Synthetic analogs of vitamin D_3 like 25-hydroxyvitamin D_3, is known to exhibit reduced calcemic activity and can elicit antiproliferative effects and other biological actions in LNCaP cells [16]. Taken together, epidemiological and dietary data suggest that both genistein and vitamin D_3 play important role in protecting against prostate cancer [17,18].

Like Genistein, another recently known anti tumor phytochemical is Topotecan, which also induces apoptosis in cancer cells. Topotecan is an FDA approved chemotherapy agent under the commercial name of Hycamtin, also referred to as topotecan hydrochloride (topotecan HCl). It was approved for treatment of multiple types of cancer, but it is most widely used for cervical and small cell lung cancer [19,20]. It is a phytochemical and a semisynthetic derivate of camptothecin. It was isolated from the bark and stem of *Camptotheca acuminata* (Camptotheca, Happy tree), a tree native to China [21,22]. Camptothecin, an alkaloid phytochemical is able to arrest cell growth and proliferation in several carcinoma cell lines. However, it has low solubility and early clinical trials reported adverse side effects, including bone marrow suppression [22-24].

Topotecan has anti-neoplastic properties and is a cytotoxic quinoline-based alkaloid which, during the S phase of the cell cycle, inhibits the standard function of topoisomerase I, an enzyme that deals with replication, repair, and recombination of the double stranded DNA by relieving torsional stress with reversible single strands cuts to the molecule. Topotecan interferes with the usual DNA-Topoisomerase I binding, thus preventing the re-annealing and repair of the single strands back to the stable double bond conformation. As a result the continuity of the replication fork is disrupted, replication is halted, and lethal double strand fragments are made which ultimately initiate apoptosis signaling. Intercalation displaces the downstream DNA, thus preventing religation of the cleaved strand. By specifically binding to the enzyme-substrate complex, Topotecan acts as an uncompetitive inhibitor [25-27].

Topotecan has low solubility and it has shown to have a high toxicity level for the optimum working concentration and dosage. Therapeutically, topotecan is mainly used as second line treatment against prostate cancer or recurring prostate tumors [28]. Cytotoxic side effects include: myelosuppression, low blood counts and suppression of the immune system leading to an increase in susceptibility to infection [25,26]. However, topotecan has been associated with inducing oxidative stress and increase in ROS (Reactive oxygen species) and nitrite. Increase in ROS causes irreversible damage to protein by forming protein carbonyl derivative and also elevates DNA stress and damage, ultimately leading to the induction of apoptotic cell death [29-31].

Genistein-Topotecan (Gn-TPT) combination has been shown to induce significant levels of apoptosis in prostate cancer [32]. Synthetic analogs of Vitamin D_3 like cholecalciferol (25-hydroxyvitamin D_3) has low hypercalcemic activity and low toxicity [16]. Here we investigated the potential additive effect of 25-hydroxyvitamin D_3 in 25-hydroxyvitamin D_3-Topotecan (Vit D-TPT) combination compared to the Gn-TPT combination in apoptosis induction in LNCaP prostate cancer cells. The potential mechanism of apoptosis induction was also determined.

2. Materials and Methods

2.1. Cell Line

LNCaP (complements of Rumbaugh-Goodwin Institute for Cancer Research [RGI], Plantation, Fl) cells were cultured and maintained in complete RPMI 1640 (Sigma-Aldrich Chemical Co., St Louis, Mo, USA) media with 10% Fetal Bovine Serum and 1% penicillin/streptomycin.

2.2. Test Agents

Genistein isoflavone (Gn) (Sigma-Aldrich, St. Louis, MO) and **25-hydroxyvitamin D_3 [Cholecalciferol] (Vit D)** (Sigma-Aldrich, St. Louis, MO) were dissolved with dimethylsulfoxide (DMSO) and diluted with RPMI-media to produce aliquots of 20 μM concentration.

Topotecan Hydrochloride (TPT) (Drummond Scientific Co., Broomall, PA) was diluted into a stock solution with dimethylsulfoxide (DMSO). Stock solution of TPT was further diluted with RPMI-media to produce aliquots ranging in concentration from 2 - 10 μM (TPT 2 - 10 μM).

Final concentrations of DMSO for genistein, topotecan, and $25(OH)D_3$ did not exceed 0.05%.

2.3. Treatment

The LNCaP cells were cultured in 75 cm^3 flask at 37°C, 5% CO_2 and 89% - 90% humidity to achieve 80% - 90% confluence. The cells were harvested, centrifuged and reconstituted into suspension with fresh RPMI 1640 media. 5 × 10^3 cells in 100 μl of media were dispenses into each well of 96 well microtiter plates and cultured for 48 hours to allow adherence and obtained > 80% confluence. The supernatants were aspirated and the adhered cells were treated with: 1) TPT; 2) Gn-TPT combination; 3) Vit D-TPT combination.

Cell cultures were distributed into three treatment groups, with Gn-TPT combination (Gn $_{20\mu M}$ + TPT $_{2\text{-}10\mu M}$), Vit D-TPT combination (Vit D $_{20\mu M}$ + TPT $_{2\text{-}10\mu M}$) and single TPT (TPT $_{2\text{-}10\mu M}$) dosages. Experiments had control groups consisting of LNCaP cells cultured in RPMI 1640 media (with 10% FBS, 1% penicillin/streptomycin and L-glutamine) without treatment. All treatments were done in triplicates, and cultured for 24 - 48 hours in a humidified incubator at 37°C and 5% CO_2. At 24 and 48 hours post-treatment/culture, cells were subjected to assay analysis as follows:

2.4. Trypan Blue Assay

Trypan blue exclusion-hemocytometer counting was used to assess the percent viability and initial concentration of the LNCaP cells prior to treatment, according to standard procedures [33,34].

2.5. MTT Assay

MTT [3-(4,5-dimethylthiazolyl-2)-2,5-diphenyltetrazolium bromide], a tetrazolium dye, is used to determine metabolic status and cell viability. Since the resulting color intensity correlates directly with the amount of metabolically active cells in each well, cell viability can be quantitatively determined by measuring the optical density (OD) in individual wells. After briefly treating the cells as described previously, a volume of 10 μL of MTT reagent was added to each well at 24 hours and 48 hours post-treatment. The plates were further incubated at 37°C and 5% CO_2 for 4 hours in the dark, after which a volume of 100 μL of DMSO was added to the cells to solubilize the formazan. Absorbance (OD) of the resultant solution was read at 490 nm using a Multiskan microplate reader (Molecular Devices, Inc., Sunnyvale, CA). The OD obtained was graphed against the concentrations of the drugs. The IC_{50} was extrapolated from the graphs as a measure to determine the concentration required to obtain expected results and compare the data.

2.6. Acridine Orange/Ethidium Bromide Fluorescence Assay

The acridine orange (AcrO)/ethidium bromide (EtBr) fluorescence assay was used to differentiate between viable, apoptotic, and necrotic cells based on fluorescence emission characteristics. Acridine orange permeates both viable and non-viable cells, causing the nuclei to emit green fluorescence. Since absorption of ethidium bromide is based on the disruption of cell membrane integrity, ethidium bromide selectively stains the nuclei of dead (non-viable) cells to produce red fluorescence. Cells that emit orange/brown colored fluorescence are indicative of apoptosis, while necrotic cells emit red fluorescence. Briefly, Ethidium bromide (25 μL) and acridine orange (75 μL) were mixed to make a cocktail: 3 μL were added to 25 μL of the cell suspensions. Wet-mounts were prepared using 10 μL of each cell suspension and analyzed under a fluorescent microscope with a band-pass filter. Apoptotic cell death was quantified by counting a total of 150 cells per 2 to 3 fields of view.

2.7. Rhodamine 123/Ethidium Bromide Fluorescence Assay

The rhodamine 123 (Rh 123)/ethidium bromide (EtBr) fluorescence assay was also used to differentiate between viable, apoptotic, and necrotic cells, as well as to determine the possible mechanism of apoptosis, based on mitochondrial transmembrane potential. Rh 123 is a cationic fluorochrome which utilizes the transmembrane potential of active mitochondria to diffuse into cells. The intact mitochondrial membranes of living cells therefore allow absorption of the dye, resulting in the emission of bright green fluorescence. Cells, in which the integrity of the mitochondrial membrane has been impaired (non-viable cells), stain lightly with Rh123. However, ethidium bromide selectively enters the disrupted membranes of dead cells and stains the nuclei to produce red fluorescence. Cell samples were briefly washed three times in PBS and resuspended in a final volume of 25 μL. Each cell suspension received 2 μL of the Rh 123 stock solution and was incubated at 37˚C for 5 minutes. Each cell suspension then received 20 µL of EtBr stock solution and was incubated at room temperature for 5 minutes. Finally, 10 µL of each cell mixture was transferred onto a microscope slide covered with a cover slip and examined/analyzed under a fluorescent microscope with a band-pass filter. Green fluorescence is indicative of viable cells; orange/brown cells are apoptotic while necrotic cells emit red fluorescence. The percentage of apoptotic cell death was quantified from an average of 150 cells spread across two to three regions/views on each slide.

2.8. DNA Fragmentation Assay

DNA fragmentation assay distinguishes between necrotic and apoptotic cell death. Cells undergoing apoptosis will cleave their nuclear DNA into 180 - 200 bp DNA breaks, which can be detected through electrophoresis. Necrotic-induced cell death results in irregular DNA breaks, which will appear as smears in the gel. Briefly, after treating the LNCaP cells as previously described, cells were collected and washed in PBS. Following the protocol from the DNeasy Blood and Tissue Kit (Qiagen, Valencia, CA), DNA was extracted from the pellet and quantified using the NanoDrop ND-2000 spectrophotometer (Nano-Drop Technologies, Wilmington, DE). A 2% agarose gel was prepared with 2 µL EtBr and loaded with 10 ng of DNA extract per well. A 1kb DNA ladder was also loaded into one well of the gel as a marker to aid quantification and sizing of the DNA fragments. Gel electrophoresis was performed at 90 - 100 mV for 30 - 45 minutes, and bands in each channel were examined.

2.9. Statistical Analyses

Experiments were performed in triplicate and repeated thrice to confirm results. Significance of the differences in mean values was determined using Graphpad Prism and the Student's t-test. Statistical significance was defined as $p \leq 0.05$.

3. Results

3.1. Growth Inhibition

The MTT assay was utilized to test the effects of genistein, 25(OH)D_3, and topotecan on the viability/chemosensitivity of LNCaP prostate carcinoma cell lines. Data collected was analyzed based on the knowledge that the absorbance is directly correlated with the amount of viable cells. Results showed that in general the number of viable cells decreased as the concentration of the drugs increased with time of exposure, demonstrating a dose dependent as well as time dependent relationship (**Figures 1(a)-(c)**). However, for the combination treatment of Vit D-Topotecan ($VitD_{20\mu M}$-$TPT_{2-10\mu M}$) there was no significant difference in comparison to the single treatment of Topotecan ($P > 0.05$) at 48 hours (**Figure 1(d)**). The combination treatment of Genistein-Topo-tecan ($Gn_{20\mu M}$-$TPT_{2-10\mu M}$) for upto 48 hours showed a more significant dose and time dependent decrease ($P < 0.001$) in cell viability than the former two treatments (**Figures 2(a)** and **(b)**, **Figure 1(d)**). Based on MTT results from previous experiment [31], the EC_{50} dose of genistein was calculated as 30 µM. In the present study, TPT and Vit D did not show a significant decrease in cell viability at this concentration.

3.2. Acridine Orange/EtBr Assay and Apoptosis Induction

The AcrO/EtBr assay determines whether any of the treatment regimens induced some of the morphological landmark features of apoptosis in the treated cells. The result indicated that both combination and single treatments induced apoptosis. At 24 hours, the combination of $Gn_{20\mu M}$-$TPT_{10\mu M}$ induced the greatest apoptosis cell death compared to the Vit D-TPT combination at the same concentrations ($p < 0.01$). (**Figure 3**).

3.3. Rhodamine 123/Ethidium Bromide Assay

The Rh 123/EtBr assay utilized differential fluorescent staining of viable, apoptotic and necrotic cells to confirm the type of treatment-induced LNCaP cell death, as well as to determine the possible mechanism of action for apoptosis (**Figure 4(b)**). A greater number of orange/brown stained cells were observed in all the treatment groups in a dose and time dependent manner (**Figure 4(c)**). Apoptotic cells with compromised mitochondria showed a reduced transmembrane potential and were visibly stained as described earlier. This suggested the use of the intrinsic-mitochondrial apoptotic pathway and the activation of caspase proteases involved in apoptosis. Results showed that initially 2 µM dose of single treatment with Topotecan was more effective compared to the combination treatments. However, with increased concentration, Genistein-TPT proved to be more effective in cell growth inhibition at a concentration of $Gn_{20\mu M}$-$TPT_{10\mu M}$ (**Figure 4(a)**) compared to single Topotecan treatment, and VitD-TPT combination at the same concentration ($p < 0.01$).

3.4. DNA Fragmentation Assay

The presence of uniform DNA fragments (Lane 4 - 6, **Figure 5**) indicated the induction of apoptotic cell death in LNCaP cells when treated with the combination dosages and single dose of topotecan. DNA breaks of 180 - 200 bp were consistent with the morphological characteristics of apoptosis. Control samples did not demon-

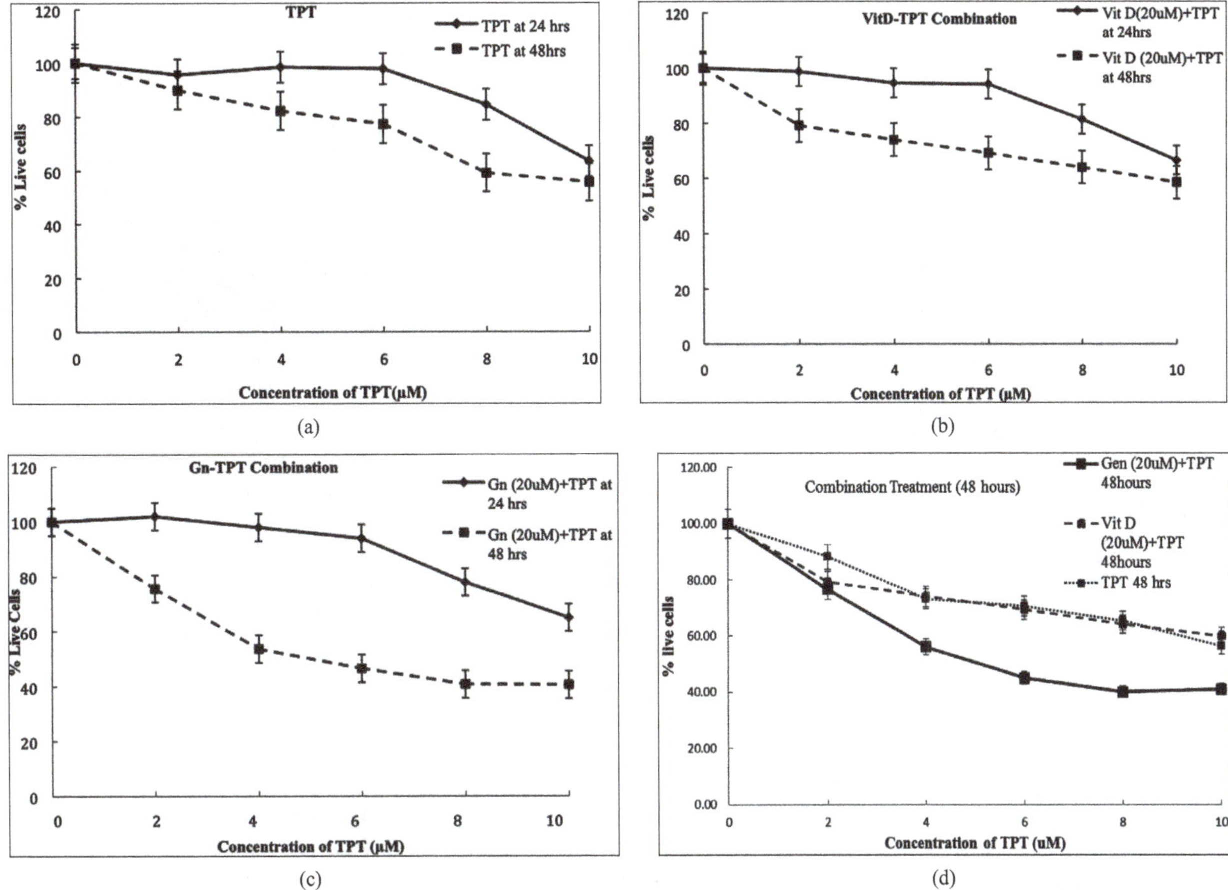

Figure 1. **(a)-(c) Growth and viability of LNCap cells was assessed using the MTT Assay. The cells were treated with varying concentrations of Topotecan($TPT_{2\text{-}10\mu M}$), Vitamin D_3-Topotecan($VitD_{20\mu M}$-$TPT_{2\text{-}10\mu M}$) and Genistein-Topotecan ($Gn_{20\mu M}$-$TPT_{2\text{-}10\mu M}$) combinations for 24 - 48 hrs; then incubated at 37°C, 5% CO_2, and 89% humidity, as previously described in the experiment. Data are the mean ± SEM(Standard Error of the Mean) of three independent experiments performed in triplicate. Bar = SEM(Standard Error of the Mean). d) Growth and viability of LNCap cells was assessed using the MTT Assay. The cells were treated with varying concentrations of Topotecan ($TPT_{2\text{-}10\mu M}$), Vitamin D_3-Topotecan($VitD_{20\mu M}$-$TPT_{2\text{-}10\mu M}$) and Genistein-Topotecan($Gn_{20\mu M}$-$TPT_{2\text{-}10\mu M}$) combinations for 48 hrs; then incubated at 37°C, 5% CO_2, and 89% humidity, as previously described in the experiment. Data are the mean ± SEM(standard Error of the Mean) of three independent experiments performed in triplicate. Bar = SEM (Standard Error of the Mean).**

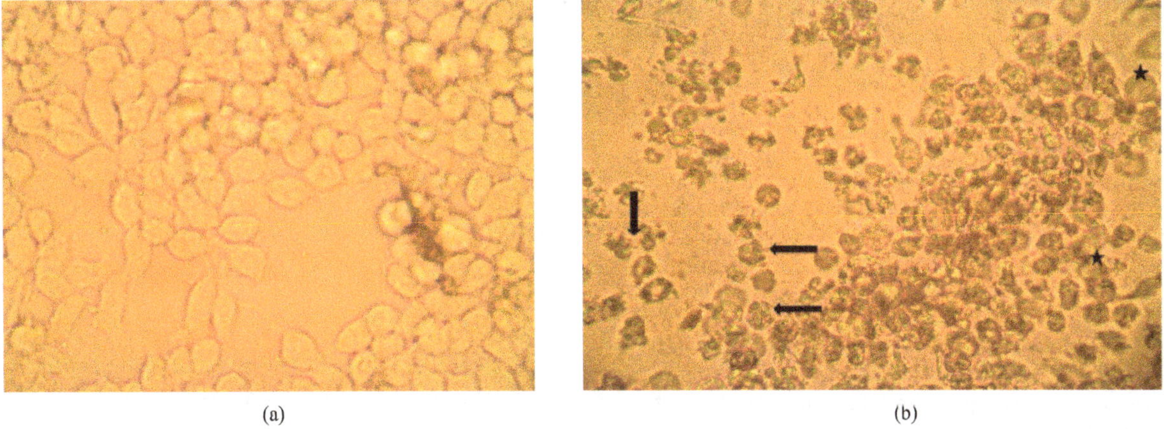

Figure 2. **(a) Growth and viability of LNCap cells without treatment after 48 hours (Brightfield imaging at 200× total magnification); (b) Fragmentation, shrinkage, and death; indicated by black arrows, of LNCap cells with ($TPT_{10\mu M}$-$Genestein_{20\mu M}$) combination treatment after 48 hours. Normal, vaible cells indicated by black stars (Brightfield imaging at 200× total magnification).**

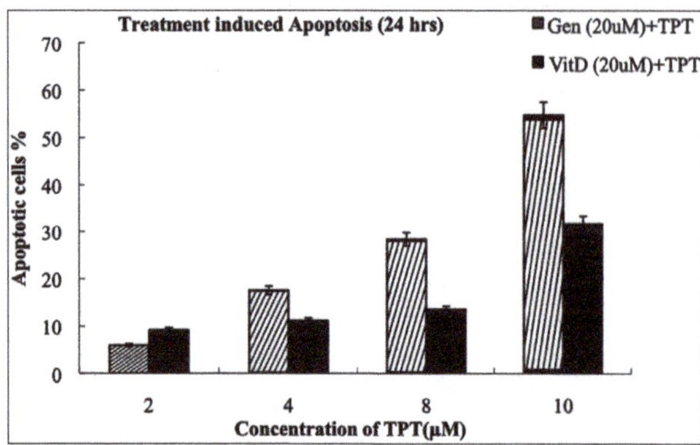

Figure 3. Treatment-induced apoptosis on LNCaP cell line was assessed using Et Br-AcrO to stain after 24 hrs. The percentage of apoptotic cells was quantified from 150 cells per 2 fields of view. At each concentration the difference between Gn + Tpt and VitD + Tpt is significant ($p < 0.01$).

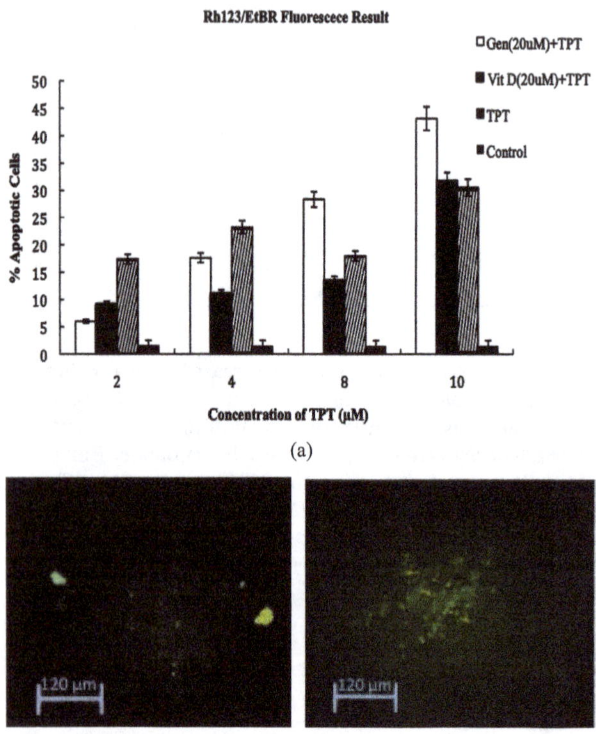

Figure 4. (a) Graph of Rh 123/EtBr fluorescence after 24-hours treatment. After 24 hours of treatment, LNCaP cells were mixed with Rh 123/EtBr and examined under the microscope for emission of fluorescent spectra indicative of apoptosis. The percentage of apoptotic cells was quantified from 150 cells per 2 fields of view. At each concentration the difference between Gn+Tpt and VitD+Tpt is significant ($p < 0.01$). (b) and (c) Images of Rh 123/EtBr fluorescence after 24-hours treatment. Green fluorescence depict viable cells, Orange/brown cells are apoptotic; (b) Control cells; (c) Cells treated with $Gn_{(20\mu M)}+TPT_{(10\ \mu M)}$.

strate cell death as shown by the single band (Lanes 2 and 7, **Figure 5**).

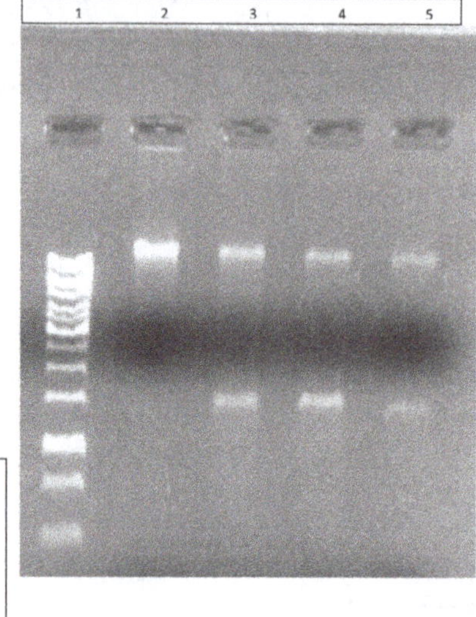

1 Marker
2 Control
3 Gn(20μM)+TPT(10μM)
4 VitD(20μM)+TPT(10μM)
5 TPT(10μM)

Figure 5. DNA Fragmentation and laddering in Gel Electrophoresis confirmed the apoptotic cell death in treatment groups. No fragments were detected in the control sample.

4. Discussion

Genistein Isoflavone is a well-known phytochemical with anti-cancer properties. Previous investigations have documented the therapeutic capacity of genistein in carcinoma treatment [4-10]. Anti-proliferative effect of cholecalciferol on prostate cancer cells has also been demonstrated [12-15]. The aim of this study was to determine and compare the potential efficacy of genistein-topotecan and Vitamin D_3-TPT combinations on LNCaP prostate cancer cells. Topotecan as a single treatment decreased LNCaP cell viability, consistent with previous reports [32]. The present data indicates that the combination treatments with Gn-TPT were significantly more cytotoxic to the cells. The results also indicate that the combination treatment with Gn-TPT was significant at a lower EC_{50} relative to VitD-TPT combination, consistent with previous reports [32]. Furthermore, the combination treatment of VitD-TPT did not show significant changes compared to the single TPT treatment.

The Rh123/Et Br and AcrO/Et Br revealed treatment-induced apoptosis to be the main mode of cell growth inhibition in LNCaP cells. The Gn-TPT combination treatment yielded the highest percentage of apoptotic cells at a low EC_{50} comparative to VitD-TPT combination. The result obtained from the Rh123/Et Br/AcrO implicates the mitochondrial pathway in the treatment-induced apoptosis and the involvement of the intrinsic pathway. The involvement of Caspases 9 and 3 in the signaling pathway of Gn-TPT combination has been demonstrated previously [32] indicating the utilization of the intrinsic apoptotic pathway. The DNA fragmentation assay reiterates the fact by displaying distinct bands for the fragmented DNA.

5. Conclusion

The overall data highlights the significance of Gn-TPT combination over Vit D-TPT combination and TPT in treatment-induced apoptosis. The Gn-TPT combination was significantly more efficacious at lower EC_{50} (lower cytotoxic level) than either compound alone, implying a greater therapeutic potential of this combination. The apoptosis induction was analyzed to be via intrinsic pathway with increased activity in a time-dependent manner; therefore, suggesting elevation of caspases 9 and 3 as a part of such pathway.

6. Acknowledgements

The authors acknowledge the equipment support of Florida Atlantic University. The collaboration and materials of Rumbaugh-Goodwin Institute for Cancer Research [RGI] at Nova Southeastern University is greatly acknowledged.

REFERENCES

[1] American Cancer Society, "Prostate Cancer Detailed Guide". http://www.cancer.org/Cancer/ProstateCancer/DetailedGuide/prostate-cancer-key-statistics

[2] M. Fukutake, K. Takahashi, H. Ishida, *et al.*, "Quantitation of Genistein and Genistin in Soyabeans and Soyabean Product," *Food and Chemical Toxicology*, Vol. 34, No. 5, 1996, pp. 457-461. doi:10.1016/0278-6915(96)87355-8

[3] C. A. Lamarliniera, "Protection against Breast Cancer with Genistein: A Component of Soy," *The American Journal of Clinical Nutrition*, Vol. 71, No. 6, 2000, pp. 1705S-1707S.

[4] J. K. Kumi-Diaka, M. Hassanhi, K. Merchant, *et al.*, "Influence of Genistein Isoflavone on Matrix Metalloproteinase-2 Expression in Prostate Cancer Cells," *Journal of Medicinal Food*, Vol. 9, No. 4, 2006, pp. 491-497.

[5] J. K. Kumi-Diaka, K. Merchant, A. Haces, *et al.*, "Genistein-Selenium Combination Induces Growth Arrest in Prostate Cancer Cells," *Journal of Medicinal Food*, Vol. 13, No. 4, 2010, pp. 842-850. doi:10.1089/jmf.2009.0199

[6] J. Kumi-Diaka, S. Saddler-Shawnette, A. Aller, *et al.*, "Potential Mechanism of Phytochemical-Induced Apoptosis in Human Prostate Adenocarcinoma Cells: Therapeutic Synergy in Genistein and Beta-Lapachone Combination Treatment," *Cancer Cell International*, Vol. 4, No. 1, 2004, p. 5. doi:10.1186/1475-2867-4-5

[7] S. Balabhadrapathruni, T. J. Thomas, E. J. Yurkow, *et al.*, "Effects of Genistein and Structurally Related Phytoestrogens on Cell Cycle Kinetics and Apoptosis in MDA-MB-468 Human Breast Cancer Cells," *Oncology Reports*, Vol. 7, No. 1, 2000, pp. 3-12.

[8] H. S. Seo, J. Ju, K. Jang *et al.*, "Induction of Apoptotic Cell Death by phyToestrogens by Up-Regulating the Levels of Phospho-p53 and p21 in Normal and Malignant Estrogen Receptor α-Negative Breast Cells," *Nutrition Research*, Vol. 31, No. 2, 2011, pp. 139-146. doi:10.1016/j.nutres.2011.01.011

[9] N. Zhou, Y. Yan, W. Li, *et al.*, "Genistein Inhibition of Topoisomerase II Alpha Expression Participated by Sp1 and Sp3 in HeLa cell," *International Journal of Molecular Sciences*, Vol. 10, No. 7, 2009, pp. 3255-3268. doi:10.3390/ijms10073255

[10] Z. Li, J. Li, W. Li, *et al.*, "Genistein Induces G2/M Cell Cycle Arrest via Stable Activation of ERK1/2 Pathway in MDA-MB-231 Breast Cancer Cells," *Cell Biology and Toxicology*, Vol. 24, No. 5, 2008, pp. 401-409. doi:10.1007/s10565-008-9054-1

[11] D. Feldman, R. J. Skowronski and D. M. Peehl, "Vitamin D and Prostate Cancer," *Advances in Experimental Medicine and Biology*, Vol. 375, 1995, pp. 53-63. doi:10.1007/978-1-4899-0949-7_5

[12] R. J. Skowronski, D. M. Peehl and D. Feldman, "Vitamin D and Prostate Cancer: 1, 25 Dihydroxy Vitamin D3 Receptors and Actions in Human Prostate Cancer Cell Lines," *Endocrinology*, Vol. 132, No. 5, 1993, pp. 1952-1960. doi:10.1210/en.132.5.1952

[13] R. J. Skowronski, D. M. Peehl and D. Feldman, "Actions of Vitamin D3, Analogs on Human Prostate Cancer Cell Lines: Comparison with 1,25-Dihydroxyvitamin D3," *Endocrinology*, Vol. 136, No. 1, 1995, pp. 20-26. doi:10.1210/en.136.1.20

[14] M. Guzey, S. Kitada and J. C. Reed, "Apoptosis Induction by 1alpha,25-Dihydroxyvitamin D3 in Prostate Cancer," *Molecular Cancer Therapeutics*, Vol. 1, No. 9, 2002, 667- pp. 677.

[15] D. M. Peeh, R. J. Skowronski, G. K. Leung, *et al.*, "Antiproliferative Effects of 1,25 Dihydroxy Vitamin D3 on Primary Cultures of Human Prostatic Cells," *Cancer Research*, Vol. 54, No. 3, 1994, pp. 805-810.

[16] R. J. Skowronski, D. M. Peehl and D. Feldman, "Actions of Vitamin D3, Analogs on Human Prostate Cancer Cell Lines: Comparison with 1,25-Dihydroxyvitamin D3," *Endocrinology*, Vol. 136, No. 1, 1995, pp. 20-26. doi:10.1210/en.136.1.20

[17] G. J. Kelloff, "Perspective on Cancer Chemoprevention Research and Drug Development," *Advances in Cancer Research*, Vol. 78, pp. 199-334. doi:10.1016/S0065-230X(08)61026-X

[18] A. Rao, R. D. Woodruff, W. N. Wade, T. E. Kute and S. D. Cramer, "Genistein and Vitamin D Synergistically Inhibit Human Prostatic Epithelial Cell Growth," *Journal of Nutrition*, Vol. 132, No. 10, 2002, pp. 3191-3194.

[19] C. M. Yashar, W. J. Spanos, *et al.*, "Potential of the Radiation Effect with Genistein in Cervical Cancer," *Gynecologic Oncology*, Vol. 99, No. 1, 2005, pp. m199-m205. doi:10.1016/j.ygyno.2005.07.002

[20] F. Lian, M. Bhuiyan, Y. W. Li, N. Wall, M. Kraut and F. H. Sarkar, "Genistein-Induced G2-M Arrest, p21WAF1 Upregulation, and Apoptosis in a Non-Small-Cell Lung Cancer Cell Line," *Nutrition and Cancer*, Vol. 31, No. 3, 1998, pp. 184-191. doi:10.1080/01635589809514701

[21] C. Law, "Basic Research Plays A Key Role in New Patient Treatments," *Journal of the National Cancer Institute*, Vol. 88, No. 13, 1996, p. 869. doi:10.1093/jnci/88.13.869

[22] C. McNeil, "Topotecan after FDA and Asco What's Next?" *Journal of the National Cancer Institute*, Vol. 88, No. 12, 1996, pp. 788-789. doi:10.1093/jnci/88.12.788

[23] H. Ulukan and P. W. Swaan, "Camptothecin: A Review of Their Chemotherapeutic Potential," *Drugs*, 62, 14, 2002, pp. 2039-2057.

[24] M. E. Wall, M. C. Wani, C. E. Cook, *et al.*, "Plant Anti-Mutagenic Agents, 1. General Bioassay and Isolation Procedures," *Journal of Natural Products*, Vol. 51, No. 5, 1988, pp. 866-873. doi:10.1021/np50059a009

[25] R. Padzur, "FDA Approval for Topotecan Hydrochloride". http://www.cancer.gov/cancertopics/druginfo/fda-topotecan-hydrochloride

[26] M. R. Redinbo, L. Stewart, P. Kuhn, *et al.*, "Crystal Structures of Human Topoisomerase I in Covalent and Non-Covalent Complexes with DNA," *Science*, Vol. 279, No. 5356, 1998, pp. 1504-1513. doi:10.1126/science.279.5356.1504

[27] D. J. Adams, M. L. Wahl, J. L. Flowers, *et al.*, "Camptothecin Analogs with Enhanced Activity against Human Breast Cancer Cells. II. Impact of the Tumor pH Gradient," *Cancer Chemotherapy and Pharmacology*, Vol. 57, No. 2, 2006, pp. 145-154. doi:10.1007/s00280-005-0008-5

[28] W. ten Bokkel Huinink, S. R. Lane and G. A. Ross, "Long-Term Survival in a Phase III, Randomized Study of Topotecan versus Paclitaxel in Advanced Epithelial Ovarian Carcinoma," *Annals of Oncology*, Vol. 15, No. 1, 2004, pp. 100-103. doi:10.1093/annonc/mdh025

[29] P. A. Ohneseit, D. Prager, R. Kehlbach and H. P. Rodeman, "Cell Cycle Effects of Topotecan Alone and in Combination with Irradiation," *Radiotherapy & Oncology*, Vol. 75, No. 2, 2005, pp. 237-245. doi:10.1016/j.radonc.2005.03.025

[30] M. Timur, S. H. Akbas and T. Ozben, "The Effect of Topotecan on Oxidative Stress in MCF-7 Human Breast Cancer Cell Line," *Acta Biochimica Polonica*, Vol. 52, No. 4, 2005, pp. 897-902.

[31] C. Erlichman, S. A. Boerner, C. G. Halgren, *et al.*, "The HER Tyrosine Kinase Inhibitor CI1033 Enhances Cytotoxicity of 7-Ethyl-10-hydroxycamptothecin and Topotecan by Inhibiting Breast Cancer Resistance Protein-Mediated Drug Efflux," *Cancer Research*, Vol. 61, No. 2, 2001, pp. 739-748.

[32] V. Hormann, J. Kumi-Diaka, M. Durity, *et al.*, "Anticancer Activities of Genistein-Topotecan Combination in Prostate Cancer Cells," *Journal of Cellular and Molecular Medicine*, Vol. 20, No. 10, 2012, pp. 1-6. doi:10.1111/j.1582-4934.2012.01576.x

[33] K. S. Louis and A. C. Siegel, "Cell Viability Analysis Using Trypan Blue: Manual and Automated Methods," *Methods in Molecular Biology*, Vol. 740, 2011, pp. 7-12. doi:10.1007/978-1-61779-108-6

[34] J. Kumi-Diaka and A. Butler, "Caspase 3 Protease Activation during the Process of Genistein-Induced Apoptosis in TM4 Testicular Cells," *Biology of the Cell*, Vol. 92, No. 2, 2000, pp. 115-124. doi:10.1016/S0248-4900(00)89019-X

9

Polyubiquitination and Proteasome Signals in Tubulobulbar Complexes of Rat Late Spermatids

Manaka Akashi, Sadaki Yokota, Hideaki Fujita
Section of Functional Morphology, Faculty of Pharmaceutical Sciences, Nagasaki International University, Nagasaki, Japan
Email: syokota123@gmail.com

ABSTRACT

To illustrate the involvement of tubulobulbar complexes (TBC) in ubiquitin-proteasome degradation of unnecessary proteins in the head cytoplasm of late spermatids, the localization of polyubiquitin and proteasome was studied by immunofluorescence and immunoelectron microscopy. Polyubiquitin localized to TBC and proteasome subunit α to dense materials surrounding the TBC in the cytoplasm of Sertoli cell enwrapping sickle-shaped spermatid heads. The results suggest that the TBC is a structural device for ubiquin-proteasome degradation of unnecessary proteins in the cytoplasm of spermatid head during rapid reduction of the head cytoplasm and nuclear compaction of late spermatids.

Keywords: Tubulobulbar Complex; Polyubiquitin Signals; Proteasome; Immunoelectron Microscopy

1. Introduction

Spermatogenic process is divided largely into three phases: 1) the proliferative phase in which spermatogonia undergo rapid successive division, 2) the meiotic phase in which recombination and segregation of chromosomes occur, 3) the differentiation phase in which spermatids transform drastically into spermatozoa which are motile cells carrying haploid genome to the egg [1,2]. The third phase is called spermiogenesis, which is divided into 19 steps in rat [3], and involves the formation of acrosome and flagellum [4-6], the condensation of nucleoplasm [7, 8] and the elimination cytoplasm [9,10]. Finally, spermatids transform to sperm which are released into the lumen of seminiferous tubule. This process, spermiation, is composed of a series of complicated phenomena in which Sertoli cells deeply involve [11-13]. Russell and Clermont [14] observed precisely the isolation process of sperm head from Sertoli cell and found that a dozen tubular processes project out from the ventral concavity of the head to indent the Sertoli cell. The tubular processes extend in the cytoplasm of Sertoli cell and their apex distends to form globule (bulbus) which assembles each other. They called the structure "tubulobulbar complexes" and supposed that the structures function as a scaffold to anchor the spermatid head to the Sertoli cell [14]. It was shown that the time course of the rapid cytoplasmic reduction in late spermatid head matched with that of the formation of tubulobulbar complexes (TBC) [15]. Moreover, the activity of acid phosphatase, a lysosomal marker enzyme, was detected in the TBC and in the cytoplasm of Sertoli cell surrounding the bulbi, suggesting that the lysosomal degradation system is involved in the elimination of cytoplasm from the head region of late spermatid [16].

The intracellular degradation system is largely divided into two categories; lysosome system [17,18] and ubiquitin-proteasome system [19,20]. In the former system, the target to be degraded is segregated by membrane to form autophagosomes which fuse later with lysosomes and the target is degraded by lysosomal enzymes [18]. In the latter system, ubiquitin molecules are covalently coupled (ubiquitinated) with the target protein which is then recognized by proteasome and degraded [19,20]. Although the relationship between the TBC and the lysosomal system is suggested [16], given the rapid and massive degradation of proteins in late spermatid head, it is also expected that the ubiquitin-proteasome system is involved in this degradation. Here, we study whether ubiquitin and proteasome localize to the TBC using immunofluorescence (IF) and immunoelectron micros- copy (IEM) techniques. The results clearly show that these two signals are present in the TBC, suggesting the ubiquitin-proteasome system participates in the degradation of unnecessary proteins in the head cytoplasm of late spermatids.

2. Materials and Methods

2.1. Animals and Antibodies

Male Wistar rats (200 g BW) were used. Testes were taken out from the animals anesthetized with ether. All experiments were performed in accordance with the Guidance for Animal Experiments issued by the Nagasaki International University. Mouse monoclonal antibody (FK2) to multi-ubiquitin chain (M-Ub) was obtained from Nippon Biotest Laboratories Inc. (Tokyo, Japan). Rabbit antibody to α subunit of 20S proteasome (P-αS) was purchased from Merck Millipore (Darmstadt, Germany). Alexa 568® or Alexa 488®-conjugated goat anti-rabbit IgG or goat anti-mouse IgG were obtained from Molecular Probes (Eugene, OR, USA). DAPI was from Hoechst (Tokyo, Japan). Protein A/G/L-gold 15-nm probe were prepared as described previously [21].

2.2. Routine Electron Microscopy

Testis tissue blocks from rats were fixed in the fixative consisting of 4% paraformaldehyde (w/v), 1% glutaraldehyde (v/v), 0.02% $CaCl_2$ (w/v) and 0.05 M Hepes-KOH (pH 7.4) overnight at 4°C. After wash in PBS, testis tissue blocks were fixed 1% reduced osmium tetroxide (w/v) for 1 h, dehydrated and embedded in Epon. Thin sections were contrasted with lead citrate and examined with a Hitachi electron microscope.

2.3. Immunofluorescence (IF) Staining

Frozen sections (6 μm thick) of testes of rats were fixed in 4% paraformaldehyde (w/v) in 0.1 M Hepes-KOH buffer (pH 7.4) for 15 min. After permeation treatment with 0.1% Triton X-100 (v/v) + 0.2% Saponin (w/v), sections were incubated in 2% fish gelatin (w/v) for 30 min to block non-specific adsorption of IgG, followed by overnight incubation with mouse FK2 antibody against M-Ub (×2000) or rabbit anti-P-αS antibody (×500). After washing with PBS, reacted IgG was visualized by Alexa 568® or Alexa 488®-conjugated goat anti-mouse IgG or anti-rabbit IgG. For dual staining of M-Ub and P-αS, sections were incubated in the mixture of mouse FK2 antibody and rabbit anti-P-αS and each reacted IgG was visualized by Alexa 568®-conjugated goat anti-mouse IgG and Alexa 488®-conjugated goat anti-rabbit IgG. For immunofluorescence control, non-immune serum was used instead of the primary specific antibodies. Nuclei were stained by 3 μM DAPI for 60 min at RT. The preparations were examined with a Nikon Eclipse E600 fluorescence microscope (Nikon, Tokyo, Japan). The images were merged using Adobe Photoshop 7.0 to determine whether each antigen colocalizes to the same area. The stage of seminiferous cycle was determined from the localization of elongating and elongated spermatids with individual tubules as previously described [22].

2.4. Immunoelectron Microscopy (IEM)

Testes of rats were cut into small blocks in the fixative consisting of 4% paraformaldehyde (w/v), 0.2% glutaraldehyde (v/v), 0.02% $CaCl_2$ (w/v) and 0.1 M Hepes-KOH (pH 7.4) and testis tissue blocks were kept in the fixative for 1h at 4°C. After wash in PBS, fixed tissue blocks were dehydrated in ethanol and embedded in LR White at −20°C, followed by resin polymerization under UV light at −20°C. Thin sections were cut with a diamond knife equipped with a Reichert Ultracut R, mounted onto nickel grids, and stored in a desiccator. Sections were treated with 2% fish gelatin in PBS (v/v) and incubated with mouse FK2 antibody (×2000) or rabbit anti-M-Ub antibody (×500) overnight at 4°C. Non-immune mouse or rabbit sera were used instead of the primary specific antibodies for control. Reacted IgG was visualized by Protein A/G/L 15-nm-gold probe. Sections were contrasted with 2% uranyl acetate (w/v) and lead citrate, coated with carbon and examined with a Hitachi H7650 electron microscope (Tokyo, Japan) at an acceleration voltage of 80 kV. Stage of seminiferous tubules and step of spermatids were judged in accordance with the stages of seminiferous cycle and step of spermatids were determined as described by Russell *et al.* [22].

3. Results

3.1. Routine Electron Microscopy of TBC

Tubulobulbar complexes (TBC) similar to those described previously [14,15] were confirmed in rat testis used in this study. The sickle-shaped heads of very late spermatids were embedded in an apical process of the Sertoli cell cytoplasm which widely spread in the ventral side but not in the dorsal side of the spermatid head. From the ventral surface of the head, very narrow tubular processes of the spermatid plasma membrane invaginated into the Sertoli cell cytoplasm (**Figure 1(A)**). The tubules extended to deeply into the cytoplasm and the apices of the tubules distended to form globules which coexisted with clear vesicles in the Sertoli cell (**Figure 1(B)**). Frequently, small protrusions of spermatid plasma membrane into the Sertoli cell were observed on the surface of the heads (**Figure 1(A)**). Many cross-sectioned tubules were noted in the section cut through plane running from the apex to base of spermatid head (**Figure 1(B)**).

3.2. IF Staining for Multi-Ubiquitin Chain

Staining for multi-ubiquitin chain (M-Ub) was observed

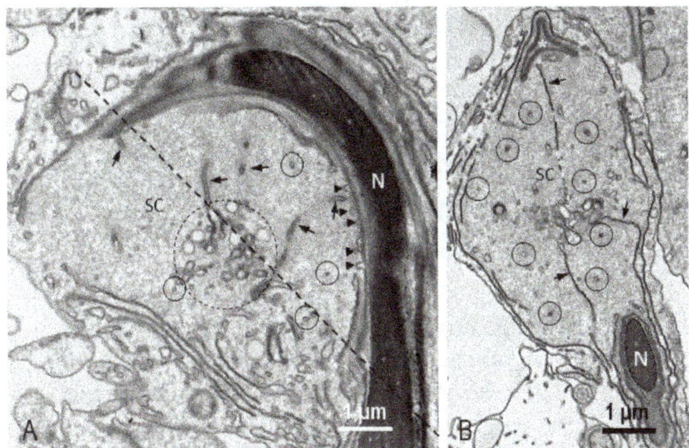

Figure 1. REM image of TBC. (A) A longitudinal section. Longitudinally-sectioned tubules (arrows) and cross-sectioned tubules (small circles) are observed in a Sertoli cell (SC). Distal ends of tubules distend and assemble to form a cluster (large circle). Small protrusions showing beginnings of the tubules are seen on the surface of sickle-shaped spermatid head (arrowheads). Oblique dotted line is a plane of section shown in B; (B) A section cut through an oblique line in A. Long tubules (arrows) extending from nucleus and from perforatorium (*) are seen. Many cross-sectioned tubules are also observed (circles). N: nucleus.

in the nuclei and cytoplasm of spermatogenic cells. We observed the staining during differentiation of spermatogenic cells. The nuclei and cytoplasm of spermatogonia were weakly and diffusely stained. Similar staining was noted in spermatocytes at stages I to III (**Figure 2(A)**). Nuclear and cytoplasmic staining increased and was highest at stage VIII (**Figures 2(B)-(D)**). Afterward, staining in the nuclei decreased rapidly whereas discrete granular staining increased. In spermatocytes at stage X, large spot in the nuclei was stained for M-Ub (**Figure 2(E)**) and at stages XII-XIII the staining intensity of this spot increased (**Figure 2(F)**). M-Ub-positive small granules were visible in the nuclei and cytoplasm of steps 1 - 8 spermatids (**Figures 2(A)-(D)**). In step 9 - 14 spermatids, the nucleoplasm was weakly stained but strong staining was noted in the cytoplasm (**Figures 2(E)** and **(F)**). The intensity of this cytoplasmic staining was highest at steps 12 - 14, afterward decreased and eventually disappeared. Instead, moderately-sized granules in the area surrounding spermatid heads were visible, and in step 19 spermatids, these granules were located in the area contact with the ventral side of the sickle-shaped heads (**Figure 2(D)**). Next, we observed these granules in steps 18 and 19 spermatids at higher magnification. The granules were noticeable in step 18 spermatids and located near the tip of head (**Figure 3(A)**) whereas in step 19 spermatids, the granules moved caudally and were located in the central region of the ventral concavity of the sickle-shaped spermatid head (**Figure 3(B)**). No staining was observed in control sections.

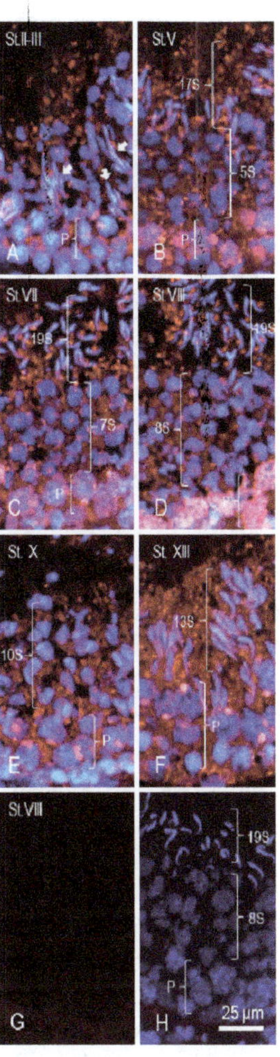

Figure 2. IF staining of M-Ub in seminiferous tubules. (A) Stage II-III. In pachytene spermatocytes (P), punctate staining in nuclei and diffuse staining in cytoplasm are observed. In elongated cytoplasm of step 16 spermatids, many granules are stained (arrows); (B) Stage V. Spotty staining in the nuclei and diffuse staining in the cytoplasm of pachytene spermatocytes (P) are observed. Dotted and diffuse staining is seen in step 5 spermatids (5S); (C) Stage VII. The nuclei and cytoplasm of pachytene spermatocytes (P) are stained. Dotted staining is seen the nuclei of step 7 spermatids (7S). Around step 19 spermatids (19S), small granules are stained; (D) Stage VIII. Pachytene spermatocytes (P) are strongly stained. In Step 8 spermatids (8S), cytoplasmic diffuse staining and granular staining are observed. Around step 19 spermatids spotty staining is seen; (E) Stage X. In the nuclei of pachytene spermatocytes (P), large spots are stained. The cytoplasm and nuclei of these cells and step 10 spermatids (10S) are also stained; (F) Stage XIII. In the nuclei of pachytene spermatocytes (P), large spot staining is seen. In the cytoplasm of step 13 spermatids (13S), numerous granules are stained; (G) Stage VIII. Control section without DAPI staining. Although very weak staining in connective tissue is seen, seminiferous tubule is completely negative for staining; (H) Control section with DAPI staining. The stage is determinable. 19S: Step 19 spermatids. 13S: Step 13 spermatids. P: Pachytene spermatocytes. No M-Ub staining is noted.

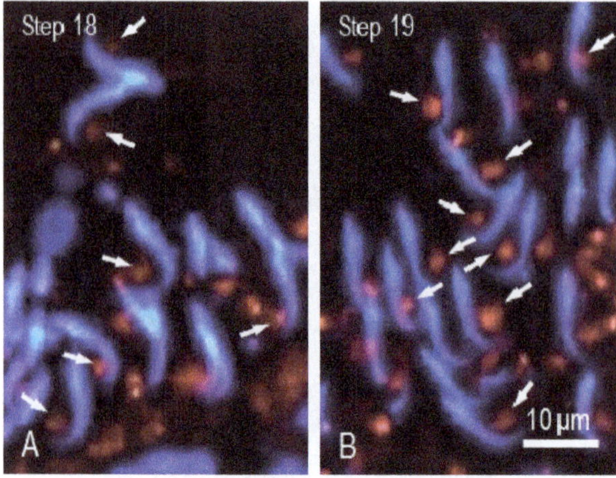

Figure 3. High power view of M-Ub staining of spermatids. (A) Sickle-shaped heads of step 18 spermatids are observed. Stained spots are located near the tip of head (arrows); (B) Heads of 19 spermatids. Stained spots moved to the center of ventral concavity of heads (arrows).

3.3. Dual IF Staining of M-Ub and Proteasome α-Subunit (P-αS)

Proteins coupled with M-Ub are recognized and degraded by proteasome. This means that both proteins might localize to the same place at the same time. Therefore, we observed the relationship between M-Ub positive spots and P-αS localization in the region of the ventral concavity of the sickle-shaped spermatid heads. P-αS staining was observed in residual bodies of step 19 spermatids and the cytoplasm of Sertoli cells (**Figure 4(B)**). M-Ub spots were also stained for P-αS (**Figures (A)-(C)**, arrowheads). In control sections, no staining for both proteins was noted (**Figure 4(D)**).

3.4. IEM Staining of M-Ub

Since IEM localization of M-Ub in spermatogenic cells during spermatogenesis was similar to the results reported previously [23], in present study, we focused on the localization of M-Ub in the IF stained M-Ub spot at the ventral concavity of step 19 spermatid heads. Strong gold labeling showing M-Ub was associated with the lumen of tubules and dense material surrounding the tubules composing the TBC (**Figures 5(A)** and **(B)**, arrowheads). Weak M-Ub labeling was observed in the cytoplasm of Sertoli cells and the labeling in the area where clear vesicles and globules assembled together, was stronger than the other area. M-Ub was detected in subacrosomal space, perforatorium and the nucleoplasm of spermatids (**Figures 5(A)** and **(B)**). In control sections, no gold labeling was observed.

3.5. IEM Localization of P-αS

Gold particles showing P-αS were observed in the cytoplasm of Sertoli cells (**Figure 6**). If carefully observed, P-αS signals were associated with dense material around the tubules and cluster of globules composing the TBC (**Figure 6**). On the contrary, the lumen of clear vesicles were almost negative for P-αS (**Figure 6**, *). In control sections, gold signals were noted.

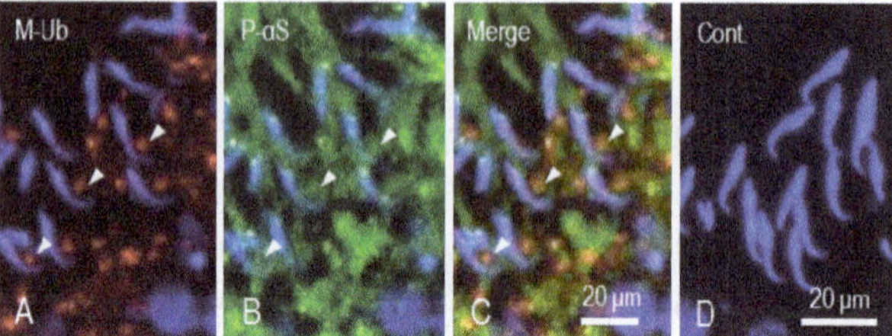

Figure 4. Dual staining of M-Ub and proteasome α subunit in step 19 spermatids. (A) M-Ub. Stained spots are seen in the area facing ventral concavity of sickle-shaped spermatid heads (arrows); (B) P-αS. Area surrounding spermatid heads are stained. Some spots are stained for both proteins (arrowheads); (C) Merged A and B. Arrowheads indicate the spots stained for both proteins; (D) Control. No staining for M-Ub and P-αS is noted.

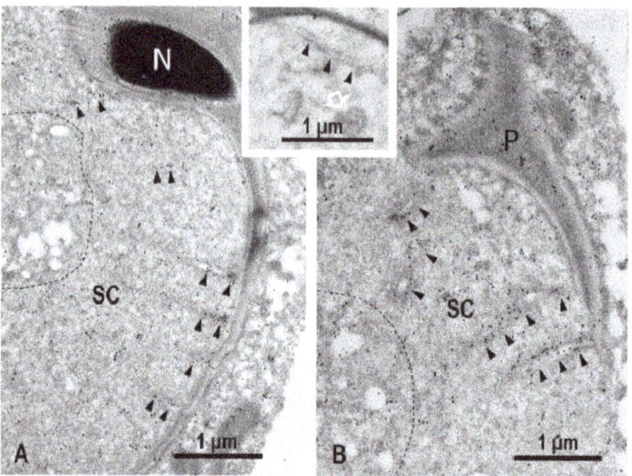

Figure 5. IEM staining of M-Ub in TBC in step 19 spermatids. (A) Nearly longitudinally sectioned sickle-shaped head. Gold particles showing M-Ub sites are seen on dense material around tubules and in the lumen of tubules (arrowheads). Globules of the distal end of tubules assemble together with clear vesicles (surrounded by dotted line). This area is also labeled. Gold particles are observed in the periphery of the nucleus of spermatid (N). SC: Sertoli cell; (B) Cross-sectioned tip of step 19 spermatid head. Gold labeling is observed in dense material surrounding tubules, in the lumen of tubules (arrowheads), and in cluster of tubule end and clear vesicles (surrounded by dotted line). Perforatorium (P) is also stained. SC: Sertoli cell. Inset. Control section. No gold labeling is noted in tubule (arrowheads) as well as tubule ends (white arrow).

4. Discussion

It was reported that approximately 25% of total cell body of spermatid is reduced when the spermatid is released to

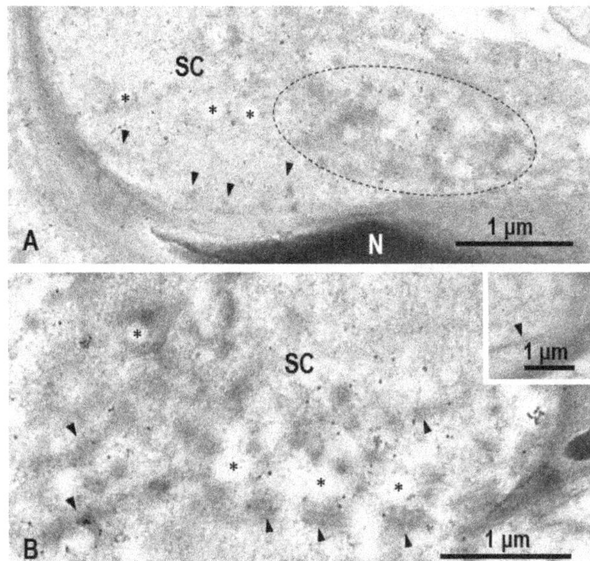

Figure 6. IEM staining of proteasome αsubunit (P-αS) in step 19 spermatid heads. (A) Longitudinal section of spermatid head. Gold particles showing P-αS are observed in the cytoplasm of Sertoli cell (SC). Some particles are associated with dense material surrounding tubules (arrowheads) and clear vesicles (*). Gold labeling is also present in cluster of tubule ends and clear vesicles (surrounded by dotted line). N: Nucleus of spermatid; (B) Cross-section of spermatid head. Gold labeling is seen on dense material along tubules (arrowheads) and around clear vesicles (*). SC: Sertoli cell. Inset. Control. No gold labeling is observed. Arrow indicates tubule.

the lumen of seminiferous tubule during the final stage of spermiogenesis [24]. It is completed by pinch-off of the cytoplasm of spermatids as residual bodies which are incorporated and degraded by Sertoli cells [25,26]. In the light of the nuclear area in step 19 spermatids, 70% of the nuclear volume is reduced as compared to that in the previous steps [16]. To accomplish such reduction, all degradation systems, including lysosomal degradation system and ubiquitin-proteasome system, are used. It is suggested that the cytoplasm casted off from spermatids is ingested by Sertoli cells and degraded in the lysosomal system [24]. Similarly, the lysosomal system of Sertoli cells is suggested to be also involved in the extensive elimination of the nuclear area through the TBC [16]. This idea is due to that the activity of acid phosphatase and a lysosomal marker enzyme is detected in the TBC [16].

In the present study, we have shown first time that multi-ubiquitin chain (M-Ub) (multi-ubiquitinated proteins) signals are associated with the TBC. The result indicates that multi-ubiquitinated proteins are present in the TBC. The multi-ubiquitinated proteins are recognized by ubiquitin receptor and trapped into proteasome to be degraded [19-20,27]. Therefore, the multi-ubiquitinated proteins in the TBC, when proteins are not clear, seem to be degraded by proteasome. In the present study we have shown that proteasome is detected in the cytoplasm of Sertoli cells and associated with dense material surrounding the TBC. The result suggests that multi-ubiquitinated proteins detected are degraded by proteasome. Since the M-Ub signals are present in the nucleoplasm as described previously [23], the multi-ubiquitinated proteins in the TBC seem to be derived also from the nucleus of step 19 spermatids. M-Ub signals are associated with not only dense material surrounding the tubules but also the lumen of them. If multi-ubiquitinated proteins are derived from the spermatid nucleus as well as cytoplasm, they should be transported across two plasma membranes of spermatids and Sertoli cell. Although the transport mechanism is quite unclear, the present results suggest that proteins which become unnecessary with rapid and extensive reduction of the cytoplasm as well as compaction of the nucleus are degraded not only by lysosomal system but also by ubiquitin-proteasome system.

5. Acknowledgements

The work was supported by the university research fund, in part by a grant-in-aid (17570158) from the Ministry of Education, Science, Culture and Sport, and by the Science Research Promotion Fund from the Promotion and Mutual Aid Corporation for Private Schools of Japan.

REFERENCES

[1] S. S. Guraya, "Biology of Spermatogenesis and Spermatozoa in Mammals," Springer-Verlag, Berlin, 1987. http://dx.doi.org/10.1007/978-3-642-71638-6

[2] D. M. de Krester and J. B. Kerr, "The Cytology of the Testis," In: E. Knobil and J. Neill, Eds., *The Physiology of Reproduction*, Raven Press, New York, 1988, pp. 837-932.

[3] Y. Clermont and A. Rambourg, "Evolution of the Endoplasmic Reticulum during Rat Spermiogenesis," *American Journal of Anatomy*, Vol. 151, No. 2, 1978, pp. 191-211. http://dx.doi.org/10.1002/aja.1001510204

[4] M. H. Burgos and D. W. Fawcett, "Studies in the Fine Structure of the Mammalian Testis. I. Differentiation of the Spermatids in the Cat *(Felis Domestica)*," *Biophysical Biochemical Cytology*, Vol. 1, No. 4, 1955, pp. 287-315. http://dx.doi.org/10.1083/jcb.1.4.287

[5] A. Abou-Halia and R. P. Tulslani, "Mammalian Sperm Acrosome: Formation, Contents, and Function," *Archive Biochemistry and Biophysics*, Vol. 379, No. 2, 2000, pp. 173-182. http://dx.doi.org/10.1006/abbi.2000.1880

[6] D. W. Fawcett and D. M. Phillips, "The Fine Structure and Development of the Neck Region of the Mammalian Spermatozoon," *Anatomical Record*, Vol. 165, No. 2, 1969, pp. 153-184.

http://dx.doi.org/10.1002/ar.1091650204

[7] W. S. Ward and D. S. Coffet, "DNA Packaging and Organization of in Mammalian Spermatozoa: Comparison with Somatic Cells," *Biology of Reproduction*, Vol. 44, No. 4, 1991, pp. 569-574. http://dx.doi.org/10.1095/biolreprod44.4.569

[8] W. S. Ward, "Function of Sperm Chromatin Structural Elements in Fertilization and Development," *Molecular Human Reproduction*, Vol. 16, No. 1, 2010, pp. 30-36. http://dx.doi.org/10.1093/molehr/gap080

[9] R. M. Sharpe, S. M. Maguire, P. T. K. Saunders, M. R. Millar, L. D. Russell, D. Garten, S. Bachmann, L. Mullins and J. J. Mullins, "Infertility in a Transgenic Rat Due to Impairment of Cytoplasmic Elimination and Sperm Release from the Sertoli Cells," *Biology of Reproduction*, Vol. 53, No. 1, 1995, pp. 214-226. http://dx.doi.org/10.1095/biolreprod53.1.214

[10] A. K. Rengan, A. Agarwal, M. van der Linde and S. S. du Plessis, "An Investigation of Excess Residual Cytoplasm in Human Spermatozoa and Its Distinction from the Cytoplasmic Droplet," *Reproductive Biology and Endocrinology*, Vol. 10, 2012, pp. 2-8. http://dx.doi.org/10.1186/1477-7827-10-92

[11] C. S. Sapsford and C. A. Rae, "Ultrastructural Studies on Sertoli Cells and Spermatids in the Bandicoot and Ram during the Movement of Mature Spermatids into Lumen of the Seminiferous Tubule," *Australian Journal of Zoology*, Vol. 17, No. 3, 1969, pp. 415-445. http://dx.doi.org/10.1071/ZO9690415

[12] J. P. Fouquet, "La Spermiation et la Formation des Corps Résiduels chez le Hamster: Role des Cellules de Sertoli," *Journal de Microscopie*, Vol. 19, No. 2, 1974, pp. 161-168.

[13] D. W. Fawcet, "Ultrastructure and Function of the Sertoli Cell," In: D. W. Hamilton and R. O. Greep, Eds., *Handbook of Physiology. Male Reproductive System, Vol. V, Section* 7, *Endocrinology*, American Physiological Society, Washington DC, Waverly Press Inc., Baltimore, 1975, pp. 21-55.

[14] L. Russell and Y. Clermont, "Anchoring Device between Sertoli Cells and Late Spermatids in Rat Seminiferous Tubules," *Anatomical Record*, Vol. 185, No. 3, 1976, pp. 259-278. http://dx.doi.org/10.1002/ar.1091850302

[15] L. D. Russell, "Spermatid-Sertoli Tubulobulbar Complexes as Devices for Elimination of Cytoplasm from the Head Region of Late Spermatids of the Rat," *Anatomical Record*, Vol. 194, No. 2, 1979, pp. 233-246. http://dx.doi.org/10.1002/ar.1091940205

[16] L. D. Russell, "Further Observations on Tubulobulbar Complexes Formed by Late Spermatids and Sertoli Cells in the Rat Testis," *Anatomical Record*, Vol. 194, No. 2, 1979, pp. 213-232. http://dx.doi.org/10.1002/ar.1091940204

[17] G. E. Mortimore and A. R. Pösö, "Lysosomal Pathway in Hepatic Protein Degradation: Regulatory Role of Amino Acids," *Federation Proceedings*, Vol. 43, No. 5, 1984, pp. 1289-1294.

[18] W. A. Dunn Jr., "Autophagy and Related Mechanisms of Lysosome-Mediated Protein Degradation," *Trends in Cell Biology*, Vol. 4, No. 4, 1994, pp. 139-143. http://dx.doi.org/10.1016/0962-8924(94)90069-8

[19] A. Hershko and A. Ciechanover, "The Ubiquitin System for Protein Degradation," *Annual Review of Biochemistry*, Vol. 61, 1992, pp. 761-807. http://dx.doi.org/10.1146/annurev.bi.61.070192.003553

[20] M. Hochstrasser, "Ubiquitin, Proteasomes, and the Regulation of Intracellular Protein Degradation," *Current Opinion in Cell Biology*, Vol. 7, No. 2, 1995, pp. 315-223. http://dx.doi.org/10.1016/0955-0674(95)80031-X

[21] S. Yokota, "Preparation of Colloidal Gold Particles and Conjugation of Protein A, IgG, F(ab')$_2$, and Streptavidin," In: S. D. Schwarzbach and T. Osafune, Eds., *Immunoelectron Microscopy. Methods and Protocols*, Springer, New York, 2012, pp. 109-119.

[22] L. D. Russell, R. A. Ettlin, A. P. S. Hikim and E. D. Clegg, "Histological and Histopathological Evaluation of the Testis," Cache River Press, Florida, 1990, pp. 59-119.

[23] C. M. Haraguchi, T. Mabuchi, S. Hirata, T. Shoda, T. Tokumoto, K. Hoshi and S. Yokota, "Possible Function of Caudal Nuclear Pocket: Degradation of Nucleoproteins by Ubiquitin-Proteasome System in Rat Spermatids and Human Sperm," *Journal of Histochemistry and Cytochemistry*, Vol. 55, No. 6, 2007, pp. 585-595. http://dx.doi.org/10.1369/jhc.6A7136.2007

[24] R. L. Sparando and L. D. Russell, "Comparative Study of Cytoplasmic Elimination in Spermatids of Selected Mammalian Species," *American Journal of Anatomy*, Vol. 178, No. 1, 1987, pp. 72-80. http://dx.doi.org/10.1002/aja.1001780109

[25] D. W. Fawcett and D. M. Phillips, "Observation on the Release of Spermatozoa and Changes in the Head during Passage through the Epididymis," *Journal of Reproduction and Fertility*, Suppl. 6, 1969, pp. 405-418.

[26] J. B. Kerr and D. M. Kretser, "The Role of the Sertoli Cell in Phagocytosis of Residual Bodies of Spermatids," *Journal of Reproduction and Fertility*, Vol. 36, 1974, pp. 439-440. http://dx.doi.org/10.1530/jrf.0.0360439

[27] K. Tanaka, "The Proteasome: Overview of Structure and Functions," *Proceedings of Japan Academy, Series B, Physical and Biological Sciences*, Vol. 85, No. 1, 2009, pp. 12-36. http://dx.doi.org/10.2183/pjab.85.12

10

Characterization of the Intracellular Distribution of Adenine Nucleotide Translocase (ANT) in *Drosophila* Indirect Flight Muscles

Vivek K. Vishnudas[1,2], Shawna S. Guillemette[1,3], Panagiotis Lekkas[1], David W. Maughan[4], Jim O. Vigoreaux[1,4*]

[1]Department of Biology, University of Vermont, Burlington, USA

[2]Berg Biosystems, Framingham, USA

[3]Department of Cancer Biology, University of Massachusetts Medical School, Worcester, USA

[4]Department of Molecular Physiology and Biophysics, University of Vermont, Burlington, USA

Email: *jvigorea@uvm.edu

ABSTRACT

Background: The high power output necessary for insect flight has driven the evolution of muscles with large myofibrils (primary energy consumers) and abundant mitochondria (primary energy suppliers). The intricate functional interrelationship between these two organelles remains largely unknown despite its fundamental importance in understanding insect flight bioenergetics. Unlike vertebrate muscle that relies on a phosphagen (creatine phosphate/creatine kinase) system to regulate high energy phosphate flux, insect flight muscle has been reported to lack mitochondrial arginine kinase (analogous to creatine kinase), a key enzyme that enables intracellular energy transport. Creatine kinase is known to interact with mitochondrial adenine nucleotide translocase (ANT) in the transfer of ADP and ATP into and out of the mitochondria. **Results:** Here, we use quantitative immunogold transmission electron microscopy to show that in *Drosophila melanogaster* indirect flight muscles (IFM), ANT is present in the mitochondria as well as throughout the myofibril. To confirm this unexpected result, we created a transgenic line that expresses a chimeric GFP-ANT protein and used an anti-GFP antibody to determine the intracellular distribution of the fusion protein in the IFM. Similar to results obtained with anti-ANT, the fusion GFP-ANT protein is detected in myofibrils and mitochondria. We confirmed the absence of arginine kinase from IFM mitochondria and show that its sarcomeric (*i.e.*, intramyofibrillar) distribution is similar to that of ANT. **Conclusions:** These results raise the possibility that direct channeling of nucleotides between mitochondria and myofibrils is assisted by an ANT protein thereby circumventing the need for a phosphagen shuttle in the IFM. The myofibrillar ANT may represent a unique adaptation in the muscles that require efficient exchange of nucleotides between mitochondria and myofibrils.

Keywords: Drosophila; Insect Flight Muscle; Adenine Nucleotide Translocase; Phosphagen System

1. Background

Insect flight is widely regarded as one of the most metabolically demanding forms of animal locomotion [1]. Respiratory rates that increase 50 - 100 fold from rest to flight have been reported for some insects [2]. In particular, insects that rely on high wing beat frequencies to achieve the power needed to sustain flight are known to have very high mass-specific metabolic rates in comparison to other animals. ATP turnover rates during flight are among the highest ever recorded [3]. The primary consumer of metabolic energy in muscle is myosin, the molecular motor that generates the forces for contraction. The *Drosophila* indirect flight muscle (IFM) myosin is especially adapted for very fast muscle speed, including a very low affinity for MgATP [4]. Thus a very high concentration of MgATP is required in the IFM to satisfy the energetic demand of flight and to compensate for myosin's weak MgATP affinity. High MgATP concentration is ensured by the large and abundant mitochondria that occupy up to 37% of the cellular space in the flight muscle [5,6].

While the role of the mitochondria as the main energy

*Corresponding author.

supplier in insect muscle is well described, the mechanism by which the intracellular energy flow is established and maintained is less clear. The adenine nucleotide translocase (ANT) is a highly abundant integral protein of the inner mitochondrial membrane that plays an essential and well established role in cellular bioenergetics [7] as well as in apoptosis [8] and cell signaling [9]. ANT catalyzes the exchange of ADP for ATP across the inner mitochondrial membrane thereby coupling oxidative phosphorylation to cell metabolism. In vertebrates and some invertebrates, ANT fulfills this important energetic function through its interaction with mitochondrial creatine kinase (Mi-CK), an enzyme that transfers high energy phosphates to creatine, thus forming the important metabolic currency, creatine phosphate (CP) [10]. Mi-CK works in conjunction with a soluble CK isoform, MM-CK, as part of the "phosphagen shuttle" responsible for transporting CP from mitochondria through the cytosol to major sites of ATP catalysis such as the myofibril and sarcoplasmic reticulum. This system creates metabolic capacitance by means of spatial and temporal buffering [6,11].

In *Drosophila* IFM, as in other insect muscles, the phosphagen system consists of arginine phosphate (AP) and arginine kinase (AK) that is assumed to operate in a manner analogous to the CK-CP system in vertebrates [12]. However, one study has questioned the existence of a mitochondrial AK isoenzyme raising the possibility that the AK-AP system may function in a manner fundamentally different from the better characterized CK-CP system [13]. Unlike vertebrate skeletal muscle, sarcoplasmic reticulum is very scarce in *Drosophila* IFM given its asynchronous mode of contraction. Thus, very little energy is expended on calcium cycling allowing the metabolic energy to be channeled preferentially to the myofibril, the site of contractile activity. How metabolic energy utilization by the myofibril is matched effectively to its production by the mitochondria remains largely unexplored.

Drosophila ANT is encoded by two tandem duplicated genes, *stress sensitive* B (*sesB*) and ANT2 [14]. Functional ANT activity in *Drosophila* adults is encoded by *sesB* [15]. Mutations in *sesB* cause paralysis and loss of synaptic transmission in the visual system [16] and lead to failure of synaptic vesicle cycling at the neuromuscular junction during periods of high demand [17]. Like the synapse, the IFM is likely to rely heavily on ANT for metabolic capacitance. Here, we have generated transgenic *Drosophila* strains that express a Green fluorescent protein-*sesB* (GFP-*sesB*) fusion protein in order to gain insight into the function of ANT in the IFM. We present new evidence of a myofibril-associated ANT that may be involved in facilitating the transport of ATP and ADP into and out of the myofibril, respectively. We also confirm earlier results that *Drosophila* IFM lack a mitochondrial AK isozyme. The study provides new molecular details of a potentially novel pathway of metabolic energy exchange involving tight functional coupling between myofibrils and mitochondria in the IFM.

2. Results

2.1. ANT Localizes to the Mitochondria and the Myofibrils of Drosophila IFM

A rabbit polyclonal antibody generated against bovine cardiac ANT was used to examine the distribution of *Drosophila* ANT in the IFM. *Drosophila* and mammalian ANT share ~80% amino acid sequence identity [18, 19]. The antibody has been reported to cross react with mouse ANT [20] and we confirmed, by western blot analysis, the specificity of the antibody for *Drosophila* *sesB*-encoded ANT (not shown, but see **Figure 3**). Studies on adult fly frozen sections revealed strong immunostaining of IFM compared to the jump muscle and other thoracic muscles (data not shown). We then performed electron microscopy on sections of the IFM to examine the intracellular distribution of ANT (**Figure 1(A)**). Gold particles were detected in the mitochondria and in the myofibril in about equal distribution, but were largely absent in the cytosol (**Figure 1(D)**). In contrast, a β ATP synthase antibody (**Figures 1(B)** and **(E)**) specifically labels the mitochondria and a flightin antibody (**Figures 1(C)** and **(F)**) labels the myofibrillar A band, as expected. Two negative controls, pre-immune rabbit serum and no primary antibody (pAG_{10} only) showed no gold distribution (data not shown).

2.2. Expression of GFP-*sesB* Fusion Protein in Drosophila Schneider Cells and Transgenic IFM

To confirm the myofibrillar localization of ANT in the IFM, we made transformation vectors encoding ANT with an N-terminal GFP tag. Zhang *et al.* reported that two adult transcripts of 1.6 KB and 1.2 KB generated from *sesB* result from alternative 3' ends [18]. We conducted RT-PCR of total mRNA from late pupae and found two similar size transcripts. DNA sequence analysis of the PCR products confirmed that the differences were only in the 3' UTRs (results not shown). To ensure that the GFP tag does not alter the mitochondrial localization of ANT, we transfected *Drosophila* Schneider cells with pCaSpeRhs GFP-*sesB* 1.6 KB and pCaSpeRhs GFP-*sesB* 1.2 KB constructs. **Figure 2** shows confocal images of tissue culture cells labeled with anti-GFP rabbit polyclonal antibodies and with Mitotracker. Both constructs express fusion proteins that are localized to the mitochondria, confirming that the addition of an N-terminal GFP tag does not affect protein targeting.

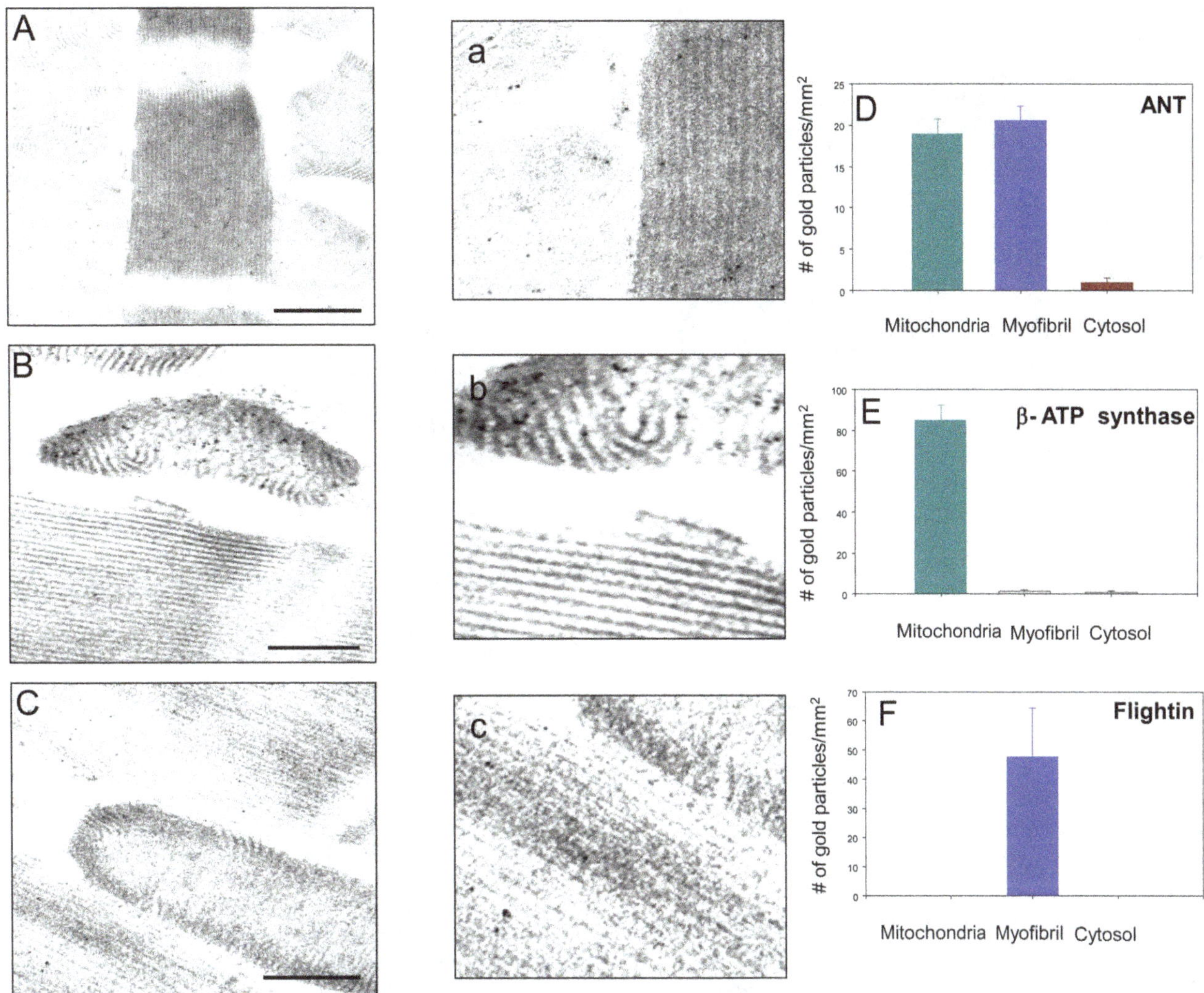

Figure 1. ANT is present in the mitochondria and myofibrils of *Drosophila* IFM. ((A), (a), and (D)) Intracellular distribution of ANT in the IFM determined with an anti-bovine cardiac ANT antibody (Bar = 200 nm). Quantification of immunogold particles revealed the dual localization of ANT (n = 12); ((B), (b), and (E)) Immunogold labeling of IFM with an anti-β ATP synthase antibody (Bar = 200 nm). Quantification of gold particle distribution shows a strictly mitochondrial localization (n = 12); ((C), (c), and (F)) Immunogold labeling with an anti-*Drosophila* flightin antibody (Bar = 200 nm). Quantification of gold particle distribution shows a strictly myofibrillar localization (n = 12). Panels a, b, and c are magnifications of a myofibril-mitochondria interface of panels (A), (B), and (C), respectively.

We confirmed the specificity of the anti-bovine cardiac ANT antisera for endogenous *Drosophila sesB*-encoded protein and GFP chimeric protein by western blot analysis of proteins extracted from the IFM of 5 transgenic lines and 2 control lines. The transgene is inserted in chromosome 1 (lines 13, 16, and 31), chromosome 2 (line 23), or chromosome 3 (line 24). The expression of the fusion protein is demonstrated by western blot with anti-GFP rabbit antibody (**Figure 3(a)**). **Figure 3(b)** is a western blot of 12% SDS-PAGE gel containing total thoracic proteins from transgenic and control lines. The anti-bovine cardiac ANT antibody cross reacts with an endogenous ~30 kD protein expressed in all transgenic and control lines, and with a ~57 kD protein expressed only in transgenic lines. The latter matches the expected size of the GFP-*sesB* fusion protein. In addition, the results show that the fusion protein expression is highest in line 31, where it is expressed at levels comparable to or higher than the endogenous ANT protein. **Figure 3(c)** is a Sypro stained 12% SDS-PAGE gel showing the expression of the fusion proteins in the five transgenic lines. Note that a band of ~57 kD appears most prominently in the heat shocked induced lane of line 31 suggesting that this line expresses the highest levels of fusion protein, in agreement with the western blot results. Line 31 was therefore selected for further study.

We characterized the localization of GFP-*sesB* fusion protein in line 31 by immunogold transmission electron

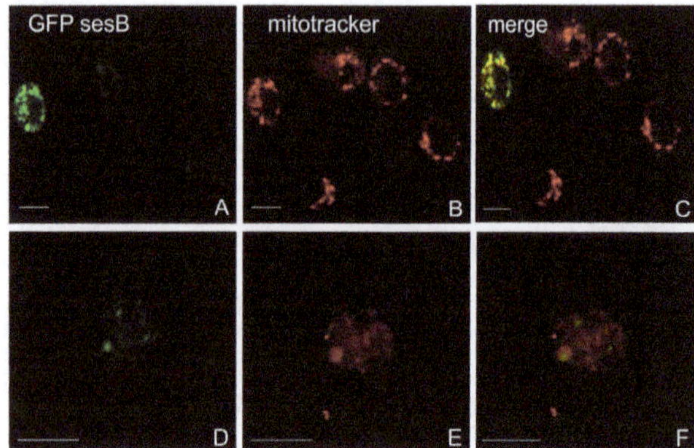

Figure 2. Confocal images of Schneider cells transfected with ANT expression vectors. ((A), (B), (C)) cells transfected with pCaSpeR hs GFP-*sesB* 1.6 KB ((D), (E), (F)) cells transfected with pCaSpeR hs GFP-*sesB* 1.2 KB. Protein expression was induced by heat shock. Cells were treated with anti-GFP antisera ((A) and (D)) or mitotracker ((B) and (E)). (C) and (F) are merged images from the two channels. Both constructs target GFP-ANT fusion protein to the mitochondria. Bar = 10 μm.

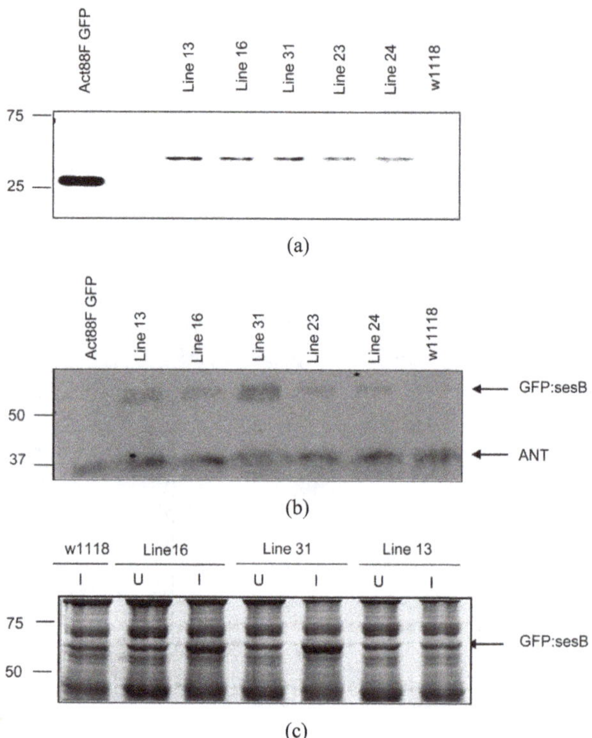

Figure 3. Expression of GFP-*sesB* fusion protein in transgenic lines. (a) Western blot showing expression levels of heat shocked induced GFP-*sesB* fusion protein in whole thorax protein extracts from transgenic and control lines using an anti-GFP antibody. The antibody detects a 30 kD band in the *Act88F*-GFP transgenic line corresponding to GFP. An ~57 kD band detected in lines 13, 16, 31, 23, and 24 correspond to the GFP-*sesB* fusion protein. No bands are detected in the non-transgenic w^{1118} sample; (b) Western blot showing expression levels of heat shocked induced GFP-*sesB* fusion protein analyzed in whole thorax protein extracts from transgenic and control lines using an anti-bovine cardiac ANT antibody. The antibody cross reacts with both endogenous ANT (30 kD band in all lanes) and GFP-*sesB* fusion protein (57 kD band in lines 13, 16, 31, 23, and 24; (c) Sypro stained gel analysis of thorax proteins from uninduced (U) and heat shock induced (I) expression of GFP-ANT transgene. Note prominent band in induced lane of lines 16 and 31 (arrow).

microscopy with an anti-GFP antibody. As controls, we used an *Act*88*F*-GFP strain, a line that expresses GFP in the IFM driven by the *Act*88*F* promoter [21], and the w^{1118} parental strain. As seen in **Figure 4(A)**, GFP labeling is found in all three sub-compartments of the muscle cell, highest in the myofibril and about equal distribution in the cytosol and mitochondria. To determine if this pattern results from a random distribution, we calculated the relative cellular area covered by each sub-compartment. In the images examined, myofibrils occupy ~70% of the area, while mitochondria and cytosol occupy ~17% and 12%, respectively (n = 8). The particle counts approximate the percent area covered by each sub-compartment (**Figure 4(D)**) suggesting that GFP distribution is random. In contrast, GFP-*sesB* shows a non-random distribution with approximately equal counts in myofibrils and mitochondria and low counts in the cytosol (**Figures 4(B)** and **(E)**). This distribution resembles the one seen with the anti-bovine cardiac ANT antibody (**Figure 1(A)**). Very few gold particles were detected in the control line w^{1118} (**Figures 4(C)** and **(F)**), and additional controls using pre-immune rabbit serum and pAG_{10} on line 31 did not show any gold distribution in the IFM (data not shown).

Several studies have shown that slow (oxidative) skeletal and cardiac muscle ANT are functionally coupled to Mi-CK [22,23]. Microfractionation studies of *Drosophila* IFM have shown that AK is found primarily in the cytosolic fraction and not associated with the mitochondria [13]. We performed immunoelectron microscopy with a rabbit anti- *Drosophila* AK polyclonal antibody to extend the study of Wyss *et al.* [13] and to determine if the myofibrillar localization of ANT is associated with AK. As seen in **Figure 5**, most of the AK localizes to the myofibril. Unlike CK in vertebrate slow skeletal and cardiac muscle, and consistent with the results of Wyss *et al.* [13], *Drosophila* AK is largely excluded from the mitochondria. We further looked at the sarcomeric distribution of ANT and AK. **Figure 6** shows that ANT and AK are primarily distributed throughout the A band. There is no difference in the gold particle distribution for the two treatments (p = 0.827, Kruskal Wallis test) suggesting that ANT and AK may co-localize in the sarcomere. In contrast, the distribution of GFP differs significantly from that of ANT ($p < 0.001$), AK ($p < 0.001$), and GFP-*sesB* ($p < 0.05$).

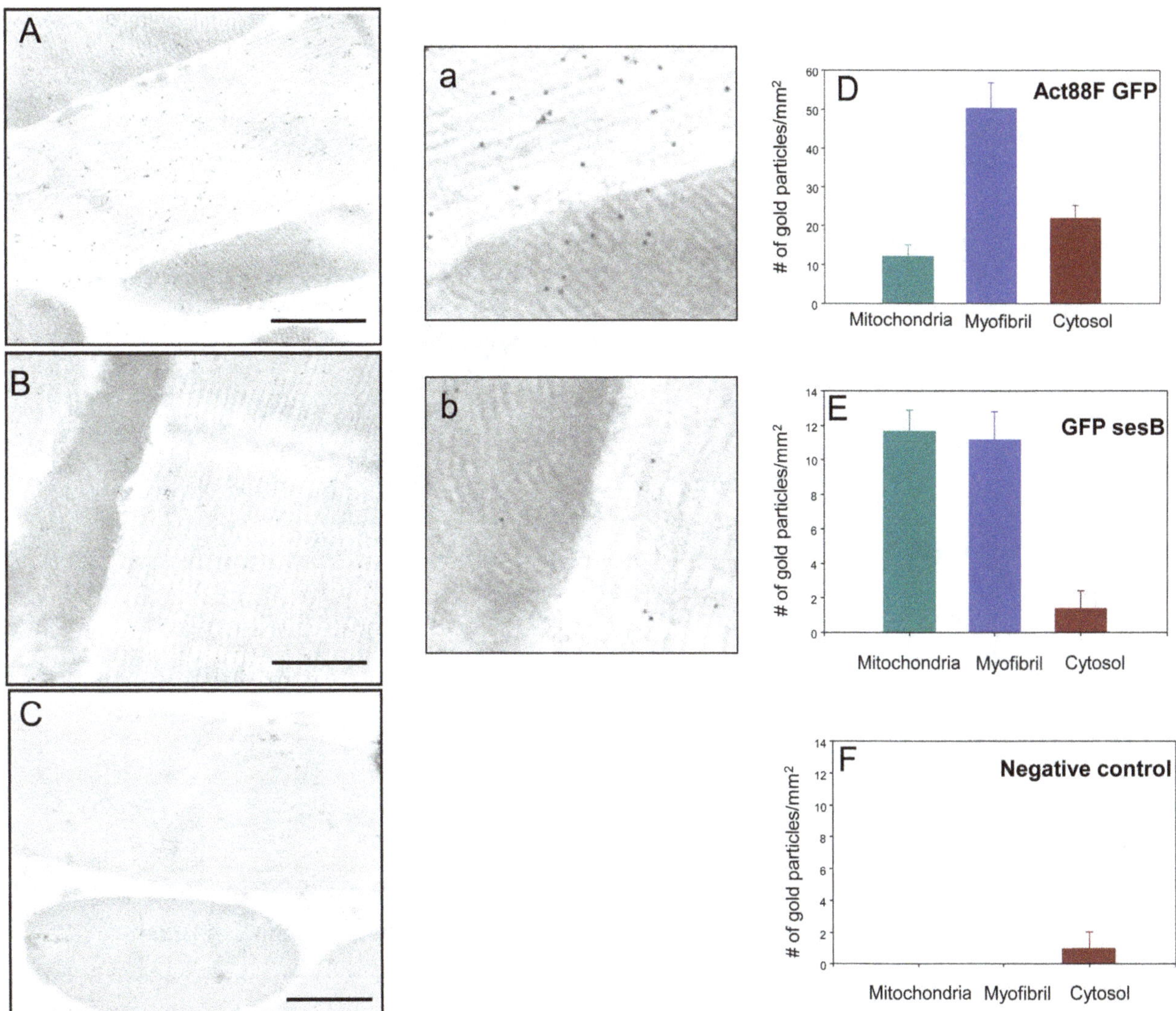

Figure 4. GFP-*sesB* is present in mitochondria and myofibrils of Drosophila IFM. Distribution of GFP signals in ((A), (a)) *Act*88F-GFP transgenic control line, ((B), (b)) transgenic Line 31, and (C) w^{1118} non-transgenic control line. All sections were probed with an anti-GFP antibody (Bar = 200 nm). (D) Quantification of gold particle distribution shows GFP is distributed randomly in transgenic control line (n = 12). (E) Quantification of gold particles shows GFP-*sesB* is preferentially targeted to mitochondria and myofibrils in transgenic line 31 (n = 12). This distribution pattern parallels that of ANT in wild-type flies (compare to Figures 1(A) and (D)). (F) No GFP signal is detected in the w^{1118} control line, confirming the specificity of the antibody. Panels a and b are magnifications of a myofibril-mitochondria interface of panels A and B, respectively.

3. Discussion

ANT is a resident protein of the inner mitochondrial membrane and research over the past several years have shown that ANT interacts with a variety of proteins in fulfilling multiple roles in energy metabolism, apoptosis, and cellular signaling [24-27]. As a component of the multiprotein permeability transition pore complex (PTPC), ANT interacts with proteins of the outer mitochondrial membrane, most notably the voltage dependent anion channel (VDAC), as well as with multiple proteins involved in metabolism and cell signaling [28]. These ANT-mediated protein interactions promote functional interplay between the mitochondria and other organelles [28]. Physical contacts have been described between the mitochondria, the endoplasmic reticulum [29,30], the sarcoplasmic reticulum [31], and adherens complexes [32], but it remains to be established if ANT forms part of the molecular connection that links mitochondria with its partners. A function for ANT outside the mitochondria has not been described. However, other mitochondrial proteins have been shown to fulfill roles outside this organelle including cytochrome c [33], ATP synthase β [34], and VDAC [35].

Here we confirmed the results of Zhang *et al.* that

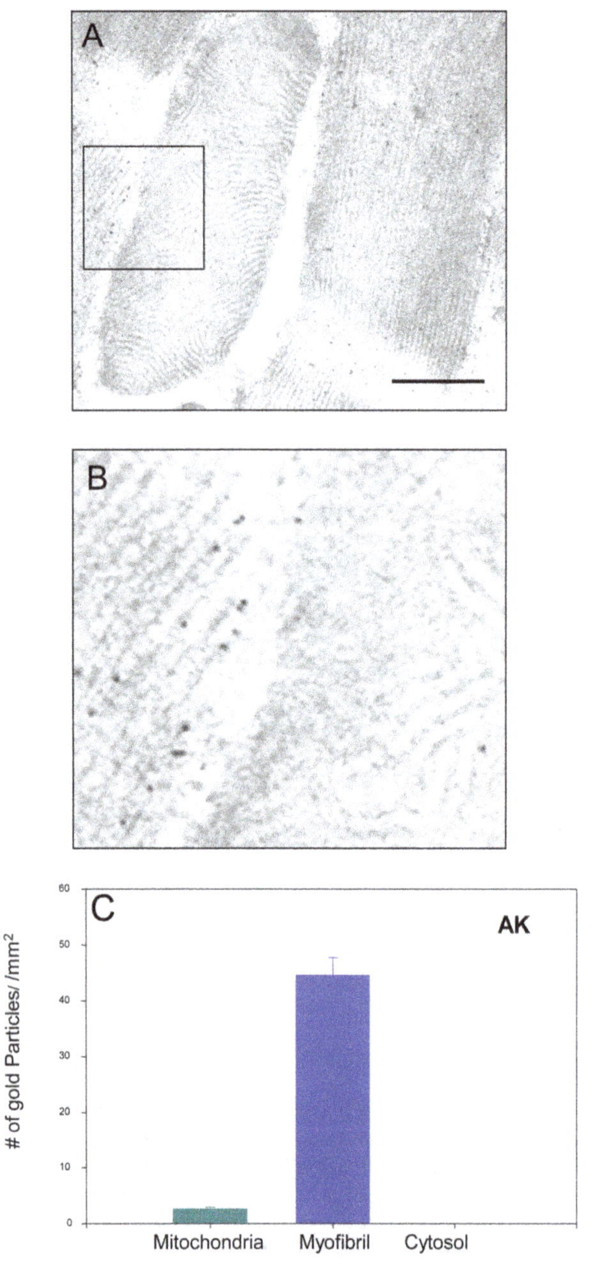

Figure 5. IFM Arginine kinase is predominantly localized to the myofibrils. (A) IFM section treated with an anti- *Drosophila* arginine kinase antibody (Bar = 200 nm); (B) Boxed area in (A) shown in greater magnification. Note the majority of gold particles localize to the myofibril; (C) Quantification of gold particle distribution demonstrates that arginine kinase is predominantly associated with the myofibril (n = 6).

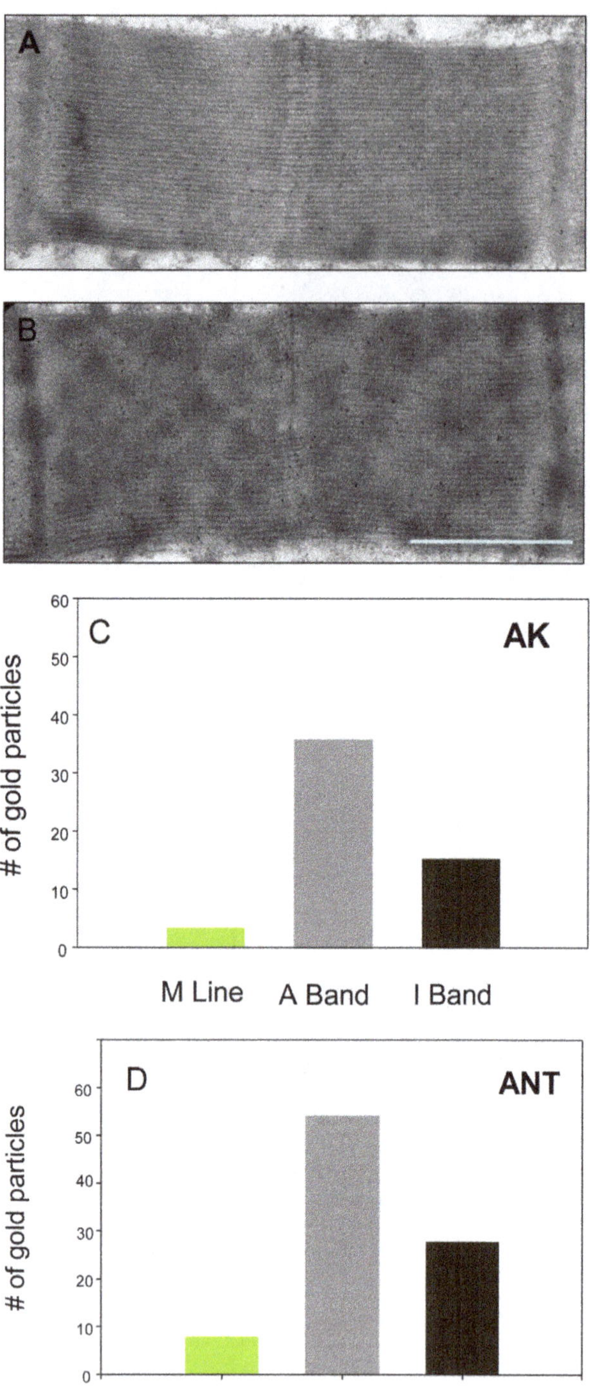

Figure 6. Sarcomeric distribution of AK ((A) and (C)) and ANT ((B) and (D)). The two proteins have similar sarcomeric distributions. Bar = 1 μm.

Drosophila sesB encodes two transcripts of ~1.2 KB and 1.6 KB that differ in their 3' UTR [14]. The functional role of the different 3' UTRs is not known but one possibility is that cis elements within the UTR determine the intracellular distribution of ANT. *sesB* transcripts, like those of other nuclear genome encoded mitochondrial proteins, rely on 3' UTR cis elements for their targeting to polysomes associated with the outer mitochondrial membrane [36]. The first 150 nucleotides of the 3' UTRs are common to both *sesB* transcripts; the 1.6 KB transcript has an additional 440 nucleotides downstream of the shared region [14]. The shared region contains a sequence element (uguaaaua) that has been identified as a mitochondrial localization motif [37]. As shown here in Schneider S2 cultured cells, both transcripts target GFP-*sesB* to the mitochondria. Preliminary secondary structure analysis revealed that the mitochondria localization motif in the 1.6 KB transcript may form base pairs with other internal sequences and therefore would not be accessible to binding proteins involved in mitochondrial targeting. Sequence motifs within the additional 440 nucleotide of the 1.6 KB transcript 3' UTR may override the mitochondrial localization signal. Of particular note is the cis element "acaccc" motif that has been identified as a binding site for MLP1 (the human homologue of *Drosophila* Muscleblind protein) [38]. This motif is required for the specific targeting of integrin alpha3 protein to distinct cytoplasmic loci [38]. Muscleblind in *Drosophila* is involved in alternative RNA splicing and is also required for terminal differentiation of muscle [39]. Further experiments will be needed to determine if sequences within the 3'UTR of the 1.6 KB transcript are responsible for an extra-mitochondrial ANT.

The functional coupling between ANT and CK in vertebrate muscles has been the subject of many studies (e.g., [10,22]. By comparison, little is known about the putative interaction between ANT and AK in non-vertebrate muscle [12]. Available evidence suggest that insect flight muscle lacks a mitochondrial AK [12,13]. However, a mitochondrial AK has been identified in the cardiac muscle of the horseshoe crab and the blue crab [40]. While a direct interaction between AK and ANT has not been demonstrated, the localization of mitochondrial AK to the intermembrane space and the outer region of the inner membrane in the crab would be conducive for such interaction to occur.

The AK/AP phosphagen system is likely to play similar physiological roles as the CK/CP phosphagen system, namely: 1) a spatio-temporal "buffering" role for adenine nucleotides, fulfilled by the cytoplasmic enzyme isoform, and 2) a '"shuttle" role for energy transfer from sites of energy production (mitochondria) to sites of energy consumption (e.g., myofibril) [23,41]. The shuttle function relies on the presence of a mitochondrial isoform at the production end and a myofibril isoform at the utilization end. Fast skeletal muscle express the cytosolic CK isoform almost exclusively, while slow skeletal muscle and heart ventricular muscle express higher proportions of mitochondrial and myofibrillar CK isoforms [23]. The functional organization of the AK/AP system in *Drosophila* IFM is not known, but given the highly oxidative nature of this muscle and the physical proximity of mitochondria to myofibrils, one would expect the system would more likely be analogous to that of vertebrate slow and cardiac muscle than to the one in skeletal muscle. However, we have confirmed the results of Wyss *et al.* [13] that *Drosophila* IFM lacks a mitochondrial AK, consistent with results of others who have found insect flight muscle lack mitochondrial AK [42]. It remains possible that a mitochondrial AK exists (as originally reported, [43]) that is not recognized by the antibodies used in this study and by Wyss *et al.* While only one AK gene has been identified in the *Drosophila* genome, this gene is predicted to encode six different polypeptides (flybase.org). Based on microfractionation studies of IFM fibers, Wyss *et al.* concluded that most of the AK in *Drosophila* IFM is cytosolic [13]. In contrast, Lang *et al.* identified AK associated with the Z line of washed myofibrils [44]. Our immunoelectron microscopy results show that most of the AK is associated with the sarcomeric A band, with no detectable distribution in the cytosol. These seemingly contradictory results can be explained if the association of AK with the myofibril is weak and fails to endure the fractionation and isolation procedures used in prior investigations. Coincidently, glycerination of *Lethocerus* IFM fibers has been shown to cause migration of zeelins from the A band to the Z band [45]. We speculate that the myofibril isolation procedure used by Lang *et al.* may have, as in the case for zeelins, lead to a redistribution of AK from the A band, as detected in this study, to the Z band given that our immunolocalization studies were conducted on IFM fixed within the intact thorax immediately following bisection. The redistribution is not unexpected assuming the interaction between AK and the myofibril is purely electrostatic, as has been shown for vertebrate CK and the sarcomeric M line [46].

The absence of a mitochondrial AK isoform would indicate that the organization of the AK/AP system in insect IFM is fundamentally different from the analogous CK/CP system in vertebrate muscle. Unlike vertebrate skeletal muscle, insect flight muscle is regarded to be functionally independent of phosphagens as evidenced by the low levels of AP in resting muscle and by low AK activity [47,48]. Nevertheless, the AK reaction plays a small role in providing metabolic capacitance during sustained flight [11,49,50]. Additionally, myofibrillar AK may be involved in phosphate (Pi) removal as increasing levels of Pi decreases work output at low ATP concentrations [4].

There is an interesting parallel between the situation in IFM and that of mice that are deficient for CK. In gastrocnemius (fast skeletal) muscle of these mice, mitochondria are more abundant than in the muscle of normal mice and are relocated toward the A band, in an apparent attempt to minimize the diffusion distance of nucleotides as a way to overcome the absence of mitochondrial and cytosolic CK [51]. Cellular remodeling presumably to enhance direct nucleotide channeling between mitochondria and organelles also has been observed in the cardiac cells of CK deficient mice [23]. We suspect that the close apposition of mitochondria and myofibrils in the IFM compensates for the lack of a cytosolic buffering phosphagen system and a mitochondrial AK, as seen in the CK deficient mice.

The IFM phosphagen system evolved in conjunction with other features of the IFM that underlie this muscle's ability for fast and sustained contractions and high power output. Among these are a myosin motor with weak affinity for ATP [4], abundant mitochondria in close apposition to the myofibrils [6], and broad myofibrils with an increased number of force producing thick filaments [5]. These modifications impose certain challenges that the cellular energetic system must overcome in order for the muscle to operate effectively. Principal among these is the large diffusion distance for metabolites in the cell and more specifically, within the confines of the myofibril. Fibers from asynchronous flight muscle have been reported to have a larger cross-sectional area than fibers from synchronous flight muscle [5]. A similar relationship is seen for myofibrils, those from asynchronous muscle have approximately twice the diameter and four times the cross-sectional area as those from synchronous muscle [5]. In crustacean tail muscle, limitations in phosphagen diffusion increase with increases in fiber size [52] and this raises the possibility that asynchronous muscles may face similar limitations. Previous studies have shown that ADP is the most diffusion constrained of all nucleotides [53]. It has been estimated that direct diffusion (i.e., in the absence of cytoplasmic CK) of ADP and ATP would compromise metabolic capacity for diffusion lengths greater than 2 μm [54]. This would be significant in the *Drosophila* IFM whose myofibrils have a diameter of nearly 2 μm. Taking these factors together we consider that diffusion of ATP and ADP alone provides a rudimentary, if any, metabolic capacitance in the IFM and therefore propose the existence of active mechanisms in these fast and powerful muscles.

The weak affinity of IFM myosin for MgATP implies that a very high concentration of this nucleotide must be present in the IFM, in the vicinity of the myosin motors. Based on skinned fiber mechanics experiments, a MgATP concentration of ~15 mM is required to attain maximal frequency of maximum work [4]. Given the limited sarcoplasmic reticulum in the IFM, and therefore limited requirement to feed SERCA pumps, most of the ATP can be channeled to the myofibril. We have no information on how ATP/ADP and AP/arginine are distributed throughout the IFM fiber but given the complexity of the cytoplasm and the high organelle occupancy, the distribution is unlikely to be homogeneous. The existence of compartmentalized pools and direct channeling of nucleotides has been demonstrated in mouse heart [55]. More recently, the existence of structural barriers that restrict the diffusion of nucleotides and metabolites in heart muscle fibers has been supported by three dimensional finite element models [56]. The large surface area and relatively large radius of IFM myofibrils could impose a significant diffusion constraint on ATP, ADP and arginine phosphate.

A myofibril-associated ANT could assist in creating and maintaining differences in ATP and ADP concentrations across compartmentalized microdomains. While mitochondria are abundant and in close proximity to the myofibril, their ATP contribution may be spatially restricted to the periphery and other means of delivering ATP deep within the confines of the myofibril may be required. In addition to weak MgATP affinity, other kinetic properties of IFM myosin appear to dictate the need for a highly efficient mechanism for nucleotide transport within the myofibril. Unlike skeletal muscle, the rate-limiting step of IFM myosin is phosphate release, *i.e.*, the rate of ADP release is very fast [4]. ADP is characterized by an inherently slow diffusion rate that in cardiac muscle is further restricted by intracellular structures and biochemical processes [57]. Thus the need to prevent pools of high ADP concentration to build up in the vicinity of the myosin motors may be as important as maintaining high ATP concentrations. While myosin motor driven fluctuations may assist in diffusing ADP, the presence of ANT may facilitate directed transfer to the mitochondria that in turn stimulates mitochondrial oxidative phosphorylation. Our observation that ANT and AK have similar cellular and myofibrillar distribution patterns raises the possibility that these proteins might directly and more efficiently buffer ADP and provide sufficient ATP required for sustaining muscle contractions at high frequencies. Further experiments will be needed to establish if and how ANT and AK, operating separately or together, contribute to nucleotide and phosphagen diffusion in the IFM. A direct transfer of nucleotides between ANT and AK would be analogous to that proposed for ANT and CK [22].

4. Conclusions

In this study we used two complementary approaches, one direct and one indirect, to examine the intracellular distribution of ANT in *Drosophila* IFM. We created a transgenic line expressing a GFP-*sesB* fusion protein and showed, using anti-GFP specific antibodies and quantitative immunoelectron microscopy, that the fusion protein is equally distributed between mitochondria and myofibrils. The distribution of the fusion protein is similar to that of the endogenous ANT observed using anti-ANT specific antibodies. The novel finding of a myofibril localized ANT is consistent with the results of a prior study that found ANT to be present among the proteins of an IFM myofibrillar preparation [58]. Furthermore, we showed that AK is primarily localized to the sarcomeric

A band and is excluded from the mitochondria.

In summary, the quest for bigger myofibrils, necessary to achieve the high levels of power output required for flight, may have propelled the evolution of energy buffering systems within the confines of the myofibril. Our results add to the growing body of evidence that phosphagen-based energy systems are highly adaptable and remarkably versatile in meeting the cellular energetic demands of specific muscle types. These findings provide a new perspective on the versatility of cellular energy transfer pathways as their adaptation to meet the energetic demands of an ultrafast, strictly oxidative and high power muscle. More studies will be needed to fully understand how energy production and utilization are functionally integrated in insect flight muscle.

5. Methods

5.1. Fly Stocks

Drosophila melanogaster w^{1118}, an otherwise wild type strain except for a *white* eye mutation, was obtained from the Bloomington Stock Center and was used as host for generation of transgenic lines and as a control line. *w/w*; *T*(2;3) ap^{xa}/ *Cyo;TM*3 *Sb e* strain was used for mapping the insertion in the transgenic lines. *Act*88F-GFP line [21], obtained from John Sparrow, was used as control line for GFP expression while Oregon R was used as the wild-type strain.

5.2. RT-PCR and Cloning

In order to generate an N-terminal GFP tag, an eGFP clone (Clonetech), was modified by Polymerase Chain Reaction (PCR) to remove the stop codon. The forward primer 5' att**accatggtg**agcaagggc 3' with a *NcoI* restriction site (shown in bold) and a reverse primer 5' attatag**cggccgc**cttgctt 3' with a *Not I* was used to amplify the coding region of the eGFP clone. The PCR product was cut with *NcoI* and *Not I* and cloned into an eGFP vector that was cut with the same combination of enzymes.

Total mRNA was isolated from wild type late stage pupae using Trizol (Invitrogen) and quantified using a Nanodrop spectrophotometer (Thermo Scientific). RT-PCR of *sesB* mRNA was performed using anchored oligo dTs [59]. The oligo dT primers 5' **actagt**-t(16)-***t*** 3', 5' **actagt**-t(16)-***a***3', 5' **actagt**-t(16)-***g***3', 5' **actcgt**-t(16)-***c***3' were used for first strand synthesis. These primers have a single nucleotide 3' anchor (bold italics) preceded by oligo dT and the *Spe I* sequence. First strand synthesis was carried out using superscript II RNase H^- Reverse transcriptase (Invitrogen) and each of the anchored oligo dTs. PCR was carried out using a forward primer that was specific to the *sesB* sequence at its 3' end and had a *Not I* restriction site at the 5' end. Using this combination of the forward primer and the anchored oligo dT, the *sesB* cDNAs that includes its 3' UTR was amplified and cloned into the modified GFP vector, immediately down stream of the GFP coding sequence using the *Not I and Spe I* restriction site so that GFP and *sesB* cDNAs are in a continuous reading frame. The GFP-*sesB* cDNA with *Sma I* and *Spe I* ends was subcloned into a modified pCaSpeR-hs, P-element transformation vector digested with *Hpa I* and *Xba I*. The pCaSpeR-hs vector was modified by *BamHI* digestion and the cut vector was religated so as to eliminate the hsp70 3' UTR. All ligations were performed using the clonables ligation/transformation kit (Novagen, Madison, WI).

5.3. Generation of Transgenic Lines and Heat Shock Treatment of Flies

The pCaSpeR-hs GFP-*sesB* vector was amplified by transformation into chemically competent *Top*10 cells (Invitrogen) and purified using the Qiagen endotoxin free Maxi prep kit (Qiagen, Valencia, CA, USA). The DNA was checked for purity and approximately 2 mg of DNA was shipped to Microinjection Services Inc (Sudbury, MA) for transformation. w^{1118} strain transformants were identified by orange eye color in the G1 generation. Homozygous strains were generated by single pair mating of siblings with the darkest eye color until no *white* eye progeny were produced. Each transgene was mapped to its resident chromosome using *w/w*; *T*(2;3) ap^{xa}/*Cyo*; *TM*3 *Sb e* strain by standard crossing techniques. Five independent lines were generated.

To induce expression of the transgene, early larvae from homozygous transgenic lines were heat shocked by placing vials in a 37°C water bath for 15 minutes followed by a five-minute incubation at room temperature. This was repeated at least 4 times every day until late pupae stages.

5.4. Preparation of Samples for Gel Electrophoresis and Western Blot Analysis

To prepare samples for gel electrophoresis, flies were placed in acetone overnight at −20°C. The next morning the acetone was removed and the flies spun for 1 hour in a speed vac after which the thorax was separated from the head, abdomen, wings and legs. The thorax was homogenized in Laemmli sample buffer with 8 $mol \cdot L^{-1}$ urea and protease inhibitors [60]. Electrophoresis on 12% polyacrylamide SDS denaturing gels were performed as described previously [60] using the discontinuous buffer system [61]. The gels were stained overnight in sypro stain (BIORAD) and viewed using an UV light box and photographed with a digital camera (Kodak Digital Science EDAS 120). For western blots, samples were separated on a 12% SDS gel and transferred onto nitrocellulose as described previously [60]. Membranes were blocked in blocking buffer (LICOR Biosciences) for one hour after which they were incubated with one of the following primary antibodies diluted in 1:1 of blocking buffer and TBST; 1) Anti-bovine cardiac ANT antibodies (Rabbit polyclonal, 1:2000), provided by Gerard Brandolin; 2) Anti-GFP polyclonal antibodies from Molecular Probes (Rabbit polyclonal, 1:2000); 3) Anti-*Drosophila* βATP synthase antibodies (Rabbit polyclonal, 1:2000) provided by Rafael Garesse. Blots were incubated in primary antibody for one hour at room temperature followed by three washes for 15 minutes each in TBST (Tris Buffer Saline 0.1% Tween). The blots were incubated in secondary anti-rabbit or anti-mouse Alexafluor 698 (Molecular Probes, Eugene, OR, USA) for one hour and then were washed in TBST and scanned on an Odyssey fluorescent scanner (LI-COR Bioscience, Lincoln, NE, USA). The images were converted from color to gray scale in Adobe Photoshop.

5.5. Insect Cell Culture, Transient Transfections and Heat Shock Treatment of Cells

Drosophila Schneider (S2) cells were cultured in 60 × 15 mm style Falcon culture plates (Becton Dickinson, Franklin Lakes, NJ, USA) in Shield and Sank–M3 Medium until 70% to 80% confluence was attained. S2 cells were grown in a 25˚C incubator. For transfection, 10 µg of pCaSpeR hs GFP-*sesB*1.6 K or pCaSpeR hs GFP-1.2 K were used. Transfections were performed using the Cellfectin reagent (Invitrogen,). Heat shock was administered by leaving cultures in a water bath at 37˚C for 30 minutes after which cells were moved back into 25˚C and allowed to recover for two hours before being processed for confocal microscopy.

5.6. Confocal Microscopy of Drosophila Schneider Cells

Cells were spun down at 4000 RPM for 5 minutes, permeabilized with 0.2% triton X-100 in 0.01 M phosphate buffer PH 7.4 for 5 minutes at room temperature, and fixed with 3.7% formaldehyde in 0.01 M phosphate buffer pH 7.4 for 15 minutes at room temperature. They were washed in 0.01 M PBS three times to remove formaldehyde and triton X-100. Coverslips overlayed with Cell Tak (BD Biosciences) in 3 M $NaHCO_3$ were prepared, cells were applied to the cover slip and incubated at room temperature for 3 minutes. The mitochondria were stained with 25 nM Mitotracker Red CMX Ros (M7512, Molecular Probes). Subsequently, cells were stained with a primary anti-GFP polyclonal antibody (Molecular Probes, 1:250 dilution) for one hour at room temperature followed by secondary goat anti- rabbit IgG (H + L), Alexafluor 488. All light sensitive reactions were performed in a light proof chamber. Cells were finally washed in PBS for one minute to remove Mitotracker and antibodies. A drop of mounting medium (Vecta shield, Vector Laboratories, Inc, Burlingame, CA 94010) was placed on the coverslip and it was inverted on to a slide. Slides were viewed using the BioRad 1024 confocal laser scanning microscope. Lasersharp 2000 version 5 software (originally from Bio-Rad Laboratories, Hercules, CA, now part of Carl Zeiss Micro imaging, Thornwood, NY) was used for image analysis and processing.

5.7. Immunogold Transmission Electron Microscopy and Image Analysis

Drosophila half thoraces were fixed for one hour at 4˚C in 0.1% glutaraldehyde and 3.0% paraformaldehyde in 0.01M Phosphate Buffered Saline (PBS), pH 7.4. After rinsing in PBS 3 × 5 min each, free aldehyde groups were quenched by washing the tissue 2 × 30 min at 4˚C in 0.05 M ammonium chloride in PBS, followed by storage overnight at 4˚C in PBS. The next day, the tissue was dehydrated in ethanol at progressively lower temperatures and embedded in Lowicryl K4M at −35˚C. Polymerization of the blocks was achieved by exposure to ultraviolet light for 24 - 48 hrs at −35˚C, followed by 48 - 72 hrs at room temperature under UV light. Semi thin sections (1 µm) were cut with a glass knive on a Reichert ultracut microtome, stained with methylene blue–azure II, and evaluated for areas of interest. Ultrathin sections (60 - 80 nm) were cut with a diamond knife, retrieved onto Carbon-Formvar coated 150 mesh nickel grids. Immunohistochemistry experiments were performed as follows: The grids were blocked with 0.5% ovalbumin solution for 20 minutes at room temperature. The following primary antibodies were used. Anti-ANT polyclonal (1:500), anti-*Drosophila* βATP synthase polyclonal (1:250), anti-*Drosophila* flightin polyclonal (1:30) [62], anti-*Drosophila* arginine kinase polyclonal (1:200) (provided by Theo Wallimann), and anti-GFP polyclonal (1:10 dilution). As a negative control, rabbit pre-immune serum was used since all primary antibodies were raised in rabbit. Sections were incubated in primary antibodies overnight at 4˚C and were washed 3 times for 5 minutes each in 1% BSA in PBS. Protein A conjugated to 10 nm gold particles (pAg_{10}) (diluted in PBS, Optical Density of 0.06 @ 525 nm) was used for detection of primary antibodies. Sections were incubated in pAg_{10} diluted in PBS for 60 minutes at room temperature followed by washes in PBS and rinses in distilled water. Sections were then contrasted with uranyl acetate (3% in deionized H_2O) and lead citrate, and examined with a JEOL 1210 TEM (JEOL USA, Inc., Peabody, Ma) operating at 60 kV. The negatives were scanned with an Epson scanner and the quantitation of gold spots using MetaMorph software (version 6.3 r7) was done as follows. The program was calibrated using standard files that take into account the magnification and size of the image. Random boxes were drawn on the image using the draw tool. Integrated morphometric analysis (IMA) function was used to calculate the area of each box (in μm^2) and the number of gold dots per area. Individual smaller boxes were drawn within the first box such that each covered mitochondria, myofibril, or cytosol. The areas of these smaller boxes and the number of gold dots per area were calculated using the IMA function. All the gold particles counted were normalized to the area and plotted using Sigma plot. Two separate flies were done for each of the treatments (five antibodies and control) and each fly was probed in duplicate, for a total of four grids per treatment. Three frames were randomly photographed from each grid totaling 12 frames per treatment.

5.8. Statistical Analysis

All analyses were performed in MATLAB using non-parametric tests. To compare distributions of gold particles in the different groups, we converted raw data (counts) into percentage data. We then performed Kruskal Wallis tests on the percent data for all groups. Further tests using the RANKSUM function (equivalent to Mann-Whitney U test) were done to compare the distribution of the GFP control to the experimental groups.

6. Authors' Contributions

VKV generated the transgenic lines and carried out the immuno EM studies and protein expression studies. He participated in the design of the study and the analysis of the data, and drafted the manuscript. SSG assisted in the characterization of the transgenic lines. PL provided overall technical assistance and participated in the immuno EM studies. DWM participated in the design and coordination of the study and helped draft the manuscript. JOV conceived of the study and had overall responsibility for its design and coordination. He assisted in data analysis and interpretation and drafting of the manuscript. All authors read and approved the final manuscript.

7. Acknowledgements

We thank Cedric Wesley for advice on the tissue culture and supplying S2 cells, and Teresa Ruiz and Michele Von Turkovich for assistance with electron microscopy. We are grateful to members of the Vigoreaux and Maughan labs for providing useful comments. We are indebted to the following individuals for supplying antibodies: Gerard Brandolin (ANT antibodies), Rafael Garesse (ATP synthase antibody) and David Sullivan (arginine kinase antibodies). Supported by NSF grants MCB-0315865, IOS-0718417 and MCB-1050834 to JOV. VKV was supported in part by a predoctoral fellowship of the Vermont Genetics Network through NIH Grant number 1 P20 RR16462 from the BRIN program of the National Center for Research Resources.

REFERENCES

[1] J. F. Harrison and S P. Roberts, "Flight Respiration and Energetics," *Annual Review of Physiology*, Vol. 62, 2000, pp. 179-205. doi:10.1146/annurev.physiol.62.1.179

[2] A. E. Kammer and B. Heinrich, "Insect Flight Metabolism," In: J. E. Treherne, M. J. Berridge and V. B. Wigglesworth, Eds., *Advances in Insect Physiology*, Vol. 13, Academic Press, London, 1978, pp. 133-228.

[3] T. M. Casey, C. P. Ellington and J. M. Gabriel, "Allometric Scaling of Muscle Performance and Metabolism: Insects," *Advances in Bioscience and Biotechnology*, Vol. 84, 1992, pp. 152-162.

[4] D. M. Swank, V. K. Vishnudas and D. W. Maughan, "An Exceptionally Fast Actomyosin Reaction Powers Insect Flight Muscle," *Proceedings of the National Academy of Sciences of the United States of America*, Vol. 103, No. 46, 2006, pp. 17543-17547. doi:10.1073/pnas.0604972103

[5] R. K. Josephson, J. G. Malamud and D. R. Stokes, "Asynchronous Muscle: A Primer," *The Journal of Experimental Biology*, Vol. 203, Pt. 18, 2000, pp. 2713-2722.

[6] V. Vishnudas and J. O. Vigoreaux, "Sustained High Power Performance: Possible Strategies for Integrating Energy Supply and Demand in Flight Muscles," In: J. O. Vigoreaux, Ed., *Nature's Versatile Engine: Insect Flight Muscle Inside and out*, Springer/Landes Bioscience, New York, 2006, pp. 188-196. doi:10.1007/0-387-31213-7_15

[7] E. Pebay-Peyroula and G. Brandolin, "Nucleotide Exchange in Mitochondria: Insight at a Molecular Level," *Current Opinion in Structural Biology*, Vol. 14, No. 4, 2004, pp. 420-425. doi:10.1016/j.sbi.2004.06.009

[8] A. Dorner and H. P. Schultheiss, "Adenine Nucleotide Translocase in the Focus of Cardiovascular Diseases," *Trends in Cardiovascular Medicine*, Vol. 17, No. 8, 2007, pp. 284-290. doi:10.1016/j.tcm.2007.10.001

[9] J. D. Sharer, J. F. Shern, H. Van Valkenburgh and D. C. Wallace and R. A. Kahn, "ARL2 and BART Enter Mitochondria and Bind the Adenine Nucleotide Transporter," *Molecular Biology of the Cell*, Vol. 13, No. 1, 2002, pp. 71-83. doi:10.1091/mbc.01-05-0245

[10] K. Guerrero, B. Wuyam, P. Mezin, I. Vivodtzev, M. Vendelin, J. C. Borel, R. Hacini, O. Chavanon, S. Imbeaud, V. Saks and C. Pison, "Functional Coupling of Adenine Nucleotide Translocase and Mitochondrial Creatine Kinase Is Enhanced after Exercise Training in Lung Transplant Skeletal Muscle," *American Journal of Physiology—Regulatory, Integrative and Comparative Physiology*, Vol. 289, No. 4, 2005, pp. R1144-R1154. doi:10.1152/ajpregu.00229.2005

[11] H. L. Sweeney, "The Importance of the Creatine Kinase Reaction: The Concept of Metabolic Capacitance," *Medicine & Science in Sports & Exercise*, Vol. 26, No. 1, 1994,

pp. 30-36. doi:10.1249/00005768-199401000-00007

[12] W. R. Ellington, "Evolution and Physiological Roles of Phosphagen Systems," *Annual Review of Physiology*, Vol. 63, 2001, pp. 289-325.

[13] M. Wyss, D. M. Maughan and T. Wallimann, "Re-Evaluation of the Structure and Physiological Function of Guanidino Kinases in Fruitfly (Drosophila), Sea Urchin (*Psammechinus miliaris*) and Man," *Biochemical Journal*, Vol. 309, Pt. 1, 1995, pp. 255-261. doi:10.1146/annurev.physiol.63.1.289

[14] Y. Q. Zhang, J. Roote, S. Brogna, A. W. Davis, D. A. Barbash, D. Nash and M. Ashburner, "Stress Sensitive B Encodes an Adenine Nucleotide Translocase in *Drosophila melanogaster*," *Genetics*, Vol. 153, No. 2, 1999, pp. 891-903.

[15] M. D. Brand, J. L. Pakay, A. Ocloo, J. Kokoszka, D. C. Wallace, P. S. Brookes and E. J. Cornwall, "The Basal Proton Conductance of Mitochondria Depends on Adenine Nucleotide Translocase Content," *Biochemical Journal*, Vol. 392, Pt. 2, 2005, pp. 353-362. doi:10.1042/BJ20050890

[16] R. Rikhy, M. Ramaswami and K. S. Krishnan, "A Temperature-Sensitive Allele of Drosophila *sesB* Reveals Acute Functions for the Mitochondrial Adenine Nucleotide Translocase in Synaptic Transmission and Dynamin Regulation," *Genetics*, Vol. 165, No. 3, 2003, pp. 1243-1253.

[17] N. Trotta, C. K. Rodesch, T. Fergestad and K. Broadie, "Cellular Bases of Activity-Dependent Paralysis in Drosophila Stress-Sensitive Mutants," *Journal of Neurobiology*, Vol. 60, No. 3, 2004, pp. 328-347. doi:10.1002/neu.20017

[18] Y. Q. Zhang, J. Roote, S. Brogna, A. W. Davis, D. A. Barbash, D. Nash and M. Ashburner, "Stress Sensitive B Encodes an Adenine Nucleotide Translocase in *Drosophila melanogaster*," *Genetics*, Vol. 153, No. 2, 1999, pp. 891-903.

[19] A. Louvi and S. G. Tsitilou, "A cDNA Clone Encoding the ADP/ATP Translocase of *Drosophila melanogaster* Shows a High Degree of Similarity with the Mammalian ADP/ATP Translocases," *Journal of Molecular Evolution*, Vol. 35, No. 1, 1992, pp. 44-50. doi:10.1007/BF00160259

[20] A. Grado, C. Manchado, R. Iglesias, M. Giralt, F. Villarroya, T. Mampel and O. Vinas, "Muscle/Heart Isoform of Mitochondrial Adenine Nucleotide Translocase (ANT1) Is Transiently Expressed during Perinatal Development in Rat Liver," *FEBS Letters*, Vol. 421, No. 3, 1998, pp. 213-216. doi:10.1016/S0014-5793(97)01563-9

[21] P. Barthmaier and E. Fyrberg, "Monitoring Development and Pathology of Drosophila Indirect Flight Muscles Using Green Fluorescent Protein," *Developmental Biology*, Vol. 169, No. 2, 1995, pp. 770-774. doi:10.1006/dbio.1995.1186

[22] M. Vendelin, M. Lemba and V. A. Saks, "Analysis of Functional Coupling: Mitochondrial Creatine Kinase and Adenine Nucleotide Translocase," *Biophysical Journal*, Vol. 87, No. 1, 2004, pp. 696-713. doi:10.1529/biophysj.103.036210

[23] R. Ventura-Clapier, A. Kaasik and V. Veksler, "Structural and Functional Adaptations of Striated Muscles to CK Deficiency," *Molecular and Cellular Biochemistry*, Vol. 256-257, No. 1-2, 2004, pp. 29-41. doi:10.1023/B:MCBI.0000009857.69730.97

[24] D. G. Brdiczka, D. B. Zorov and S. S. Sheu, "Mitochondrial Contact Sites: Their Role in Energy Metabolism and Apoptosis," *Biochimica et Biophysica Acta*, Vol. 1762, No. 2, 2006, pp. 148-163.

[25] A. P. Halestrap and C. Brennerb, "The Adenine Nucleotide Translocase: A Central Component of the Mitochondrial Permeability Transition Pore and Key Player in Cell Death," *Current Medicinal Chemistry*, Vol. 10, No. 16, 2003, pp. 1507-1525. doi:10.2174/0929867033457278

[26] S. M. Claypool, Y. Oktay, P. Boontheung, J. A. Loo and C. M. Koehler, "Cardiolipin Defines the Interactome of the Major ADP/ATP Carrier Protein of the Mitochondrial Inner Membrane," *The Journal of Cell Biology*, Vol. 182, No. 5, 2008, pp. 937-950. doi:10.2174/0929867033457278

[27] R. Liu, A. L. Strom, J. Zhai, J. Gal, S. Bao, W. Gong and H. Zhu, "Enzymatically Inactive Adenylate Kinase 4 Interacts with Mitochondrial ADP/ATP Translocase," *The International Journal of Biochemistry & Cell Biology*, Vol. 41, No. 6, 2009, pp. 1371-1380. doi:10.1016/j.biocel.2008.12.002

[28] F. Verrier, A. Deniaud, M. Lebras, D. Metivier, G. Kroemer, B. Mignotte, G. Jan and C. Brenner, "Dynamic Evolution of the Adenine Nucleotide Translocase Interactome during Chemotherapy-Induced Apoptosis," *Oncogene*, Vol. 23, No. 49, 2004, 8049-8064. doi:10.1038/sj.onc.1208001

[29] G. Csordas, C. Renken, P. Varnai, L. Walter, D. Weaver, K. F. Buttle, T. Balla, C. A. Mannella and G. Hajnoczky, "Structural and Functional Features and Significance of the Physical Linkage between ER and Mitochondria," *The Journal of Cell Biology*, Vol. 174, No. 7, 2006, pp. 915-921. doi:10.1083/jcb.200604016

[30] R. Rizzuto, P. Pinton, W. Carrington, F. S. Fay, K. E. Fogarty, L. M. Lifshitz, R. A. Tuft and T. Pozzan, "Close Contacts with the Endoplasmic Reticulum as Determinants of Mitochondrial Ca2+ Responses," *Science*, Vol. 280, No. 5370, 1998, pp. 1763-1766. doi:10.1126/science.280.5370.1763

[31] J. Dai, K. H. Kuo, J. M. Leo, C. van Breemen and C. H. Lee, "Rearrangement of the Close Contact between the Mitochondria and the Sarcoplasmic Reticulum in Airway Smooth Muscle," *Cell Calcium*, Vol. 37, No. 4, 2005, pp. 333-340. doi:10.1016/j.ceca.2004.12.002

[32] K. C. Rowland, N. K. Irby and G. A. Spirou, "Specialized Synapse-Associated Structures within the Calyx of Held," *The Journal of Neuroscience*, Vol. 20, No. 24, 2000, pp. 9135-9144.

[33] X. Jiang and X. Wang, "Cytochrome C-Mediated Apoptosis," *Annual Review of Biochemistry*, Vol. 73, 2004, pp. 87-106. doi:10.1146/annurev.biochem.73.011303.073706

[34] L. O. Martinez, S. Jacquet, J. P. Esteve, C. Rolland, E. Cabezon, E. Champagne, T. Pineau, V. Georgeaud, J. E. Walker, F. Terce, *et al.*, "Ectopic Beta-Chain of ATP

Synthase Is An Apolipoprotein A-I Receptor in Hepatic HDL Endocytosis," *Nature*, Vol. 421, No. 6918, 2003, pp. 75-79. doi:10.1038/nature01250

[35] R. Buettner, G. Papoutsoglou, E. Scemes, D. C. Spray and R. Dermietzel, "Evidence for Secretory Pathway Localization of a Voltage-Dependent Anion Channel Isoform," *Proceedings of the National Academy of Sciences of the United States of America*, Vol. 97, No. 7, 2000, pp. 3201-3206. doi:10.1073/pnas.97.7.3201

[36] J. Sylvestre, A. Margeot, C. Jacq, G. Dujardin and M. Corral-Debrinski, "The Role of the 3' Untranslated Region in mRNA Sorting to the Vicinity of Mitochondria Is Conserved from Yeast to Human Cells," *Molecular Biology of the Cell*, Vol. 14, No. 9, 2003, pp. 3848-3856. doi:10.1091/mbc.E03-02-0074

[37] R. Shalgi, M. Lapidot, R. Shamir and Y. Pilpel, "A Catalog of Stability-Associated Sequence Elements in 3' UTRs of Yeast mRNAs," *Genome Biology*, Vol. 6, 2005, p. R86. doi:10.1186/gb-2005-6-10-r86

[38] Y. Adereth, V. Dammai, N. Kose, R. Li and T. Hsu, "RNA-Dependent Integrin Alpha3 Protein Localization Regulated by the Muscleblind-Like Protein MLP1," *Nature Cell Biology*, Vol. 7, No. 12, 2005, pp. 1240-1247.

[39] L. Machuca-Tzili, H. Thorpe, T. E. Robinson, C. Sewry and J. D. Brook, "Flies Deficient in Muscleblind Protein Model Features of Myotonic Dystrophy with Altered Splice Forms of Z-Band-Associated Transcripts," *Human Genetics*, Vol. 120, No. 4, 2006, pp. 487-499. doi:10.1007/s00439-006-0228-8

[40] A. J. Pineda and W. R. Ellington, "Immunogold Transmission Electron Microscopic Localization of Arginine Kinase in Arthropod Mitochondria," *Journal of Experimental Zoology*, Vol. 281, No. 2, 1998, pp. 73-79. doi:10.1002/(SICI)1097-010X(19980601)281:2<73::AID-JEZ1>3.0.CO;2-7

[41] T. Wallimann, M. Wyss, D. Brdiczka, K. Nicolay and H. M. Eppenberger, "Intracellular Compartmentation, Structure and Function of Creatine Kinase Isoenzymes in Tissues with High and Fluctuating Energy Demands: The 'Phosphocreatine Circuit' for Cellular Energy Homeostasis," *The Biochemical Journal*, Vol. 281, Pt. 1, 1992, pp. 21-40.

[42] W. R. Ellington and A. C. Hines, "Mitochondrial Activities of Phosphagen Kinases Are Not widely Distributed in the Invertebrates," *Biological Bulletin*, Vol. 180, No. 3, 1991, pp. 505-507. doi:10.2307/1542352

[43] L. R. Munneke and G. E. Collier, "Cytoplasmic and Mitochondrial Arginine Kinases in *Drosophila*: Evidence for a Single Gene," *Biochemical Genetics*, Vol. 26, No. 1-2, 1988, pp. 131-141. doi:10.1007/BF00555494

[44] A. B. Lang, C. Wyss and H. M. Eppenberger, "Localization of Arginine Kinase in Muscles Fibres of *Drosophila melanogaster*," *Journal of Muscle Research & Cell Motility*, Vol. 1, No. 2, 1980, pp. 147-161. doi:10.1007/BF00711796

[45] C. Ferguson, A. Lakey, A. Hutchings, G. W. Butcher, K. R. Leonard and B. Bullard, "Cytoskeletal Proteins of Insect Muscle: Location of Zeelins in Lethocerus Flight and Leg Muscle," *Journal of Cell Science*, Vol. 107, Pt. 5, 1994, pp. 1115-1129.

[46] T. Hornemann, M. Stolz and T. Wallimann, "Isoenzyme-Specific Interaction of Muscle-Type Creatine Kinase with the Sarcomeric M-Line Is Mediated by Nh$_2$-Terminal Lysine Charge-Clamps," *The Journal of Cell Biology*, Vol. 149, 2000, pp. 1225-1234. doi:10.1083/jcb.149.6.1225

[47] T. Wallimann, M. Wyss, D. Brdiczka, K. Nicolay and H. M. Eppenberger, "Intracellular Compartmentation, Structure and Function of Creatine Kinase Isoenzymes in Tissues with High and Fluctuating Energy Demands: The 'Phosphocreatine Circuit' for Cellular Energy Homeostasis," *Biochemical Journal*, Vol. 281, Pt. 1, 1992, pp. 21-40.

[48] P. W. Hochachka, "Muscles as Molecular and Metabolic Machines," CRC Press, Boca Raton, 1994.

[49] B. Sacktor, "Utilization of Fuels by Muscle," In: D. J. Candy and B. A. Kilby, Eds., *Insect Biochemistry and Function*, Chapman and Hall, London, 1975, pp. 1-81. doi:10.1007/978-94-009-5853-1_1

[50] B. Saktor, "Utilization of Fuels by Muscle," Chapman and Hall, London, 1975.

[51] M. Novotova, M. Pavlovicova, V. I. Veksler, R. Ventura-Clapier and I. Zahradnik, "Ultrastructural Remodeling of Fast Skeletal Muscle Fibers Induced by Invalidation of Creatine Kinase," *American Journal of Physiology Cell Physiology*, Vol. 291, No. 6, 2006, pp. C1279-C1285. doi:10.1152/ajpcell.00114.2006

[52] A. G. Jimenez, B. R. Locke and S. T. Kinsey, "The Influence of Oxygen and High-Energy Phosphate Diffusion on Metabolic Scaling in Three Species of Tail-Flipping Crustaceans," *The Journal of Experimental Biology*, Vol. 211, 2008, pp. 3214-3225. doi:10.1242/jeb.020677

[53] K. Yoshizaki, H. Watari and G. K. Radda, "Role of Phosphocreatine in Energy Transport in Skeletal Muscle of Bullfrog Studied by ^{31}P-NMR," *Biochimica et Biophysica Acta*, Vol. 1051, No. 2, 1990, pp. 144-150. doi:10.1016/0167-4889(90)90186-H

[54] R. A. de Graaf, A. van Kranenburg and K. Nicolay, "*In Vivo* ^{31}P-NMR Diffusion Spectroscopy of ATP and Phosphocreatine in Rat Skeletal Muscle," *Biophysical Journal*, Vol. 78, No. 4, 2000, pp. 1657-1664. doi:10.1016/S0006-3495(00)76717-8

[55] A. Kaasik, V. Veksler, E. Boehm, M. Novotova, A. Minajeva and R. Ventura-Clapier, "Energetic Crosstalk between Organelles: Architectural Integration of Energy Production and Utilization," *Circulation Research*, Vol. 89, 2001, pp. 153-159. doi:10.1161/hh1401.093440

[56] H. R. Ramay and M. Vendelin, "Diffusion Restrictions Surrounding Mitochondria: A Mathematical Model of Heart Muscle Fibers," *Biophysical Journal*, Vol. 97, No. 2, 2009, pp. 443-452. doi:10.1016/j.bpj.2009.04.062

[57] V. Saks, A. Kuznetsov, T. Andrienko, Y. Usson, F. Appaix, K. Guerrero, T. Kaambre, P. Sikk, M. Lemba and M. Vendelin, "Heterogeneity of ADP Diffusion and Regulation of Respiration in Cardiac Cells," *Biophysical Journal*, Vol. 84, No. 5, 2003, pp. 3436-3456.

doi:10.1016/S0006-3495(03)70065-4

[58] K. Ashman, T. Houthaeve, J. Clayton, M. Wilm, A. Podtelejnikov, O. N. Jensen and M. Mann, "The Application of Robotics and Mass Spectrometry to the Characterisation of the *Drosophila melanogaster* Indirect Flight Muscle Proteome," *Letters in Peptide Science*, Vol. 4, No. 2, 1997, pp. 57-65. doi:10.1007/BF02443516

[59] D. K. Nam, S. Lee, G. Zhou, X. Cao, C. Wang, T. Clark, J. Chen, J. D. Rowley and S. M. Wang, "Oligo (dT) Primer Generates a High Frequency of Truncated c DNAs through Internal Ploy(A) Priming during Reverse Transcription," *Proceedings of the National Academy of Sciences of the United States of America*, Vol. 99, No. 9, 2002, pp. 6152-6156. doi:10.1073/pnas.092140899

[60] J. O. Vigoreaux, J. D Saide and M. L. Pardue, "Structurally Different *Drosophila* Striated Muscles Utilize Distinct Variants of Z-Band-Associated Proteins," *Journal of Muscle Research & Cell Motility*, Vol. 12, No. 4, 1991, pp. 340-354. doi:10.1007/BF01738589

[61] U. K. Laemmli, "Cleavage of Structural Proteins during the Assembly of the Head of Bacteriophage T4," *Nature*, Vol. 277, 1970, pp. 680-685. doi:10.1038/227680a0

[62] G. Ayer and J. O. Vigoreaux, "Flightin Is a Myosin Rod Binding Protein," *Cell Biochemistry and Biophysics*, Vol. 38, No. 1, 2003, pp. 41-54. doi:10.1385/CBB:38:1:41

11

Uptake of Cystatin by Melanoma Cells in Culture

Lauren Deady[1], James L. Cox[2*]
[1]Department of Biology, Truman State University, Kirksville, USA
[2]Department of Biochemistry, AT Still University, Kirksville, USA
Email: *jcox@atsu.edu

ABSTRACT

The cystatins are a super family of cysteine protease inhibitors which are ubiquitous in their biologic occurrence. Cystatin C, a type II cystatin, is primarily a secreted protein found in most biological fluids. Besides acting as inhibitors of cathepsin, the cystatins have been found to have some non-inhibitor related functions and multiple physiological roles. Much interest has been generated for the cystatins as metastasis "suppressor-like" proteins, as they have been shown to inhibit metastasis for multiple cancer types. The sites and actions of the cystatins related to tumor suppressor actions are still unclear, however. In this work, we have examined the uptake of cystatin by metastatic melanoma cells in culture. Our results indicate cystatin uptake is mediated by a non-canonical endocytotic pathway in B16 murine melanoma cells.

Keywords: Cystatin; Melanoma; Endocytosis; Fluorescent

1. Introduction

The cystatins are a super family of cysteine protease inhibitors that primarily inhibit cathepsin-type cysteine proteases (family C1) [1]. The cystatins serve to block excess or inadvertent cysteine protease activity that may occur in cells or tissues. Additional physiological roles for the cystatins are expanding and include inhibition of microbial cell invasion and replication, modulation of immune function, and growth factor effect son cells [2-4]. The type I cystatins are primarily intracellular proteins. However, low levels of extra cellular type I cystatins have also been reported and they may therefore influence other cells in the vicinity of the cystatin producing cells [5]. Type II cystatins are primarily secreted inhibitor proteins, which may also have some intracellular actions as uptake by cellular endocytosis has been observed in vivo and in cells in culture [6]. As secreted inhibitors, they may interact with a wide variety of different cell types. For instance, uptake of cystatin C into neuroretinal ganglion cells has recently been reported [7]. Cystatins are able to suppress the metastasis of a variety of cancer cell types when over expressed by cancer cells [8,9]. Although a number of studies have examined changes in cystatin over-expression effects on cancer cells, exact sites of cellular action of the cystatins are still unclear [10-12]. The cellular sites of cystatin action are particularly important as cystatins have recently been shown to not only interact with non-cathepsin proteins but also influence gene expression in cancer cells [13,14]. As many cancers types over-express and secrete cathepsins, which are also found plasma membrane bound and even in nuclei, there are both extra cellular and intracellular sites where cystatins may interfere with cathepsin activity and function in tumor cells [15]. To explore questions related to cystatin sites of action and cellular entry we have examined the process of cystatin uptake by murine melanoma cells in cell culture.

*Corresponding author.

2. Materials and Methods

2.1. Cell Culture and Antibodies

B16F10 melanoma cells were purchased from the ATCC. Cells were cultured at 37˚C at 5% CO_2 in RPMI 1640 medium (Atlanta Biological, Inc.) containing 10% fetal bovine serum and antibiotics (100 I.U./ml penicillin, 100 μg/ml streptomycin, and 0.25 μg/ml amphotericin) (MP Biomedicals). Caveolin-1 polyclonal antibody was purchased from Santa Cruz Biotechnology, Inc. and ezrin mouse monoclonal antibody was from DSHB, University of Iowa, Iowa City, Iowa. Anti-Cystatin C polyclonal antibody was purchased from Thermo Scientific.

2.2. FITC Labeling and Uptake Conditions

Egg white chicken cystatin (Calbiochem) was labeled with FITC for cell uptake experiments. Labeling of cystatin (100 μg) was carried out in 100 μl of 50 mM Na_2CO_3, 0.85% NaCl buffer containing 0.8 mM FITC (Sigma-Aldrich) on a tube rotator for 1 hour at 4oC. Following the reaction, the cystatin solution was dialyzed overnight against 1 liter of phosphate buffered saline (PBS) at 4°C. B16F10 melanoma cells were seeded on acidpre-treated glass coverslips overnight. FITC-labeled cystatin (5 μl (~5 μg cystatin) per 50 μl phenol red-free RPMI media plus 10% FBS) was added directly onto cells on coverslips and incubation followed for 1 hour at 37°C under cell culture conditions. The cells were then washed 3× with PBS and fixed with ice cold methanol for 10 minutes at −20°C. In some experiments, 5 μl Texas Red dextran, 1 mg/ml (Molecular Probes, Eugene, Oregon) was added per 50 μl media at the same time as labeled cystatin with parallel incubation conditions as for labeled cystatin. Inhibitors were obtained from Sigma-Aldrich and used to pretreat the cells as follows: Chlorpromazine pretreatment was for 1 hour at 50 μM before cystatin uptake. Blebbistatin pretreatment was for 1 hour at 100 μM before cystatin uptake. Colchicine and wortmannin were used at 50 μM and 100 nM respectively in 1 hour pretreatment before cystatin uptake experiments.

2.3. Time Course of Cystatin Uptake into Melanoma Cells

Melanoma cells were plated in 96-well plates at 2 × 104 cells/well and allowed to attach overnight. FITC-labeled labeled cystatin was added to the cells at 1:25 dilution in culture medium for various times. Labeled cystatin solution was then removed and cells were washed with PBS. The plate was then read with a BioTek Fl×-800 plate reader for recording of the relative fluorescence of the wells.

2.4. Fluorescent Antibody Staining

The method followed for fluorescent antibody staining was from Chan *et al.* [16]. B16F10 melanoma cells were attached to glass coverslips overnight under cell culture conditions. After cystatin uptake protocol, cells were fixed for 10 minutes with cold methanol at −20°C. Cells were then washed twice with PBS and permeabilized with 0.1% Triton × 100 in PBS for 2 minutes followed by three additional washes with PBS. Permeabilized cells were then blocked with 5% horse serum in PBS for 30 minutes. Primary antibodies (1:100 diluted in blocking solution) were spotted on parafilm in 50 μl drops and the coverslips were inverted onto drops for 30 minutes. Following primary antibody, coverslips were washed 3 times for 5 minutes each with PBS. Fluorescent labeled secondary antibodies were then applied as for primary with incubation for 30 minutes at room temperature. Three PBS washes were again applied, and a brief rinse with deionized water was carried out before mounting the coverslips on glass slides with Prolong Gold mounting medium (Invitrogen).

2.5. Fluorescent and Confocal Microscopy

Fluorescent microscopy was imaged with a Zeiss Axiovert 200 M microscope (63× oil with 1.4 NA Plan apo lens and an Axiocam MRM camera. Images were captured with Axiovision version 4.7 software. Confocal microscopic analysis and image capture was with a Leica SP5 microscope using a 63× oil objective (Leica Microsystems, Heidelberg, Germany).

3. Results and Discussion

Cystatins are synthesized by virtually all eukaryotic cells, and the type II cystatins are secreted cysteine protease inhibitors that are found in all body fluids at various levels [17]. As cystatins are the major cysteine protease inhibitors found in biological fluids, it was of interest to examine the interaction and uptake of cystatin into cells in culture. The levels of cystatin present in serum, for example, might be taken up by endothelial cells and influence the behavior of these cells. The uptake of many different labeled proteins by cells in culture has been examined. However, there are very few reports on the process of cystatin uptake in metastatic cancer cells. Our goal in this study was to probe the process of cystatin uptake in a metastatic cancer cells because overexpression of cystatin in cancer cells or delivery of cystatin to cancer cells *in vivo* has been shown to be effective in blocking metastasis [18]. We wished to probe the pathway of cystatin uptake in cancer cells through the use of inhibitors and antibody labeling of potential cellular uptake components.

The direct uptake of labeled cystatin from cell culture media into metastatic murine B16F10 melanoma cells as a model system was investigated. We focused on B16F10 melanoma because our previous work showed cystatin C could significantly inhibit metastasis in mice when overexpressed in these cells [8]. The type II cystatins are found in all biological fluids and therefore are able to interact and be taken up by cells these fluids come into contact with. Fluorescently-labeled (FITC) chicken cystatin was prepared and added to melanoma cells in culture such that cellular uptake could be imaged with fluorescent microscopy. We added the FITC-labeled chicken egg white cystatin to B16 melanoma cells for a brief period of time (10 minutes) and found pretreatment of the cells with unlabeled cystatin could partially block cystatin uptake by the cells (**Figure 1**). This result suggested a potential receptor mediated uptake of cystatin by the melanoma cells. A time course of cystatin uptake by B16 melanoma cells showed a rapid (minutes) cell associated cystatin fluorescence followed by a lower level fluores cence by 30 minutes (**Figure 2**). We next wanted to look

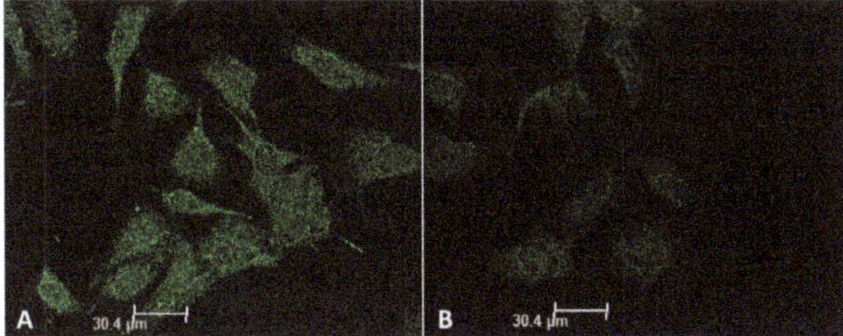

Figure 1. Uptake of FITC-labeled cystatin into B16F10 melanoma cells. Cells were incubated with FITC-labeled cystatin for 10 minutes, washed twice with PBS, fixed with methanol, mounted on coverslips for confocal microscopy. (A) without or (B) with unlabeled cystatin (20 ug/ml) pretreatment of cells.

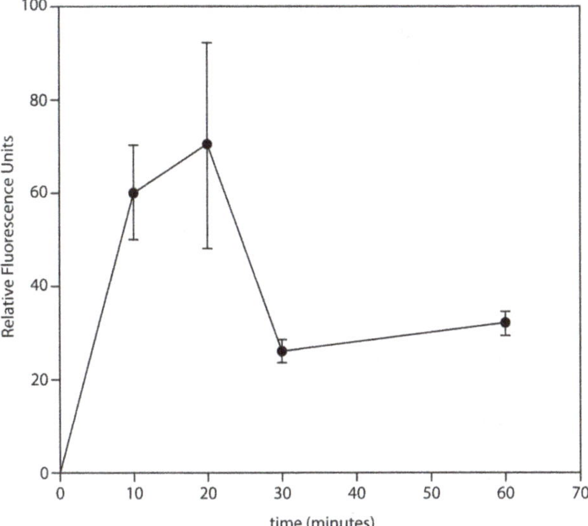

Figure 2. Time course of cystatin uptake into B16 melanoma cells. FITC-labeled cystatin was added to B16F10 melanoma cells for various times. Following timed cystatin exposure, the cells were washed with PBS and fluorescence was measured with a fluorescent plate reader. Relative fluorescence was recorded +/− SEM for each time point.

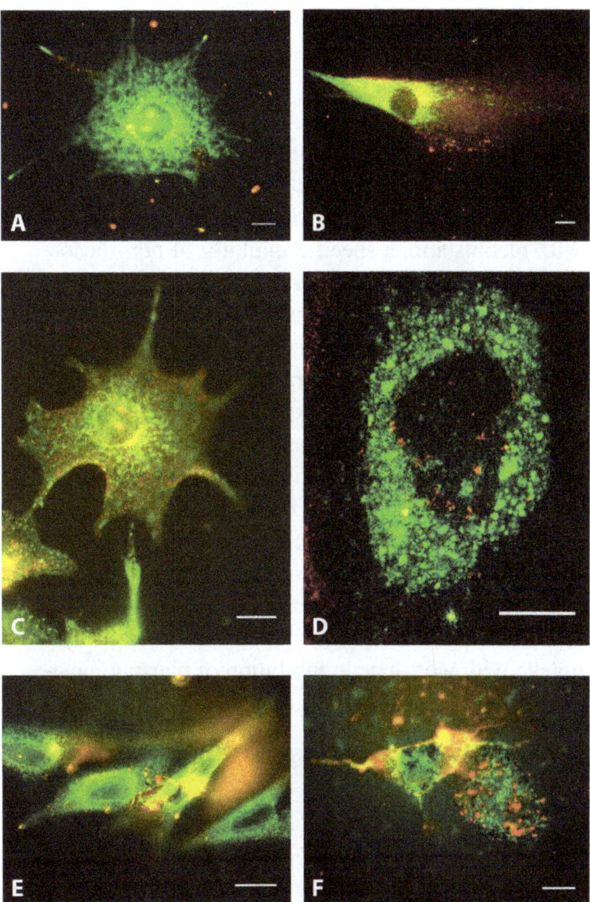

Figure 3. Uptake of FITC-labeled cystatin into B16F10 melanoma cells.Melanoma cells were incubated with both labeled cystatin and Texas red dextran beads for one hour at 37˚C. Cells were then methanol fixed and in some cases stained with fluorescent antibody. ((A), (B)) control uptake of cystatin (green) and dextran (red); (C) cystatin uptake (green) plus caveolin-1 stained (red); (D) same as c, plus additional 2× magnification; (E) control cystatin uptake (green) and dextran (red); (F) chlorpromazine pretreated for 1 hour. Representative images of cystatin uptake experiments performed in triplicate. Scale bar = 15 μm.

at cystatin uptake under different conditions and test certain inhibitors for their actions on cystatin uptake by the melanoma cells. When chicken cystatin was added to B16F10 melanoma cells for 60 minutes under cell culture conditions there was uptake of cystatin into cytoplasmic vesicles which ranged in cellular location from general cytoplasmic to perinuclear (**Figure 3**). Exposure of the cells to Texas Red-labeled dextran together with the FITC-labeled cystatin appeared to label distinct intracellular vesicle populations (**Figure 3**). This observation suggested that each fluorescent label followed a distinct endocytotic pathway and that uptake of cystatin is not simply macropinocytosis as has been described for dextran particle endocytosis. It was also noted that perinuclear regions of the melanoma cells tended to show co-localization of both labeled cystatin and dextran. Since lysosomes are generally perinuclearin location and represent a final destination for many endocytosed proteins, it appears the bulk of the endocytosed cystatin is destined for lysosomes.

A major endocytotic pathway is through caveolae mediated uptake [19]. To examine this possibility, we used a primary antibody to caveolin-1 and a secondary TRITC-labeled antibody to stain cells which had been fixed and permeabilized following one hour exposure to labeled cystatin. We found no co-localization of labeled cystatin vesicles and caveolin-1 in the melanoma cell periphery near the membrane region (**Figure 3**). This finding suggests uptake of cystatin is not mediated by caveosomes or caveolin-1 associated vesicles. To examine potential clathrin mediated uptake of cystatin, we pre-incubated melanoma cells in culture with chlorpromazine, a known inhibitor of clathrin mediated uptake [20]. We found no inhibition of cystatin uptake with chlorpromazine treatment, suggesting cystatin uptake is non-clathrin mediated (**Figure 1**).

The involvement of the cellular cytoskeleton in endocytosis is well known. We tested the involvement of myosin II in the uptake of cystatin by pretreatment of the cells with blebbistatin, a specific inhibitor of non-muscle myosin [21]. No major inhibition of labeled cystatin uptake was observed with blebbistatin pretreatment of the melanoma cells (**Figure 4**). In separate experiments, colchicine, a microtubule inhibitor, was used to pre-treat cells prior to cystatin uptake experiments. No dramatic inhibition of cystatin uptake was noted with colchicine pretreatment (**Figure 4**). We did not expect inhibition with colchicine as other reports found vesicular trafficking of macropinocytosis most closely linked to microtubules. Since our earlier dextran uptake showed macropinocytosis was not involved the lack of colchicine effect was not surprising. We did find partial inhibition of cystatin uptake with wortmannin inhibition of phosphoinositide 3-kinase (PI3K) (**Figure 4**). It was noted that intracellular vesicles were enlarged which might indicate PI3K inhibition interferes with endocytotic vesicle trafficking. PI3K is involved in macrophage and other cell phagocytosis through FcγR receptors [22]. Although of interest, more work will have to be undertaken to look at possible involvement of phagocytosis through a receptor mediated pathway in melanoma cells.

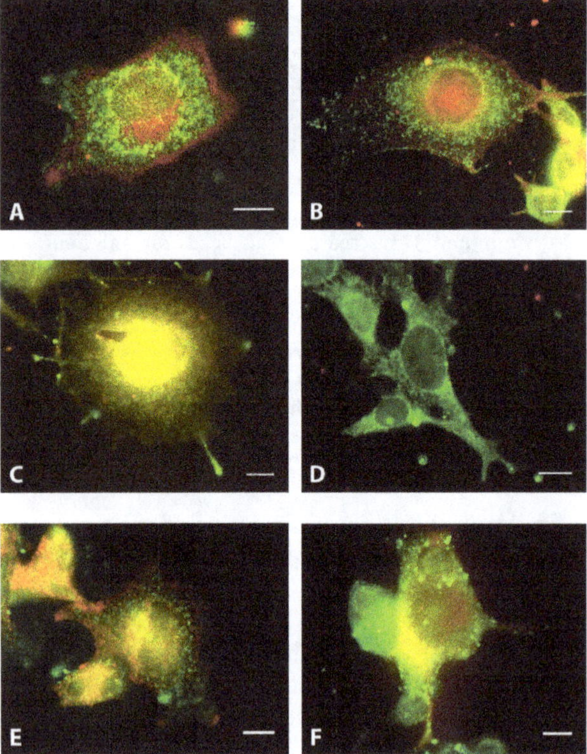

Figure 4. Uptake of FITC-labeled cystatin into B16F10 melanoma cells, inhibitor treatment. Melanoma cells were pre-incubated at 37°C for 1 hour prior to labeled-cystatin uptake experiments. (A) control blebbistatin; (B) blebbistatin treated; (C) control colchicines; (D) colchicine treated; (E) control wortmannin; (F) wortmannin treated. Representative images of cystatin uptake experiments performed in triplicate. Scale bar = 15 μm.

Researchers have described another phagocytic pathway in melanoma and some other cancer cell types through ezrin linked vacuoles [23]. This phagocytic pathway is dependent on the actin cytoskeleton, and vacuoles found in this pathway also express lamp1, which might be expected for lysosomal destination of the vacuole cargo. Our approach to test for ezrin-linked cystatin phagocytosis was to allow melanoma cell uptake of labeled cystatin and then stain the cells for ezrin. We found no clear evidence of ezrin co-localization with fluorescent cystatin which had been taken up (**Figure 5**). Future work should also examine lamp1 and rab5 which are endosomal markers linked with ezrin, particularly in light of the blockade of metastasis seen with ezrin mutants seen in metastatic melanoma [24]. In this work we have focused on a general description of the cystatin uptake process in metastatic melanoma cells. Information on the pathway of cystatin uptake and utilization by cancer cells may lead to new ways to exploit influence of cystatinas an anti-metastatic agent in the multiple cancer types previously studied.

4. Conclusion

Work shown here on cystatin and other recent work had shown that uptake of cystatin by cancer cells is rapid, selective, and saturable, all pointing towards an active, specific receptor mediated process [25]. First, we have demonstrated the bulk of labeled cystatin uptake by melanoma cells is not through macropinocytosis. We have also extended previous work to show that cystatin uptake by melanoma cells is mediated by a non-clathrin and non-caveolin endocytotic pathway. The uptake of cystatin into melanoma cells requires neither microtubule involvement nor myosin II dependence, but the process showed some inhibition with wortmannin treatment,

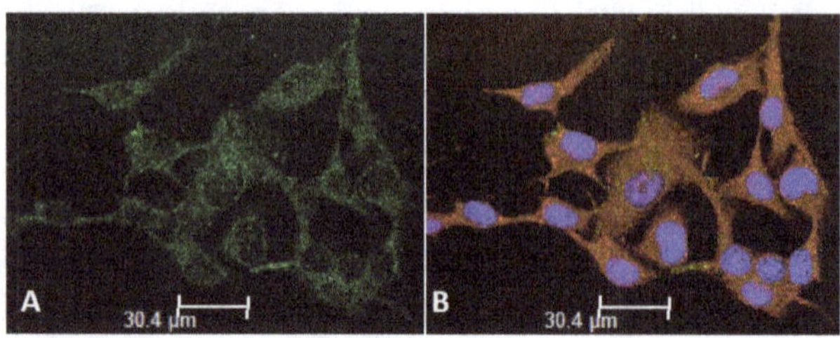

Figure 5. Confocal image of labeled cystatin uptake (green) plus anti-ezrinantibody stained (red) B16 melanoma cells. (A) cystatin uptake, green channel, (B) cystatin uptake and anti-ezrin (red secondary). Nuclear stain (blue), Draq5.

suggesting PI3 kinase involvement. Ezrin staining did not localize with cystatin C uptake vesicles. Future work will define the membrane receptor for cystatin, define further requirements for cystatin uptake, and examine more closely possible involvement of phagocytic pathways for cystatin uptake in metastatic cells.

REFERENCES

[1] V. Turk, V. Stoka and D. Turk, "Cystatins: Biochemical and Structural Properties, and Medical Relevance," *Frontiers in Bioscience: A Journal and Virtual Library*, Vol. 13, 2008, pp. 5406-5420.

[2] M. F. Blankenvoorde, W. van't Hof, E. Walgreen-Weterings, T. J. van Steenbergen, H. S. Brand, E. C. Veerman and A. V. N. Amerongen, "Cystatin and Cystatin-Derived Peptides Have Antibacterial Activity against the Pathogen *Porphyromonas gingivalis*," *Biological Chemistry*, Vol. 379, No. 11, 1998, pp. 1371-1375.

[3] N. Kopitar-Jerala, "The Role of Cystatins in Cells of the Immune System," *FEBS Letters*, Vol. 580, No. 27, 2006, pp. 6295-6301. doi:10.1016/j.febslet.2006.10.055

[4] Q. Sun, "Growth Stimulation of 3T3 Fibroblasts by Cystatin," *Experimental Cell Research*, Vol. 180, No. 1, 1989, pp. 150-160. doi:10.1016/0014-4827(89)90219-X

[5] M. Abrahamson, A. J. Barrett, G. Salvesen and A. Grubb, "Isolation of Six Cysteine Proteinase Inhibitors from Human Urine. Their Physicochemical and Enzyme Kinetic Properties and Concentrations in Biological Fluids," *The Journal of Biological Chemistry*, Vol. 261, No. 24, 1986, pp. 11282-11289.

[6] H. Wallin, M. Bjarnadottir, L. K. Vogel, J. Wassélius, U. Ekström and M. Abrahamson, "Cystatins—Extra- and Intracellular Cysteine Protease Inhibitors: High-Level Secretion and Uptake of Cystatin C in Human Neuroblastoma Cells.," *Biochimie*, Vol. 92, No. 11, 2010, pp. 1625-1634. doi:10.1016/j.biochi.2010.08.011

[7] J. Wassélius, K. Johansson, K. Håkansson, M. Abrahamson and B. Ehinger, "Cystatin C Uptake in the Eye," *Graefe's Archive for Clinical and Experimental Ophthalmology = Albrecht von Graefes Archiv für klinische und experimentelle Ophthalmologie*, Vol. 243, No. 6, 2005, pp. 583-592.

[8] H. Ervin and J. L. Cox, "Late Stage Inhibition of Hematogenous Melanoma Metastasis by Cystatin C Over- Expression," *Cancer Cell International*, Vol. 5, No. 1, 2005, p. 14. doi:10.1186/1475-2867-5-14

[9] W. Li, F. Ding, L. Zhang, Z. Liu, Y. Wu, A. Luo, M. Wu, M. Wang, Q. Zhan and Z. Liu, "Overexpression of Stefin A in Human Esophageal Squamous Cell Carcinoma Cells Inhibits Tumor Cell Growth, Angiogenesis, Invasion, and Metastasis," *Clinical Cancer Research : An Official Journal of the American Association for Cancer Research*, Vol. 11, No. 24, 2005, pp. 8753-8762.

[10] S. D. Konduri, N. Yanamandra, K. Siddique, A. Joseph, D. H. Dinh, W. C. Olivero, M. Gujrati, G. Kouraklis, A. Swaroop, A. P. Kyritsis and J. S. Rao, "Modulation of Cystatin C Expression Impairs the Invasive and Tumorigenic Potential of Human Glioblastoma Cells," *Oncogene*, Vol. 21, No. 57, 2002, pp. 8705-8712. doi:10.1038/sj.onc.1205949

[11] B. Wegiel, T. Jiborn, M. Abrahamson, L. Helczynski, L. Otterbein, J. L. Persson and A. Bjartell, "Cystatin C Is Downregulated in Prostate Cancer and Modulates Invasion of Prostate Cancer Cells via MAPK/Erk and Androgen Receptor Pathways," *PloS One*, Vol. 4, No. 11, 2009, e7953. doi:10.1371/journal.pone.0007953

[12] S. Alvarez-Díaz, N. Valle, J. M. García, C. Peña, J. M. P. Freije, V. Quesada, A. Astudillo, F. Bonilla, C. López-Otín and A. Muñoz, "Cystatin D Is a Candidate Tumor Suppressor Gene Induced by Vitamin D in Human Colon Cancer Cells.," *The Journal of Clinical Investigation*, Vol. 119, No. 8, 2009, pp. 2343-2358. doi:10.1172/JCI37205

[13] J. P. Sokol and W. P. Schiemann, "Cystatin C Antagonizes Transforming Growth Factor Beta Signaling in Normal and Cancer Cells," *Molecular Cancer Research: MCR*, Vol. 2, No. 3, 2004, pp. 183-195.

[14] J. Song, C. Jie, P. Polk, R. Shridhar, T. Clair, J. Zhang, L. Yin and D. Keppler, "The Candidate Tumor Suppressor CST6 Alters the Gene Expression Profile of Human Breast Carcinoma Cells: Down-Regulation of the Potent Mitogenic, Motogenic, and Angiogenic Factor Autotaxin," *Biochemical and Biophysical Research Communications*, Vol. 340, No. 1, 2006, pp. 175-182. doi:10.1016/j.bbrc.2005.11.171

[15] D. Cavallo-Medved and B. F. Sloane, "Cell-Surface Cathepsin B: Understanding Its Functional Significance," *Current Topics in Developmental Biology*, Vol. 54, 2003, pp. 313-341. doi:10.1016/S0070-2153(03)54013-3

[16] K. T. Chan, C. L. Cortesio and A. Huttenlocher, "Integrins in Cell Migration," *Methods in Enzymology*, Vol. 426, 2007, pp. 47-67. doi:10.1016/S0076-6879(07)26003-3

[17] J. Ochieng and G. Chaudhuri, "Cystatin Superfamily," *Journal of Health Care for the Poor and Underserved*, Vol. 21, No. 1, 2010, pp. 51-70. doi:10.1353/hpu.0.0257

[18] J. L. Cox, "Cystatins and Cancer," *Frontiers in Bioscience: A Journal and Virtual Library*, Vol. 14, 2009, pp. 463-474.

[19] G. Bathori, L. Cervenak and I. Karadi, "Caveolae—An Alternative Endocytotic Pathway for Targeted Drug Delivery," *Critical Reviews in Therapeutic Drug Carrier Systems*, Vol. 21, No. 2, 2004, pp. 67-95. doi:10.1615/CritRevTherDrugCarrierSyst.v21.i2.10

[20] K. M. Hussain, K. L. J. Leong, M. M.-L. Ng and J. J. H. Chu, "The Essential Role of Clathrin-Mediated Endocytosis in the Infectious Entry of Human Enterovirus 71," *The Journal of Biological Chemistry*, Vol. 286, No. 1, 2011, pp. 309-321. doi:10.1074/jbc.M110.168468

[21] M. Rey, A. Valenzuela-Fernández, A. Urzainqui, M. Yáñez-Mó, M. Pérez-Martínez, P. Penela, F. Mayor and F. Sánchez-Madrid, "Myosin IIA Is Involved in the Endocytosis of CXCR4 Induced by SDF-1alpha.," *Journal of Cell Science*, Vol. 120, No. 6, 2007, pp. 1126-1133.

[22] M. Bohdanowicz, G. Cosío, J. M. Backer and S. Grinstein, "Class I and Class III Phosphoinositide 3-Kinases Are Required for Actin Polymerization That Propels Phago-

somes," *The Journal of Cell Biology*, Vol. 191, No. 5, 2010, pp. 999-1012. doi:10.1083/jcb.201004005

[23] L. Lugini, P. Matarrese, A. Tinari, F. Lozupone, C. Federici, E. Iessi, M. Gentile, F. Luciani, G. Parmiani, L. Rivoltini, W. Malorni and S. Fais, "Cannibalism of Live Lymphocytes by Human Metastatic but Not Primary Melanoma Cells.," *Cancer Research*, Vol. 66, No. 7, 2006, pp. 3629-3638. doi:10.1158/0008-5472.CAN-05-3204

[24] C. Federici, D. Brambilla, F. Lozupone, P. Matarrese, A. de Milito, L. Lugini, E. Iessi, S. Cecchetti, M. Marino, M. Perdicchio, M. Logozzi, M. Spada, W. Malorni and S. Fais, "Pleiotropic Function of Ezrin in Human Metastatic Melanomas," *International Journal of Cancer*, Vol. 124, No. 12, 2009, pp. 2804-2812.

[25] U. Ekström, H. Wallin, J. Lorenzo, B. Holmqvist, M. Abrahamson and F. X. Avilés, "Internalization of Cystatin C in Human Cell Lines," *The FEBS Journal*, Vol. 275, No. 18, 2008, pp. 4571-4582. doi:10.1111/j.1742-4658.2008.06600.x

12

Manifestation of Key Molecular Genetic Markers in Pharmacocorrection of Endogenous Iron Metabolism in MCF-7 and MCF-7/DDP Human Breast Cancer Cells

Vasyl' Chekhun, Natalia Lukianova, Dmytro Demash, Tetiana Borikun, Svyatoslav Chekhun, Yulia Shvets

Department of Mechanisms of Antitumor Therapy, R.E. Kavetsky Institute of Experimental Pathology, Oncology and Radiobiology, NAS of Ukraine, Kyiv, Ukraine

Email: chekhun@onconet.kiev.ua

ABSTRACT

Effects of the nanocomposite and its components (magnetic fluid, cisplatin) on the level of endogenous iron exchange and the key links of genetic and epigenetic regulation of apoptotic program of sensitive and resistant MCF-7 cells were examined. We showed genetic and epigenetic mechanisms of action of nanocomposite of magnetic fluid and cisplatin. Nanocomposite caused elevation of number of cells in apoptosis in sensitive and especially resistant MCF-7 cells compared to cisplatin alone. It was proved that impact of nanocomposite on MCF-7/S and MCF-7/DDP cells caused more significant changes in expression of apoptosis regulators p53, Bcl-2 and Bax. We also suggested that changes in endogenous iron homeostasis and activation of free radical processes caused significant impact on apoptosis. Those changes included changes in methylation and expression of transferrin, its receptors, ferritin heavy and light chains (predominantly in resistant cell line), which caused activation of free radical synthesis and development of oxidative stress. We also showed that nanocomposite impact resulted into significant changes in expression of miRNA-34a and miRNA-200b, which regulated apoptosis, cell adhesion, invasion and activity of ferritin heavy chains gene. Thus, use of nanocomposite containing cisplatin and ferromagnetic as exogenous source of Fe ions caused changes of endogenous iron levels in sensitive and resistant cells allowing to increase specific activity of cytostatics and overcome factors, which promoted MDR development. Pharmacocorrection of endogenous iron metabolism allowed increasing antitumor activity of cisplatin and overcoming drug resistance.

Keywords: Nanocomposite; Iron Metabolism; Apoptosis; ROS; Drug Resistance; Breast Cancer Cells

1. Introduction

A new round of research in the field of tumor cells' biology significantly changed the common view not only about the latent period of malignancy, its clinical and morphological classification, characteristics of the proliferative and metastatic potential, but also the selection of treatment, the criteria for disease-free and overall patients' survival.

If, for example, we consider only the cellular heterogeneity and microenvironment in tumor tissue, then modern immunohistochemical subtypes from basal and luminal A/B to "triple negative" breast cancer are just like surrogate analogues of real molecular genetic characteristics that can be significant for clinical oncology [1,2].

One of the basic reasons for the heterogeneity of tumor tissue is the genetic and genomic instability caused by DNA sequence violations and rearrangement of chromosomes that determine the level of oncogenes and expression of tumor suppressors [3,4].

Increased expression of certain genes and reduction of others constantly change their dynamic balance, which determines not only the degree of heterogeneity, but also the level of sensitivity to therapy [5].

The emergence of these changes is predominantly associated with violation of the program of epigenetic regulation of tumor cell, which is accompanied by general hypomethylation and hypermethylation of promoters of individual genes. The rapid progress in studying of epigenetic regulation mechanisms of gene expression revealed regulatory role of small miRNA molecules, which regulate post-translational process at the level of protein synthesis [6-8].

However, significant progress in understanding of the nature of the malignant process does not produce the expected results in terms of early diagnostics, effective treatment and life quality improvement of cancer patients.

The need to improve clinical outcomes dictates a paradigm shift in the system of accumulation, analysis and interpretations of our knowledge.

Recent fundamental studies on metabolism of endogenous and exogenous iron in normal and tumor cells showed high degree of Fe ions integration into biological systems of human organism that offered a prospect of new markers and targets identification for development of antineoplastic agents. Iron is an essential element for cell growth and division because it exists in the structure of active sites of various proteins and takes part in many different reactions. Iron ions property to change quickly its valence from Fe^{2+} to Fe^{3+} made them vitally important in such reactions as DNA synthesis, oxygen transport, energy metabolism, cell cycle control, etc. At the same time its high reactivity makes it rather toxic compound, which causes hydroxyl radicals production. H. Fenton showed that during interaction with different reductants inside, the cell Fe^{3+} turned into Fe^{2+} [9]. It is known that formed reactive oxygen species interact with lipids, proteins, DNA, causing mutations and cell damage, which can initiate malignant transformation. Such unfavorable scenario for normal cell might become an optimal finding for selective damage of tumor cell. One of the unique properties of tumor cells is their increased affinity to Fe^{2+} ions [10-12]. It is important that intracellular Fe depot exists in forms of free/bound iron pools and ferritin. In healthy adults about 25% - 30% of Fe ions are connected with ferritin, thus preventing the development of toxic effects caused by this metal [13]. In normal cells in cases of supplement needs, iron is escaped from depot and is transported out of the cells by ferroportin and serum transferrin [14]. High intracellular Fe level causes ferritin translation, while its deficiency blocks this process [15]. Due to potential toxicity and possibility of reactive oxygen species (ROS) generation, the Fe level in organism is strictly controlled [16]. Instead, this is not typical for tumor cell. For activation of DNA synthesis it elevates ferritin levels, including multidrug resistance (MDR) formation [12,17]. And this is the second paradox. The fact of immunological identity between ferritin and autocrine growth factor secreted by leucosis cells is very interesting [18].

Today it is well known that besides the role in DNA synthesis, Fe ions affect the expression of proteins that are involved in cell cycle regulation, cell proliferation and angiogenesis and participate in the formation of cells' metastatic potential. Numerous studies have proved the role of iron in the anti-tumor effect of cytotoxics [19,20]. A new wave of interest in the investigation of Fe ions and iron-containing proteins greatly expanded understanding of their dominant role in vital functions of both normal and tumor cells [12]. Today, it is clearly possible to speak about a whole family of iron-containing proteins, which define the system of cell activity and can be considered, on the one hand, as biological markers of tumor process and, on the other hand, as promising targets for anti-tumor therapy. It is known that Fe ions, like most cytotoxic agents, are capable to initiate apoptosis in cell lines of different origin. At the same time different chelators and donators of iron ions are being discovered and actively studied [21,22].

At the present stage of development of theoretical, experimental and clinical oncology considerable interest is focused on the sources of iron in the form of nanomaterials, first, as a potential delivery vector for cytostatics; second, as an exogenous modifier of endogenous iron metabolism; and third, as a factor of selective accumulation and overcoming of resistance to anti-tumor drugs (ATDR).

The aim of this study was to examine the effect of the nanocomposite (NC) and its components (magnetic fluid (MF), cisplatin (DDP)) on the level of endogenous iron exchange and the key links of genetic and epigenetic regulation of apoptotic program of sensitive and resistant MCF-7 cells.

2. Materials and Methods

The investigation was carried out *in vitro* on cultures of MCF-7 human breast cancer cells, sensitive (MCF-7/S) and resistant to DDP (MCF-7/DDP). The cells of original line (sensitive to ATDR) were cultivated in modified culture medium Dulbecco ISCOVE (Sigma, Germany) with adding of 10% of fetal bovine serum (Sangva, Ukraine) at the temperature of 37˚C and CO_2 concentration of 5%. The cells have been reseeded twice a week with density 2 - 4 × 10^4 cells/cm^2. We obtained a variant of this line resistant to DDP (MCF-7/DDP) by growing of cells in culture medium with adding of rising concentrations of DDP. Every two months studied cells were analyzed in order to determine the level of their resistance with 3-(4,5-dimethylthiazol-2-1)-2,5-diphenyltetrazolium bromide (MTT viability test). On the moment of investigation the level of resistance of MCF-7/DDP cells was 4.

2.1. MTT Assay

After incubation period 10 μl of MTT were added into every well of 96-well plate. Cells were incubated at 37˚C in humid atmosphere for 3 hours and then centrifuged (1500 rpm for 5 min). Violet crystals of formazan were visually detected on the bottom of wells. After removal of the supernatant 50 μl of DMSO were added into wells to dissolve formazan crystals. After 20 min of incubation at room temperature crystals totally dissolved. Optical density in wells was measured by multiwell spectrophotometer (Labsystems Multiscan PLUS, USA) at 540 nm [23].

We studied cells sensitivity and cytotoxic effect of NC and its components. NC and DDP were used in doses equal to IC_{20} (by DDP) for each studied cell line (sensitive and resistant). Fe_3O_4 concentration in medium was equal to 12 μg/ml for MCF-7/S and 48 μg/ml for MCF-7/DDP cells, respectively. After addition of these agents into cultivation medium cells were cultivated for 48 hours in standard conditions.

After cultivation we studied expression of endogenous iron metabolism proteins, apoptosis regulators, CpG-sites methylation in promoters of ferritin heavy chains and transferrin receptor 1 genes, pool of reactive oxygen species (ROS), number of cells in apoptosis and necrosis, and expression of miRNA which regulate apoptosis, cell adhesion and activity of ferritin heavy chains.

2.2. Immunocytochemistry

The cells for the immunocytochemical analysis were grown on cover glasses, incubated in fixative solution (methanol:acetone 1:1) during 2 hours at t -20°C, than with 1% solution of bovine serum albumin (BSA) for 20 minutes. The monoclonal antibodies for defying of proteins were against transferrin receptor 1 (TFR1) (Bioworld Technology, USA), transferrin (Tf), ferritin light chains (FTL) (Abcam, USA), ferritin heavy chains (FTH) (GeneTex, USA), p53, Bcl-2 and Bax (DakoCytomation, Denmark) were applied in standard conditions of cultivation with according to algorithm of methodical approaches [24].

The estimation of results was made with the help of optical microscope (×100, oil immersion) with usage of classical method of H-Score:

$$S = 1 \times N_{1+} + 2 \times N_{2+} + 3 \times N_{3+}$$

where S-"H-Score" index, N_{1+}, N_{2+} and N_{3+} -numbers of cells with low, medium or high expression of the marker [25].

2.3. Methylation-Specific PCR (MSP)

Bisulfite conversion involves the deamination of unmodified cytosine residues to uracil under the influence of hydrosulfite ion from water solution of sodium bisulfite. Such treatment doesn't affect 5-mC and in further amplification uracils are amplified as thymines, whereas 5-mC residues get amplified as cytosines. Bisulfite conversion was performed using EZ DNA Methylation Gold-Kit (Zymo Research, USA) according to manufacturer's protocol. Aliquots of bisulfite-modified DNA were stored at −20°C and were used for MSP. Methylation-specific PCR was performed using the standard protocols; primer sequences are available in **Table 1**.

MSP-mixture contained 12.5 μl of Master Mix (Applied Biosystem, USA), 50 pM forward and reverse primers (IDT, USA), 3 ml (~50 ng) of bisulfite-converted DNA and deionized water to a final reaction volume of 25 μl. PCR reaction consisted of the following steps: denaturation of DNA at 95°C during 10 min, 35 cycles of amplification (denaturation at 95°C for 30 s), annealing of primers (56°C for 20 s), polymerization (72°C for 30 s) and final polymerization at 72°C for 10 sec.

PCR products were analyzed by agarose gel electrophoresis in 1.2% agarose "Low EEO, Type 1-A" ("Sigma", USA). Samples were loaded in amount of 6 ml PCR-product in well for each sample. For electrophoresis was used Tris-acetate buffer (40 mM Tris-acetate, pH 8.0; 1 mM EDTA). After electrophoresis the results was visualized by ethidium bromide, photographed under UV light and evaluated by a computer program TotalLab v2.01.

2.4. MicroRNA Expression Analysis

Total RNA extraction from the cells was performed using a commercial set of "Rhibo-zol" (Amplisense, Russia). Concentration of isolated RNA was determined using spectrophotometer "NanoDrop 2000c" (ThermoScientific, USA). Purity of isolated RNA is controlled by E_{260}/E_{280} ratio. RNA then was dissolved in TE buffer and stored at −20°C.

For the reverse transcriptase PCR (RT-PCR) we used the reaction mixture, listed in **Table 2**. RT-PCR was performed "Tertsyk" thermocycler ("DNA technology", Russia) using the following reaction parameters (**Table 3**). After the RT-PCR reaction product was added to a mixture of reagents (**Table 4**). AmpliTaq Gold Enzyme was activated and PCR was performed (40 cycles, denature-

Table 1. Primer sequences used in MSP.

Name	Nucleotide sequence	Final product size
FTH_M (forward)	5'-cgagggtttttagcggtc-3'	128 bp
FTH_M (reverse)	5'-atctcttataaccgcgtcgac-3'	
FTH_U (forward)	5'-gtgagggtttttagtggtt-3'	128 bp
FTH_U (reverse)	5'-aatctcttataaccacatcaac-3'	
TFR1_M (forward)	5'-gtagttgggattataggcgc-3'	180 bp
TFR1_M (reverse)	5'-taattaccaaacgcgataactc-3'	
TFR1_U (forward)	5'-tgagtagttgggattataggtgt-3'	180 bp
TFR1_U (reverse)	5'-taattaccaaacacaataactcac-3'	

Table 2. Composition of the reaction mixture for RT-PCR.

Components of the reaction mixture	Volume (μL)
100 mM dNTP mix	0.05
Reverse transcriptase multiscribe, 50 U/μL	0.33
10× RT buffer	0.50
RNAse inhibitor 20 U/μL	0.06
Nuclease-free water	1.56
5× TaqMan miRNA primer	1.00
Total RNA sample	1.50
Total	5.0

Table 3. Temperature parameters for RT-PCR.

Temperature (°C)	Time (min)
16	30
42	30
85	5
4	∞

Table 4. Composition of the reaction mixture for PCR.

Reagent	Volume (μL)
2× TaqMan 2 universal PCR master mix, no AmpErase UNG	10.00
Nuclease-free water	4.00
20× TaqMan MicroRNA assays mix	1.00
Product from RT reaction	5.00
Total	20.00

tion at 95°C for 15 sec, annealing and extension of primers at 65°C, 60 sec).

Horizontal gel electrophoresis was performed as described above. The values of miRNA expression were normalized by calculating the miRNA/miRNA-u87 (positive control) expression ratios for each miRNA.

2.5. Measurement of Intracellular ROS

CM-H2DCFDA, a lipid soluble membrane permeable dye upon entering cells undergoes deacetylation by intracellular esterases and forms the more hydrophilic, non-fluorescent dye Dichlorodihydrofluorescein (DCFH2). This is subsequently oxidized by ROS with formation of a highly fluorescent oxidation product, Dichlorofluorescein (DCF). The generated fluorescence is directly proportional to the amount of ROS. Fluorescence was analyzed by flow cytometry. The effect of DDP, MF and NC on generation of ROS was measured in MCF-7/S and MCF-7/DDP cell lines (2.5×10^5 cells). After centrifugation (1500 rpm for 5 minutes) cells were resuspended in PBS, incubated for 30 minutes at 37°C with CM-H2DCFDA (10 mM) for measurement of ROS. Positive control with 25 μM H_2O_2 was also made (data not presented). Fluorescence was acquired in the log mode and expressed as geometrical mean fluorescence channel (GMean). Acquisition was performed on 10,000 gated events.

2.6. Measurement of Annexin V Positivity

Translocation of phosphatidylserine from the inner to the outer leaflet of the plasma membrane occurs during apoptosis and can be assessed by exploiting the high binding affinity of Annexin V, a Ca^{2+}-dependent phospholipid binding protein to phosphatidylserine. To examine whether cell death occurred via apoptosis or necrosis, propidium iodide (PI), a non-permeable stain with affinity towards nucleic acids, which selectively enters necrotic or late apoptotic cells, was used. Therefore, co-staining of Annexin V and PI helps discriminate between live cells (PI and Annexin V negative), cells in early apoptosis (Annexin V positive, PI negative), cells undergoing late apoptosis (both Annexin V and PI positive) or necrotic cells (PI positive, Annexin V negative).

For detection of apoptotic cells we used apoptosis detection kit (Annexin V-FITC kit, Beckman Coulter, USA). Briefly, MCF/S and MCF/DDP (2.5×10^5/ml) were incubated with DDP, MF and NC as described above. After two washes, cells were resuspended in Annexin V binding buffer (10 mM HEPES/NaOH, pH 7.4, 140 mM NaCl, 2.5 mM $CaCl_2$) and Annexin V-FITC was added according to the manufacturers' instructions. The cells were incubated for 10 minutes in the dark at 37°C and just 5 min prior to acquisition, PI (0.1 mg/ml) was added and cells were washed and placed then in a flow cytometer.

2.7. Statistical Processing of the Results

Statistical processing of the obtained results was carried out with the help of mathematical program of medical and biological statistics STATISTICA 6.0 and in the environment of Microsoft Excel. Calculation and comparison of the significance of differences between the average values was carried out with usage of Student's t-criterion. Differences were considered significant with the probability not less than 95% ($P < 0.05$).

3. Results

Currently it has been clearly demonstrated that the effectiveness of most ATDR depends not only on the depth of DNA violation or other cellular targets damages, but it is also associated with the apoptotic program induction, cells' oxidative stress, impaired structural and functional state of their cytoskeleton, and changes in transport and metabolic processes, including iron metabolism. At the same time one of the biggest obstacles in the treatment of patients with malignant tumors, including breast cancer, is resistance to ATDR. According to the literature data and the results of our research, the development of resistance to cytostatic agents is accompanied by changes in the level of apoptosis regulatory-proteins, proliferation, transport, detoxication system, intercellular adhesion and receptor status [26].

At the same time, it was shown that the formation of the phenotype of drug resistance to DDP is accompanied by impaired protein metabolism of endogenous iron: increased expression of regulatory-proteins of transmembrane transport (TFR1), cell influx of iron (Tf) and ferritin (FTL, FTH) as a protein of intracellular iron storage and deposition (**Figure 1**, **Table 5**). Development of resistance to anticancer drugs also was accompanied by the shift of ferritin's light chains from cytoplasm to cell nucleus.

It should be noted that the development of DDP-resistance in MCF-7 cells led to the increase of TFR1 expression by 200%, Tf level more than 80%, and FTL and FTH—by 50% and 75%, respectively.

Based on the data presented above, and in contrast to the searching for binding factors of Fe ions in tumor cells, we investigated the anti-tumor effect of NC containing DDP and mixture Fe_2O_3-Fe_3O_4 nanoparicles, as delivery vector for cytostatic. To elucidate the possible mechanisms of death of original and resistant cells caused by NC and its components we held a wide range of studies of molecular genetic and epigenetic markers that might be associated with iron regulatory-proteins and survival of MCF-7 and MCF-7/DDP cells.

In our previous studies with use of light and electron microscopy we showed that MF and NC were able to penetrate and accumulate in both MCF-7/S and MCF-7/DDP cells [26] using histochemical staining with potassium ferro- and ferricyanide. We found that NC incorporation into MCF-7/DDP cells compared to MCF-7/S cells is much more active. Cytomorphology and electron microscopy of resistant cells with high levels of accumulated NC showed its higher cytotoxic effects in particular dystrophy, necrobiosis, necrosis and apoptosis.

It was established that cultivation of original and resistant MCF-7 cells with MF and NC was accompanied by

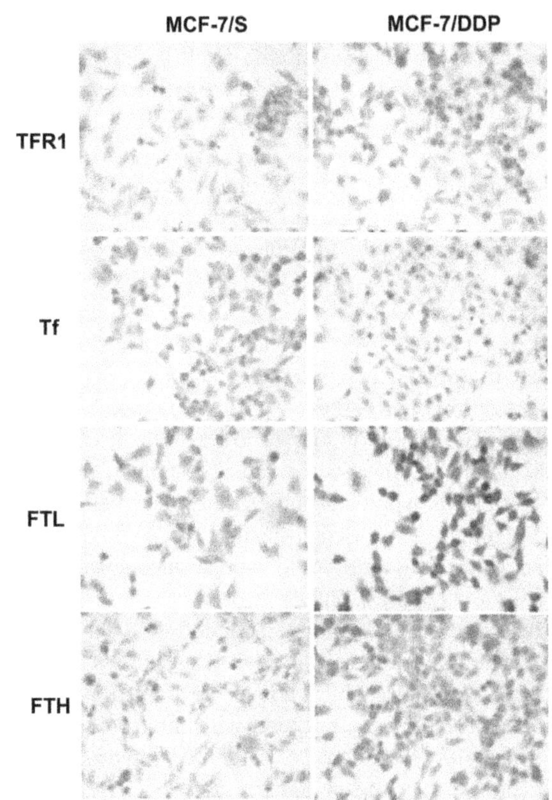

Figure 1. Expression of iron metabolism proteins in MCF-7/S and MCF-7/DDP cells.

Table 5. Expression of iron-regulatory proteins in the MCF-7/S and MCF-7/DDP cells.

Marker	Cell line	Expression of studied markers, H-score values			
		Control	MF	DDP	NC
TFR1	MCF-7/S	72 ± 1.8	101 ± 3.6	72 ± 1.6	110 ± 3.5
	MCF-7/DDP	215 ± 5.5	235 ± 3.6	215 ± 3.4	288 ± 3.5
Tf	MCF-7/S	160 ± 5.4	184 ± 2.5	156 ± 2.4	197 ± 3.6
	MCF-7/DDP	293 ± 6.0	296 ± 3.5	293 ± 4.3	299 ± 0.3
FTL	MCF-7/S	195 ± 4.5	209 ± 2.8	193 ± 3.5	214 ± 3.1
	MCF-7/DDP	296 ± 2.8	299 ± 2.8	294 ± 5.8	299 ± 0.5
FTH	MCF-7/S	125 ± 2.7	160 ± 3.2	125 ± 2.5	169 ± 2.0
	MCF-7/DDP	220 ± 6.1	277 ± 3.6	223 ± 4.3	298 ± 2.0

significant changes in the expression of the investigated iron regulatory-proteins (**Table 5**). Herewith, the most significant changes in original cells were observed in the levels of Tf and TFR1 expression, as well as FTL and FTH. Under the influence of MF and NC TFR1 expression level increased by 40.3% and 52.8%, and the expression level of FTH—by 28.0% and 35.2%, respectively. Significant Tf expression increase in sensitive cells was observed only under NC treatment. In resistant cells the most significant changes were found in TFR1 expression under the influence of NC and in FTH expression when MF and NC have been used (**Table 5**). It is noteworthy that the level of FTL expression under the influence of MF and NC did not change significantly in original and resistant cells.

At the same time we found that under the influence of NC and its components the disturbances of epigenetic regulation of some regulatory genes of endogenous iron metabolism occur. It was shown that cultivation of original and resistant MCF-7 cells with DDP, MF and NC was accompanied by hypomethylation of CpG-sites of TFR1 gene promoters and hypermethylation of CpG-sites of FTH gene promoters (**Figure 2**).

Free intracellular iron is known to be associated with free radicals generation. So we had investigated alterations of ROS level in parental and DDP resistant cell lines after the treatment with NC and its derivatives. It should be stressed here that ROS could induce oncogenesis in normal cells and apoptosis in cancer cells.

Significance of ROS activity isn't definitely characterized to date. Development of DDP resistance is associated with 1.83-fold free oxygen level elevation (**Table 6, Figure 3**) and could be associated with failure of antioxidant system or rise of content of iron keeping proteins.

Significant increase of free oxygen radical activity was detected after NC treatment in MCF-7 cells (by 15%) and MCF-7/DDP (by 25%). Elevation of ROS level after NC treatment could be important factor on the one hand in DDP antitumor activity increase and on the other hand

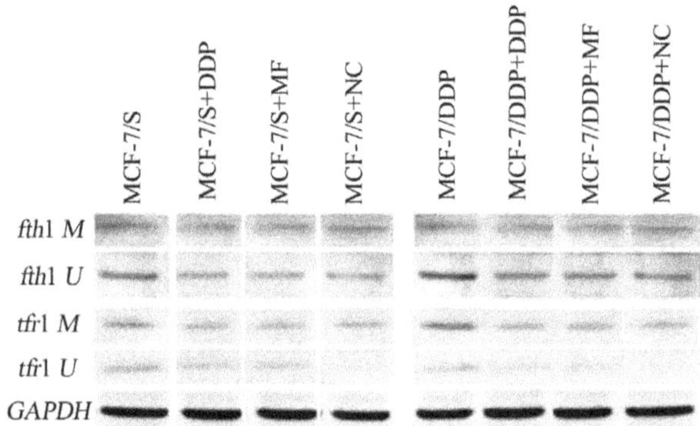

Figure 2. Peculiarities of methylation CpG-sites in the genes promoters of FTH and TFR1 in original and resistant MCF-7 cells under the influence of MF, DDP and NC.

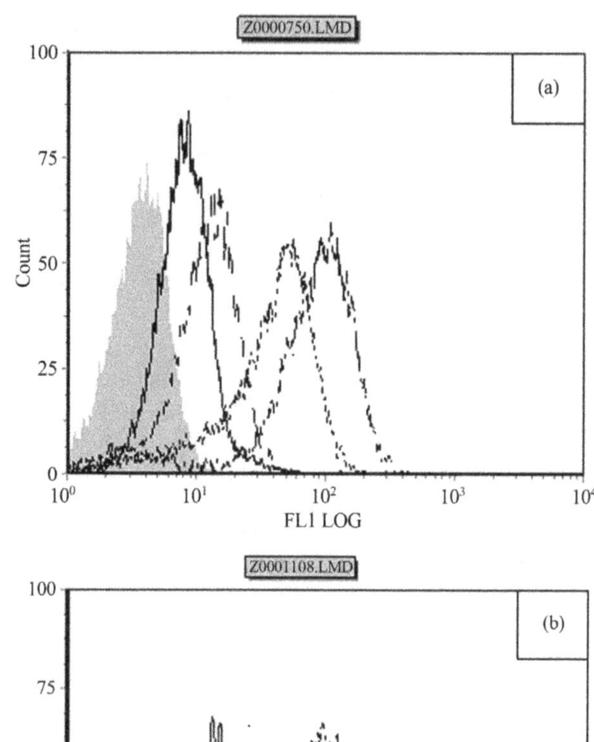

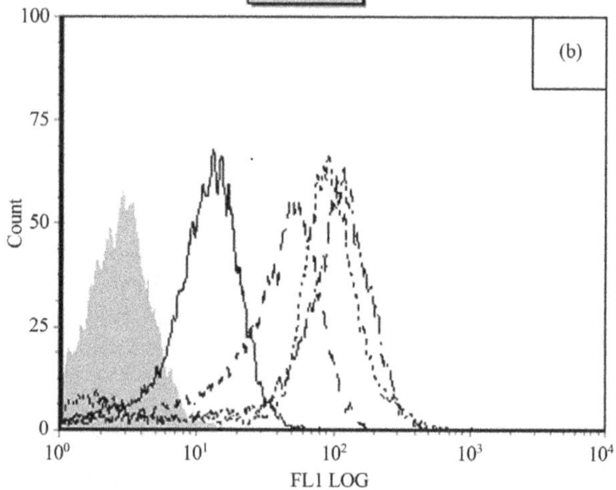

Figure 3. Flow cytometry study of ROS expression in (a) MCF-7/S and (b) MCF-7/DDP cells. Filled histogram is isotype control, ROS in control cells, ROS in cells treated with DDP, MF and NC.

Table 6. Flow cytometry studies of intracellular ROS (GMean values).

Cell lines	ROS level			
	Control	DDP	MF	NC
MCF-7/S	7.27 ± 0.65	8.61 ± 1.55	13.88 ± 1.96	35.18 ± 5.02
MCF-7/DDP	13.30 ± 1.24	18.96 ± 2.48	35.48 ± 3.45	53.67 ± 4.44

could be a key mechanism in overcoming DDP resistance DDP is known to induce tumor cell death by the apoptotic mechanism. Meanwhile last year's investigations suggest that NCs of different origin can also induce apoptosis in tumor cells. Thus next direction of our experiments was to elucidate the NC role in apoptosis initiation in MCF-7/S and MCF-7/DDP cells.

Table 7 and **Figure 4** show that DDP induced apop-

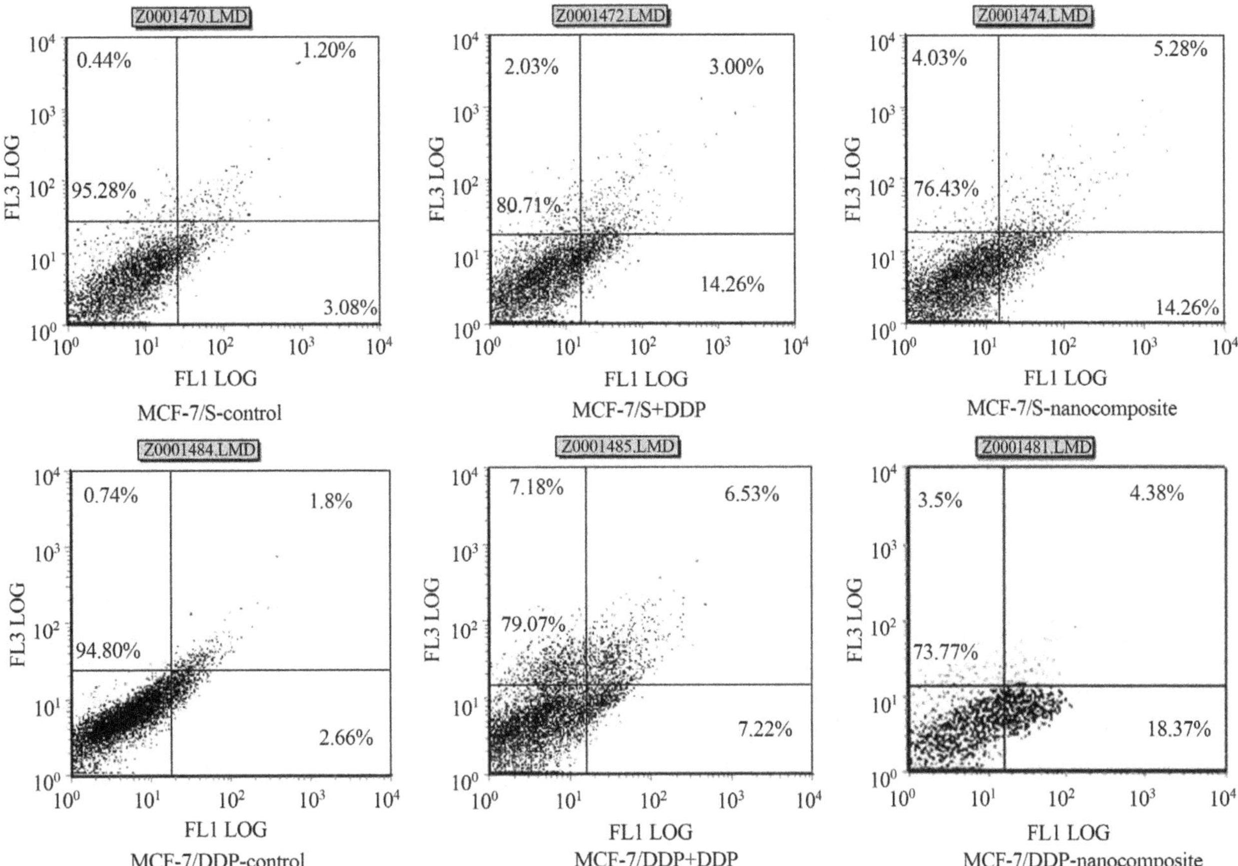

Figure 4. Changes in percent of MCF-7/S (top) and MCF-7/DDP cells (bottom) in necrosis and apoptosis after treatment with DDP and NC.

Table 7. Changes in percentage of MCF-7/S and MCF-7/DDP cells in necrosis and apoptosis after treatment with DDP, MF and NC.

Cell lines	Parameters	Studied factors and number of cells in apoptosis and necrosis (%)			
		Control	DDP	MF	NC
MCF-7/S	Live cells	95.28 ± 1.44	80.71 ± 2.35	93.46 ± 2.66	76.43 ± 3.44
	Necrosis	0.44 ± 0.06	2.03 ± 0.09	2.35 ± 0.12	4.03 ± 0.21
	Apoptosis	4.28 ± 0.53	17.26 ± 1.06	4.19 ± 0.61	19.54 ± 1.46
MCF-7/DDP	Live cells	97.80 ± 1.56	79.07 ± 1.96	89.83 ± 1.65	73.77 ± 2.65
	Necrosis	0.74 ± 0.08	7.18 ± 0.11	3.96 ± 0.31	3.50 ± 0.74
	Apoptosis	4.46 ± 0.63	19.75 ± 1.85	5.21 ± 0.55	22.73 ± 2.77

tosis in 17.3% and necrosis in 2% of MCF-7/S cells. Quantity of apoptotic and necrotic MCF-7 cells after MF treatment was at the level of control cells. NC induced increase of number of cells in apoptosis and necrosis to 19.9% and 3.5% for MCF-7/S cell line. MF treatment slightly increased percent of apoptotic and necrotic MCF-7/DDP cells. It must be noted, that NC treatment did not reduce but, on the contrary, increased rates of apoptotic and necrotic cells in sensitive and resistant to DDP cell lines. Efficiency of apoptosis induction with antitumor drugs is known to be dependent on the expression level of a number of key regulatory molecules and the degree of activity of signaling cascades formation. So the next stage of our research was to determine the expression level of apoptosis regulating proteins in human breast cancer cell lines MCF-7/S and MCF-7/DDP.

Previously presented data [27] showed that generation of DDP resistance in MCF-7/DDP cell line is accompanied by slight decrease of p53 expression and significant changes in expression of antiapoptotic proteins Bcl-2 and Bax.

However, recently p53 subunit (R2), which has Fe

binding sites and provides nucleotides for DNA reparation was discovered [28-30]. Thus, investigations of modification role of exogenous iron source in key links of apoptotic program arouse scientific interest.

While studying changes of expression of proteins, associated with apoptosis, we found out that NC influences expression of p53, Bcl-2 and Bax proteins (**Table 8**). In parental MCF-7 cells DDP and NC induced decrease of Bcl-2 expression on 23.5% and on 22.6%, respectively, and increased expression of Bax on 14.9% and 24.3% respectively. In DDP resistant cells MCF-7/DDP similar changes in proteins expression were detected. However, it is important to note that NC significantly increased expression of proapoptotic protein p53 in MCF-7/DDP cells. Obtained data indicate that NC is able to initiate apoptotic program not only in MCF-7/S cells, but more significantly in MCF-7/DDP cells due to reduction of antiapoptotic protein Bcl-2 expression and elevation of proapoptotic p53 and Bax expression. Our results indicate the impact of exogenous iron, which is the part of NC, on the changes of molecular genetic markers expression level, which determines sensitivity to apoptosis induction, the degree of adhesion and invasion of MCF-7/S and MCF-7/DDP cells. At the same time, in the available literature there are no data on the influence of exogenous sources of iron on expression of miRNAs.

In particular, our attention was attracted by miR-34 and miR-200b, which are involved in the regulation of apoptosis, cell-cell adhesion, invasion and activity of ferritin heavy chains gene [12,31].

We have found that another link in the chain of mechanisms, that causes the increase in the percentage of cells dying by apoptosis in sensitive and resistant MCF-7 cells during their cultivation with DDP, MF and NCs, is the miR-34 and miR-200b expression elevation (**Figure 5**).

Obtained results are also evidence of increased sensitivity of the investigated cells to antitumor drugs, which is particularly important fact concerning the resistant cells.

4. Discussion

So, *in vitro* we showed genetic and epigenetic mechanisms of action of NC of MF and DDP. NC caused elevation of number of cells in apoptosis in sensitive and especially resistant MCF-7 cells compared to DDP alone. It was proved that impact of NC on MCF-7/S and MCF-7/DDP cells caused more significant changes in p53, Bcl-2 and Bax expression, which regulated apoptosis.

We also suggested that changes in endogenous iron homeostasis and activation of free radical processes caused significant impact on apoptosis because of oxidative stress development.

It is known that elevation of iron concentration in the cytoplasm might cause oxidative stress due to ROS generation in Fenton reactions [9]. DDP itself is also able to cause ROS generation by induction of NADPH oxidadases expression [31]. In our study we noted elevation of ROS content in both cell lines after impact of NC.

It is also known that FTH expression during oxidative stress increased by IRE/IRP system to transfer iron ions into the bound state and decreased number of ROS generation sites [32].

IRP usually connects with IRE on mRNA by SH-groups in cysteine residues. It was shown that accumula-

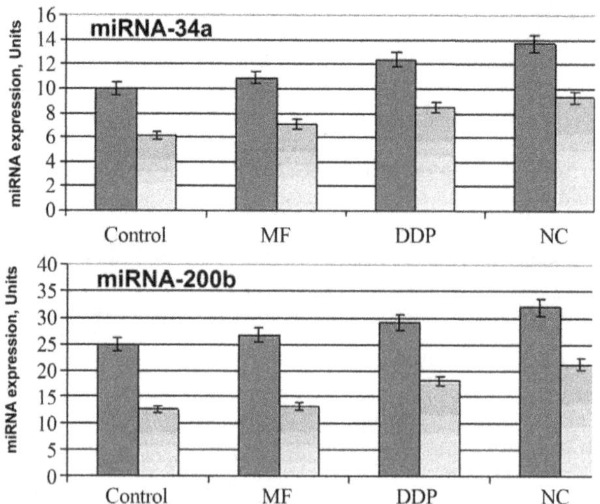

Figure 5. miRNA-34a and miRNA-200b expression in MCF-7/S cells (dark) and MCF-7/DDP cells (light) after impact of MF, DDP and NC.

Table 8. Changes in expression of proteins which regulate apoptosis in MCF-7/S and MCF-7/DDP cells after impact of NC and its components.

Marker	Cell lines	Studied markers expression (H-score values)			
		Control	MF	DDP	NC
p53	MCF-7/S	183 ± 4.7	180 ± 3.6	208 ± 4.6	179 ± 2.6
	MCF-7/DDP	164 ± 4.1	135 ± 2.6	166 ± 3.9	180 ± 4.1
Bcl-2	MCF-7/S	179 ± 2.9	169 ± 2.3	137 ± 2.5	126 ± 3.0
	MCF-7/DDP	36 ± 0.9	36 ± 2.5	35 ± 2.5	18 ± 1.7
Bax	MCF-7/S	181 ± 2.4	178 ± 2.1	208 ± 2.8	225 ± 3.1
	MCF-7/DDP	62 ± 1.4	62 ± 2.8	64 ± 3.4	121 ± 2.7

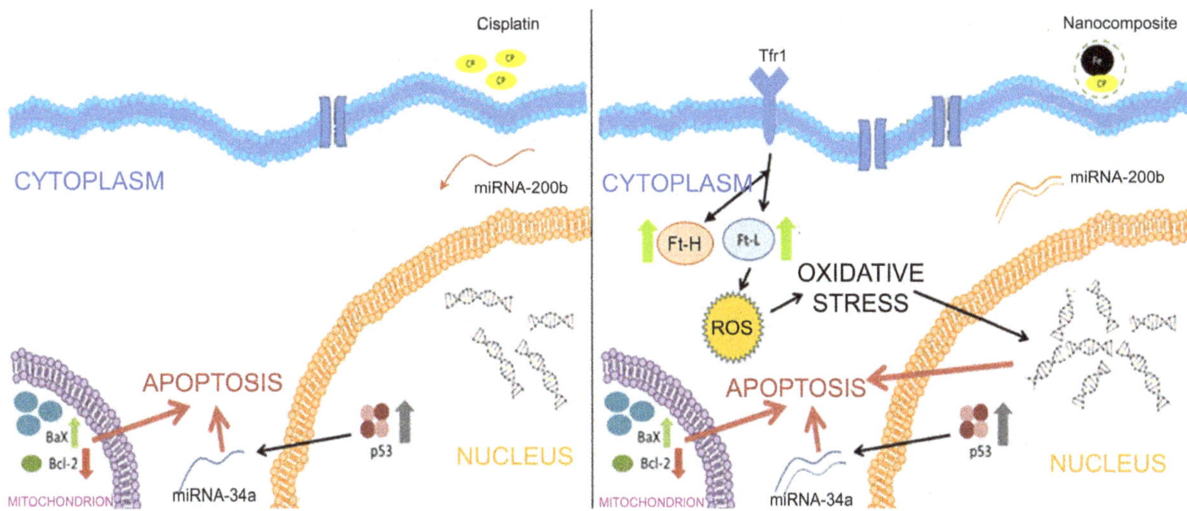

Figure 6. Schematic illustration of mechanisms of action of DDP and NC (based on ferromagnetic and DDP).

tion of ROS-generating substances (ATDR, iron-containing nanoparticles) by cells caused oxidation of these residues, formation of disulfide bonds and resulted in disturbances in protein conformation and interaction with mRNA causing increase of TFR1 expression [33]. Such phenomenon of sinchronous changes in FTH and TFR1 expression as a result of oxidative stress was observed by different groups of scientists on different cell lines [34, 35].

On the other hand, we suggest that there is another mechanism of TFR1 expression increase associated with hypomethylation of its gene. It is known that under oxidative stress DNA methyltransferases 1 and 3B are able to change their localization in chromatin causing changes in global methylation patterns [36]. This mechanism probably is the basis of increase of TFR1 expression.

We also showed that NC impact resulted into significant changes of expression of miRNA-34a and miRNA-200b, which regulated apoptosis, cell adhesion, invasion and activity of ferritin heavy chains gene.

Thus, the use of NC containing DDP and ferromagnetic as exogenous source of Fe ions causes changes of endogenous iron levels in sensitive and resistant cells allowing to increase specific activity of cytostatic agent and overcome factors, which promote MDR development (**Figure 6**).

REFERENCES

[1] V. Almendro, A. Marusyk and K. Polyak, "Cellular Heterogeneity and Molecular Evolution in Cancer," *Annual Review of Pathology—Mechanisms of Disease*, Vol. 8, 2013, pp. 277-302.

[2] M. H. Barcellos-Hoff, "Does Microenvironment Contribute to the Etiology of Estrogen Receptor-Negative Breast Cancer?" *Clinical Cancer Research*, Vol. 19, No. 3, 2013, pp. 541-548. http://dx.doi.org/10.1158/1078-0432.CCR-12-2241

[3] V. F. Chekhun, S. D. Sherban and Z. D. Savtsova, "Tumor Heterogeneity—Dynamical State," *Oncology*, Vol. 14, No. 1, 2012, pp. 4-12.

[4] V. F. Chekhun, "From System Cancer Biology to Personalized Treatment," *Oncology*, Vol. 14, No. 2, 2012, pp. 84-88.

[5] N. A. Saunders, F. Simpson, E. W. Thompson, M. M. Hill, L. Endo-Munoz, G. Leggatt, R. F. Minchin and A. Guminski, "Role of Intratumoural Heterogeneity in Cancer Drug Resistance: Molecular and Clinical Perspectives," *EMBO Molecular Medicine*, Vol. 4, No. 8, 2012, pp. 675-684. http://dx.doi.org/10.1002/emmm.201101131

[6] T. V. Bagnyukova, I. P. Pogribny and V. F. Chekhun, "MicroRNAs in Normal and Cancer Cells: A New Class of Gene Expression Regulators," *Experimental Oncology*, Vol. 28, No. 4, 2006, pp. 263-269.

[7] K. R. Kutanzi, O. V. Yurchenko, F. A. Beland, V. F. Checkhun and I. P. Pogribny, "MicroRNA-Mediated Drug Resistance in Breast Cancer," *Clinical Epigenetics*, Vol. 2, No. 2, 2011, pp. 171-185. http://dx.doi.org/10.1007/s13148-011-0040-8

[8] T. A. Farazi, J. I. Hoell, P. Morozov and T. Tusch, "MicroRNAs in Human Cancer," *Advances in Experimental Medicine and Biology*, Vol. 774, 2013, pp. 1-20. http://dx.doi.org/10.1007/978-94-007-5590-1_1

[9] S. Toyokuni, "Iron and Carcinogenesis: From Fenton Reaction to Target Genes," *Redox Report*, Vol. 7, No. 4, 2002, pp. 189-197. http://dx.doi.org/10.1179/135100002125000596

[10] P. Karihtala and Y. Soini, "Reactive Oxygen Species and Antioxidant Mechanisms in Human Tissues and Their Relation to Malignancies," *APMIS*, Vol. 115, No. 2, 2007, pp. 81-103. http://dx.doi.org/10.1111/j.1600-0463.2007.apm_514.x

[11] H. Wiseman and B. Halliwell, "Damage to DNA by Reactive Oxygen and Nitrogen Species: Role in Inflammatory Disease and Progression to Cancer," *Biochemical Journal*, Vol. 313, No. 1, 1996, pp. 17-29.

[12] S. I. Shpyleva, V. P. Tryndyak, O. Kovalchuk, A. Starlard-Davenport, V. F. Chekhun, F. A. Beland and I. P. Pogribny, "Role of Ferritin Alterations in Human Breast Cancer Cells," *Breast Cancer Research and Treatment*, Vol. 126, No. 1, 2011, pp. 63-71. http://dx.doi.org/10.1007/s10549-010-0849-4

[13] K. Fan, L. Gao and X. Yan, "Human Ferritin for Tumor Detection and Therapy," *Wiley Interdisciplinary Reviews Nanomedicine and Nanobiotechnology*, Vol. 5, No. 4, 2013, pp. 287-298. http://dx.doi.org/10.1002/wnan.1221

[14] C. Datz, T. K. Felder, D. Niederseer and E. Aigner, "Iron Homeostasis in the Metabolic Syndrome," *European Journal of Clinical Investigation*, Vol. 43, No. 2, 2013, pp. 215-224. http://dx.doi.org/10.1111/eci.12032

[15] E. C. Theil, R. K. Behera and T. Tosha, "Ferritins for Chemistry and for Life," *Coordination Chemistry Reviews*, Vol. 257, No. 2, 2013, pp. 579-586. http://dx.doi.org/10.1016/j.ccr.2012.05.013

[16] M. Geppert, M. C. Hohnholt, S. Nürnberger and R. Dringen, "Ferritin Up-Regulation and Transient ROS Production in Cultured Brain Astrocytes after Loading with Iron Oxide Nanoparticles," *Acta Biomaterialia*, Vol. 8, No. 10, 2012, pp. 3832-3839. http://dx.doi.org/10.1016/j.actbio.2012.06.029

[17] V. F. Chekhun, N. Yu. Lukianova and N. O. Bezdenezhnykh, "Features of Iron Metabolism Regulatingproteins Expression in Sensitive and Resistant to Antitumor Drugs Breast Cancer Cells *in Vitro*," *Clinical Oncology* (*SE*), *Proceedings of XII Ukrainian Oncologists Meeting*, Sudak, 2011, p. 227.

[18] N. Kikyo, M. Suda, N. Kikyo, K. Hagiwara, K. Yasukawa, M. Fujisawa, Y. Yazaki and T. Okabe, "Purification and Characterization of a Cell Growth Factor from a Human Leukemia Cell Line: Immunological Identity with Ferritin," *Cancer Research*, Vol. 54, No. 1, 1994, pp. 268-271.

[19] E. Laqué-Rupérez, M. J. Ruiz-Gómez, L. de la Peña, L. Gil and M. Martínez-Morillo, "Methotrexate Cytotoxicity on MCF-7 Breast Cancer Cells Is Not Altered by Exposure to 25 Hz, 1.5 mT Magnetic Field and Iron (III) Chloride Hexahydrate," *Bioelectrochemistry*, Vol. 60, No. 1-2, 2003, pp. 81-86. http://dx.doi.org/10.1016/S1567-5394(03)00054-9

[20] J. F. Head, F. Wang and R. L. Elliott, "Antineoplastic Drugs That Interfere with Iron Metabolism in Cancer Cells," *Advances in Enzyme Regulation*, Vol. 37, 1997, pp. 147-169. http://dx.doi.org/10.1016/S0065-2571(96)00010-6

[21] S. V. Torti and F. M. Torti, "Cellular Iron Metabolism in Prognosis and Therapy of Breast Cancer," *Critical Reviews in Oncogenesis*, Vol. 18, No. 5, 2013, pp. 435-448. http://dx.doi.org/10.1615/CritRevOncog.2013007784

[22] J. L. Heath, J. M. Weiss, C. P. Lavau and D. S. Wechsler, "Iron Deprivation in Cancer-Potential Therapeutic Implications," *Nutrients*, Vol. 5, No. 8, 2013, pp. 2836-2859. http://dx.doi.org/10.3390/nu5082836

[23] M. Niks and M. Otto, "Towards an Optimized MTT Assay," *Journal of Immunological Methods*, Vol. 130, No. 1, 1990, pp. 149-151. http://dx.doi.org/10.1016/0022-1759(90)90309-J

[24] Y. L. Chao, C. R. Shepard and A. Wells, "Breast Carcinoma Cells Re-Express E-Cadherin during Mesenchymal to Epithelial Reverting Transition," *Molecular Cancer*, Vol. 9, No. 1, 2010, pp. 179-197. http://dx.doi.org/10.1186/1476-4598-9-179

[25] R. A. McCelland, D. Wilson and R. Leake, "A Multicentre Study into the Reliability of Steroid Receptor Immunocyto-Chemical Assay Quantification," *European Journal of Cancer*, Vol. 27, 1991, pp. 711-715. http://dx.doi.org/10.1016/0277-5379(91)90171-9

[26] V. F. Chekhun, O. V. Yurchenko, L. A. Naleskina, D. V. Demash, N. Yu. Lukianova and Yu. V. Lozovska, "*In Vitro* Modification of Cisplatin Cytotoxicity with Magnetic Fluid," *Experimental Oncology*, Vol. 35, No. 1, 2013, pp. 15-19.

[27] N. Yu. Lukyanova, N. V. Rusetskaya and N. A. Tregubova, "Molecular Profile and Cell Cycle in MCF-7 Cells Resistant to Cisplatin and Doxorubicin," *Experimental Oncology*, Vol. 31, No. 2, 2009, pp. 87-92.

[28] A. Jordan and P. Reichard, "Ribonucleotide Reductases," *Annual Review of Biochemistry*, Vol. 67, 1998, pp. 71-98. http://dx.doi.org/10.1146/annurev.biochem.67.1.71

[29] H. Tanaka, H. Arakawa, T. Yamaguchi, K. Shiraishi, S. Fukuda, K. Matsui, Y. Takei and Y. Nakamura, "A Ribonucleotide Reductase Gene Involved in a p53-Dependent Cellcycle Checkpoint for DNA Damage," *Nature*, Vol. 404, 2000, pp. 42-49. http://dx.doi.org/10.1038/35003506

[30] D. S. Byun, K. S. Chae, B. K. Ryu, M. G. Lee and S. G. Chi, "Expression and Mutation Analyses of P53R2, a Newly Identified p53 Target for DNA Repair in Human Gastric Carcinoma," *International Journal of Cancer*, Vol. 98, No. 5, 2002, pp. 718-723. http://dx.doi.org/10.1002/ijc.10253

[31] N. Yu. Lukianova, L. A. Naleskina, N. O. Bezdenezhnykh, L. M. Kunskaya, D. V. Demash, Yu. V. Yanish, I. M. Todor and V. F. Chekhun, "Reactive Changes of Cytophysiological Properties, Molecular-Biological Profile and Functional Metabolic Status of Cells *in Vitro* with Different Sensitivity to Cytostatic Agents under the Influence of Magnetic Fluid," *Journal of Cancer Research*, Vol. 1, No. 1, 2013, pp. 7-14. http://dx.doi.org/10.11648/j.crj.20130101.12

[32] H.-J. Kim, J.-H. Lee, S.-J. Kim, G. S. Oh, H.-D. Moon, K.-B. Kwon, C. Park, B. H. Park, H.-K. Lee, S.-Y. Chung, R. Park and H.-S. So, "Roles of NADPH Oxidases in Cisplatin-Induced Reactive Oxygen Species Generation and Ototoxicity," *The Journal of Neuroscience*, Vol. 30, No. 11, 2010, pp. 3933-3946. http://dx.doi.org/10.1523/JNEUROSCI.6054-09.2010

[33] A. Cozzi, B. Corsi, S. Levi, P. Santambrogio, G. Biasiotto and P. Arosi, "Analysis of the Biologic Functions of H- and L-Ferritins in HeLa Cells by Transfection with siRNAs and cDNAs: Evidence for a Proliferative Role of L-Ferritin," *Blood*, Vol. 103, No. 6, 2004, pp. 2377-2383. http://dx.doi.org/10.1182/blood-2003-06-1842

[34] X. Xu, H. L. Persson and D. R. Richardson, "Molecular Pharmacology of the Interaction of Anthracyclines with

Iron," *Molecular Pharmacology*, Vol. 68, No. 2, 2005, pp. 261-271.

[35] E. Pawelczyk, A. S. Arbab, S. Pandit, E. Hu and J. A. Frank, "Expression of Transferrin Receptor and Ferritin Following Ferumoxides-Protamine Sulfate Labeling of Cells: Implications for Cellular Magnetic Resonance Imaging," *NMR in Biomedicine*, Vol. 19, 2006, pp. 581-592. http://dx.doi.org/10.1002/nbm.1038

[36] H. M. O'Hagan, W. Wang, S. Sen, C. D. Shields, S. S. Lee, Y. W. Zhang, E. G. Clements, Y. Cai, L. Van Neste, H. Easwaran, R. A. Casero, C. L. Sears and S. B. Baylin, "Oxidative Damage Targets Complexes Containing DNA Methyltransferases, SIRT1, and Polycomb Members to Promoter CpG Islands," *Cancer Cell*, Vol. 20, No. 5, 2011, pp. 606-619. http://dx.doi.org/10.1016/j.ccr.2011.09.012

Natural Antimicrobial Peptides: Pleiotropic Molecules in Host Defense

Mercedes Leonor Sánchez[1,2], Melina María Belén Martínez[3], Paulo César Maffia[1,3*]
[1]National Council for Scientific Research (CONICET), Buenos Aires, Argentina
[2]School of Medicine, University of Buenos Aires, Buenos Aires, Argentina
[3]Laboratory of Molecular Microbiology, National University of Quilmes, Buenos Aires, Argentina
Email: *paulo.maffia@unq.edu.ar

ABSTRACT

Natural antimicrobial peptides (AMPs) are small cationic molecules that display antimicrobial activity against a wide range of bacteria, fungi and viruses. AMPs are multifunctional molecules that have an essential activity in infection and inflammation: they play an important role in the innate immune response, not only as antimicrobial agents, but also as immunomodulating molecules and as an important link between the innate and adaptive immune response. In this article, we will discuss the antimicrobial activity, together with the novel properties of some of these molecules as immune modulators on the innate and adaptive immune response.

Keywords: Antimicrobial Peptides; Immune Response; Cathelicidins; Defensins; SLPI

1. Introduction

Adaptive immune response is considered critical to prevent the establishment or progression of infections by parasites, bacteria and viruses. But it is the innate immunity that is responsible for the rapid initial defense against the pathogen.

The co-evolution of hosts and pathogens has led to a diverse group of peptides that the host produces in order to kill or reduce the infective microbes. These peptides, called antimicrobial peptides (AMPs) can be found in almost all forms of life, in organisms like bacteria or plants and also in invertebrate and vertebrate species, including mammals. Among the latter, humans have several cell types that synthesize and secrete AMPs, such as epithelial cells, epidermal keratinocytes, neutrophils, macrophages, and natural killer cells. In mammals, these AMPs can be considered as part of the innate immune system.

Most of the AMPs are generally expressed as propeptides that undergo subsequent proteolytic process to release the biologically active and mature host defense peptide. Some AMPs are constitutively expressed but others are synthesized upon infection signals such as various exogenous and endogenous inflammatory mediators [1,2].

Classically, AMPs are short, amphiphilic, and mostly cationic polypeptides with a diverse repertoire of activities within the innate immunity system. These multifaceted molecules may be able to enhance phagocytosis, stimulate prostaglandin release, neutralize the septic effects of LPS, and promote recruitment and accumulation of various immune cells at inflammatory sites among other functions [3,4].

In recent years, these molecules have been the center of attention, because of their promising uses as alternative approaches to infection management.

AMPs are mainly cationic and amphipathic peptides; two characteristics together with their conformational flexibility and secondary structure allow them to interact with and insert into biomembranes, leading to disruption of cytoplasmic membrane integrity. AMPs can also interact with intracellular bacterial targets, resulting in microbial killing [5]. In general, cationic peptides interfere with bacterial membrane integrity such as membrane wrinkling and the formation of ion-permeable channels that probably increase membrane permeability and finally lead to bacterial cell lysis.

It is also worthy to notice that it has recently been shown that bacteria are capable of adapting and resisting host AMPs, by means of production of peptidases and

*Corresponding author.

proteases that degrade antimicrobial peptides, or compounds that inhibit their action, and also by reduction of net anionic charge of the bacterial cell envelope. The relationship between antibacterial peptides and cell wall targets may shed light in strategies to challenge antibiotic resistance.

However, even though these host defense peptides originally gained prominence through initial descriptions of their direct antimicrobial functions [5], not all protective cationic peptides are necessarily working through direct microbicidal action. In contrast, under such conditions, a wide range of functions have been demonstrated for these peptides in the context of host immunity, as will be discussed in this review.

Some AMPs are involved in the transition to the adaptive immune response, as they were shown to be chemotactic for human monocytes [6], T cells [7], modulating dendritic cell differentiation and dendritic cell-induced T cell polarization [8]. However, some of them can also function as negative feedback regulators facilitating the resolution of inflammation [9]. Therefore, it is possible that a malfunction of the feedback control, *i.e.* through a decreased secretion of AMPs that participates as negative feedback regulator, can amplify and perpetuate the inflammatory process.

As we mentioned, the role of AMPs in innate immunity and the pathogen target has been very well studied, but it is less known the ability of them to induce not only a T helper cell response, but also an appropriate T cells response. This is a very important issue, since an appropriate T cell response, *i.e.* Th1 and Th2 for intracellular and extracellular pathogens respectively, will aid the resolution or perpetuation of the infection. In this review, we have chosen only three AMPs: cathelicidins, defensins and SLPI, to describe the typical functions and to illustrate their ability to modulate the immune response.

2. AMPs Categories

Host defense peptides can be organized into three basic categories regarding the target: 1) plasma membrane-active peptides are thought to act in a multistage process where they electrostatically bind to a membrane surface, aggregate to form superstructures, and disrupt membrane integrity; 2) a second group of peptides act on intracellular targets to inhibit transcriptional, translational or other processes; and 3) cell wall-active peptides target precursors, mechanisms, and/or essential intermediates in peptidoglycan, lipopolysaccharide (LPS) or other biosynthetic pathways interfering with functional cell wall synthesis and ensuing bacterial replication.

3. Eukariotic Peptides

When the first eukaryotic defensin peptides were characterized nearly 30 years ago, mechanism of action studies indicated that these molecules were able to rapidly and efficiently permeabilize artificial membrane bilayers [10].

Eukaryotic peptides are typically small and cationic, and are structurally classified into five major groups: 1) cysteine-stabilized (e.g., defensins); 2) cysteine-stabilized loop (e.g., protegrins, tachyplesins); 3) linear α-helical (e.g., LL-37, kinocidin helices); 4) enriched in one or more specific amino acid residues (*i.e.*, Bac5); or 5) combinations of the above. Although certain families of host defense peptides, particularly those of higher eukaryotes, show evidence of extensive gene duplication, and in some cases positive selection, their structural (cysteine stabilization, α-helicity) and biophysical (amphipathicity, cationicity) features are often highly conserved.

4. Defensins

Defensins are cationic AMPs produced at a variety of epithelial surfaces by various cells as a component of the innate host defense. They are cyclic peptides which are categorized into three subfamilies on the basis of the disulfide pairings between their six conserved cysteine residues (α- and β-defensins) or their macrocyclic nature (θ-defensins). The α- and β-defensins are widely distributed in vertebrate species, whereas θ-defensins have so far been identified only in Old World monkeys and apparently only in neutrophils and monocytes [11,12]. In mammalian species, around 50 α-defensins and 90 β-defensins have been identified which are either stored in the granules of neutrophils and Paneth cells, or are generated by monocytes/macrophages, keratinocytes or epithelial cells of the respiratory, digestive, urinary and reproductive systems [13,14].

α- and β-defensins modify cell migration and maturation, induce cytokines and trigger histamin and prostaglandin D2 release from mast cells. β-defensins are chemoattractive for immature dendritic cells and memory T cells.

The production of defensins is regulated by pathogens or by inflammatory cytokines released at the site of inflammation. For example, alpha-defensins can be released upon microbial invasion or up-regulated by stimulation with lipopolisaccharides and tumor necrosis factor-α [15]. The β-defensins, found in epithelial cells as well as monocytes/macrophages and dendritic cells, is up-regulated in bacterial infections through the recognition of bacterial components by the Toll-like receptor (TLR) [16-19]. The production and secretion can be also modulated by other AMPs. For example, the human cathelicidin LL37 can synergize with IL-1β to increase the production of IL-6, IL-8, IL-10, CCL2, as well as to increase the synthesis and release of α-defensins [20,21].

On the contrary, downregulation of defensins can be induced by endogenous glucocorticoids leading to increased severity of group A *Streptococcus pyogenes* skin infection [22]. It was also demonstrated that the inactivation of the antimicrobial activity of β-defensins contributes with the recurrent airways infections in patients with cystic fibrosis [23,24].

The antimicrobial activity of defensins has been determined in *Staphylococcus aureus*, *Pseudomonas aeruginosa*, *Salmonella* and certain enveloped viruses [25,26]. This antimicrobial function is supported by experiments performed in knocking-out β-defensin 1 mice [27].

AMPs constitute a substantial part of the mucosal barrier, and β-defensins are the principal type secreted by the epithelium. The four well-characterized human β defensins, hBD-1-4, encoded by β-defensin 1 (DEFB1), DEFB4, DEFB103, and DEFB104, are small (30 - 47 amino acid), cationic, cysteine-rich peptides that possess broad antimicrobial activity [28].

The antimicrobial activity is potentiated under reducing conditions that exist in the hypoxic gut lumen, whereas other antimicrobial peptides, such as hBD-3, are diminished in the reduced state.

The expression of hBD-1 is constitutive, whereas other defensins are expressed in response to microbial and inflammatory stimuli. Defective expression of hBD-1 is associated with mucosal diseases, such as inflammatory bowel disease, candidia carriage, periodontitis, and dental carries.

DEFB114 is a β defensin that exhibited a broad spectrum of antimicrobial activity against typical pathogenic microbes. This AMP also demonstrated to have LPS binding activity while inhibiting the release of TNF-α in RAW264.7 culture when challenged with LPS through the inhibition of MAP kinase p42/44. This peptide also displayed antimicrobial activity against typical pathogens, *i.e.*, *Escherichia coli*, *Staphylococcus aureus*, and *Candida albicans*, while showing low cytotoxicity toward human erythrocytes. All these features are indicative of its potential therapeutic use in the treatment of LPS-induced inflammation [29].

Besides inhibiting microbial growth, an additional function of some of these AMPs is their influence on the immune response by recruiting leukocytes through the induction of chemokines such as CXCL8 (IL-8), CCL2 (MCP-1), IP-10, MIP3α, RANTES and cytokines such as IL-6, IL-10, IL-18, IFNγ and IFN-α. Cells that can be recruited and activated by defensins are neutrophils, monocytes, macrophages, immature dendritic cells, mast cells and T cells [30-32]. Therefore, defensins can tailor the adaptive immune response even by inducing the expression of the costimulatory molecules CD80, CD86 and CD40 on monocytes and dendritic cells [31].

Antigen specific cytotoxic T lymphocytes can be induced by α-defensins and murine β-defensins, thereby enhancing Th1-dependent cellular responses, and potentially anti-tumour immunity [33,34].

However, systemic injections of human α defensins into mice result in augmentation of both Th1 and Th2 immune responses [35]. Whether human β defensin also induce a Th1 and Th2 immune response it is less known since human β defensins are inactive in mice. Experiments with murine β defensins (MBDs) in mice suggest that these defensins induce mainly a Th1 response [36].

On the other hand, it was shown that human defensins co-administered with antigens in mice resulted in an increased production of antibodies, providing evidence of the role of these peptides in humoral response [37]. Other functions of defensins have been deeply discussed by Oppenheim [36]. Based on these findings, it seems that the action of defensins on T cell responses is not specific, since these peptides enhance both cellular and humoral cytokine production and immune responses.

As it was mentioned above, a less beneficial effect of β-defensins is also suggested, since β-defensins could amplify inflammation by binding CCR6 expressed by Th17 cells [38,39].

In 2008, a study by Sass *et al.* [40], evaluated the antibiotic mode of action of β-defensin (hβD-3) against *Staphylococcus aureus*. They showed that hβD-3 induces alterations in the *S. aureus* core cell wall stress regulon in a similar fashion to the cell wall-active antibiotic vancomycin. This observation is consistent with the concept that hβD-3 perturbs specific component(s) of the cell wall, triggering upregulation of cell wall stress response pathways. Further, hβD-3 inhibits purified enzymes required for cell wall synthesis. Finally, transmission electron microscopy revealed that exposure to hβD-3 causes apparent fissures in the cell wall that allowed for protrusions of membrane bound cytoplasmic contents of *S. aureus*.

5. Cathelicidin

The name cathelicidin comes from the highly conserved N-terminal region of these proteins known as the cathelin domain. This protein contains two disulfide bonds between cysteine residues C85-C96 and C107-C124 [41] and was given its name based on an ability to inhibit the protease cathepsin-L. The cathelicidin LL-37 is the only member of the cathelicidin family expressed in humans [42].

The main source of cathelicidin in the human body is azurophilic granules of neutrophil, but it can also be found in mucosal surfaces and in keratinocytes during inflammation [42]. Another study also demonstrated the increased expression of LL-37 in macrophages, endothelial cells and T cells in atherosclerotic lesions [43].

Unlike defensins, cathelicidin expression in humans is less directly modulated by TLRs or cytokines, but rather relies on the action and modulation of Vitamin D in specific tissues [44].

Like other AMPs, cathelicidins kill bacteria through permeabilization of bacterial cell membranes and binding to LPS [45]. The microbicidal activity of cathelicidin was further supported by *in vivo* experiments with cathelicidin knockout mice [46]. Furthermore, cathelicin LL-37 acts on mast cells and increases the expression of TLR4, while releasing histamine, prostaglandins and the cytokines IL-4, IL-5, IL-1β [47] and IL-8 in airway epithelial cells and keratinocytes [48]. Thus, it acts as a chemokine for neutrophils, monocytes, mast cells and T cells but not for dendritic cells [49]. Cathelicidin also suppresses neutrophil apoptosis, and on the other hand promotes apoptosis of epithelial cells [49].

It has been demonstrated that LL-37 is a potent modifier of DC differentiation. LL-37-derived DC displayed significantly up-regulated endocytic capacity, modified phagocytic receptor expression and function, up-regulated co stimulatory molecule expression, enhanced secretion of Th-1 inducing cytokines, and promoted Th1 responses *in vitro* [50].

However, Kandler *et al.* showed that LL-37 suppressed the maturation and activation of human dendritic cells in response to a number of TLR ligands and thus the *ex vivo* costimulation of T cells [51]. Furthermore, LL-37 increased the level of TLR4 and induced the release of Th2 cytokines IL-4, IL-5 and IL-1β from mast cells [47]. Besides, LL37 synergistically enhances the IL-1β induced production of cytokines (IL-6, IL-10) and chemokines such as MCP-1 and MCP-3 in human peripheral blood monocytes [20]. These findings suggest that LL-37 induces mainly a Th2 response.

So far, there is no data on LL-37 and the induction of Th17, although Peric, M. *et al.* [52] reported that human keratinocytes could increase the expression of cathelicidin when stimulated in the presence of vitamin D(3) and this increase was signaled through the IL-17RA.

The hCAP-18 prepropeptide is mainly produced in leucocytes and epithelial cells. After secretion, processing to the active AMP occurs by local proteases. As the proteolytic activity of various cells and tissues differs, hCAP-18 can be processed either to the full-length active peptide LL-37 found in the exocytosed material of neutrophils or to multiple smaller peptides such as RK-31 and KS-30, which occur on the skin surface and display increased antimicrobial activity.

In neutrophils, hCAP-18 is stored in the specific granules and, upon degranulation, is processed to LL-37 by protease 3 released from the azurophilic granules. In the skin, RK-31 and KS-30 are produced by the action of SCTE (stratum corneum tryptic enzyme, kallikrein 5/hK5) and SCCE (stratum corneum chymotryptic enzyme, kallikrein 7/hK7). High concentrations of hCAP-18 are present in seminal plasma. Proteolytic cleavage by semen-derived gastricsin occurs when the pH drops in the vagina generating the active peptide ALL-38. Similar to the defensins, cathelicidin peptides exert various immunomodulatory functions. As a striking example for the importance of cathelicidin for human health, LL-37 deficiency was found to be associated with chronic peridontal disease. High levels of hCAP-18/LL-37 are produced in healing human skin suggesting an important role in re-epithelialization during wound healing.

Consistently, healing was inhibited by LL-37 specific antibodies in an organ-cultured human skin model and in chronic ulcers, hCAP-18/LL-37 was found to be absent in ulcer edge epithelium.

Although not a direct measure of microbicidal activity, numerous studies have demonstrated that LL-37 binds LPS with high affinity [53-55] and hence may limit its endotoxic properties as well as downstream immunological signaling events.

In Gram negative bacteria, resistance to cationic antimicrobial peptides is conferred via activation of PhoP/PhoQ and PmrA/PmrB two component regulatory systems, which lead to increased incorporation of positively charged l-4-aminoarabinose subunits into LPS and other adaptive responses [56]. This molecular modification results in a more neutral or even cationic surface charge, and hence reduced electrostatic attraction for cationic antimicrobial peptides, affording increased survival of organisms.

Although the above studies are not a direct measure of LL-37's mechanism of action, they do suggest that members of this classical α-helical host defense peptides family may have specific interactions with certain Gram-negative cell wall component structures.

6. SLPI

Secretory Leukocyte Protease Inhibitor (SLPI) is a serine protease inhibitor of cathepsin G, trypsin and chymotrypsin, but primarily against neutrophil elastase [57]. SLPI was first isolated from bronchial secretions and was later found to be also produced by many mucosal surfaces, keratinocytes, neutrophils and macrophages.

This polycationic non-glycosylated peptide display antimicrobial properties *in vivo* and *in vitro* [58,59]. The antimicrobial activity of human SLPI has been described for *Escherichia coli*, *Pseudomonas aeruginosa*, *Staphylococcus aureus*, *Staphylococcus epidermidis*, *Streptococcus sp.*, *Aspergillus fumigatus*, *Candida albicans*, and HIV [58]. It has been recently reported that mouse and even human SLPI shows antibacterial activity against mycobacteria and it constitutes a pattern recognition receptor for mycobateria that not only kills bacteria but facilitates their phagocytosis by murine and human macrophages [60]. The antimycobacterial activity of SLPI resides in the WAP (Whey Acidic Protein) domains of the molecule and is quite similar to other cationic peptides [61]. The antimicrobial activity clearly does not depend on inhibition of serine proteases. Instead, disruption of the membranes of target organisms is observed, most likely depending of the cationic nature of these small proteins.

As it was mentioned before, SLPI is constitutively produced by epithelial cells and neutrophils and alveolar macrophages. Its expression is upregulated by diverse inflammatory stimuli such as TNFα and *M. tuberculosis* and downmodulated by adenoviral infection, TGFβ1 cytokine production and during chronic obstructive lung disease [62-67]. In a recent review, the multifaceted roles of SLPI and elafin/trappin-2 were examined in the context of their possible use as inhaled drugs for treating chronic lung diseases such as CF (cystic fibrosis) and COPD (chronic obstructive pulmonary disease) [68].

6.1. SLPI as a Biomarker

Jendeberg *et al.* measured the antimicrobial peptide concentration in plasma from patients with community-acquired pneumonia (CAP), in this work the authors found that in subjects with CAP, mean plasma concentrations of SLPI and bactericidal/permeability-increasing protein (BPI) were significantly higher than in healthy control subjects, but less markedly increased in patients with non-respiratory tract infections. The finding of higher SLPI levels in male subjects with CAP implies that there are sex-dependent immunological differences in SLPI turnover [69].

The relationships between rhinovirus infection and bacterial infection and the role of antimicrobial peptides in COPD exacerbations were also investigated. In this studies the antimicrobial peptides SLPI, elafin, pentraxin, LL-37, α-defensins and β-defensin-2, and the protease neutrophil elastase were evaluated. The authors concluded that neutrophil elastase was significantly increased and SLPI and elafin significantly reduced after rhinovirus infection exclusively in subjects with COPD with secondary bacterial infections, and SLPI and elafin levels correlated inversely with bacterial load. Rhinovirus infections are frequently followed by secondary bacterial infections in COPD and cleavage of the antimicrobial peptides SLPI and elafin by virus-induced neutrophil elastase may precipitate in these secondary bacterial infections. Therapy targeting neutrophil elastase or enhancing innate immunity may be useful novel therapies for prevention of secondary bacterial infections in virus-induced COPD exacerbations [70].

6.2. SLPI as Anti-Inflammatory Molecule

Besides the antimicrobial activity, SLPI major function is to inhibit inflammation by blocking the proteolitic activity of serine proteinases released by leukocytes and also through downmodulating several cytokines up-regulated via LPS such as TNFα, MCP-1 and IL-6 [71-75].

The anti-inflammatory activity is also mediated by inhibition of proteolitic degradation of IkB and the activation of the transcription factor nuclear factor NFkB [75, 76]. Moreover, SLPI knock out mice show impaired cutaneous wound healing with increased activation of local TGFβ [77].

The role of SLPI in adaptive immunity is less clear. Samsom *et al.* proposed that SLPI expression in dendritic cells located in cervical lymph node contributes to mucosal tolerance [78]. Furthermore, SLPI modulates antibody class switching by dampening IgG and IgA class switching recombination without affecting B cell proliferation [79]. However, splenic B cells from SLPI knock out mice have higher proliferation rates and produce higher levels of IgM [80]. Overall these data indicate that SLPI may affect adaptive immune response, however, it is not known whether SLPI might affect directly the classical Th1 or Th2 pattern of cytokines.

Another serine protease inhibitor of the same family of SLPI, named ELAFIN, can activate lung dendritic cells and produce a bias to Th1 response, in an *in vivo* murine model [81]. Moreover, it was suggested that SLPI could also favour a Th1 response, since SLPI can inhibit LPS-induced production of PGE2 from monocytes and PGE2 skews T-helper lymphocyte cytokine production towards a Th2 immune response [82]. However, this data does not agree with the suppression of inflammation observed in an arthritis mouse model [83]. The activity of SLPI on adaptive immune response seems to be complex. It was observed that SLPI-treated human peripheral mononuclear cells may release Th2 cytokines. Therefore, further studies are required to clarify the role of SLPI in T helper cells immune response.

Although not yet understood in detail, SLPI was demonstrated to be a pivotal endogenous factor necessary for optimal cutaneous and mucosal wound healing, most likely due to its antiprotease activity [77,84,85].

6.3. SLPI in HPV and HIV Infection

Hoffmann, M., *et al* demonstrated that exposure of human cervical epithelial cells to high-risk human papillomaviruses results in a reduction in the expression of SLPI [86]. In head and neck squamous cell carcinoma (HNSCC) tissue, HPV DNA was analyzed and correlated with SLPI expression. A possible role of smoking on SLPI expression in clinically normal mucosa was also investigated, 19 patients treated for non-malignant diseases (non-HNSCC) were analyzed for SLPI expression and correlated with smoking habits. In HNSCC patients, SLPI expression showed a significant inverse correlation with HPV status. For the first time, a correlation between SLPI downregulation and HPV infection was demonstrated, suggesting that high levels of SLPI, possibly induced by environmental factors such as tobacco smoking, correlate with protective effects against HPV infection [87].

Depressed cervicovaginal SLPI levels have been correlated with *Trichomonas vaginalis* infection. SLPI levels were lower in females with a positive *T. vaginalis* antigen test result, a vaginal pH > 4.5, vaginal leukocytosis, and recurrent *T. vaginalis* infection. SLPI level was reduced by >50% in a *T. vaginalis* load-dependent manner. The SLPI level could be used as a vaginal-health marker to evaluate interventions and vaginal products [88].

SLPI was identified as a factor responsible for selective anti-HIV activity of human saliva [89]. It appears that this activity results from binding of SLPI to host cell membrane associated proteins such as scramblase and/or annexin II, rather than direct interaction with the virus particle [90,91]. SLPI have also been associated with anti-HIV activity of vaginal fluid [92,93]. Mechanistically, SLPI has been shown to inhibit HIV-1 infection of macrophages by binding to and blocking cell surface annexin A2 [34]. Annexin A2 is found at the cell surface as the annexin A2 heterotetramer (A2t) consisting of two annexin A2 monomers and an S100A10 dimer [94,95], which are co-expressed by basal epithelial cells [96]. It was also demonstrated that the annexin A2 heterotetramer (A2t) contributes to HPV16 (the most common high-risk genotype) infection and co-immunoprecipitates with HPV16 particles on the surface of epithelial cells in an L2-dependent manner. Inhibiting A2t with an endogenous annexin A2 ligand, SLPI, or with an annexin A2 antibody significantly reduces HPV16 infection [97].

6.4. SLPI as a PRR

The innate immune system of the host is able to detect "pathogen-associated molecular patterns" (PAMPs), conserved molecular structures produced exclusively by microorganisms that are essential for the physiology of microbes [98]. PAMPs are recognized by pattern recognition receptors (PRRs) of the host. Some of them, like TLR2 and TLR4, are present on the cell surface of cells from the innate immune system, in contrast, other PRRs, like TLR3 and TLR7, are found inside the cells. Finally, the last group of known PRRs, such as surfactants and C-reactive protein, are soluble and present in extracellular body fluids. The recognition of PAMPs by PRRs facilitates the uptake of the pathogen or the signaling needed for the induction of the appropriate immune response of the host [98,99]. AMPs are able to bind and kill a pathogen but it is not clear whether most of these peptides could be considered as a PRR, specifically binding to PAMPs and facilitating the clearing by macrophages. Gomez *et al.* [59] demonstrated that human recombinant SLPI not only kills mycobacteria but also acts as a pattern recognition receptor for the host immune system.

7. Conclusions

There is no doubt that AMPs have a microbicidal activity and a role in modulating the immune response. Moreover, some of them, like SLPI, acts like pattern recognition receptors that facilitate the clearing of the pathogens by immune cells.

Therefore, in the era of antibiotic resistance they look as excellent candidates to develop new antimicrobial agents. However, they may elicit an array of different actions on adaptive immune response that it has to be determined before the new drugs become available. The activities of these peptides should be analyzed under different physiological and pathological settings, since the activity may vary depending on the immune response elicited by the pathogen.

Understanding these pleiotropic molecules that function as "natural antibiotics" and "immune regulators" has great promise for yielding new strategies in the control of human disease.

REFERENCES

[1] R. A. Dorschner, *et al.*, "Cutaneous Injury Induces the Release of Cathelicidin Anti-Microbial Peptides Active against Group A Streptococcus," *Journal of Investigative Dermatology*, Vol. 117, No. 1, 2001, pp. 91-97. http://dx.doi.org/10.1046/j.1523-1747.2001.01340.x

[2] M. Zasloff, "Inducing Endogenous Antimicrobial Peptides to Battle Infections," *Proceedings of the National Academy of Sciences of the United States of America*, Vol. 103, No. 24, 2006, pp. 8913-8914. http://dx.doi.org/10.1073/pnas.0603508103

[3] P. Elsbach, "What Is the Real Role of Antimicrobial Polypeptides That Can Mediate Several Other Inflammatory Responses?" *Journal of Clinical Investigation*, Vol. 111, No. 11, 2003, pp. 1643-1645.

[4] D. Yang, *et al.*, "Multiple Roles of Antimicrobial Defensins, Cathelicidins, and Eosinophil-Derived Neurotoxin in Host Defense," *Annual Review of Immunology*, Vol. 22, 2004, pp. 181-215. http://dx.doi.org/10.1146/annurev.immunol.22.012703.104603

[5] R. E. Hancock, "Cationic Peptides: Effectors in Innate Immunity and Novel Antimicrobials," *Lancet Infectious Diseases*, Vol. 1, No. 3, 2001, pp. 156-164. http://dx.doi.org/10.1016/S1473-3099(01)00092-5

[6] M. C. Territo, *et al.*, "Monocyte-Chemotactic Activity of

Defensins from Human Neutrophils," *Journal of Clinical Investigation*, Vol. 84, No. 6, 1989, pp. 2017-2020. http://dx.doi.org/10.1172/JCI114394

[7] O. Chertov, *et al.*, "Identification of Defensin-1, Defensin-2, and CAP37/Azurocidin as T-Cell Chemoattractant Proteins Released from Interleukin-8-Stimulated Neutrophils," *Journal of Biological Chemistry*, Vol. 271, No. 6, 1996, pp. 2935-2940. http://dx.doi.org/10.1074/jbc.271.6.2935

[8] D. M. Bowdish, *et al.*, "The Human Cationic Peptide LL-37 Induces Activation of the Extracellular Signal-Regulated Kinase and p38 Kinase Pathways in Primary Human Monocytes," *Journal of Immunology*, Vol. 172, No. 6, 2004, pp. 3758-3765.

[9] J. Shi, *et al.*, "A Novel Role for Defensins in Intestinal Homeostasis: Regulation of IL-1beta Secretion," *Journal of Immunology*, Vol. 179, No. 2, 2007, pp. 1245-1253.

[10] B. L. Kagan, T. Ganz and R. I. Lehrer, "Defensins: A Family of Antimicrobial and Cytotoxic Peptides," *Toxicology*, Vol. 87, No. 1-3, 1994, pp. 131-149. http://dx.doi.org/10.1016/0300-483X(94)90158-9

[11] H. Jenssen, P. Hamill and R. E. Hancock, "Peptide Antimicrobial Agents," *Clinical Microbiology Reviews*, Vol. 19, No. 3, 2006, pp. 491-511. http://dx.doi.org/10.1128/CMR.00056-05

[12] Y. Q. Tang, *et al.*, "A Cyclic Antimicrobial Peptide Produced in Primate Leukocytes by the Ligation of Two Truncated Alpha-Defensins," *Science*, Vol. 286, No. 5439, 1999, pp. 498-502. http://dx.doi.org/10.1126/science.286.5439.498

[13] A. Fahlgren, *et al.*, "Increased Expression of Antimicrobial Peptides and Lysozyme in Colonic Epithelial Cells of Patients with Ulcerative Colitis," *Clinical & Experimental Immunology*, Vol. 131, No. 1, 2003, pp. 90-101. http://dx.doi.org/10.1046/j.1365-2249.2003.02035.x

[14] Y. Lai and R. L. Gallo, "AMPed up Immunity: How Antimicrobial Peptides Have Multiple Roles in Immune Defense," *Trends in Immunology*, Vol. 30, No. 3, 2009, pp. 131-141. http://dx.doi.org/10.1016/j.it.2008.12.003

[15] D. Yang, *et al.*, "Beta-Defensins: Linking Innate and Adaptive Immunity through Dendritic and T Cell CCR6," *Science*, Vol. 286, No. 5439, 1999, pp. 525-528. http://dx.doi.org/10.1126/science.286.5439.525

[16] J. Harder, *et al.*, "Differential Gene Induction of Human Beta-Defensins (hBD-1, -2, -3, and -4) in Keratinocytes Is Inhibited by Retinoic Acid," *Journal of Investigative Dermatology*, Vol. 123, No. 3, 2004, pp. 522-529. http://dx.doi.org/10.1111/j.0022-202X.2004.23234.x

[17] D. Proud, S. P. Sanders and S. Wiehler, "Human Rhinovirus Infection Induces Airway Epithelial Cell Production of Human Beta-Defensin 2 both *in Vitro* and *in Vivo*," *Journal of Immunology*, Vol. 172, No. 7, 2004, pp. 4637-4645.

[18] P. Vora, *et al.*, "Beta-Defensin-2 Expression Is Regulated by TLR Signaling in Intestinal Epithelial Cells," *Journal of Immunology*, Vol. 173, No. 9, 2004, pp. 5398-5405.

[19] J. Harder, *et al.*, "A Peptide Antibiotic from Human Skin," *Nature*, Vol. 387, No. 6636, 1997, p. 861. http://dx.doi.org/10.1038/43088

[20] J. Yu, *et al.*, "Host Defense Peptide LL-37, in Synergy with Inflammatory Mediator IL-1beta, Augments Immune Responses by Multiple Pathways," *Journal of Immunology*, Vol. 179, No. 11, 2007, pp. 7684-7691.

[21] Y. Zheng, *et al.*, "Cathelicidin LL-37 Induces the Generation of Reactive Oxygen Species and Release of Human Alpha-Defensins from Neutrophils," *British Journal of Dermatology*, Vol. 157, No. 6, 2007, pp. 1124-1131. http://dx.doi.org/10.1111/j.1365-2133.2007.08196.x

[22] K. M. Aberg, *et al.*, "Psychological Stress Downregulates Epidermal Antimicrobial Peptide Expression and Increases Severity of Cutaneous Infections in Mice," *Journal of Clinical Investigation*, Vol. 117, No. 11, 2007, pp. 3339-3349. http://dx.doi.org/10.1172/JCI31726

[23] J. J. Smith, *et al.*, "Cystic Fibrosis Airway Epithelia Fail to Kill Bacteria Because of Abnormal Airway Surface Fluid," *Cell*, Vol. 85, No. 2, 1996, pp. 229-236. http://dx.doi.org/10.1016/S0092-8674(00)81099-5

[24] M. J. Goldman, *et al.*, "Human Beta-Defensin-1 Is a Salt-Sensitive Antibiotic in Lung That Is Inactivated in Cystic Fibrosis," *Cell*, Vol. 88, No. 4, 1997, pp. 553-560. http://dx.doi.org/10.1016/S0092-8674(00)81895-4

[25] L. C. Huang, *et al.*, "*In Vitro* Activity of Human Beta-Defensin 2 against *Pseudomonas aeruginosa* in the Presence of Tear Fluid," *Antimicrobial Agents and Chemotherapy*, Vol. 51, No. 11, 2007, pp. 3853-3860. http://dx.doi.org/10.1128/AAC.01317-06

[26] R. I. Lehrer and T. Ganz, "Antimicrobial Peptides in Mammalian and Insect Host Defence," *Current Opinion in Immunology*, Vol. 11, No. 1, 1999, pp. 23-27. http://dx.doi.org/10.1016/S0952-7915(99)80005-3

[27] G. Morrison, *et al.*, "Characterization of the Mouse Beta Defensin 1, Defb1, Mutant Mouse Model," *Infection and Immunity*, Vol. 70, No. 6, 2002, pp. 3053-3060. http://dx.doi.org/10.1128/IAI.70.6.3053-3060.2002

[28] C. J. Kelly, *et al.*, "Fundamental Role for HIF-1Alpha in Constitutive Expression of Human Beta Defensin-1," *Mucosal Immunology*, Vol. 6, 2013, pp. 1110-1118. http://dx.doi.org/10.1038/mi.2013.6

[29] H. Yu, *et al.*, "The Novel Human Beta-Defensin 114 Regulates Lipopolysaccharide(LPS)-Mediated Inflammation and Protects Sperm from Motility Loss," *Journal of Biological Chemistry*, Vol. 288, No. 17, 2013, pp. 12270-12282. http://dx.doi.org/10.1074/jbc.M112.411884

[30] D. Yang, *et al.*, "Mammalian Defensins in Immunity: More Than Just Microbicidal," *Trends in Immunology*, Vol. 23, No. 6, 2002, pp. 291-296. http://dx.doi.org/10.1016/S1471-4906(02)02246-9

[31] N. Funderburg, *et al.*, "Human-Defensin-3 Activates Professional Antigen-Presenting Cells via Toll-Like Receptors 1 and 2," *Proceedings of the National Academy of Sciences of the United States of America*, Vol. 104, No. 47, 2007, pp. 18631-18635. http://dx.doi.org/10.1073/pnas.0702130104

[32] F. Niyonsaba, *et al.*, "The Human Beta-Defensins (-1, -2, -3, -4) and Cathelicidin LL-37 Induce IL-18 Secretion through p38 and ERK MAPK Activation in Primary Human Keratinocytes," *Journal of Immunology*, Vol. 175, No. 3, 2005, pp. 1776-1784.

[33] J. W. Lillard, Jr., *et al.*, "Mechanisms for Induction of Acquired Host Immunity by Neutrophil Peptide Defensins," *Proceedings of the National Academy of Sciences of the United States of America*, Vol. 96, No. 2, 1999, pp. 651-656. http://dx.doi.org/10.1073/pnas.96.2.651

[34] A. Biragyn, *et al.*, "Mediators of Innate Immunity that Target Immature, But Not Mature, Dendritic Cells Induce Antitumor Immunity When Genetically Fused with Nonimmunogenic Tumor Antigens," *Journal of Immunology*, Vol. 167, No. 11, 2001, pp. 6644-6653.

[35] K. Tani, *et al.*, "Defensins Act as Potent Adjuvants That Promote Cellular and Humoral Immune Responses in Mice to a Lymphoma Idiotype and Carrier Antigens," *International Immunology*, Vol. 12, No. 5, 2000, pp. 691-700.

[36] J. J. Oppenheim, A. Biragyn, L. W. Kwak and D. Yang, "Roles of Antimicrobial Peptides such as Defensins in Innate And Adaptive Immunity," *Annals of the Rheumatic Diseases*, Vol. 62, Suppl. 2, 2003, pp. ii17-ii21. http://dx.doi.org/10.1136/ard.62.suppl_2.ii17

[37] K. A. Brogden, M. Heidari, R. E. Sacco, D. Palmquist, J. M. Guthmiller, G. K. Johnson, H. P. Jia, B. F. Tack and P. B. McCray Jr., "Defensin-Induced Adaptive Immunity in Mice And Its Potential in Preventing Periodontal Disease," *Oral Microbiology and Immunology*, Vol. 18, No. 2, 2003, pp. 95-99. http://dx.doi.org/10.1034/j.1399-302X.2003.00047.x

[38] J. K. Kolls, P. B. McCray Jr. and Y. R. Chan, "Cytokine-Mediated Regulation of Antimicrobial Proteins," *Nature Reviews. Immunology*, Vol. 8, No. 11, 2008, pp. 829-835.

[39] E. V. Acosta-Rodriguez, L. Rivino, J. Geginat, D. Jarrossay, M. Gattorno, A. Lanzavecchia, F. Sallusto and G. Napolitani, "Surface Phenotype and Antigenic Specificity of Human Interleukin 17-Producing T Helper Memory Cells," *Nature Immunology*, Vol. 8, No. 6, 2007, pp. 639-646.

[40] V. Sass, U. Paga, A. Tossib, G. Bierbaumc and H. G. Sahl, "Mode of Action of Human *Beta-Defensin* 3 against *Staphylococcus aureus* and Transcriptional Analysis of Responses to Defensin Challenge," *International Journal of Medical Microbiology*, Vol. 298, No. 7-8, 2008, pp. 619-633. http://dx.doi.org/10.1016/j.ijmm.2008.01.011

[41] J. F. Sanchez, F. Wojcik, Y.-S. Yang, M.-P. Strub, J. M. Strub, A. Van Dorsselaer, M. Martin, R. Lehrer, T. Ganz, A. Chavanieu, B. Calas and A. Aumelasa, "Overexpression and Structural Study of the Cathelicidin Motif of the Protegrin-3 Precursor," *Biochemistry*, Vol. 41, No. 1, 2002, pp. 21-30. http://dx.doi.org/10.1021/bi010930a

[42] U. H. Durr, U. S. Sudheendra and A. Ramamoorthy, "LL-37, the Only Human Member of the Cathelicidin family of Antimicrobial Peptides," *Biochimica et Biophysica Acta (BBA)—Biomembranes*, Vol. 1758, No. 9, 2006, pp. 1408-1425. http://dx.doi.org/10.1016/j.bbamem.2006.03.030

[43] K. Edfeldt, B. Agerberth, M. E. Rottenberg, G. H. Gudmundsson, X. B. Wang, K. Mandal, Q. B. Xu and Z. Q. Yan, "Involvement of the Antimicrobial Peptide LL-37 in Human Atherosclerosis," *Arteriosclerosis, Thrombosis, and Vascular Biology*, Vol. 26, No. 7, 2006, pp. 1551-1557. http://dx.doi.org/10.1161/01.ATV.0000223901.08459.57

[44] S. Yim, P. Dhawan, C. Ragunath, S. Christakos and G. Diamond, "Induction of Cathelicidin in Normal and CF Bronchial Epithelial Cells by 1,25-Dihydroxyvitamin D_3," *Journal of Cystic Fibrosis*, Vol. 6, No. 6, 2007, pp. 403-410. http://dx.doi.org/10.1016/j.jcf.2007.03.003

[45] M. Behuliak, R. Pálffy, R. Gardlík, J. Hodosy, L. Halčák and P. Celec, "Variability of Thiobarbituric Acid Reacting Substances in Saliva," *Disease Markers*, Vol. 26, No. 2, 2009, pp. 49-53. http://dx.doi.org/10.1155/2009/175683

[46] M. Chromek, Z. Slamová, P. Bergman, L. Kovács, L. Podracká, I. Ehrén, T. Hökfelt, G. H Gudmundsson, R. L Gallo, B. Agerberth and A. Brauner, "The Antimicrobial Peptide Cathelicidin Protects the Urinary Tract against Invasive Bacterial Infection," *Nature Medicine*, Vol. 12, No. 6, 2006, pp. 636-641. http://dx.doi.org/10.1038/nm1407

[47] M. Yoshioka, N. Fukuishi, Y. Kubo, H. Yamanobe, K. Ohsaki, Y. Kawasoe, M. Murata, A. Ishizumi, Y. Nishii, N. Matsui and M. Akagi, "Human Cathelicidin CAP18/LL-37 Changes Mast Cell Function toward Innate Immunity," *Biological and Pharmaceutical Bulletin*, Vol. 31, No. 2, 2008, pp. 212-216. http://dx.doi.org/10.1248/bpb.31.212

[48] G. S. Tjabringa, J. Aarbiou, D. K. Ninaber, J. W. Drijfhout, O. E. Sørensen, N. Borregaard, K. F. Rabe and P. S. Hiemstra, "The Antimicrobial Peptide LL-37 Activates Innate Immunity at the Airway Epithelial Surface by Transactivation of the Epidermal Growth Factor Receptor," *Journal of Immunology*, Vol. 171, No. 12, 2003, pp. 6690-6696.

[49] A. Nijnik and R. E. Hancock, "The Roles of Cathelicidin LL-37 in Immune Defences and Novel Clinical Applications," *Current Opinion in Hematology*, Vol. 16, No. 1, 2009, pp. 41-47. http://dx.doi.org/10.1097/MOH.0b013e32831ac517

[50] D. J. Davidson, A. J. Currie, G. S. Reid, D. M. Bowdish, K. L. MacDonald, R. C. Ma, R. E. Hancock and D. P. Speert, "The Cationic Antimicrobial Peptide LL-37 Modulates Dendritic Cell Differentiation and Dendritic Cell-Induced T Cell Polarization," *Journal of Immunology*, Vol. 172, No. 2, 2004, pp. 1146-1156.

[51] K. Kandler, R. Shaykhiev, P. Kleemann, F. Klescz, M. Lohoff, C. Vogelmeier and R. Bals, "The Anti-Microbial Peptide LL-37 Inhibits the Activation of Dendritic Cells by TLR Ligands," *International Immunology*, Vol. 18, No. 12, 2006, pp. 1729-1736. http://dx.doi.org/10.1093/intimm/dxl107

[52] M. Peric, S. Koglin, S. M. Kim, S. Morizane, R. Besch, J. C. Prinz, T. Ruzicka, R. L. Gallo and J. Schauber "IL-17A Enhances Vitamin D3-Induced Expression of Cathelicidin Antimicrobial Peptide in Human Keratinocytes," *Journal of Immunology*, Vol. 181, No. 12, 2008, pp. 8504-8512.

[53] Y. Rosenfeld, N. Papo and Y. Shai, "Endotoxin (Lipopolysaccharide) Neutralization by Innate Immunity Host-De-

fense Peptides. Peptide Properties and Plausible Modes of Action," *The Journal of Biological Chemistry*, Vol. 281, No. 3, 2006, pp. 1636-1643.
http://dx.doi.org/10.1074/jbc.M504327200

[54] C. Junkes, R. D. Harvey, K. D. Bruce, R. Dölling, M. Bagheri and M. Dathe, "Cyclic Antimicrobial R-, W-Rich Peptides: The Role of Peptide Structure and *E. coli* Outer and Inner Membranes in Activity and the Mode of Action," *European Biophysics Journal*, Vol. 40, No. 4. 2011, pp. 515-528.
http://dx.doi.org/10.1007/s00249-011-0671-x

[55] P. N. Domadia, A. Bhunia, A. Ramamoorthy and S. Bhattacharjya, "Structure, Interactions, and AntiBacterial Activities of MSI-594 Derived Mutant Peptide MSI-594F5A in Lipopolysaccharide Micelles: Role of the Helical Hairpin Conformation in Outer-Membrane Permeabilization," *Journal of American Chemical Society*, Vol. 132, No. 51, 2010, pp. 18417-18428.
http://dx.doi.org/10.1021/ja1083255

[56] M. R. Yeaman and N. Y. Yount, "Mechanisms of Antimicrobial Peptide Action and Resistance," *Pharmacological Reviews*, Vol. 55, No. 1, 2003, pp. 27-55.
http://dx.doi.org/10.1124/pr.55.1.2

[57] S. E. Williams, T. I. Brown, A. Roghanian and J. M. Sallenave, "SLPI and Elafin: One Glove, Many Fingers," *Clinical Science*, Vol. 110, No. 1, 2006, pp. 21-35.
http://dx.doi.org/10.1042/CS20050115

[58] J. M. Sallenave, "Antimicrobial Activity of Antiproteinases," *Biochemical Society Transactions*, Vol. 30, No. 2, 2002, pp. 111-115.
http://dx.doi.org/10.1042/BST0300111

[59] S. A. Gomez, C. L. Argüelles, D. Guerrieri, N. L. Tateosian, N. O. Amiano, R. Slimovich, P. C. Maffia, E. Abbate, R. M. Musella, V. E. Garcia and H. E. Chuluyan, "Secretory Leukocyte Protease Inhibitor: A Secreted Pattern Recognition Receptor for Mycobacteria," *American Journal of Respiratory and Critical Care Medicine*, Vol. 179, No. 3, 2009, pp. 247-253.
http://dx.doi.org/10.1164/rccm.200804-615OC

[60] J. Nishimura, H. Saiga, S. Sato, M. Okuyama, H. Kayama, H. Kuwata, S. Matsumoto, T. Nishida, Y. Sawa, S. Akira, Y. Yoshikai, M. Yamamoto and K. Takeda, "Potent Antimycobacterial Activity of Mouse Secretory Leukocyte Protease Inhibitor," *Journal of Immunology*, Vol. 180, No. 6, 2008, pp. 4032-4039.

[61] C. Verma, S. Seebah, S. M. Low, L. Zhou, S. P. Liu, J. Li and R. W. Beuerman, "Defensins: Antimicrobial Peptides for Therapeutic Development," *Biotechnology Journal*, Vol. 2, No. 11, 2007, pp. 1353-1359.
http://dx.doi.org/10.1002/biot.200700148

[62] H. Saitoh, T. Masuda, S. Shimura, T. Fushimi and K. Shirato, "Secretion and Gene Expression of Secretory Leukocyte Protease Inhibitor by Human Airway Submucosal Glands," *American Journal of Physiology. Lung Cellular and Molecular Physiology*, Vol. 280, No. 1, 2001, pp. L79-L87.

[63] S. van Wetering, A. C. van der Linden, M. A. van Sterkenburg, K. F. Rabe, J. Schalkwijk and P. S. Hiemstra, "Regulation of Secretory Leukocyte Proteinase Inhibitor (SLPI) Production by Human Bronchial Epithelial Cells: Increase of Cell-Associated SLPI by Neutrophil Elastase," *Journal of Investigative Medicine*,. Vol. 48, No. 5, 2000, pp. 359-366.

[64] B. L. Luo, R. C. Niu, J. T. Feng, C. P. Hu, X. Y. Xie and L. J. Ma, "Downregulation of Secretory Leukocyte Proteinase Inhibitor in Chronic Obstructive Lung Disease: The Role of TGF-beta/Smads Signaling Pathways," *Archives of Medical Research*, Vol. 39, No. 4, 2008, pp. 388-396.
http://dx.doi.org/10.1016/j.arcmed.2008.02.002

[65] Y. Higashimoto, Y. Yamagata, T. Iwata, T. Ishiguchi, M. Okada, M. Masuda, H. Satoh and H. Itoh, "Adenoviral E1A Suppresses Secretory Leukoprotease Inhibitor and Elafin Secretion in Human Alveolar Epithelial Cells and Bronchial Epithelial Cells," *Respiration*, Vol. 72, No. 6, 2005, pp. 629-635.

[66] F. Jaumann, A. Elssner, G. Mazur, S. Dobmann and C. Vogelmeier, "Transforming Growth Factor-Beta1 Is a Potent Inhibitor of Secretory Leukoprotease Inhibitor exPression in a Bronchial Epithelial Cell Line. Munich Lung Transplant Group," *European Respiratory Journal*, Vol. 15, No. 6, 2000, pp. 1052-1057.
http://dx.doi.org/10.1034/j.1399-3003.2000.01513.x

[67] A. H. Ding, H. W. Yu, J. X. Yang, S. P. Shi and S. Ehrt, "Induction of Macrophage-Derived SLPI by Mycobacterium Tuberculosis Depends on TLR2 but Not MyD88," *Immunology*, Vol. 116, No. 3, 2005, pp. 381-389.
http://dx.doi.org/10.1111/j.1365-2567.2005.02238.x

[68] M. L. Zani, A. Tanga, A. Saidi, H. Serrano, S. Dallet-Choisy, K. Baranger and T. Moreau, "SLPI and Trappin-2 as Therapeutic Agents to Target Airway Serine Proteases in Inflammatory Lung Diseases: Current and Future Directions," *Biochemical Society Transactions*, Vol. 39, No. 5, 2011, pp. 1441-1446.
http://dx.doi.org/10.1042/BST0391441

[69] A. L. Jendeberg, K. Stralin and O. Hultgren, "Antimicrobial Peptide Plasma Concentrations in Patients with Community-Acquired Pneumonia," *Scandinavian Journal of Infectious Diseases*, Vol. 45, No. 6, 2013, pp. 432-437.
http://dx.doi.org/10.3109/00365548.2012.760844

[70] P. Mallia, *et al.*, "Rhinovirus Infection Induces Degradation of Antimicrobial Peptides and Secondary Bacterial Infection in Chronic Obstructive Pulmonary Disease," *American Journal of Respiratory and Critical Care Medicine*, Vol. 186, No. 11, 2012, pp. 1117-1124.
http://dx.doi.org/10.1164/rccm.201205-0806OC

[71] T. Moreau, K. Baranger, S. Dadé, S. Dallet-Choisy, N. Guyot and M. L. Zan, "Multifaceted Roles of Human Elafin and Secretory Leukocyte Proteinase Inhibitor (SLPI), Two Serine Protease Inhibitors of the Chelonianin Family," *Biochimie*, Vol. 90, No. 2, 2008, pp. 284-295.
http://dx.doi.org/10.1016/j.biochi.2007.09.007

[72] S. Weldon and C. C. Taggart, "Innate Host Defense Functions of Secretory Leucoprotease Inhibitor," *Experimental Lung Research*, Vol. 33, No. 10, 2007, pp. 485-491.
http://dx.doi.org/10.1080/01902140701756547

[73] J. X. Yang, J. Zhu, D. X. Sun and A. H. Ding, "Suppression of Macrophage Responses to Bacterial Lipopolysac-

charide (LPS) by Secretory Leukocyte Protease Inhibitor (SLPI) Is Independent of Its Anti-Protease Function," *Biochimica et Biophysica Acta*, Vol. 1745, No. 3, 2005, pp. 310-317.
http://dx.doi.org/10.1016/j.bbamcr.2005.07.006

[74] F. Jin, C. F. Nathan, D. Radzioch and A. Ding, "Lipopolysaccharide-Related Stimuli Induce Expression of the Secretory Leukocyte Protease Inhibitor, a Macrophage-Derived Lipopolysaccharide Inhibitor," *Infection and Immunity*, Vol. 66, No. 6, 1998, pp. 2447-2452.

[75] A. B. Lentsch, J. A Jordan, B. J. Czermak, K. M. Diehl, E. M. Younkin, V. Sarma and P. A. Ward, "Inhibition of NF-kappaB Activation and Augmentation of IkappaBbeta by Secretory Leukocyte Protease Inhibitor during Lung Inflammation," *The American Journal of Pathology*, Vol. 154, No. 1, 1999, pp. 239-247.
http://dx.doi.org/10.1016/S0002-9440(10)65270-4

[76] C. C. Taggart, S. A. Cryan, S. Weldon, A. Gibbons, C. M. Greene, E. Kelly, T. B. Low, S. J. O'Neill and N. G. McElvaney, "Secretory Leucoprotease Inhibitor Binds to NF-kappaB Binding Sites in Monocytes and InHibits p65 Binding," *The Journal of Experimental Medicine*, Vol. 202, No. 12, 2005, pp. 1659-1668.
http://dx.doi.org/10.1084/jem.20050768

[77] G. S. Ashcroft, K. J. Lei, W. W. Jin, G. Longenecker, A. B. Kulkarni, T. Greenwell-Wild, H. Hale-Donze1, G. McGrady, X. Y. Song and S. M. Wahl, "Secretory Leukocyte Protease Inhibitor Mediates Non-Redundant Functions Necessary for Normal Wound Healing," *Nature Medicine*, Vol. 6, No.10, 2000, pp. 1147-1153.
http://dx.doi.org/10.1038/80489

[78] J. N. Samsom, A. P. van der Marel, L. A. van Berkel, J. M. van Helvoort, Y. Simons-Oosterhuis, W. Jansen, M. Greuter, R. L. Nelissen, C. M. Meeuwisse, E. E. Nieuwenhuis, R. E. Mebius and G. Kraal, "Secretory Leukoprotease Inhibitor in Mucosal Lymph Node Dendritic Cells Regulates the Threshold for Mucosal Tolerance," *Journal of Immunology*, Vol. 179, No. 10, 2007, pp. 6588-6595.

[79] W. Xu, A. Chiu, A. Chadburn, M. Shan, M. Buldys, A. Ding, D. M. Knowles, P. A. Santini and A. Cerutti, "Epithelial Cells Trigger Frontline Immunoglobulin Class Switching through a Pathway Regulated by the Inhibitor SLPI," *Nature Immunology*, Vol. 8, No. 3, 2007, pp. 294-303.

[80] A. Nakamura, Y. Mori, K. Hagiwara, T. Suzuki, T. Sakakibara, T. Kikuchi, T. Igarashi, M. Ebina, T. Abe, J. Miyazaki, T. Takai and T. Nukiwa, "Increased Susceptibility to LPS-Induced Endotoxin Shock in Secretory Leukoprotease Inhibitor (SLPI)-Deficient Mice," *Journal of Experimental Medicine*, Vol. 197, No. 5, 2003, pp. 669-674.
http://dx.doi.org/10.1084/jem.20021824

[81] A. Roghanian, S. E. Williams, T. A. Sheldrake, T. I. Brown, K. Oberheim, Z. Xing, S. E. M. Howie and J. M. Sallenave, "The Antimicrobial/Elastase Inhibitor Elafin Regulates Lung Dendritic Cells and Adaptive Immunity," *American Journal of Respiratory Cell and Molecular Biology*, Vol. 34, No. 5, 2006, pp. 634-642.
http://dx.doi.org/10.1165/rcmb.2005-0405OC

[82] Y. Zhang, D. L. DeWitt, T. B. McNeely, S. M. Wahl and L. M. Wahl, "Secretory Leukocyte Protease Inhibitor Suppresses the Production of Monocyte Prostaglandin H Synthase-2, Prostaglandin E2, and Matrix Metalloproteinases," *Journal of Clinical Investigation*, Vol. 99, No. 5, 1997, pp. 894-900.
http://dx.doi.org/10.1172/JCI119254

[83] X. Y. Song, Li. Zenga, W. W. Jina, J. Thompsona, D. E. Mizela, K. J. Leia, R. C. Billinghurstb, A. R. Pooleb and S. M. Wahl, "Secretory Leukocyte Protease Inhibitor Suppresses the Inflammation and Joint Damage of Bacterial Cell Wall-Induced Arthritis," *Journal of Experimental Medicine*, Vol. 190, No. 4, 1999, pp. 535-542.
http://dx.doi.org/10.1084/jem.190.4.535

[84] J. Zhu, C. Nathan, W. W. Jin, D. Sim, G. S. Ashcroft, S. M. Wahl, L. Lacomis, H. Erdjument-Bromage, P. Tempst, C. D. Wright and A. H. Ding, "Conversion of Proepithelin to Epithelins: Roles of SLPI and Elastase in Host Defense and Wound Repair," *Cell*, Vol. 111, No. 6, 2002, pp. 867-878.
http://dx.doi.org/10.1016/S0092-8674(02)01141-8

[85] N. Angelov, N. Moutsopoulos, M. J. Jeong, S. Nares, G. Ashcroft and S. M. Wahl, "Aberrant Mucosal Wound Repair in the Absence of Secretory Leukocyte Protease Inhibitor," *Thrombosis and Haemostasis*, Vol. 92, No. 2, 2004, pp. 288-297.

[86] E. Fakioglu, S. S. Wilson, P. M. M. Mesquita, E. Hazrati1, N. Cheshenko, J. A. Blaho and B. C. Herold, "Herpes Simplex Virus Downregulates Secretory Leukocyte Protease Inhibitor: A Novel Immune Evasion Mechanism," *Journal of Virology*, Vol. 82, No. 19, 2008, pp. 9337-9344.
http://dx.doi.org/10.1128/JVI.00603-08

[87] M. Hoffmann, E. S. Quabius, S. Tribius, L. Hebebrand, T. Görögh, G. Halec, T. Kahn, J. Hedderich, C. Röcken, J. Haag, T. Waterboer, M. Schmitt, A. R. Giuliano and W. M. Kast, "Human Papillomavirus Infection in Head and Neck Cancer: The Role of the Secretory Leukocyte Protease Inhibitor," *Oncology Reports*, Vol. 29, No. 5, 2013, pp. 1962-1968.

[88] J. S. Huppert, B. Huang, C. Chen, H. Y. Dawood and R. N. Fichorova, "Clinical Evidence for the Role of Trichomonas Vaginalis in Regulation of Secretory Leukocyte Protease Inhibitor in the Female Genital Tract," *The Journal of Infectious Diseases*, Vol. 207, No. 9, 2013, pp. 1462-1470.
http://dx.doi.org/10.1093/infdis/jit039

[89] T. B. McNeely, D. C. Shugars, M. Rosendahl, C. Tucker, S. P. Eisenberg and S. M. Wahl, "Inhibition Of Human Immunodeficiency Virus Type 1 Infectivity by Secretory Leukocyte Protease Inhibitor Occurs Prior to Viral Reverse Transcription," *Blood*, Vol. 90, No. 3, 1997, pp. 1141-1149.

[90] C. C. Tseng and C. P. Tseng, "Identification of a Novel Secretory Leukocyte Protease Inhibitor-Binding Protein Involved in Membrane Phospholipid Movement," *FEBS Letters*, Vol. 475, No. 3, 2000, pp. 232-236.
http://dx.doi.org/10.1016/S0014-5793(00)01700-2

[91] G. Ma, T. Greenwell-Wild, K. J. Lei, W. W. Jin, J. Swisher, N. Hardegen, C. T. Wild and S. M. Wahl, "Secretory Leukocyte Protease Inhibitor Binds to Annexin II,

a Cofactor for Macrophage HIV-1 Infection," *Journal of Experimental Medicine*, Vol. 200, No. 10, 2004, pp. 1337-1346. http://dx.doi.org/10.1084/jem.20041115

[92] K. Pillay, A. Coutsoudis, A. K. Agadzi-Naqvi, L. Kuhn, H. M. Coovadia and E. N. Janoff, "Secretory Leukocyte Protease Inhibitor in Vaginal Fluids And Perinatal Human Immunodeficiency Virus Type 1 Transmission," *The Journal of Infectious Diseases*, Vol. 183, No. 4, 2001, pp. 653-656. http://dx.doi.org/10.1086/318535

[93] S. M. Iqbal, *et al.*, "Elevated Elafin/Trappin-2 in The Female Genital Tract Is Associated with Protection Against HIV Acquisition," *AIDS*, Vol. 23, No. 13, 2009, pp. 1669-1677. http://dx.doi.org/10.1097/QAD.0b013e32832ea643

[94] D. M. Waisman, "Annexin II Tetramer: Structure and Function," *Molecular and Cellular Biochemistry*, Vol. 149-150, No. 1, 1995, pp. 301-322. http://dx.doi.org/10.1007/BF01076592

[95] U. Rescher and V. Gerke, "S100A10/p11: Family, Friends and Functions," *Pflügers Archiv*, Vol. 455, No. 4, 2008, pp. 575-582. http://dx.doi.org/10.1007/s00424-007-0313-4

[96] E. Pena-Alonso, *et al.*, "Annexin A2 Localizes to the Basal Epithelial Layer and Is Down-Regulated in Dysplasia and Head and Neck Squamous Cell Carcinoma," *Cancer Letters*, Vol. 263, No. 1, 2008, pp. 89-98. http://dx.doi.org/10.1016/j.canlet.2007.12.029

[97] A. W. Woodham, D. M. Da Silva, J. G. Skeate, A. B. Raff, M. R. Ambroso, H. E. Brand, J. M. Isas, R. Langen and W. M. Kast, "The S100A10 Subunit of the Annexin A2 Heterotetramer Facilitates L2-Mediated Human Papillomavirus Infection," *PLoS ONE*, Vol. 7, No. 8, 2012, Article ID: e43519. http://dx.doi.org/10.1371/journal.pone.0043519

[98] R. Medzhitov and C. A. Janeway Jr., "Decoding the Patterns of Self and Nonself by the Innate Immune System," *Science*, Vol. 296, No. 5566, 2002. pp. 298-300. http://dx.doi.org/10.1126/science.1068883

[99] T. Andrews and K. E. Sullivan, "Infections in Patients with Inherited Defects in Phagocytic Function," *Clinical Microbiology Reviews*, Vol. 16, No. 4, 2003, pp. 597-621. http://dx.doi.org/10.1128/CMR.16.4.597-621.2003

14

The Knee Joint Tissues Differ Significantly in TGFβ1 Expression and Its Sensitivity

Sadanand Fulzele[1*], Monte Hunter[1], Rajnikumar Sangani[1], Norman Chutkan[1], Carlos Isales[1], Mark W. Hamrick[2]

[1]Department of Orthopaedic Surgery, Medical College of Georgia, Augusta, USA
[2]Department of Cellular Biology and Anatomy, Medical College of Georgia, Augusta, USA
Email: [*]sfulzele@mail.mcg.edu

ABSTRACT

The knee joint is the largest and most complex joint in the human body. In this study, we investigated TGFβ1 expression in the outer meniscus, inner meniscus and articular cartilage of rabbit and human knee tissue (outer and inner menisci) in order to determine the potential role of this factor in normal meniscal function. We also examined the potential of TGF-β1 stimulation to promote tissue regeneration in the two different regions of rabbit knee meniscus tissue. Immunohistochemical investigations of TGF-β1 were performed on rabbit and human knee tissue. The rabbit outer, inner and articular cartilage cells were culture and stimulated with TGF-β1 followed by cell proliferation assay and extracellular matrix analysis. Regulatory studies were performed using TGF-β1 inhibitors SB-431542 and PD98059. Gene expression was analyzed by quantitative polymerase chain reaction. We found marked regional variation in the expression of TGF-β1 in rabbit and human knee. TGF-β1 expressions are relatively greater in the outer meniscus than inner meniscus. Furthermore, we found that exogenous TGF-β1 stimulation increased cell proliferation and aggrecan synthesis more so in the outer than in the inner meniscus. Articular cartilage tissue shows moderate levels of cell proliferation and ECM synthesis when compared with outer and inner meniscus. These findings suggest that growth factors used to enhance the repair and regeneration of meniscal tissue should be tailored to enhance region-specific variation in cell proliferation and extracellular matrix synthesis.

Keywords: Meniscus; Outer Meniscus; Inner Meniscus; TGFβ1; Articular Cartilage

1. Introduction

The knee joint is one of the largest articulations in the body. Menisci within the knee are crucial to its proper function. The meniscus of the knee is functionally a two-component connective tissue that distributes compressive load and acts as a lubricated bearing surface for rotation and sliding of the femoral condyles upon the tibial plateau [1]. The two components are the outer and inner menisci, which differ in the predominant collagen and proteoglycan isomers that constitute them. The inner meniscus is an avascular zone that contains primarily type II collagen and higher glycosaminoglycans (GAGs), whereas outer meniscus contains mostly type I and less GAGs [2]. A torn meniscus can result from any activity that causes forcefully twist or rotate knee, such as aggressive pivoting or sudden stops and turns. Torn menisci causes significant pain and disability and thus, require expeditious management. Failure of the meniscus to withstand the high stresses applied to it results in the common clinical condition of a meniscal tear. Treatment of tears is confounded by a limited blood supply, which effectively ends at the transition of the inner and outer regions [3]. To aid healing of torn menisci, investigators are now examining the potential of growth factor therapy. Primary candidates among these are basic fibroblast growth factor bFGF [4], TGF-β [5,6] and platelet derived growth factor-AB. Previous studies have indicated a dramatic, order-of-magnitude effect on meniscal cell proliferation with exogenous bFGF in monolayer culture [4].

Other than meniscus, articular cartilage is also important tissue of knee. Damaged articular cartilage has a limited regenerative potential and is responsible for considerable disability in the form of arthritis and joint

[*]Corresponding author.

trauma. Once articular cartilage substance is lost, the damage is generally permanent and is often progressive [7]. In the normal articular joint, cartilage homeostasis is maintained by a balance between the synthesis and degradation of articular cartilage composed of proteoglycans and type II collagen [8]. However, in Osteoarthritis (OA), the balance shifts toward catabolism, leading to cartilage destruction because of excessive production of proteolytic enzymes. On the other hand, the articular cartilage is hard to regenerate during the development of OA. Several investigator have used Insulin-like growth factor I (IGF-I) [9], fibroblast growth factor-2 (FGF-2) [10] and TGFβ [11], the cell-regulatory molecules that promote anabolic and mitogenic activities by articular chondrocytes which may possess therapeutic potential.

The growth factor TGF-β is an important factor for cartilage development (chondrogenesis), its maintenance and regeneration [12]. Transforming growth factor-β (TGF-β) superfamily, composed of TGF-β, bone morphogenic protein (BMP), activin and cartilage-derived growth factor (CDGF) subfamilies, regulates a variety of cellular processes including embryonic differentiation, extracellular matrix formation, cell proliferation and apoptosis [13,14]. TGF-β stimulates chondrocyte differentiation by accumulating chondrocyte-specific gene expression such as type II collagen and aggrecan. In addition, TGF-β can potentially inhibit the release of catalytic factors, which are elevated in osteoarthritis [15]. Based on the available evidence obtained from various *in vitro* and *in vivo* studies, TGF-β is considered a potentially useful agent for the treatment of arthritic conditions.

The aim of this study was to assess the effects of recombinant TGF-β1 on the activity of different region of knee tissue particularly articular cartilage and meniscal cells harvested from the inner and outer zones of the meniscus. We hypothesized that there is regional variation in the expression and stimulatory effect of TGF-β1 on knee tissue. This is the first report showing the regional variation in expression of TGF-β1 in human and rabbit knee tissue as well as differential stimulatory effect of TGFβ1 on the different region of the knee tissue in rabbit model. The study was designed to 1) analyze the steady-state level of TGFβ1 expression and its effect on cell proliferation in different region of knee tissue 2) to determine the TGFβ1 signaling pathway using transcription inhibitor, and 3) to analysis the effect of TGFβ1 on knee chondrocytes specific gene expression such as collagen type II and proteoglycan. This study would also enable the investigators to assess whether cells from the avascular regions of the meniscus have the ability to proliferate and produce extracellular matrix (ECM) in a similar manner as cells from the vascular region, when exposed to TGF-β1 *in vitro*.

2. Material and Method

2.1. Primary Cell Culture

Four month-old New Zealand white rabbits (n = 6) were sacrificed and articular Cartilage, Inner and Outer menisci were harvested. The meniscus was divided approximately at the radial midpoint to separate the inner and outer portions. The cells were isolated by 2-hr digestion at 37°C in 0.05% pronase (Roche Diagnostics, Indianapolis) followed by overnight digestion at 37°C in 0.2% collagenase (type II, Worthington Biochemical, Lakewood, NJ) using F12 medium (Mediatech, Herndon, VA) modified with 4.8 mM $CaCl_2$ (Sigma, St. Louis, MO) and 40 mM HEPES buffer (Sigma). The cells were washed in phosphate buffered saline (PBS, Fisher Biotech, Fair Lawn, NJ) and plated at 2.0×10^4 cells/cm^2 in 100-mm tissue culture plates (Becton Dickinson Labware, Franklin Lakes, NJ), then grown for 10 days with 3× /week changes of supplemented Hams F12 medium (Mediatech) containing 50 U/ml penicillin, 50 ug/ml streptomycin (Mediatech), 1% l-glutamine (Hyclone), and 10% fetal bovine serum (FBS, Hyclone). Cells were treated with 0.25% trypsin (Mediatech) for 5 minutes on a rotating table to ease detachment when plating the test samples.

2.2. Human Patients' Samples

Meniscus tissues from patients (n = 5) were acquired after joint replacement surgeries. We obtained informed consent from each patient. The experimental protocol was approved by the Institutional Review Board. The outer and inner meniscus were separated and used for immunohistochemistry. The tissue were embedded in OCT, snap frozen in liquid nitrogen, and cryostatin sections cut at 6 - 8 um for immunohistochemistry.

2.3. Chondrocytes Culture and Treatment with TGF-β1 and Its Inhibitors

Fibroblasts were passage upon reaching 70% confluence. After the last passage, fibroblasts were expanded until reaching 70% confluence on 12 well culture plates for RNA and 96 well plates for cell proliferation and inhibitory study. Chondrocytes fibroblasts were then starved of serum for 24 h before treatment by replacing serum containing media with DMEM plus 1% Antibiotic plus 1× insulin-transferrin-selenium supplement (Invitrogen). Recombinant TGF-β1 (R & D Systems) was dissolved into the serum-free media at a final concentration of 5 or 10 ng/ml. Stock solutions of inhibits TGF-βRI activity (SB-431542) and inhibitor of MEK1/2 activation (PD98059) were prepared by dissolving these solid anhydrous compounds in dimethyl sulfoxide (DMSO). These stock solutions were then added to serum-free media containing at a final concentration of 20 μm for SB-431542 and 30 μm concentration of PD98059. Fibroblasts were pretreated with SB-431542 and PD98059 for 1 h before treatment with TGF-β1.

2.4. Proliferation Assays

The number of viable cultured cells in proliferation was determined using a Promega CellTiter 96® AQueous One MTS Cell Proliferation Assay. Briefly, cells were plated in triplicate at an initial density of 5000/cm^2 in 96-well plates (BD Labware) using supplemented Hams F12 medium containing 5% FBS to support overnight attachment. The following day, fibroblasts were starved of serum for 24 h before treatment by replacing serum containing media with Ham F12 media plus containing 50 U/ml penicillin, 50 ug/ml streptomycin plus 1× ITS (Insulin Transferrin Selenium supplement (BD Biosciences, Bedford, MA). The next day cells were fed with fresh supplemented Hams F12, substituting the FBS with 1% ITS and adding 5 and 10 ng/ml of recombinant TGF-β1 for 24 hr, 48 hrs and 72 hrs. Cells were washed with PBS twice and add 100 μl of Media and 20 μl of MTS (CellTiter 96® AQueous One Solution Reagent, Promega) assay buffer for 3 hr and incubate at 37˚C in a humidified, 5% CO_2 incubator. Optical density (OD) was read at 490 ηm.

2.5. mRNA Determination by Real-Time Polymerase Chain Reaction

After 16 hrs of TGFb1 stimulation, ribonucleic acid (RNA) was extracted by TRIzol® (Invitrogen), following manufacturer's instructions, and assayed for absorbance at 260 and 280 nm (Helios-Gamma, Thermo Spectronic, Rochester, NY). The RNA was reverse-transcribed into complementary deoxyribonucleic acid (cDNA) using iScript reagents from Bio-Rad on a programmable thermal cycler (PCR-Sprint, Thermo Electron, Milford, MA). 50 ng of cDNA was amplified in each real-time polymerase chain reaction using a Bio-Radi Cycler, ABgene reagents (Fisher scientific) and custom-designed primers for the ECM genes specific (**Table 1**) to the rabbit. An glyceraldehyde-3-phosphate dehydrogenase (GAPDH) threshold cycles was used to normalize the expression of the target genes to the constitutive transcriptional activity.

2.6. Immunohistochemistry

Articular cartilage and portions of the outer and inner meniscus were embedded in OCT, snap frozen in liquid nitrogen, and cryostat in sections cut at 6 - 8 um. Sections were fixed with cold acetone for 5 minutes, blocked in normal donkey serum, and incubated with primary TGFβ1 antibody (Santa Cruz, Inc.) for 2 hrs at room temperature then washed and incubated with FITC-labeled goat anti-mouse secondary antibody. Sections were counterstained with DAPI and mounted using aqueous medium.

2.7. Statistical Analysis

Data are expressed as the mean SD. Differences in measured variables between experimental and control groups were assessed using Student's t-test. A p-value $<$ 0.05 was considered statistically significant in between sample comparisons.

3. Results

3.1. Differential Expression of TGFβ in Rabbit and Human Knee Tissue

The immunostaining results show that TGF-β is present in all different type of rabbit knee tissue. Outer meniscus shows the most abundant amount of TGF-β and inner meniscus the least in rabbit knee tissue (**Figure 1(a)**). The rabbit articular cartilage showed the moderate level of TGF-β (**Figure 1(a)**). Human knee tissue also showed similar type of results as rabbit knee tissue. Outer meniscus showed most expression of TGF-β1 than inner meniscus (**Figure 1(b)**).

3.2. Effect of TGF-β1 on Knee Chondrocytes Proliferation

The rabbit meniscus and articular cartilage cells were grown on 96 well cell culture plates and stimulate with and without TGF-β1. The morphology of cells exposure

Table 1. Nucleotide sequences of rabbit gene primers used for real time-PCR.

Gene	Primer Sequence	Product size in base pair	Annealing temperature (˚C)	Reference
GAPDH	5'GTC GTC TCC TGC GAC TTC AAC 3' 5'TAC CAG GAA ATG AGC TTC ACA AAG 3'	100	60	L23961
COL-2	5'GCT CTG AAC AGC CAA AGG AC 3' 5'TCT GCC CAG TTC AGG TCT CT 3'	191	60	S83370
Aggrecan	5'CTG GGT GTC AGG ACC GTG TA 3' 5'TTC GCC TGT GTA GCA GAT GG 3'	90	60	L38480
Biglycan	5'CCT CCA GGT GGT CTA TCT GC 3' 5'GAG GCT GAT GCC GTT GTA GT 3'	75	60	AF020290

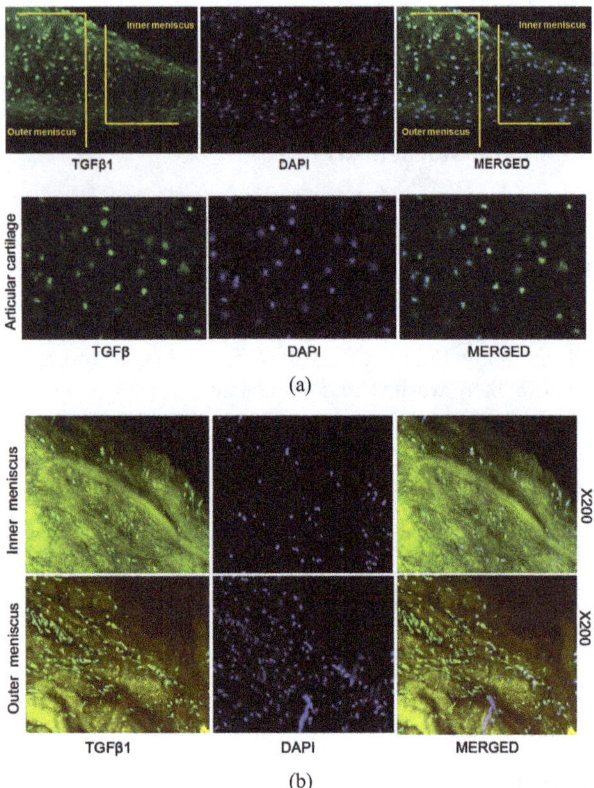

Figure 1. Immunofluorescent staining of (a) inner meniscus, outer meniscus, and articular cartilage tissue using TGF-β1 antibodies for rabbit knee tissue; (b) Immunofluorescent staining of inner and outer meniscus tissue using antibodies specific for human TGFβ1.

to TGF-β1 showed no striking difference compared with the control group. MTT proliferation assay showed that TGF-β1 significantly stimulated the proliferation of all different type of chondrocytes in a dose-dependent manner (**Figure 2**). We found that 10 ηg/ml concentrations stimulate more cell proliferation than 5 ng/ml TGF-β1 (**Figure 3**). So we carried out all experiment with concentration with 10 ng/ml concentration unless it mention. To determine whether TGF-β1 induced the proliferation of chondrocytes, chondrocytes were cultured on 96 well plates and subsequently treated with TGF-β1 for 24, 48 and 72 hrs.

TGF-β1 treatment with concentration of 10 ng/ml increased chondrocytes cell proliferation in outer, inner and articular cartilage cell. Outer meniscus showed more cell proliferation than inner meniscus whereas articular cartilage showed moderate level of cell proliferation.

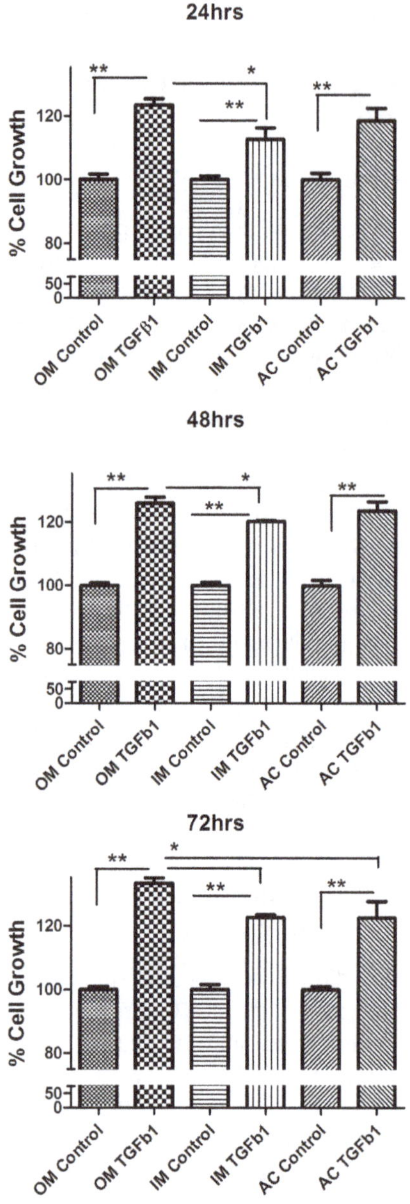

Figure 2. MTT proliferation assay of (a) outer (OM), (b) inner (IM) and (c) articular cartilage (AC) chondrocytes treated with concentrations (10 ng/ml) of recombinant TGF-β1. Data were recorded 24, 48, and 72 hours following treatment (n = 4). Data were analyzed by ANOVA followed by Bonferroni post hoc test ($^{*}p < 0.05$; $^{**}p < 0.01$).

3.3. Influence of Pathway Inhibitors Blocked Cell Growth under TGF-β1 Stimulation

MEK1/2 are critical members of the MAPK pathway that have been shown to be involved in the growth and cell proliferation of cells. PD98059 is a potent and specific cell-permeable inhibitor of MEK1/2 activation. As shown in figure (**Figure 3**), 10 ng/ml TGF-β1 treatment increased cell proliferation in all different types of chondrocytes. Pretreatment with the 30 μm concentration of PD98059 for one hour prior to TGF-β1 treatment, an inhibitor of extracellular signal regulated kinase (ERK1/2) significantly decrease the cell proliferation by 90% for outer, inner and AC cells when compared to with the TGF-β1-treated group (**Figure 3**).

SB-431542 is a novel small molecule that potently in

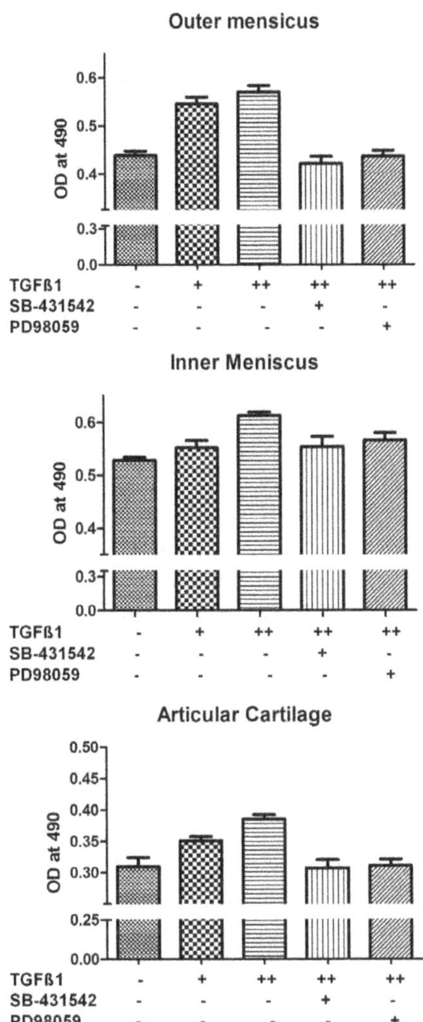

Figure 3. TGF-β1 inhibitors inhibit cell proliferation. MTT assay of inner, outer and articular cartilage chondrocytes treated with concentrations $^{+}$5 ng/ml TGF-β1, $^{++}$10 ng/ml TGF-β1, *20 uM SB-431542 and $^{\#}$PD98059 30 uM (n = 4).

hibits TGF-β type I receptor activity [16,17]. The cells pretreated with 20 μm SB-431542 and followed by 10 ng/ml TGF-β1 treatment showed significant decrease in cell proliferation by 90% - 100% in outer, Inner and AC cells compared to TGF-β1 treated group (**Figure 3**). Taken together, these results suggest that 30 μm concentration of PD98059 and 20 μm SB-431542 can effectively inhibit TGF-β1 signal transduction mediated growth.

3.4. TGF-β1 Induces the Expression of Type II Collagen and Proteoglycan in Chondrocytes

The next step was to determine whether TGF-β1 regulated the expression of collagen type II and proteoglycan like aggrecan and biglycan. We measure the mRNA level of Collagen type II, aggrecan and biglycan in outer, inner and AC chondrocytes after 24 hrs of 10 ng/ml TGF-β1 treatment. TGF-β1 stimulated type II collagen synthesis in outer, inner and AC chondrocytes but most up-regulation in inner meniscus than outer and articular cartilage. The mRNA level of aggrecan and biglycan was also upregulatedin outer, inner and AC chondrocytes after treatment of TGF-β1. Outer meniscus showed most aggrecan up-regulation whereas inner meniscus showed biglycan compare to others (**Figure 4**).

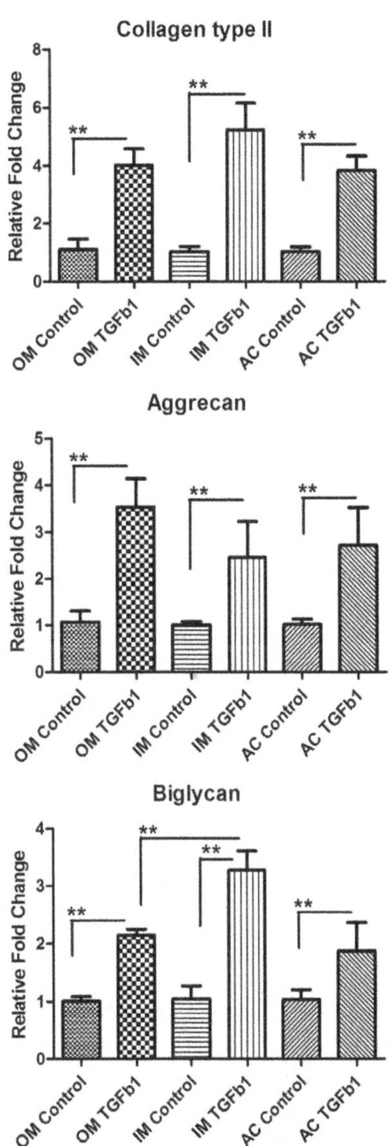

Figure 4. Real-time PCR analysis of (a) type II collagen, (b) Aggrecan, and (c) Biglycan expression in inner (IM), outer (OM) and articular cartilage (AC) chondrocytes treated with 10 ng/ml concentrations of recombinant TGF-β1. Data were recorded after 16 hrs hours following treatment. Data for each sample were normalized with glyceraldehyde-3-phosphate dehydrogenase (GAPDH) mRNA. Data (means ± SD, n = 6) are represented as the fold change in expression compared to control. *P < 0.05. Data were analyzed by ANOVA followed by Bonferroni post hoc test (*p < 0.05; **p < 0.01).

4. Discussion

Although TGF-β1 is a potent inhibitor of growth in most cell types, it has been shown to stimulate growth of certain cells in culture, such as mouse bone marrow mesenchymal stem cells [18], rat and avian articular chondrocytes [19,20], human nasal septal chondrocytes [21] and cells with an osteoblastic phenotype from rat parietal bone [22]. We first did immunostaining of the rabbit knee tissue to know the regional difference in expression of TGF-β level. Our results show that TGF-β express in all different region of rabbit knee with regional variation in expression. The outer meniscus cells shows most staining and then gradually staining goes down toward inner meniscus and this may be the one of the reasons inner meniscus poorly heal when compare to outer meniscus. The articular cartilage shows the moderate level of TGF-β expression when compare to meniscus tissue. Interesting results in rabbit meniscus made us curious to analyze level of TGFβ1 in human meniscus. Interestingly, TGFβ1 expression in inner meniscus is lower than outer meniscus.

The next step was to assess the effect of recombinant TGF-β1 on these different tissues of same rabbit knee joint. Our results indicate that outer, inner meniscus and articular cartilage chondrocytes when cultured in TGF-β1 enriched medium displayed increased cell number compared to controls. In addition, the dose-dependent effect of TGF-β1 on cell proliferation was found between the concentration ranges of 5 - 10 ng/mL. This range is very important in the future therapeutic applications of the TGF-β1 in cartilage healing. This is consistent with previous findings reported by other researcher [5,23]. We determine the regional differences in the cell proliferation response to TGF-β1. The cell proliferation is higher in outer meniscal cells than inner and articular cartilage. This varying response between the tissues of same joints may be due to number of TGFβ receptors per cell. We did not observed any tremendous change in cell morphology in samples cultured with medium containing TGF-β1. It is known that cells in culture have a tendency to change their phenotype and behavior especially in cell lines. Tissue handling is an important factor in securing cell viability and may also affect cell behavior [6]. In this study, the experiments were carried out on meniscus and articular cartilage cells from primary cultures that have been passage only twice to reduce their tendency to dedifferentiate.

To ascertain the effects of TGF-β1, we examined the cell cycle regulatory effect of TGF-β1 in rabbit meniscus and articular cartilage cells *in vitro*. We show that pretreatment with PD98059 significantly blocked the mitogenic and cell cycle promotive effects of TGF-β1 (MTT assay). PD98059 is an inhibitor for MAP kinase kinases 1 and 2 (MKK), also called MAP/ERK kinases (MEK), the upstream activator of ERK1/2. These results suggest that phosphorylated ERK1 is necessary to maintain and promote cell cycle progression under TGF-β1 stimulation. Our results agree with earlier reports showing that ERK1/2 plays a crucial mediating role in mitogenic signaling of TGF-β1 in rat articular chondrocytes [24] and airway smooth muscle cell [25]. We used another small molecule inhibitor SB-431542 that potently inhibits TGF-β RI activity at nanomolar concentrations [16,17]. Our data showed SB-431542 molecule counteract TGF-β1 mediated growth leading to cell inhibition on inner, outer and articular cartilage cell. There are also reports that SB-431542 inhibited TGF-β1 mediated proliferation of osteosarcoma cell line that is growth stimulated in response to TGF-β1 [26].

Chondrocytes and their extracellular matrix (ECM) are the two major components in cartilage biology, both playing critical roles in maintaining tissue integrity and function. We analysis the ECM genes and showed that the presence of type II collagen was enhanced in TGF-β1 treated cultures in outer, inner and articular cartilage. Similar type of finding was reported in meniscus [27] and articular cartilage cell [28] when stimulated by TGF-β1 exogenous treatment or gene transfer. The inner meniscus showed more collagen type II than outer and articular cartilage. This may be due to anatomical variation in the tissue, it is well known that inner meniscus predominantly contain collagen type II and outer meniscus contain collagen type I [2]. Our data also showed that there is increase in proteoglycan synthesis such as aggrecan and biglycan in the presence of TGF-β1. We also found regional variation in biglycan synthesis following TGF-β1 stimulation. The inner meniscus showed significantly up-regulation of biglycan synthesis than outer and articular cartilage. There are reports that TGF-β1 enhanced biglycan synthesis, and increased the length of the GAG chains on all secreted Proteoglycan [5,29] but little is known about the significance of biglycan up regulation following TGF-β1 stimulation. There is no significant different in the regional variation of the aggrecan synthesis of knee joint after TGF-β1 stimulation. To considering clinical application of our results, we should collect more histological and biomechanical data on knee tissues using the current approach in human and larger animal models.

The signal pathways activated by growth factors are complex. It will be interesting to analysis the regional variation in expression of other growth factors (IGF, FGF and platelet derived growth factor) and their receptors which play important roles in cell proliferation and ECM synthesis. In summary, these experiments highlight the regional variation in expression and stimulatory activity of TGF-β1 in articular cartilage and meniscal fibrochondrocytes. This study also indicates that meniscus cells from the avascular zone (inner meniscus) of the meniscus are capable of responding favorably to the addition of TGF-β1. Although this was an *in vitro* study, we made encouraging observations that can form the basis for *in vivo* research aimed at enhancing articular cartilage and meniscal repair, even within the avascular zone, following surgical repair. Such type of study has immense clinical significance because it will give important information about which region of organ is more benefited from particular growth factor that could potentially help to develop better and more effective treatment strategies.

REFERENCES

[1] P. Ghosh, Y. Numata, S. Smith, R. Read, S. Armstrong and K. Johnson, "The Metabolic Response of Articular Cartilage to Abnormal Mechanical Loading Induced by Medial or Lateral Meniscectomy," *Agents and Actions Supplement*, Vol. 39, 1993, pp. 89-93.

[2] J. Sanchez-Adams and K. Athanasiou, "The Knee Meniscus: A Complex Tissue of Diverse Cells," *Cellular and Molecular Bioengineering*, Vol. 2, No. 3, 2009, pp. 332-340. http://dx.doi.org/10.1007/s12195-009-0066-6

[3] S. P. Arnoczky, "Building a Meniscus. Biologic Considerations," *Clinical Orthopaedics and Related Research*, No. 367, 1999, pp. S244-S253. http://dx.doi.org/10.1097/00003086-199910001-00024

[4] M. Cucchiarini, S. Schetting, E. F. Terwilliger, D. Kohn and H. Madry, "rAAV-Mediated Overexpression of FGF-2 Promotes Cell Proliferation, Survival, and Alpha-SMA Expression in Human Meniscal Lesions," *Gene Therapy*, Vol. 16, No. 11, 2009, pp. 1363-1372. http://dx.doi.org/10.1038/gt.2009.91

[5] S. Collier and P. Ghosh, "Effects of Transforming Growth Factor Beta on Proteoglycan Synthesis by Cell and Explant Cultures Derived from the Knee Joint Meniscus," *Osteoarthritis and Cartilage*, Vol. 3, No. 2, 1995, pp. 127-138. http://dx.doi.org/10.1016/S1063-4584(05)80045-7

[6] T. Tanaka, K. Fujii and Y. Kumagae, "Comparison of Biochemical Characteristics of Cultured Fibrochondrocytes Isolated from the Inner and Outer Regions of Human Meniscus," *Knee Surgery, Sports Traumatology, Arthroscopy*, Vol. 7, No. 2, 1999, pp. 75-80. http://dx.doi.org/10.1007/s001670050125

[7] J. A. Buckwalter, H. J. Mankin and A. J. Grodzinsky, "Articular Cartilage and Osteoarthritis," *Instructional Course Lectures*, Vol. 54, 2005, pp. 465-480.

[8] M. W. Lark, E. K. Bayne, J. Flanagan, C. F. Harper, L. A. Hoerrner, N. I. Hutchinson, I. I. Singer, S. A. Donatelli, J. R. Weidner, H. R. Williams, R. A. Mumford and L. S. Lohmander, "Aggrecan Degradation in Human Cartilage. Evidence for Both Matrix Metalloproteinase and Aggrecanase Activity in Normal, Osteoarthritic, and Rheumatoid Joints," *Journal of Clinical Investigation*, Vol. 100, No. 1, 1997, pp. 93-106. http://dx.doi.org/10.1172/JCI119526

[9] L. A. Fortier, H. O. Mohammed, G. Lust and A. J. Nixon, "Insulin-Like Growth Factor-I Enhances Cell-Based Repair of Articular Cartilage," *Journal of Bone & Joint Surgery*, Vol. 84, No. 2, 2002, pp. 276-288. http://dx.doi.org/10.1302/0301-620X.84B2.11167

[10] S. B. Trippel, M. C. Whelan, M. Klagsbrun and S. R. Doctrow, "Interaction of Basic Fibroblast Growth Factor with Bovine Growth Plate Chondrocytes," *Journal of Orthopaedic Research*, Vol. 10, No. 5, 1992, pp. 638-646. http://dx.doi.org/10.1002/jor.1100100506

[11] F. D. Shuler, H. I. Georgescu, C. Niyibizi, R. K. Studer, Z. Mi, B. Johnstone, R. D. Robbins and C. H. Evans, "Increased Matrix Synthesis Following Adenoviral Transfer of a Transforming Growth Factor Beta1 Gene into Articular Chondrocytes," *Journal of Orthopaedic Research*, Vol. 18, No. 4, 2000, pp. 585-592. http://dx.doi.org/10.1002/jor.1100180411

[12] E. N. Blaney Davidson, E. L. Vitters, P. M. van der Kraan and W. B. van den Berg, "Expression of Transforming Growth Factor-Beta (TGFbeta) and the TGF-Beta Signalling Molecule SMAD-2P in Spontaneous and Instability-Induced Osteoarthritis: Role in Cartilage Degradation, Chondrogenesis and Osteophyte Formation," *Annals of the Rheumatic Diseases*, Vol. 65, No. 11, 2006, pp. 1414-1421. http://dx.doi.org/10.1136/ard.2005.045971

[13] H. L. Moses and R. Serra, "Regulation of Differentiation by TGF-Beta," *Current Opinion in Genetics & Development*, Vol. 6, No. 5, 1996, pp. 581-586. http://dx.doi.org/10.1016/S0959-437X(96)80087-6

[14] P. A. Hoodless and J. L. Wrana, "Mechanism and Function of Signaling by the TGF Beta Superfamily," *Current Topics in Microbiology and Immunology*, Vol. 228, 1998, pp. 235-272. http://dx.doi.org/10.1007/978-3-642-80481-6_10

[15] S. H. Tsai, M. T. Sheu, Y. C. Liang, H. T. Cheng, S. S. Fang and C. H. Chen, "TGF-Beta Inhibits IL-1beta-Activated PAR-2 Expression through Multiple Pathways in Human Primary Synovial Cells," *Journal of Biomedical Science*, Vol. 16, 2009, p. 97. http://dx.doi.org/10.1186/1423-0127-16-97

[16] G. J. Inman, F. J. Nicolás, J. F. Callahan, J. D. Harling, L. M. Gaster, A. D. Reith, N. J. Laping and C. S. Hill, "SB-431542 Is a Potent and Specific Inhibitor of Transforming Growth Factor-Beta Superfamily Type I Activin Receptor-Like Kinase (ALK) Receptors ALK4, ALK5, and ALK7," *Molecular Pharmacology*, Vol. 62, No. 1, 2002, pp. 65-74. http://dx.doi.org/10.1124/mol.62.1.65

[17] N. J. Laping, E. Grygielko, A. Mathur, S. Butter, J. Bomberger, C. Tweed, W. Martin, J. Fornwald, R. Lehr, J. Harling, L. Gaster, J. F. Callahan and B. A. Olson, "Inhibition of Transforming Growth Factor (TGF)-Beta1-Induced Extracellular Matrix with a Novel Inhibitor of the TGF-Beta Type I Receptor Kinase Activity: SB-431542," *Molecular Pharmacology*, Vol. 62, No. 1, 2002, pp. 58-64. http://dx.doi.org/10.1124/mol.62.1.58

[18] L. Longobardi, L. O'Rear, S. Aakula, B. Johnstone, K. Shimer, A. Chytil, W. A. Horton, H. L. Moses and A. Spagnoli, "Effect of IGF-I in the Chondrogenesis of Bone Marrow Mesenchymal Stem Cells in the Presence or Absence of TGF-Beta Signaling," *Journal of Bone and Mineral Research*, Vol. 21, No. 4, 2006, pp. 626-636.

http://dx.doi.org/10.1359/jbmr.051213

[19] T. Tsukazaki, T. Usa, T. Matsumoto, H. Enomoto, A. Ohtsuru, H. Namba, K. Iwasaki and S. Yamashita, "Effect of Transforming Growth Factor-Beta on the Insulin-Like Growth Factor-I Autocrine/Paracrine Axis in Cultured Rat Articular Chondrocytes," *Experimental Cell Research*, Vol. 215, No. 1, 1994, pp. 9-16. http://dx.doi.org/10.1006/excr.1994.1307

[20] K. T. Rousche, B. C. Ford, C. A. Praul and R. M. Leach, "The Use of Growth Factors in the Proliferation of Avian Articular Chondrocytes in a Serum-Free Culture System," *Connective Tissue Research*, Vol. 42, No. 3, 2001, pp. 165-174. http://dx.doi.org/10.3109/03008200109005647

[21] J. D. Richmon, A. B. Sage, V. W. Wong, A. C. Chen, C. Pan, R. L. Sah and D. Watson, "Tensile Biomechanical Properties of Human Nasal Septal Cartilage," *American Journal of Rhinology*, Vol. 19, No. 6, 2005, pp. 617-622.

[22] M. Centrella, T. L. McCarthy and E. Canalis, "Transforming Growth Factor Beta Is a Bifunctional Regulator of Replication and Collagen Synthesis in Osteoblast-Enriched Cell Cultures from Fetal Rat Bone," *Journal of Biological Chemistry*, Vol. 262, No. 6, 1987, pp. 2869-2874.

[23] M. K. Akens and M. B. Hurtig, "Influence of Species and Anatomical Location on Chondrocyte Expansion," *BMC Musculoskeletal Disorders*, Vol. 6, 2005, p. 23. http://dx.doi.org/10.1186/1471-2474-6-23

[24] A. Yonekura, M. Osaki, Y. Hirota, T. Tsukazaki, Y. Miyazaki, T. Matsumoto, A. Ohtsuru, H. Namba, H. Shindo and S. Yamashita, "Transforming Growth Factor-Beta Stimulates Articular Chondrocyte Cell Growth through p44/42 MAP Kinase (ERK) Activation," *Endocrine Journal*, Vol. 46, No. 4, 1999, pp. 545-553. http://dx.doi.org/10.1507/endocrj.46.545

[25] G. Chen and N. Khalil, "TGF-Beta1 Increases Proliferation of Airway Smooth Muscle Cells by Phosphorylation of Map Kinases," *Respiratory Research*, Vol. 7, 2006, p. 2. http://dx.doi.org/10.1186/1465-9921-7-2

[26] S. Matsuyama, M. Iwadate, M. Kondo, M. Saitoh, A. Hanyu, K. Shimizu, H. Aburatani, H. K. Mishima, T. Imamura, K. Miyazono and K. Miyazawa, "SB-431542 and Gleevec Inhibit Transforming Growth Factor-Beta-Induced Proliferation of Human Osteosarcoma Cells," *Cancer Research*, Vol. 63, No. 22, 2003, pp. 7791-7798.

[27] H. Goto, F. D. Shuler, C. Niyibizi, F. H. Fu, P. D. Robbins and C. H. Evans, "Gene Therapy for Meniscal Injury: Enhanced Synthesis of Proteoglycan and Collagen by Meniscal Cells Transduced with a TGFbeta(1)Gene," *Osteoarthritis and Cartilage*, Vol. 8, No. 4, 2000, pp. 266-271. http://dx.doi.org/10.1053/joca.1999.0300

[28] P. Galéra, D. Vivien, S. Pronost, J. Bonaventure, F. Rédini, G. Loyau and J. P. Pujol, "Transforming Growth Factor-Beta 1 (TGF-Beta 1) Up-Regulation of Collagen Type II in Primary Cultures of Rabbit Articular Chondrocytes (RAC) Involves Increased mRNA Levels without Affecting mRNA Stability and Procollagen Processing," *Journal of Cellular Physiology*, Vol. 153, No. 3, 1992, pp. 596-606. http://dx.doi.org/10.1002/jcp.1041530322

[29] K. G. Vogel and D. J. Hernandez, "The Effects of Transforming Growth Factor-Beta and Serum on Proteoglycan Synthesis by Tendon Fibrocartilage," *European Journal of Cell Biology*, Vol. 59, No. 2, 1992, pp. 304-313.

15

Expression Patterns of CAPN1 and CAPN8b Genes during Embryogenesis in *Xenopus laevis*

Lucie Abrouk-Vérot, Claire Brun, Jean-Marie Exbrayat
Université de Lyon, UMRS 449, Biologie Générale, Université Catholique de Lyon, Reproduction et Développement Comparé, Ecole Pratique des Hautes Etudes, Lyon, France
Email: labrouk@univ-catholyon.fr, cbrun@univ-catholyon.fr, jmexbrayat@univ-catholyon.fr

ABSTRACT

Calpains are a superfamily of Ca^{2+}-dependent cysteine proteases, implicated in various cellular processes and thus probably necessary in all the stages of cell life. The first extended report of quantification of total RNAs within the developmental stages of *Xenopus laevis* was described in this study. Decreases of total RNAs were positively associated with waves of apoptotic cell death (onset of gastrulation, and morphogenesis). Using qPCR, the temporal expression pattern of CAPN1 and CPAN8b (XCL-2) were characterized during the *Xenopus laevis* embryogenesis. Transcripts of the CAPN1 and CAPN8 genes were detectable from gastrula stage and their levels oscillated throughout development. The expression of the CAPN1 (mu/I) gene was observed in earliest stage, indicating a maternal origin, while expression of the CAPN8b gene was detectable after midblastula transition. The levels of the two transcripts then started to rise again obviously as a result of zygotic expression (stage 11). The CAPN1 gene expression was particularly expressed at tailbud stage, while the CAPN8 transcripts were found at gastrula, neurula and tailbud stages. This is the first report of quantification of mRNAs CAPN8b and CAPN1 (mu/I) within the developmental stages of *Xenopus laevis* by qPCR.

Keywords: Calpain; Embryonic Development; Cell Death; *Xenopus laevis*

1. Introduction

Found in all eucaryotes and some bacteria, the multigenic calpain family encodes calcium-dependent cysteine proteases that vary in their domain structure, expression and substrate specificity [1]. Calpains have a limited proteolytic activity and function in order to modulate their substrates activities. There are typical and atypical proteases according to the presence or absence of the calmodulin-like domain [2,3]. Moreover, some typical calpains are ubiquitous, while others are tissue-specific.

The ubiquitously expressed calpain 1 (CAPN1) is well characterized [4]. It is a heterodimeric enzyme consisting of a large catalytic subunit (80 kDa) and a small regulatory subunit (30 kDa), common with the calpain 2. These two enzymes have mostly indistinguishable substrate and they are ubiquitously expressed, but they differ in the *in vitro* Ca^{2+} requirement (μM versus mM) for the proteolytic activity. The large and small subunits contain four (I-IV) and two (V-VI) domains respectively: the regulatory N-terminal domain (I), the protease domain (II), the C2-domain-like Ca^{2+}/phospholipid-binding domain (III), the penta-EF-hand domain (IV and VI) and the Gly-clustering hydrophobic domain (V). In the presence of Ca^{2+}, the binding of Ca^{2+} to domains II, III, IV and the domain VI induces some conformational changes that allow domain II to form a single active domain [5]. Ubiquitous calpains are synthesized under inactive form, activated after translocation near the plasmic membrane [6]. Their activity is strictly regulated.

The functions of the ubiquitous calpains are fundamental; they are implicated in a large range of biological processes, such as regulation of cell cycle, cell migration, apoptosis or intracellular signaling [7-9]. The calpains are thus probably necessary in all the stages of cell life. Calpain genes were particularly indispensable for early development of vertebrates [10-13]. The CAPN1 (NP_ 001080485.1) distribution was performed during the development of *Xenopus laevis* and its essential role demonstrated [14].

The mammalian typical calpain 8 (CAPN8), previously known as nCL-2, is specifically expressed in the stomach and weakly in the intestine [15,16]. An ortholog of human CAPN8/nCL-2, XCL-2, was identified during embryogenesis of *Xenopus laevis* and it is actually referred by the official symbol CAPN-8b. The role of CAPN-8b was demonstrated during the cell movements of gastrulation and neurulation [17].

Only a few studies regarding the calpains have been realized in amphibian [18]. The aim of the present paper is to compare during the embryogenesis of *Xenopus laevis* the expression profiles of a well-known calpain, the CAPN1, with a calpain that is less studied, the CAPN-8b. The objective is to show whether there are quantitative differences between their expressions. Quantification is carried out by a relative approach comparing the amount of the target mRNA with those of a reference gene, the fixB gene.

2. Materials and Methods

2.1. Embryo Culture

Three independent *in vitro* fertilizations (FIV) (with three different females) were realized following standard procedures. *Xenopus laevis* females were stimulated by hCG (human chorionic gonadotrophin) injection. For each one, *in vitro* fertilized eggs were incubated at 23°C. Only eggs that turned round with the animal pole on the top were harvested. For each FIV, different sets of few embryos from developmental stages 0 (unfertilized eggs), 1 (fertilized eggs), 5 (morula, 2 h 45 pf at 23°C), 8 - 9 (blastula, 5 h pf at 23°C), 11 (gastrula, 11 h 45 pf at 23°C), 20 (neurula, 21 h 45 pf at 23°C), 36 (tailbud, 50 h pf at 23°C), 48 and 58 (metamorphosis, about 2 months) were collected, anesthesized with MS222 (Sigma, St Louis MO), and immediately immersed in RNA*later* RNA stabilization Reagent (Qiagen) with β mercaptoethanol (10 µl β mercaptoethanol for 1 ml of RNA*later* RNA stabilization Reagent) and stored at −70°C. The embryos were staged according to [19].

2.2. RNA Extraction

Total RNAs from each sample were extracted using RNeasy^R Lipid Tissue Mini kit (Qiagen) according to the manufacturer's instruction, including on column DNase treatment. Total RNAs were eluted into 30 µl of elution buffer. The RNAs yield and purity were assessed by spectrophotometric measurements on Nanodrop ND-1000 (NanoDrop technologies). Each sample was measured three times and the average value was determined. The quality of total RNAs was analyzed by 1% ethidium bromide agarose gel electrophoresis. RNAs integrity was estimated using AGILENT RNA 6000 Nano kit on an Agilent 2100 Bioanalyzer (Agilent technologies).

2.3. Reverse Transcription

Total RNAs (1 µg) were reverse transcribed with anchored Oligo-dT. Then, cDNAs were amplified using VersoTM cDNA kit (Thermoscientific) according to the manufacturer's instruction. 50 pg fixB (bacterial gene, spike control) were added to a total volume of 20 µl and incubated at 42°C for 45 min. Enzyme was inactivated at 95°C for 2 min. The reactions were subsequently diluted to 80 µl and frozen at −20°C.

2.4. Primer Design

Sequences of primers for *X. laevis* CAPN1 (mu/I) and CAPN8b were designed by using the Primer3 software (Whitehead Institute/MIT, USA) and purchased from Eurogentec (Seraing, Belgium). These primers have Tms between 52°C and 60°C and the products were 160 and 137 bp. Primers used for qPCR detection of are shown in **Table 1**.

2.5. Real-Time Quantitative PCR Using SYBR Green I

LightCycler DNA Master SYBR Green I (Roche Diagnostics GmbH, Mannheim, Germany) combined with LightCycler Instrument (Roche Diagnostics GmbH, Mannheim, Germany) and suitable primers were used for qRT-PCR, according to the manufacturer's instruction. The LightCycler experimental run protocol consisted of an initial Taq activation at 95°C for 8 min, followed by a "touch down" PCR step of 45 cycles consisting of denaturation at 95°C for 15 s, annealing at 68°C for 5 s, and elongation at 72°C for 8 s. After completed qPCR, melting curves were collected. A slower melt rate of 0.1 degrees C/s was effective for detecting weak amplicons, and improved resolution of the T(m) of amplicons amplified simultaneously. Relative quantification was carried by the standard curve method out using the LightCycler® Relative quantification Software (Roche Diagnostics GmbH, Mannheim, Germany, version 1.01). A series of dilutions of calibrator sample (external standard = pool of four sets of embryos from stage 48 = 2 µg) was included in each experiment in order to generate an external standard curve. That curve was used for quantification of both target and FixB gene (spike control) in each sample. The calculation of data was based on the cross

Table 1. Primers for qPCR detection.

CAPN1 mu/I forward	5'-TAGCTTTGGAGTCTGCAGGT-3'
CAPN1 mu/I reverse	5'-TGTGTGTCCATACCCTGGAA-3'
CAPN8b forward	5'-TAAAGGCAGACTCGGATCGT-3'
CAPN8b reverse	5'-AAGAGCCAGGTCATTGGATG-3'

Primers for bacterial gene fixB (internal standard, spike control) were obtained with Spike Control Kit (Codelink™ iExpress iAmplify cRNA prep and hybrid kit, GE healthcare).

ing point (Cp) values obtained by the lightCycler® Software. Results were calculated as the target/reference ratio of the sample divided by the target/reference ratio of the calibrator. This corrected for sample in homogeneity and variability of detection.

Accession numbers of all sequences used in this study are given in **Table 2**.

3. Results

3.1. Quantification of Total RNAs

Amount of total RNAs per embryo in the course of early *Xenopus laevis* embryogenesis stayed stable up to about stage 36 (tailbud). A decrease was then observed so that the amount of total RNAs at stage 48 was about two-and-half times lower than that at earlier stages. Then the amount of total RNAs began to increase to a maximum at stage 58 (**Figure 1**).

RNA integrity was estimated using AGILENT RNA 6000 Nano kit on an Agilent 2100 Bioanalyzer. The method provided total RNA extracts of good quantity and quality (**Figure 2**).

3.2. Expression Pattern of CAPN8b and CAPN1 (mu/I) during Development. Temporal Expression during Embryogenesis

Using real time PCR, which is currently the most sensitive and reliable technique for quantitative mRNA analy-

Table 2. Accession numbers of sequences used.

Definition	Official symbol	Accession
calpain 1, (mu/I) large subunit	CAPN1	NM_001087016.1
calpain 8	CAPN8-b	NM_001088543.1

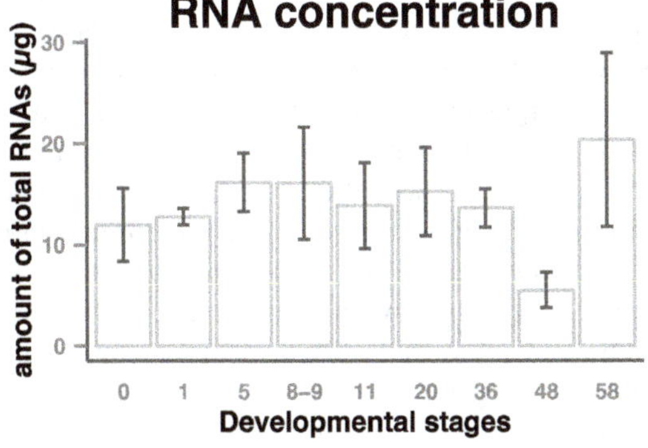

Figure 1. Total RNAs amount in *Xenopus laevis* early developmental stages. RNA was extracted from eggs/embryos and diluted to 30 µl as described in the Materials and Methods section. Amount of total RNAs (columns) per embryo and standard deviations (bars) from 3 or 4 sets of eggs/embryos from stages indicated on the horizontal axes are shown.

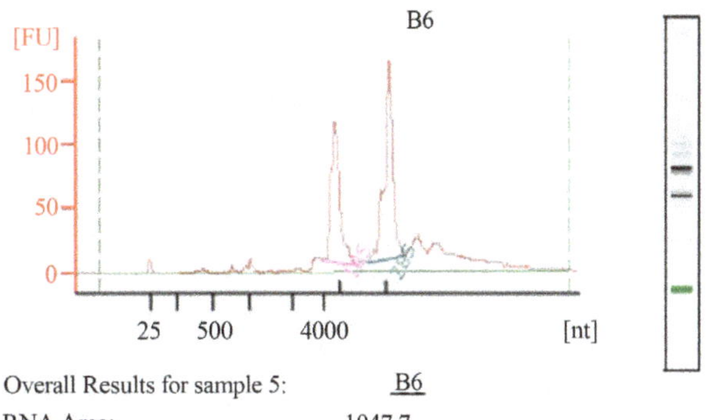

Overall Results for sample 5: B6
RNA Area: 1047.7
RNA Concentration: 674 ng/µl
rRNA Ratio [28s / 18s]: 1.3
RNA Integrity Number (RIN): 10.0 (B.02.05)
Result Flagging Color:
Result Flagging Label: RIN:10
Fragment table for sample 5: B6

Name	Start Size [nt]	End Size [nt]	Area	% of total Area
18S	3.801	5.901	253.4	24.2
28S	6.433	8.701	318.1	30.4

Figure 2. RNAs integrity using AGILENT RNA 6000 Nano kit on an Agilent 2100 Bioanalyzer (Agilent technologies).

sis, the expressions levels of both two calpain genes have been examined and compared during the period of early development of *Xenopus laevis*. The quantitative expression of CAPN8b and CAPN1 (mu/I) mRNAs determined by RT-qPCR during the period of early development of *Xenopus laevis* (9 developmental stages) are shown in **Figure 3**.

CAPN1 (mu/I) mRNAs were already detectable in *Xenopus* one-cell embryos, contrary to CAPN8b mRNA. Following fertilization, the levels of two mRNA stayed stable up to about stage 8 (blastula). Then, the level of two transcripts began to increase to a maximum. It was reached at stage 22 - 43 for RNA CAPN1 or 13 - 21 for RNA CAPN8b. The CAPN1 (mu/I) and CAPN8b mRNA levels gradually decreased up to about stage 58, the last stage tested in this study. The levels of CAPN8b mRNA were kept about 10 times greater than was the level of CAPN1 (mu/I) mRNA from gastrula until the metamorphosis.

4. Discussion

This is the first extended report of quantification of total RNAs within the developmental stages of *Xenopus laevis*. The quantification of total RNA profile presented here complements and correlates the partial quantification data previously published [20]. A light decrease of the concentration of total RNA was observed at stage 11 (gastrula) and was positively associated with a maternal cell death program which is set up at fertilization and abruptly activated at the onset of gastrulation (10.5) [21]. The second wave of apoptotic cell death has been recog-

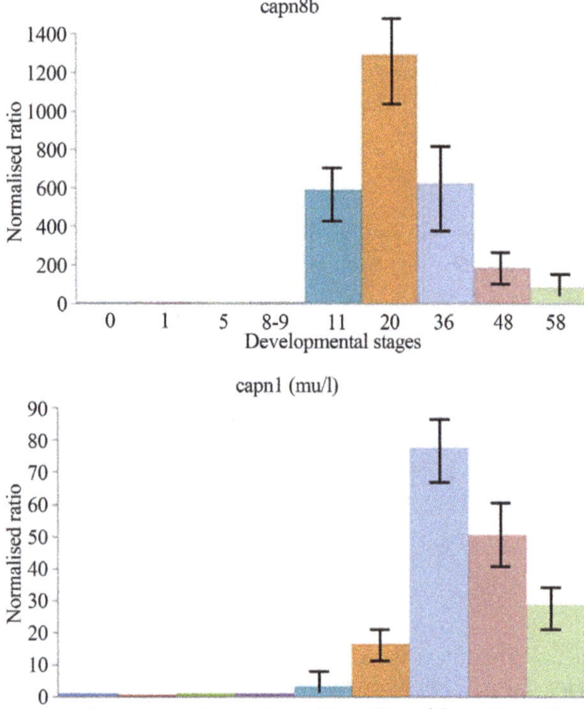

Figure 3. The mRNA expression profiles of *Xenopus laevis* CAPN1 (mu/I) and CAPN8b genes, normalized to total RNA and stage 0 (unfertilized eggs) and expressed in arbitrary units. The numbers on the vertical axis represent the ratio between the average amount of copies of a mRNA at a particular developmental stage and stage zero normalized to the same amount of input RNA (means ± SD, n = 3 replicates). The numbers on the horizontal axis represent the *Xenopus* developmental stages determined according to Niewkoop and Faber [19].

nized to occur throughout neurogenesis (stages 13 to 21) [22] and was correlated with a light increase of total RNA. Stages 35 to 46 represent the end of morphogenesis leading to the development of organs necessary for free floating life. A significant decrease of the concentration of total RNA was introduced at stage 36 to stage 48 and was positively associated with a large wave of apoptotic cell death in several organs (ectoderm, digestive tracts, somites) [23].

qPCR data are frequently normalized by one of several options, including the expression of reference genes, number of cells, weight of tissue, DNA/RNA spike and total RNAs concentration. The RT-qPCR normalization of mRNA expression patterns to reference genes such as ODC, GAPDH, EF-1α, H4 or L8, widely used in *Xenopus* RT-qPCR experiments, is not particulary suitable, because their levels vary during *Xenopus* development. The normalization to total RNAs is more appropriate [20]. Therefore, the mRNA expression profiles of CAPN8b and CAPN1 (mu/I) in stage series were normalized to total RNA of each embryonic stage and to stage 0, and are presented in arbitrary units. This is the first report of quantification of mRNAs CAPN8b and CAPN1 (mu/I) within the developmental stages of *Xenopus laevis* by qPCR.

The CAPN1 (mu/I) mRNAs were expressed maternally contrary to CAPN8b mRNA. Expression of these two genes was remained at a constant level until the onset of gastrulation. CAPN1 (mu/I) and CAPN8b genes display distinct embryonic specific expression patterns, suggesting potentially different developmental roles for these tow genes, during amphibian early embryogenesis.

The levels of two transcripts then started to rise again obviously as a result of zygotic expression. Zygotic expression of CAPN8b mRNA was detectable greatly at stage 11 (gastrulation) and its level of expression increases until stages 13 - 21 (neurulation). Zygotic expression of CAPN1 (mu/I) mRNA was detectable weakly at stage 11 (gastrulation) and its level of expression increased until stages 22 - 43 (tailbud). The CAPN1 (mu/I) and CAPN8b mRNA levels gradually decreased up to about stage 50 (metamorphosis), without nevertheless reached to maternal levels.

While the separate expression patterns of CAPN1 and CAPN8b genes particularly highlight similarity in their expression profiles in the course of early development, we observed quantitative differences in the expression of these genes. The levels of CAPN8b mRNA were kept about 10 times greater than was the level of CAPN1 (mu/I) mRNA from gastrula until the metamorphosis.

The CAPN8b temporal expression profile presented here complemented the partial CAPN8b expression data previously published [17] and also corroborated very well with the RT-PCR expression analysis described in [17]. The temporal expression profile of CAPN8b transcripts determined by RT-qPCR was consistent with the spatial expression of XCL-2 examined by whole-mount *in situ* hybridization and described in [17]. Signals were first detected in the area close to the ventral blastoporal lip at stage 12.5 (gastrula). In late gastrulae and neurulae, XCL-2 gene was expressed exclusively in the ventral circumblastoporal collar and the mesoderm-free zone at the most anterior tip of neural fold where the stomodeum and cement gland form. From the late neurula stage, it was expressed significantly in the cement gland and proctodeum, and after stage 34, expression was only found in the cement gland.

The temporal expression of CAPN1 (mu/I) and CAPN8b transcripts determined by RT-qPCR was correlated with the temporal expression of calpains 1 and 2 examined by immunohistochemistry with antibodies directed against human isoforms and described in [14]. Calpains expression was weak in the early stages, strong between neurulation and growth period and decreased in the second part of metamorphosis. The mRNA expression profiles of CAPN8b and CAPN1 (mu/I) were correlated with the expression of calpains. The high expression of mRNA and calpains between the gastrulation and the metamorphosis suggest their fundamental role in organogenesis and at the onset of metamorphosis. Our data suggest that CAPN1 (mu/I) and CAPN8b are a prerequisite for morphogenetic movements during embryogenesis in *Xenopus laevis*.

The temporal changes in gene expression are a key mechanism in embryo development. Three major changes in the cellular activities take place at the midblastula stage (just after 12 cleavages). This midblastula transition (MBT) was assumed to involve the coordinated initiation of transcription, acquisition of cell motility and lengthening of cell giving rise to G1 and G2 phases [24].

REFERENCES

[1] H. Sorimachi, S. Hata and Y. Ono, "Calpain Chronicle—An Enzyme Family under Multidisciplinary Characterization," *Proceedings of Japan Academy Ser B Physiological and Biological Sciences*, Vol. 87, No. 6, 2011, pp. 287-327.

[2] H. Sorimachi and K. J. Suzuki, "The Structure of Calpain," *Biochemistry*, Vol. 129, No. 5, 2001, pp. 653-664. http://dx.doi.org/10.1093/oxfordjournals.jbchem.a002903

[3] D. E. Goll, V. F. Thompson, H. Li, W. Wei and J. Cong, "The Calpain System," *Physiological Reviews*, Vol. 83, No. 3, 2003, pp. 731-801.

[4] K. Suzuki, S. Hata, Y. Kawabata and H. Sorimachi, "Structure, Activation, and Biology of Calpain," *Diabetes*, Vol. 53, Suppl. 1, 2004, pp. S12-S18. http://dx.doi.org/10.2337/diabetes.53.2007.S12

[5] R. L. Campbell and P. L. Davies, "Structure-Function Relationships in Calpains," *Biochemical Journal*, Vol. 447, No. 3, 2012, pp. 335-351. http://dx.doi.org/10.1042/BJ20120921

[6] K. Suzuki and H. Sorimachi. "A Novel Aspect of Calpain Activation," *FEBS Letters*, Vol. 433, No. 1-2, 1998, pp. 1-4. http://dx.doi.org/10.1016/S0014-5793(98)00856-4

[7] M. Noguchi, A. Sarin, M. J. Aman, H. Nakajima, E. W. Shores, P. A. Henkart and W. J. Leonard, "Functional Cleavage of the Common Cytokine Receptor Gamma Chain (Gammac) by Calpain," *Proceedings of National Academy of Sciences USA*, Vol. 94, No. 21, 1997, pp. 11534-11539. http://dx.doi.org/10.1073/pnas.94.21.11534

[8] L. Leloup, H. Shao, Y. H. Bae, B. Deasy, D. Stolz, P. Roy and A. Wells, "m-Calpain Activation Is Regulated by Its Membrane Localization and by Its Binding to Phosphatidylinositol 4,5-Bisphosphate," *The Journal of Biological Chemistry*, Vol. 285, No. 43, 2010, pp. 33549-334566. http://dx.doi.org/10.1074/jbc.M110.123604

[9] N. N. Danial and S. J. Korsmeyer, "Cell Death: Critical Control Points," *Cell*, Vol. 116, No. 2, 2004, pp. 205-219. http://dx.doi.org/10.1016/S0092-8674(04)00046-7

[10] J. S. Arthur, J. S. Elce, C. Hegadorn, K. Williams and P. A. Greer, "Disruption of the Murine Calpain Small Subunit Gene, CAPN4: Calpain Is Essential for Embryonic Development, but Not for Cell Growth and Division," *Molecular and Cellular Biology*, Vol. 20, No. 12, 2000, pp. 4474-4481. http://dx.doi.org/10.1128/MCB.20.12.4474-4481.2000

[11] U. J. Zimmerman, L. Boring, J. H. Pak, N. Mukerjee and K. K. Wang, "The Calpain Small Subunit Gene Is Essential: Its Inactivation Results in Embryonic Lethality," *International Union of Biochemistry and Molecular Biology Life*, Vol. 50, No. 1, 2000, pp. 63-68.

[12] P. Dutt, D. E. Croall, J. S. Arthur, T. D. Veyra, K. Williams, J. S. Elce and P. A. Greer, "m-Calpain Is Required for Preimplantation Embryonic Development in Mice," *BMC Developmental Biology*, Vol. 24, No. 6, 2006, p. 3. http://dx.doi.org/10.1186/1471-213X-6-3

[13] S. E. Lepage and A. E. Bruce, "Characterization and Comparative Expression of Zebrafish Calpain System Genes during Early Development," *Developmental Dynamics*, Vol. 237, No. 3, 2008, pp. 819-829. http://dx.doi.org/10.1002/dvdy.21459

[14] E. N. Moudilou, N. Mouterfi, J. M. Exbrayat and C. Brun, "Calpains Expression during *Xenopus laevis* Development," *Tissue and Cell*, Vol. 42, No. 5, 2010, pp. 275-281. http://dx.doi.org/10.1016/j.tice.2010.07.001

[15] H. Sorimachi, S. Ishiura and K. Suzuki, "A Novel Tissue-Specific Calpain Species Expressed Predominantly in the Stomach Comprises Two Alternative Splicing Products with and without Ca(2+)-Binding Domain," *Journal of Biological Chemistry*, Vol. 268, No. 26, 1993, pp. 19476-19482.

[16] S. Hata, S. Koyama, H. Kawahara, N. Doi, T. Maeda, N. Toyama-Sorimachi, K. Abe, K. Suzuki and H. Sorimachi, "Stomach-Specific Calpain, nCL-2, Localizes in Mucus Cells and Proteolyzes the Beta-Subunit of Coatomer Complex, Beta-COP," *Journal of Biological Chemistry*, Vol. 281, No. 16, 2006, pp. 11214-11224. http://dx.doi.org/10.1074/jbc.M509244200

[17] Y. Cao, H. Zhao and H. Grunz, "XCL-2 Is a Novel m-Type Calpain and Disrupts Morphogenetic Movements during Embryogenesis in *Xenopus laevis*," *Development, Growth and Differentiation*, Vol. 43, No. 5, 2001, pp. 563-571. http://dx.doi.org/10.1046/j.1440-169X.2001.00592.x

[18] J. M., Exbrayat, E. A. Moudilou, L. Abrouk and C. Brun, "Apoptosis in Amphibian Development," *Advances in Bioscience and Biotechnology*, Vol. 3, No. 6A, 2012, pp. 669-678. http://dx.doi.org/10.4236/abb.2012.326087

[19] P. D. Nieuwkoop and J. Faber, "Normal Table of *Xenopus laevis*," North Holland Publishing Company, Amsterdam, 1967.

[20] R. Sindelka, Z. Ferjentsik and J. Jonák, "Developmental Expression Profiles of *Xenopus laevis* Reference Genes," *Developmental Dynamics*, Vol. 235, No. 3, 2006, pp. 754-758. http://dx.doi.org/10.1002/dvdy.20665

[21] C. Hensey and J. Gautier, "A Developmental Timer that Regulates Apoptosis at the Onset of Gastrulation," *Mechanisms of Development*, Vol. 69, No. 1-2, 1997, pp. 183-195.

http://dx.doi.org/10.1016/S0925-4773(97)00191-3

[22] C. Hensey and J. Gautier, "Programmed Cell Death during *Xenopus* Development: A Spatio-Temporal Analysis," *Developmental Biology*, Vol. 203, No. 1, 1998, pp. 36-48. http://dx.doi.org/10.1006/dbio.1998.9028

[23] J. Estabel, A. Mercer, N. König and J. M. Exbrayat, "Programmed Cell Death in *Xenopus laevis* Spinal Cord, Tail and Other Tissues, Prior to, and during, Metamorphosis," *Life Sciences*, Vol. 73, No. 25, 2003, pp. 3297-3306. http://dx.doi.org/10.1016/j.lfs.2003.06.015

[24] J. Newport and M. Kirschner, "A Major Developmental Transition in Early *Xenopus* Embryos: I. Characterization and Timing of Cellular Changes at the Midblastula Stage," *Cell*, Vol. 30, No. 3, 1982, pp. 675-686. http://dx.doi.org/10.1016/0092-8674(82)90272-0

16

Characteristics of Mesenchymal Stem Cells under Hypoxia

Bruna Amorin[1,2], Ana Paula Alegretti[1,2], Vanessa de Souza Valim[1,2], Annelise Martins Pezzi da Silva[1,2], Maria Aparecida Lima da Silva[1,2], Felipe Sehn[1], Lucia Silla[1,2]

[1]Cell Technology Center, Experimental Research Center, Hospital de Clínicas de Porto Alegre-Porto Alegre, Rio Grande do Sul, Brazil

[2]Graduate Program in Medicine: Medical Sciences, Universidade Federal do Rio Grande do Sul-Porto Alegre, Rio Grande do Sul, Brazil

Email: lsilla@hcpa.ufrgs.br

ABSTRACT

Mesenchymal stem cells (MSC) are considered non-hematopoietic multipotent stem cells with self-renewal properties and the ability to differentiate into a variety of mesenchymal tissues. Optimal conditions for the culture of these cells have been the subject of investigation for several years. In particular, ideal oxygen tension levels have not been established in the literature. In physiological environments, oxygen tension may vary from 12% in peripheral blood to 1% in the deep zone of cartilage regions. In any case, oxygen tension is considerably lower *in vivo* when compared with the normal atmosphere of standard cell culture conditions (21%). The objective of this study was to review the literature available on MSC characteristics (cell cycle, survival, proliferation, differentiation, morphology, immunophenotype, cytogenetics) when cultured under hypoxic conditions. Our focus on optimal culture conditions is justified by the key role currently played by these cells in regenerative medicine.

Keywords: Mesenchymal Stem Cells; Culture; Hypoxia

1. Introduction

Mesenchymal stem cells (MSCs), also referred to mesenchymal stromal cells [1], are considered non-hematopoietic multipotent stem cells with self-renewal properties and the ability to differentiate into mesoderm tissues [2]. MSCs were first described by Friedenstein *et al.* in the 1970s, as cells morphologically similar to fibroblasts and with a high ability to adhere to plastic surfaces [3]. Several subsequent studies have reported the multipotent nature of these cells, *i.e.*, their ability to differentiate into embryonic mesoderm-derived cells, namely, osteocytes, chondroblasts, and adipocytes [4-6].

The first studies designed to assess the effects of different oxygen (O_2) tension levels in MSC culture date back to 1958, when Cooper *et al.* and Zwartouw & Westwood observed that some cells proliferated more rapidly under low O_2 tension levels when compared with normal atmospheric levels [7,8].

In recent years, studies have evidenced that MSC are recruited to areas of tissue damage, such as fractures, myocardial infarction, and ischemic brain lesions, where they become involved in both the regulation of inflammatory response and tissue repair [9,10], and hypoxia appears to be an important regulator of MSC recruitment, migration, and differentiation [11,12], In an animal model, Rochefort *et al.* have shown that MSCs, but not hematopoietic progenitor cells, were mobilized from the bone marrow into peripheral blood through hypoxia [13]. Additionally, low O_2 tension levels have been implicated in the maintenance of stem-cell quiescence and plasticity in general [14,15].

Based on the above, studies on optimal *in vitro* culture conditions have focused on ideal O_2 tension levels [16]. In particular, it has been hypothesized that the survival and proliferation of MSCs can be improved by maintaining cells at low O_2 tension levels although it remains unclear whether different *in vitro* concentrations of O_2 over long periods of time change typical features of MSC [14]. Within this scenario, a relevant fact is that cultures under standard conditions (21% O_2) are exposed to a significantly higher amount of O_2 when compared with physiological *in vivo* conditions [17]. In fact, approximately 1% - 1.5% of the genome appears to be regulated by hypoxia [18].

Recently, with the growing interest in the potential application of MSCs in regenerative medicine, the possibility to obtain a higher rate of proliferation, and the availability of more appropriate methods to change O_2 tension levels in culture,has motivated the publication of several studies on the effects of low O_2 tension levels on MSC behavior and function [16,17,19,21]. It is important to emphasize that subtle differences in culture conditions, e.g., medium supplementation with different growth factors, probably account for the heterogeneity of results found in the literature. The present study discusses several described effects of hypoxia on MSC.

2. Bone Marrow Stromal Cells and MSC Niches

Stromal cells, together with extracellular matrix and soluble regulatory factors, once regarded as secondary components [22], are currently believed to be essential to maintain hematopoiesis [23]. In addition to MSC, stromal cells, both *in vitro* and *in vivo*, are formed by a heterogeneous population of cells including macrophages, fibroblasts, adipocytes, osteoblasts, and endothelial cells. These cells are considered to be the main components of the niche and seem to play a critical role in the regulation of the hematopoietic stem cells [24,25].

The term niche was introduced in 1980 to describe the spatial structure that lodges stem cells [26]. The hematopoietic stem cell niche is located in the bone marrow, and currently there are models that advocate two superimposed populations: the endosteal niche-close to the bone surface, where quiescent hematopoietic stem cells are located and maintained; and the perivascular niche-associated with sinusoidal endothelial cells, where hematopoietic stem cells primarily divide and self-renew. MSCs are present in these two niches, and they participate in hematopoiesis and ontogeny [23].

3. MSC Survival in a Hypoxic Environment

In vivo O_2 tension levels have been described to range between 4% and 7% in the bone marrow, sometimes reaching as low as 1% - 2% [27-30]. Therefore, MSCs and all other stroma cells have to be able to live in a hypoxic microenvironment [31].

MSCs, as all cells, have the ability to effectively change metabolic pathways from aerobic to anaerobic, an essential aspect for the survival of these cells under hypoxic conditions [16]. An experimental study performed with rats on the changes in MSCs under serum-deprivation and hypoxia conditions, concluded that serum deprivation was the main reason leading to ischemia-induced apoptosis of MSCs. However, it also showed that prolonged exposure to hypoxia leaded to mitochondrial dysfunction and Caspase-3 activation, a key factor in apoptosis [32]. Although it has been shown that MSCs are able to withstand hypoxia (e.g., $O_2 < 1\%$) for at least 48 hours (37), the accumulation of lactate resulting from glycolysis could become an inhibitory factor in the long term [16-33].

When bone marrow-derived MSCs are cultured under hypoxic conditions, intracellular signaling pathways associated with cell survival are stimulated such as hypoxia-inducible factor-1 alpha (HIF-1α) that when stabilized migrates into the cell nucleus and combines with hypoxia-inducible factor-1 beta (HIF-1β). Subsequently, these dimeric structures bind to the promoter region of hypoxia-responsive genes, including glucose-6-phosphate transporter (G6PT), which controls gluconeogenesis. The increased level of glucose resulting from gluconeogenesis appears to contribute to MSC survival under hypoxic or serum-deprivation conditions (37-39). Additionally, an increased survival rate under hypoxia, when compared to normoxia, can be attributed not only to the overexpression of HIF-1, but also to an increase in erythropoietin receptors, and anti-apoptotic factors Bcl-2 and Bcl-XL, followed by decreased Caspase-3 levels (70,71). Moreover, interleukin IL-6 and vascular endothelial growth factor (VEGF)-two proangiogenic factors are also stimulated during hypoxia, further contributing to cell survival (70). These beneficial effects are regulated by a complex array of signaling pathways, including Akt and ERK pathways (26,72,73).

Hypoxia and the HIF-1α stabilization cause phosphorylation of the Akt signaling pathway which is degraded under normoxia. When Akt is activated, the expression of pro-apoptotic factor Bax reduces, and the expression of anti-apoptotic factor Bcl-2 increases. Such overexpression may interact with the Bax accumulated in the mitochondrion, triggering apoptosis or further stabilization of HIF-1α causing, as mentioned above, its translocation into the cell nucleus and the activation of hypoxia-responsive genes such as G6PT and angiogenesis-related factors such as VEGF and IL-6 (18).

In addition to O_2 tension levels, the optimal culture time of MSCs under hypoxic conditions also remains to be determined. Wang *et al.* used short time hypoxic preconditioning in MSCs and observed favorable effects on cell viability and angiogenic properties after 10 minutes of culture; stronger effects were observed after longer culture times [34].

4. MSC Proliferation and Cell Cycle under Hypoxic Conditions

Despite the fact that physiological levels of O_2 tension, even in healthy tissues, are significantly below 21%, cells are most frequently cultured at this O_2 tension level. Culture under normal physiological O_2 tension levels might affect the proliferation rates of several types of cells [16]. Lenon *et al.* showed that the culture of MSCs derived from the bone marrow of mice at 5% O_2 resulted in approximately 40% more cells at first passage when compared with cells cultivated at 21% O_2 [28]. Similar findings were reported for human MSCs by Grayson *et al.*, that showed an increased cell proliferation rate under low O_2 conditions (2%) for 7 passages, resulting in a 30 times higher number of cells when compared with cultures under normoxia [35]. Hung *et al.* [20] also observed a higher proliferation capacity of MSCs after 7 days of culture under hypoxia (1% O_2). In the same study, an *in vitro* migration assay was performed and showed that hypoxia enhanced the migration capacity of MSCs [36]. In spite of a short exposure time (24 hours) at 1.5% O_2, Matin-Rendon *et al.* also observed increased MSC proliferation [37]. Accordingly, D'Ippolito *et al.*, showed that a low O_2 tension level decreased the time necessary for the cell population to double when cultured at 3% O_2 [14], and Ren *et al.* showed an increase in the number of cells in the G2/S/M phase during hypoxia [38].

Conversely, a study by Holzwarth *et al.* analyzing MSCs proliferation at 21%, 5%, 3%, and 1% O_2 observed reduced rates of proliferation after 7 days of culture under hypoxia. In their study, cultures at 21% O_2, considered to be hyperoxic in comparison with the physiological environment in which MSCs reside, showed robust proliferation rates. Assessing MSC cell cycle after 7 days of culture only 1.37% of the cells entered the G2/M phase in hypoxic cell cultures (1% O_2) compared with 2.50% at an O_2 concentration of 21%. The authors concluded that the reduced number of cells in the G2/M phase confirms the inhibitory effect on cell proliferation under reduced O_2 concentrations [17]. An inhibitory effect on MSC proliferation rate cultured under hypoxia (1% O_2) in a medium containing 17% of fetal bovine serum (FBS) has also been reported [39].

5. MSC Plasticity under Hypoxic Conditions

The multilineage potential of MSCs is one of the reasons underlying their use in regenerative medicine [40]. According to several studies, MSC differentiation into other lineages can either increase or decrease under hypoxia [17,20,41]. Some *in vitro* studies have shown that cultures with low O_2 concentrations stimulated differentiation processes, inducing cells to differentiate into adipogenic, osteogenic, or chondrogenic cells [28,38,42]. Conversely, some others have reported suppressive effects of low O_2 tension levels on the plasticity of MSCs [43,44].

5.1. Osteogenic Differentiation

HIF-1α and VEGF are MSCs key elements in bone development and regeneration. VEGF is a transcriptional target of HIF-1α and has an important role in angiogenesis to which osteogenesis-bone development and regeneration, is strongly associated [45].

A study conducted by Huang *et al.* investigated the effects of hypoxia (5% O_2) in relation to the biological capacity of MSCs obtained from rabbits. That study found that hypoxia significantly increased the proliferation of MSCs, and the expression of messenger RNA (mRNA) for core binding factor alpha-1 (Cbfα-1) increased 1 hour after hypoxia. These results indicate that hypoxia increases differentiation rates in osteogenic lineages and suggest that Cbfα-1 may be positivity influenced by HIF1-α [41]. Some studies point to an improved osteogenic differentiation capacity of MSCs, however with conflicting results. Lennon *et al.* cultivated cells for several passages under hypoxia (5% O_2) and also at 21% O_2. Subsequently, the osteogenic differentiation capacity of these cells was assessed, suggesting that 5% O_2 was associated with a better differentiation response [28]. Valorani *et al.* also observed that, under hypoxic conditions (2% O_2), MSCs obtained from human adipose tissue showed an increase in their potential to differentiate into osteocytes [46]. Another study showed that different rates of O_2 (in hypoxia) were associated with different results: cells cultured at 1% O_2 showed a negative effect on osteogenic differentiation, whereas an increase in O_2 tension to 3% caused recovery of osteogenic differentiation [17]. In contrast, D'Ippolito *et al.* [14], Hung *et al.* [39], Muller *et al.* [47], Salim *et al.* [48], and Martin-Rendon *et al.* [37] observed either a reduced capacity or no effect on the differentiation capacity of MSCs into osteoblasts under hypoxic conditions or after exposure to hypoxia.

5.2. Chondrogenic Differentiation

Although Scherer *et al.* have shown that 5% O_2 promoted chondrogenic differentiation in the presence of a chondrogenic medium [49] and chondrocytes are known to develop in an extremely hypoxic environment [16] there are few articles assessing hypoxia effects on MSCs chondrogenic differentiation. There are, however, some indirect evidences that hypoxia might play a key role in *in vivo* MSCs chondrogenic differentiation since Sox9, an important transcription factor involved in chondrogenesis, was observed to be upregulated and to involve HIF-1α and p38MAPK/Akt pathways under hypoxic conditions, similarly to the MSC survival mechanisms in a hypoxic environment mentioned above [50,51].

5.3. Adipogenic Differentiation

Although several studies have suggested that hypoxia can increase the MSCs differentiation in adipocytes [38,42, 46,52,53] there are some that suggested hypoxia has no effect [35-54] or even suppress adipogenic MSCs differentiation [39]. Again, several different O_2 tension and culture conditions have been utilized in these studies.

6. MSC Morphology under Hypoxic Conditions

Some studies have described the morphology of MSCs

and correlated it with the "quality" of these cells, reporting that smaller cells have higher self-renewal capacity and an enhanced differentiation potential. Grayson *et al.* observed some differences in cells cultured under hypoxia in terms of their cellular and nuclear morphology, as well as in the formation of reinforced extracellular matrix when compared with MSCs in normoxia [35]. Holzwarth *et al.* microscopically analyzed the morphology of MSCs cultured under both hypoxic and in normoxic conditions after 1 and 3 weeks of culture. The authors observed that cell morphology under hypoxia was donor-dependent-some samples did not show differences between the two O_2 tension conditions, and others died after exposure to hypoxia [17].

7. Immunophenotypic Characteristics of MSCs under Hypoxic Conditions

MSC immunophenotype is characterized by the expression of CD73, CD90, CD105, CD106, CD146, and MHC class I molecules, and the absence of markers such as CD45 and CD34 or MHC class II molecules [1,55]. According to one study by Holzwarth *et al.*, there were no significant differences in the expression of cell surface markers after 14 days of culture at 1% when compared to 20% of O_2 [17].

8. Cytogenetic Characteristics of MSCs under Hypoxic Conditions

Senescence is a typical phenomenon of the *in vitro* cells cultures thought to interrupt cell proliferation, and attributed to everal factors, including progressive telomere shortening secondary to loss of telomerase activity [56]. In normal O_2 tension, time in culture has been linked to an increase number of mutations and to malignant transformation in murine mesenchymal cells [57] and in human MSC cultures, after more than 50 *in vitro* passages [58,59]. Holzwarth *et al.* did not see cytogenetic alterations in cultures of MSCs either under hypoxic or normoxic conditions during six weeks of culture [17].

9. Gene Expression in MSCs Cultured under Hypoxia

As mentioned above, several studies have described genes that may be under-or over-expressed in MSCs during hypoxia. These genes are involved in different functions, such as DNA repair, cell cycle, chromosome segregation, apoptosis, glycolysis, angiogenesis, proliferation, and adhesion [33,35,60,64]. A study conducted by Onishi *et al.* reported over-expression of several genes in rat MSCs cultivated under hypoxia vs. normoxia. That study also revealed that most genes analyzed were upregulated after 24 hours even in cells cultured at 10% O_2 (moderate hypoxia) [64].

The **Table 1** describes some of the genes identified in different studies involving human MSCs in relation to their expression during hypoxia.

10. Immunomodulatory Effects and Homing under Hypoxia

In addition to the easy isolation and culture of MSCs, their differentiation potential and the associated production of growth factors and cytokines, these cells have also become the focus of attention due to their immunomodulatory properties [65,66]. In this context, hypoxia

Table 1. Expression of genes in human MSCs cultured under hypoxic conditions.

Gene	Function
Upregulated	
HIF[a]	Hypoxia-related transcription factor
CXCR4[bc]	Migration
CX3CR1[b]	Migration
RRM2[d]	Ribonucleotide reductase
XRCC2[d]	DNA repair
KIF24[d]	Chromatid assembly
POLQ[d]	DNA polymerase
E2F8[d]	Cell cycle progression
FANCD2[d]	DNA repair
ESCO 2[d]	Sister chromatid cohesion
AURKB[d]	Chromosome segregation
CENPN[d]	Centromere binding
MKI67[d]	Cell proliferation
Downregulated	
TFC1[d]	Hepatic transcription factor
LEP[d]	Metabolism, apoptosis, angiogenesis
ANGPT2[d]	Antagonist on vascular remodeling
ZP1[d]	Sperm binding in the pellucid zone
VWF[d]	Platelet binding to endothelium
GIMAP4[d]	Development of T cells
CD93[d]	Intercellular adhesion, clearance of apoptotic cells
PLVAP[d]	Cell adhesion
ESAM[d]	Endothelial cell adhesion
PCDH17[d]	Cell-cell connections within the brain

References: [a]Holzwarth *et al.* [17]; [b]Hung *et al.* [39]; [c]Liu F., [d]Basciano [60].

seems to regulate the levels of soluble factors (such as VEGF, fibroblast growth factor-2 [FGF2], hepatocyte growth factor, and insulin-like growth factor 1 [IGF-1]), as well as the levels of tumor necrosis factor alpha (TNF-α), based on nuclear factor kappa β-dependent mechanisms [67]. These findings suggest an influence of hypoxia on the immunoregulatory properties of MSCs.

Wu *et al.* used human MSCs and observed that most genes were regulated after 24 hours of hypoxia. However, in less than 4 hours of hypoxia, an increased secretion of VEGF and membrane type 1-matrix metalloproteinase (MT1-MMP) was observed, as well as reduced levels of matrix metalloproteinase-2 (MMP2) [68]. Muir *et al.* also confirmed an increased expression of VEGF under hypoxic conditions [69]. Potier *et al.* cultivated human MSCs with FBS in hypoxia and observed a decreased expression of TGF-β3, as well as increased expression of FGF2 and VEGF after 48 hours. In that study, IL-6, IL-8, and MPC1 levels were not affected [44]. Conversely, in a study by Hung *et al.*, who cultivated MSCs in a FBS-free medium, the expression of IL-6, macrophage chemotactic protein (MCP1) and VEGF was found to be upregulated under hypoxia [39].

In addition to influencing the secretion of soluble factors, hypoxia also regulates chemokine receptors CX3CR1 and CXCR4 [39-70], and hepatocyte growth factor receptor cMet [21]. These receptors increase MSCs' migration and homing potential to cell-damaged areas [16]. Rosova *et al.* found a possible role of hepatocyte growth factor and its receptor cMet, whose expression is upregulated during hypoxia [21]. One study conducted by Hung *et al.* showed that hypoxia increased the migration capacity of MSCs [20], whereas Wang *et al.* showed an improved migration potential of MSCs in brain lesion. In the latter study, the CXCR4 receptor was shown to be involved in ability of these cells to migrate to damaged tissue under hypoxic conditions [70].

Hypoxia has been shown to influence the secretion of trophic factors and membrane markers associated with MSC migration and homing and, in animal models, MSCs cultured under hypoxia appears to have improved immune regulatory performance suggesting a possible role of hypoxia in cellular therapy strategies [71,72]. Migration may involve several cytokines, chemokines, and integrins. Among chemokines, the stromal cell-derived factor-1 (SDF-1/CXCR4 receptor) is expressed in a wide variety of tissues. In this context, SDF-1 is a potent progenitor cell mobilization agent present in the bone marrow niche, and its receptor CXCR4 is essential in the migration and homing of stem cells [73]. Therefore, the interaction between this axis and CXCR4 expression on the surface of MSCs cultured under hypoxia has an important role in the migration of transplanted cells [37].

11. Final Considerations

The recent interest in the potential application of MSCs in regenerative medicine and the availability of more appropriate methods for cell cultures were followed by the conduction and publication of several studies assessing the effects of low O_2 levels on the behavior and function of MSCs. This condition of cell culture appears to enhance MSCs migration and immune regulatory performance in damaged tissues with relevant consequences for cellular therapy. However, optimal conditions for the culture of MSCs have not yet been clearly defined and it is extremely important to accurately determine whether cells cultured under hypoxia are affected in terms of their differentiation, proliferation, and morphology, among other aspects, including the secretion of trophyc factors and membrane markers associated with MSCs migration and homing. Although the most of studies suggest a beneficial effect of hypoxia on MSCs, some authors have also reported opposite and negative effects. In this context, further research is essential so that a consensus can be reached, especially because of the great interest in standardizing the culture of MSCs for use in cell therapy, a promising tool for the treatment of different malignant conditions that are untreatable today.

12. Acknowledgements

The authors are grateful for the financial support provided by Fundo de Incentivo à Pesquisa e Eventos (FIPE/HCPA).

REFERENCES

[1] M. Dominici, K. Le Blanc, I. Mueller, I. Slaper-Cortenbach, F. Marini, D. Krause, R. Deans, A. Keating, D. Prockop and E. Horwitz, "Minimal Criteria for Defining Multipotent Mesenchymal Stromal Cells. The International Society for Cellular Therapy Position Statement," *Cytotherapy*, Vol. 8, No. 4, 2006, pp. 315-317. doi:10.1080/14653240600855905

[2] S. P. Bruder, K. H. Kraus, V. M. Goldberg and S. Kadiyala, "The Effect of Implants Loaded with Autologous Mesenchymal Stem Cells on the Healing of Canine Segmental Bone Defects," *The Journal of Bone Joint Surgery*, Vol. 80, No. 7, 1998, pp. 985-996.

[3] A. J. Friedenstein, R. K. Chailakhjan and K. S. Lalykina, "The Development of Fibroblast Colonies in Monolayer Cultures of Guinea-Pig Bone Marrow and Spleen Cells," *Cell and Tissue Kinetics*, Vol. 3, No. 4, 1970, pp. 393-403.

[4] D. J. Prockop, "Marrow Stromal Cells as Stem Cells for Nonhematopoietic Tissues," *Science*, Vol. 276, No. 5309, 1997, pp. 71-74. doi:10.1126/science.276.5309.71

[5] A. I. Caplan, "All MSCs Are Pericytes?" *Cell Stem Cell*, Vol. 3, No. 3, 2008, pp. 229-230. doi:10.1016/j.stem.2008.08.008

[6] M. F. Pittenger, A. M. Mackay, S. C. Beck, R. K. Jaiswal, R. Douglas, J. D. Mosca, M. A. Moorman, D. W. Simonetti, S. Craig and D. R. Marshak, "Multilineage Potential of Adult Human Mesenchymal Stem Cells," *Science*, Vol. 284, No. 5411, 1999, pp. 143-147. doi:10.1126/science.284.5411.143

[7] P. D. Cooper, A. M. Burt and J. N. Wilson, "Critical Effect of Oxygen Tension on Rate of Growth of Animal Cells in Continuous Suspended Culture," *Nature*, Vol. 182, No. 4648, 1958, pp. 1508-1509. doi:10.1038/1821508b0

[8] H. T. Zwartouw and J. C. Westwood, "Factors Affecting Growth and Glycolysis in Tissue Culture," *British Journal of Experimental Pathology*, Vol. 39, No. 5, 1958, pp. 529-539.

[9] S. Aggarwal and M. F. Pittenger, "Human Mesenchymal Stem Cells Modulate Allogeneic Immune Cell Responses," *Blood*, Vol. 105, No. 4, 2005, pp. 1815-1822. doi:10.1182/blood-2004-04-1559

[10] C. T. van Velthoven, A. Kavelaars, F. van Bel and C. J. Heijnen, "Mesenchymal Stem Cell Treatment after Neonatal Hypoxic-Ischemic Brain Injury Improves Behavioral Outcome and Induces Neuronal and Oligodendrocyte Regeneration," *Brain Behavior Immunity*, Vol. 24, No. 3, 2010, pp. 387-393. doi:10.1016/j.bbi.2009.10.017

[11] L. F. Raheja, D. C. Genetos and C. E. Yellowley, "The Effect of Oxygen Tension on the Long-Term Osteogenic Differentiation and MMP/TIMP Expression of Human Mesenchymal Stem Cells," *Cells Tissues Organs*, Vol. 191, No. 3, 2010, pp. 175-184. doi:10.1159/000235679

[12] L. F. Raheja, D. C. Genetos, A. Wong and C. E. Yellowley, "Hypoxic Regulation of Mesenchymal Stem Cell Migration: The Role of RhoA and HIF-1Alpha," *Cell Biology International*, Vol. 35, No. 10, 2011, pp. 981-989. doi:10.1042/CBI20100733

[13] G. Y. Rochefort, B. Delorme, A. Lopez, O. Herault, P. Bonnet, P. Charbord, V. Eder and J. Domenech, "Multipotential Mesenchymal Stem Cells Are Mobilized into Peripheral Blood by Hypoxia," *Stem Cells*, Vol. 24, No. 10, 2006, pp. 2202-2208. doi:10.1634/stemcells.2006-0164

[14] G. D'Ippolito, S. Diabira, G. A. Howard, B. A. Roos and P. C. Schiller, "Low Oxygen Tension Inhibits Osteogenic Differentiation and Enhances Stemness of Human MIAMI Cells," *Bone*, Vol. 39, No. 3, 2006, pp. 513-522. doi:10.1016/j.bone.2006.02.061

[15] L. Liu, Q. Yu, J. Lin, X. Lai, W. Cao, K. Du, Y. Wang, K. Wu, Y. Hu, L. Zhang, H. Xiao, Y. Duan and H. Huang, "Hypoxia-Inducible Factor-1alpha Is Essential for Hypoxia-Induced Mesenchymal Stem Cell Mobilization into the Peripheral Blood," *Stem Cells Development*, Vol. 20, No. 11, 2011, pp. 1961-1971. doi:10.1089/scd.2010.0453

[16] R. Das, H. Jahr, G. J. van Osch and E. Farrell, "The Role of Hypoxia in Bone Marrow-Derived Mesenchymal Stem Cells: Considerations for Regenerative Medicine Approaches," *Tissue Engineering Part B Reviews*, Vol. 16, No. 2, 2010, pp. 159-168. doi:10.1089/ten.teb.2009.0296

[17] C. Holzwarth, M. Vaegler, F. Gieseke, S. M. Pfister, R. Handgretinger, G. Kerst and I. Muller, "Low Physiologic Oxygen Tensions Reduce Proliferation and Differentiation of Human Multipotent Mesenchymal Stromal Cells," *BMC Cell Biology*, Vol. 11, 2010, p. 11. doi:10.1186/1471-2121-11-11

[18] R. P. Hill, D. T. Marie-Egyptienne and D. W. Hedley, "Cancer Stem Cells, Hypoxia and Metastasis," *Seminars in Radiation Oncology*, Vol. 19, No. 2, 2009, pp. 106-111.

[19] A. Efimenko, E. Starostina, N. Kalinina and A. Stolzing, "Angiogenic Properties of Aged Adipose Derived Mesenchymal Stem Cells after Hypoxic Conditioning," *Journal of Translational Medicine*, Vol. 9, 2011, p. 10. doi:10.1186/1479-5876-9-10

[20] S. P. Hung, J. H. Ho, Y. R. Shih, T. Lo and O. K. Lee, "Hypoxia Promotes Proliferation and Osteogenic Differentiation Potentials of Human Mesenchymal Stem Cells," *Journal of Orthopaedic Research*, Vol. 30, No. 2, 2012, pp. 260-266. doi:10.1002/jor.21517

[21] I. Rosova, M. Dao, B. Capoccia, D. Link and J. A. Nolta, "Hypoxic Preconditioning Results in Increased Motility and Improved Therapeutic Potential of Human Mesenchymal Stem Cells," *Stem Cells*, Vol. 26, No. 8, 2008, pp. 2173-2182. doi:10.1634/stemcells.2007-1104

[22] N. B. Nardi and S. Meirelles Lda, "Mesenchymal Stem-Cells: Isolation, *in Vitro* Expansion and Characterization," *Handbook of Experimental Pharmacology*, Vol. 174, 2006, pp. 249-282. doi:10.1007/3-540-31265-X_11

[23] K. Le Blanc and D. Mougiakakos, "Multipotent Mesenchymal Stromal Cells and the Innate Immune System," *Nature Reviews Immunology*, Vol. 12, No. 5, 2012, pp. 383-396. doi:10.1038/nri3209

[24] C. Lo Celso, J. W. Wu and C. P. Lin, "*In Vivo* Imaging of Hematopoietic Stem Cells and Their Microenvironment," *Journal of Biophotonics*, Vol. 2, No. 11, 2009, pp. 619-631. doi:10.1002/jbio.200910072

[25] Y. Shen and S. K. Nilsson, "Bone, Microenvironment and Hematopoiesis," *Current Opinion in Hematology*, Vol. 19, No. 4, 2012, pp. 250-255.

[26] R. Schofield, "The Stem Cell System," *Biomedicine Pharmacotherapy*, Vol. 37, No. 8, 1983, pp. 375-380.

[27] M. G. Cipolleschi, P. Dello Sbarba and M. Olivotto, "The Role of Hypoxia in the Maintenance of Hematopoietic Stem Cells," *Blood*, Vol. 82, No. 7, 1993, pp. 2031-2037.

[28] D. P. Lennon, J. M. Edmison and A. I. Caplan, "Cultivation of Rat Marrow-Derived Mesenchymal Stem Cells in Reduced Oxygen Tension: Effects on *in Vitro* and *in Vivo* Osteochondrogenesis," *Journal of Cellular Physiology*, Vol. 187, No. 3, 2001, pp. 345-355. doi:10.1002/jcp.1081

[29] T. Ma, W. L. Grayson, M. Frohlich and G. Vunjak-Novakovic, "Hypoxia and Stem Cell-Based Engineering of Mesenchymal Tissues," *Biotechnology Progress*, Vol. 25, No. 1, 2009, pp. 32-42. doi:10.1002/btpr.128

[30] Z. Ivanovic, "Hypoxia or *in Situ* Normoxia: The Stem Cell Paradigm," *Journal of Cellular Physiology*, Vol. 219, No. 2, 2009, pp. 271-275. doi:10.1002/jcp.21690

[31] I. G. Winkler, V. Barbier, R. Wadley, A. C. Zannettino, S. Williams and J. P. Levesque, "Positioning of Bone Mar-

row Hematopoietic and Stromal Cells Relative to Blood Flow *in Vivo*: Serially Reconstituting Hematopoietic Stem Cells Reside in Distinct Nonperfused Niches," *Blood*, Vol. 116, No. 3, 2010, pp. 375-385.

[32] W. Zhu, J. Chen, X. Cong, S. Hu and X. Chen, "Hypoxia and Serum Deprivation-Induced Apoptosis in Mesenchymal Stem Cells," *Stem Cells*, Vol. 24, No. 2, 2006, pp. 416-425. doi:10.1634/stemcells.2005-0121

[33] S. Lord-Dufour, I. B. Copland, L. C. Levros Jr., M. Post, A. Das, C. Khosla, J. Galipeau, E. Rassart and B. Annabi, "Evidence for Transcriptional Regulation of the Glucose-6-Phosphate Transporter by HIF-1Alpha: Targeting G6PT with Mumbaistatin Analogs in Hypoxic Mesenchymal Stromal Cells," *Stem Cells*, Vol. 27, No. 3, 2009, pp. 489-497. doi:10.1634/stemcells.2008-0855

[34] J. A. Wang, T. L. Chen, J. Jiang, H. Shi, C. Gui, R. H. Luo, X. J. Xie, M. X. Xiang and X. Zhang, "Hypoxic Preconditioning Attenuates Hypoxia/Reoxygenation-Induced Apoptosis in Mesenchymal Stem Cells," *Acta Pharmacologica Sinica*, Vol. 29, No. 1, 2008, pp. 74-82. doi:10.1111/j.1745-7254.2008.00716.x

[35] W. L. Grayson, F. Zhao, B. Bunnell and T. Ma, "Hypoxia Enhances Proliferation and Tissue Formation of Human Mesenchymal Stem Cells," *Biochemical and Biophysical Research Communications*, Vol. 358, No. 3, 2007, pp. 948-953.

[36] S. P. Hung, J. H. Ho, Y. R. Shih, T. Lo and O. K. Lee, "Hypoxia Promotes Proliferation and Osteogenic Differentiation Potentials of Human Mesenchymal Stem Cells," *Journal of Orthopaedic Research*, Vol. 30, No. 2, 2012, pp. 260-266. doi:10.1002/jor.21517

[37] E. Martin-Rendon, S. J. Hale, D. Ryan, D. Baban, S. P. Forde, M. Roubelakis, D. Sweeney, M. Moukayed, A. L. Harris, K. Davies and S. M. Watt, "Transcriptional Profiling of Human Cord Blood CD133+ and Cultured Bone Marrow Mesenchymal Stem Cells in Response to Hypoxia," *Stem Cells*, Vol. 25, No. 4, 2007, pp. 1003-1012. doi:10.1634/stemcells.2006-0398

[38] H. Ren, Y. Cao, Q. Zhao, J. Li, C. Zhou, L. Liao, M. Jia, H. Cai, Z. C. Han, R. Yang, G. Chen and R. C. Zhao, "Proliferation and Differentiation of Bone Marrow Stromal Cells under Hypoxic Conditions," *Biochemical and Biophysical Research Communications*, Vol. 347, No. 1, 2006, pp. 12-21. doi:10.1016/j.bbrc.2006.05.169

[39] S. C. Hung, R. R. Pochampally, S. C. Hsu, C. Sanchez, S. C. Chen, J. Spees and D. J. Prockop, "Short-Term Exposure of Multipotent Stromal Cells to Low Oxygen Increases Their Expression of CX3CR1 and CXCR4 and Their Engraftment *in Vivo*," *PLoS One*, Vol. 2, No. 5, 2007, p. e416. doi:10.1371/journal.pone.0000416

[40] D. Baksh, L. Song and R. S. Tuan, "Adult Mesenchymal Stem Cells: Characterization, Differentiation, and Application in Cell and Gene Therapy," *Journal of Cellular and Molecular Medicine*, Vol. 8, No. 3, 2004, pp. 301-316. doi:10.1111/j.1582-4934.2004.tb00320.x

[41] J. Huang, F. Deng, L. Wang, X. R. Xiang, W. W. Zhou, N. Hu and L. Xu, "Hypoxia Induces Osteogenesis-Related Activities and Expression of Core Binding Factor Alpha1 in Mesenchymal Stem Cells," *The Tohoku Journal of Experimental Medicine*, Vol. 224, No. 1, 2011, pp. 7-12.

[42] T. Fink, L. Abildtrup, K. Fogd, B. M. Abdallah, M. Kassem, P. Ebbesen and V. Zachar, "Induction of Adipocyte-Like Phenotype in Human Mesenchymal Stem Cells by Hypoxia," *Stem Cells*, Vol. 22, No. 7, 2004, pp. 1346-1355. doi:10.1634/stemcells.2004-0038

[43] C. Fehrer, R. Brunauer, G. Laschober, H. Unterluggauer, S. Reitinger, F. Kloss, C. Gully, R. Gassner and G. Lepperdinger, "Reduced Oxygen Tension Attenuates Differentiation Capacity of Human Mesenchymal Stem Cells and Prolongs Their Lifespan," *Aging Cell*, Vol. 6, No. 6, 2007, pp. 745-757. doi:10.1111/j.1474-9726.2007.00336.x

[44] E. Potier, E. Ferreira, R. Andriamanalijaona, J. P. Pujol, K. Oudina, D. Logeart-Avramoglou and H. Petite, "Hypoxia Affects Mesenchymal Stromal Cell Osteogenic Differentiation and Angiogenic Factor Expression," *Bone*, Vol. 40, No. 4, 2007, pp. 1078-1087. doi:10.1016/j.bone.2006.11.024

[45] C. Wan, J. Shao, S. R. Gilbert, R. C. Riddle, F. Long, R. S. Johnson, E. Schipani and T. L. Clemens, "Role of HIF-1Alpha in Skeletal Development," *Annals of New York Academy of Science*, Vol. 1192, 2010, pp. 322-326. doi:10.1111/j.1749-6632.2009.05238.x

[46] M. G. Valorani, E. Montelatici, A. Germani, A. Biddle, D. D'Alessandro, R. Strollo, M. P. Patrizi, L. Lazzari, E. Nye, W. R. Otto, P. Pozzilli and M. R. Alison, "Pre-Culturing Human Adipose Tissue Mesenchymal Stem Cells under Hypoxia Increases Their Adipogenic and Osteogenic Differentiation Potentials," *Cell Proliferation*, Vol. 45, No. 3, 2012, pp. 225-238. doi:10.1111/j.1365-2184.2012.00817.x

[47] I. Muller, M. Vaegler, C. Holzwarth, N. Tzaribatchev, S. M. Pfister, B. Schutt, P. Reize, J. Greil, R. Handgretinger and M. Rudert, "Secretion of Angiogenic Proteins by Human Multipotent Mesenchymal Stromal Cells and Their Clinical Potential in the Treatment of Avascular Osteonecrosis," *Leukemia*, Vol. 22, No. 11, 2008, pp. 2054-2061. doi:10.1038/leu.2008.217

[48] A. Salim, R. P. Nacamuli, E. F. Morgan, A. J. Giaccia and M. T. Longaker, "Transient Changes in Oxygen Tension Inhibit Osteogenic Differentiation and *Runx*2 Expression in Osteoblasts," *Journal of Biological Chemistry*, Vol. 279, No. 38, 2004, pp. 40007-40016. doi:10.1074/jbc.M403715200

[49] K. Scherer, M. Schunke, R. Sellckau, J. Hassenpflug and B. Kurz, "The Influence of Oxygen and Hydrostatic Pressure on Articular Chondrocytes and Adherent Bone Marrow Cells *in Vitro*," *Biorheology*, Vol. 41, No. 3-4, 2004, pp. 323-333.

[50] M. Kanichai, D. Ferguson, P. J. Prendergast and V. A. Campbell, "Hypoxia Promotes Chondrogenesis in Rat Mesenchymal Stem Cells: A Role for AKT and Hypoxia-Inducible Factor (HIF)-1Alpha," *Journal of Cellular Physiology*, Vol. 216, No. 3, 2008, pp. 708-715. doi:10.1002/jcp.21446

[51] J. C. Robins, N. Akeno, A. Mukherjee, R. R. Dalal, B. J. Aronow, P. Koopman and T. L. Clemens, "Hypoxia Induces Chondrocyte-Specific Gene Expression in Mesen-

chymal Cells in Association with Transcriptional Activation of Sox9," *Bone*, Vol. 37, No. 3, 2005, pp. 313-322. doi:10.1016/j.bone.2005.04.040

[52] W. L. Grayson, F. Zhao, R. Izadpanah, B. Bunnell and T. Ma, "Effects of Hypoxia on Human Mesenchymal Stem Cell Expansion and Plasticity in 3D Constructs," *Journal of Cellular Physiology*, Vol. 207, No. 2, 2006, pp. 331-339. doi:10.1002/jcp.20571

[53] L. A. Mylotte, A. M. Duffy, M. Murphy, T. O'Brien, A. Samali, F. Barry and E. Szegezdi, "Metabolic Flexibility Permits Mesenchymal Stem Cell Survival in an Ischemic Environment," *Stem Cells*, Vol. 26, No. 5, 2008, pp. 1325-1336. doi:10.1634/stemcells.2007-1072

[54] S. Carrancio, N. Lopez-Holgado, F. M. Sanchez-Guijo, E. Villaron, V. Barbado, S. Tabera, M. Diez-Campelo, J. Blanco, J. F. San Miguel and M. C. Del Canizo, "Optimization of Mesenchymal Stem Cell Expansion Procedures by Cell Separation and Culture Conditions Modification," *Experimental Hematology*, Vol. 36, No. 8, 2008, pp. 1014-1021. doi:10.1016/j.exphem.2008.03.012

[55] E. M. Horwitz, K. Le Blanc, M. Dominici, I. Mueller, I. Slaper-Cortenbach, F. C. Marini, R. J. Deans, D. S. Krause and A. Keating, "Clarification of the Nomenclature for MSC: The International Society for Cellular Therapy Position Statement," *Cytotherapy*, Vol. 7, No. 5, 2005, pp. 393-395. doi:10.1080/14653240500319234

[56] M. Kassem, M. Kristiansen and B. M. Abdallah, "Mesenchymal Stem Cells: Cell Biology and Potential Use in Therapy," *Basic & Clinical Pharmacology & Toxicology*, Vol. 95, No. 5, 2004, pp. 209-214. doi:10.1111/j.1742-7843.2004.pto950502.x

[57] M. Miura, Y. Miura, H. M. Padilla-Nash, A. A. Molinolo, B. Fu, V. Patel, B. M. Seo, W. Sonoyama, J. J. Zheng, C. C. Baker, W. Chen, T. Ried and S. Shi, "Accumulated Chromosomal Instability in Murine Bone Marrow Mesenchymal Stem Cells Leads to Malignant Transformation," *Stem Cells*, Vol. 24, No. 4, 2006, pp. 1095-1103. doi:10.1634/stemcells.2005-0403

[58] A. S. Grigorian, P. V. Kruglyakov, U. A. Taminkina, O. A. Efimova, A. A. Pendina, A. V. Voskresenskaya, T. V. Kuznetsova and D. G. Polyntsev, "Alterations of Cytological and Karyological Profile of Human Mesenchymal Stem Cells during *in Vitro* Culturing," *Bulletin of Experimental Biology and Medicine*, Vol. 150, No. 1, 2010, pp. 125-130. doi:10.1007/s10517-010-1086-x

[59] D. Rubio, J. Garcia-Castro, M. C. Martin, R. de la Fuente, J. C. Cigudosa, A. C. Lloyd and A. Bernad, "Spontaneous Human Adult Stem Cell Transformation," *Cancer Research*, Vol. 65, No. 8, 2005, pp. 3035-3039.

[60] L. Basciano, C. Nemos, B. Foliguet, N. de Isla, M. de Carvalho, N. Tran and A. Dalloul, "Long Term Culture of Mesenchymal Stem Cells in Hypoxia Promotes a genetic Program Maintaining Their Undifferentiated and Multipotent Status," *BMC Cell Biology*, Vol. 12, 2011, p. 12. doi:10.1186/1471-2121-12-12

[61] Y. Jin, T. Kato, M. Furu, A. Nasu, Y. Kajita, H. Mitsui, M. Ueda, T. Aoyama, T. Nakayama, T. Nakamura and J. Toguchida, "Mesenchymal Stem Cells Cultured under Hypoxia Escape from Senescence via Down-Regulation of p16 and Extracellular Signal Regulated Kinase," *Biochemical and Biophysical Research Communications*, Vol. 391, No. 3, 2010, pp. 1471-1476. doi:10.1016/j.bbrc.2009.12.096

[62] H. Liu, W. Xue, G. Ge, X. Luo, Y. Li, H. Xiang, X. Ding, P. Tian and X. Tian, "Hypoxic Preconditioning Advances CXCR4 and CXCR7 Expression by Activating HIF-1Alpha in MSCs," *Biochemical and Biophysical Research Communications*, Vol. 401, No. 4, 2010, pp. 509-515. doi:10.1016/j.bbrc.2010.09.076

[63] C. C. Tsai, Y. J. Chen, T. L. Yew, L. L. Chen, J. Y. Wang, C. H. Chiu and S. C. Hung, "Hypoxia Inhibits Senescence and Maintains Mesenchymal Stem Cell Properties through Down-Regulation of E2A-p21 by HIF-TWIST," *Blood*, Vol. 117, No. 2, 2011, pp. 459-469. doi:10.1182/blood-2010-05-287508

[64] S. Ohnishi, T. Yasuda, S. Kitamura and N. Nagaya, "Effect of Hypoxia on Gene Expression of Bone Marrow-Derived Mesenchymal Stem Cells and Mononuclear Cells," *Stem Cells*, Vol. 25, No. 5, 2007, pp. 1166-1177. doi:10.1634/stemcells.2006-0347

[65] M. Shi, Z. W. Liu and F. S. Wang, "Immunomodulatory Properties and Therapeutic Application of Mesenchymal Stem Cells," *Clinical and Experimental Immunology*, Vol. 164, No. 1, 2011, pp. 1-8. doi:10.1111/j.1365-2249.2011.04327.x

[66] C. D. Wan, R. Cheng, H. B. Wang and T. Liu, "Immunomodulatory Effects of Mesenchymal Stem Cells Derived from Adipose Tissues in a Rat Orthotopic Liver Transplantation Model," *Hepatobiliary and Pancreatic Diseases International*, Vol. 7, No. 1, 2008, pp. 29-33.

[67] P. R. Crisostomo, Y. Wang, T. A. Markel, M. Wang, T. Lahm and D. R. Meldrum, "Human Mesenchymal stem Cells Stimulated by TNF-Alpha, LPS, or Hypoxia Produce Growth Factors by an NF Kappa B, but Not JNK-Dependent Mechanism," *American Journal of Physiology: Cell Physiology*, Vol. 294, No. 3, 2008, pp. C675-C682. doi:10.1152/ajpcell.00437.2007

[68] E. H. Wu, H. S. Li, T. Zhao, J. D. Fan, X. Ma, L. Xiong, W. J. Li, L. L. Zhu and M. Fan, "Effect of hypoxia on the gene profile of human bone marrow-derived mesenchymal stem cells," *Acta Physiologica Sinica*, Vol. 59, No. 2, 2007, pp. 227-232.

[69] C. Muir, L. W. Chung, D. D. Carson and M. C. Farach-Carson, "Hypoxia Increases VEGF-A Production by Prostate Cancer and Bone Marrow Stromal Cells and Initiates Paracrine Activation of Bone Marrow Endothelial Cells," *Clinical and Experimental Metastasis*, Vol. 23, No. 1, 2006, pp. 75-86. doi:10.1007/s10585-006-9021-2

[70] Y. Wang, Y. Deng and G. Q. Zhou, "SDF-1Alpha/CXCR4-Mediated Migration of Systemically Transplanted Bone Marrow Stromal Cells towards Ischemic Brain Lesion in a Rat Model," *Brain Research*, Vol. 1195, No. 2008, pp. 104-112. doi:10.1016/j.brainres.2007.11.068

[71] X. Hu, S. P. Yu, J. L. Fraser, Z. Lu, M. E. Ogle, J. A. Wang and L. Wei, "Transplantation of Hypoxia-Preconditioned Mesenchymal Stem Cells Improves Infracted Heart Function via Enhanced Survival of Implanted Cells and Angiogenesis," *Journal of Thoracic and Cardiovascu-*

lar Surgery, Vol. 135, No. 4, 2008, pp. 799-808. doi:10.1016/j.jtcvs.2007.07.071

[72] X. Mao, Q. Zeng, X. Wang, L. Cao and Z. Bai, "Angiogenic Potency of Bone Marrow Stromal Cells Improved by *ex Vivo* Hypoxia Prestimulation," *Journal of Huazhong University of Science and Technology* (*Medical Science*), Vol. 24, No. 6, 2004, pp. 566-568. doi:10.1007/BF02911356

[73] H. Liu, S. Liu, Y. Li, X. Wang, W. Xue, G. Ge and X. Luo, "The Role of SDF-1-CXCR4/CXCR7 Axis in the Therapeutic Effects of Hypoxia-Preconditioned Mesenchymal Stem Cells for Renal Ischemia/Reperfusion Injury," *PLoS One*, Vol. 7, No. 4, 2012, p. e34608. doi:10.1371/journal.pone.0034608

17

Protective Effects of Flavonoid Baicalein against Menadione-Induced Damage in SK-N-MC Cells

Maryam Moslehi, Razieh Yazdanparast*
Institute of Biochemistry and Biophysics, University of Tehran, Tehran, Iran
Email: *yazdan@ibb.ut.ac.ir

ABSTRACT

Oxidative damage and redox metal homeostasis loss are two contributing factors in brain aging and widely distributed neurodegenerative diseases. Oxidative species in company with excessive amounts of intracellular free iron result in Fenton-type reaction with subsequent production of highly reactive hydroxyl radicals which initiate peroxidation of biomolecules and further formation of non-degradable toxic pigments called lipofuscin that amasses in long-lived post-mitotic cells such as neurons. Dietary flavonoid baicalein can counteract the detrimental consequences through exertion of a multiplicity of protective actions within the brain including direct ROS scavenging activity and iron chelation. In this study, we evaluated the neuroprotective effects of baicalein in menadione (superoxide radical generator)-treated SK-N-MC neuroblastoma cell line. Our results showed that treatment of cells with menadione led to lipofuscin formation due to elevated intracellular iron contents and accumulation of oxidative products such as MDA and PCO. Also, menadione caused apoptotic cell death in SK-N-MC cells. However, pretreatment with baicalein (40 μM) reversed the harmful effects by chelating free iron and preventing biomolecules peroxidations. Moreover, baicalein prevented cell death through modulation of key molecules in apoptotic pathways including suppression of Bax and caspase-9 activities and induction of bcl2 expression. Key structural features such as presence of hydroxyl groups and iron-binding motifs in baicalein make it the appropriate candidate in antioxidant-based therapy in age-related neurodegenerative diseases.

Keywords: Aging; Baicalein; Lipofuscin; Menadione; Neurodegenerative Disease; Oxidative Stress

1. Introduction

The key precept of the oxidative stress theory of aging is that senescence-related loss of function is due to the progressive and irreparable accrual of molecular oxidative damage which is brought about by powerful pro-oxidant species including reactive oxygen species (ROS) [1,2]. ROS include a broad range of partially reduced metabolites of oxygen (e.g. superoxide, hydrogen peroxide and hydroxyl radical) having higher reactivity than molecular oxygen [3]. Their raison d'être remains unclear. Putative explanations for their occurrence range from inadvertent by-products of aerobic metabolism to highly regulated and intricate signaling mechanisms [4]. Free radical or oxidative stress theory of aging was first proclaimed by Denham Harman demonstrating the role of oxidative species in aging process acceleration and cell death [5,6]. This theory can explain many of the senescent changes including accumulation of brown-yellow, electron-dense, autofluorescent bodies in cells called lipofuscin pigments or age pigments [5,7,8]. Correlation of lipofuscin with aging is not only because the amount of lipofuscin elevates with age, but also, more significantly because the rate of lipofuscin accumulation negatively correlates to longevity. High consumption of oxygen via brain makes it susceptible to oxidative damage [9,10]. Reactive oxygen species which are generated by mitochondria through different ways, diffuse into lysosomes which encompass a variety of macromolecules under degradation as well as redox-active low molecular iron which would be released from different sorts of metalloproteins. Based on Fenton reaction, hydroxyl radical can be generated through the reaction of hydrogen peroxide with iron, bringing about the cross-linking of adjacent macromolecules and resultant lipofuscin formation [7,8,10-12]. There is a debate on the function of lipofuscin pigments formed during exposure of cells to oxidative agents. Some researchers believe that lipofuscin formation does not have any serious effects on normal func-

*Corresponding author.

tion of cells. On the contrary, some scientists state that although lipofuscin cannot react directly with extralysosomal constituents because of the lysosomal membrane, the high content of iron within lipofuscin granules may promote generation of ROS, sensitizing cells to oxidative injury through lysosomal destabilization. Destabilization of lysosomal membrane results in leaking of hydrolytic enzymes into the cytosol. Hence, oxidative species and redox-active transition metals homeostasis impairments which facilitate further formation of active and hazardous reactive oxygen species might be two main characteristics of age-related neurodegenerative diseases [13].

Human's aspiration for greater longevity has long been a strong motivation for a lot of studies in the field of aging and age-related disorders. Escalating body of evidence implies that lifestyle factors, and specially the diet, may counteract oxidative damage [2,14]. Dietary flavonoids with blood-brain barrier ability were shown to have potential anti-aging and brain-protective activities [5, 15-18]. Baicalein (5,6,7-trihydroxy-2-phenyl-4H-1-benzopyran-4-one), one of the naturally occurring flavonoids in *Scutellaria baicalensis* GEORGI known as "Huang qin" in China and "Ogon" in Japan, is prescribed for oxidative stress-related diseases [19]. Numerous studies have shown that baicalein protects neurons from oxidative damage via multiple bio-effects ranging from classic radical scavenging activities to modulation of signaling pathways involved in stress-associated diseases. Moreover, recent studies have denoted that baicalein mitigates formation of hydroxyl radical through its iron-binding (anti-Fenton) and strong chelation properties [20-23].

In this study, we scrutinize the effect of baicalein on menadione (superoxide anion generator)-induced lipofuscin formation in human neuroblastoma SK-N-MC cell line to comprehend the mechanism by which baicalein protect SK-N-MC cells against oxidative damages.

2. Materials and Methods

2.1. Materials

The cell culture medium (RPMI-1640), penicillin-streptomycin and fetal bovine serum (FBS) were purchased from Gibco BRL (Life technology, Paisely, Scotland). The culture plates were purchased from Nunc (Brand products, Denmark). dimethyl sulfoxide (DMSO), $FeCl_3$ and $KMnO_4$ were obtained from Merck (Darmstadt, Germany). Ethidium bromide, acridine orange, Baicalein and Triton X-100 were purchased from Pharmacia LKB Biotechnology (Sweden). MTT [3-(4,5-dimethyl tiazol-2, 5-diphenyl tetrazolium bromide], phenylmethylsulphonyl fluoride (PMSF), leupeptin, pepstatin, aprotinin, monochlorobimane (mBCL), dithionitrobenzoic acid (DTNB), GSH, ascorbic acid, ferrozine and pan-caspase inhibitor (Z-VAD-fmk) were purchased from Sigma Chem. Co. (Germany). 2',7'-dichlorofluorescein diacetate (DCFH-DA) was obtained from Molecular Probe (Eugene, Oregon, USA). Ethylenediaminetetraacetic acid (EDTA) was from Aldrich (Germany). Human SK-N- MC neuroblastoma cells were obtained from Pasteur Institute (Tehran, Iran). All antibodies including anti-Bax, anti-Bcl-2, anti-cleaved caspase-9, anti-tubulin and mouse/rabbit horseradish peroxidase-conjugated second-dary antibodies were purchased from Biosource (Nivelles, Belgium). Chemiluminescence detection system was purchased from Amersham-Pharmacia (Piscataway, NJ, USA).

2.2. Cell Culture

Human neuroblastoma cell line SK-N-MC was cultured in RPMI-1640 medium supplemented with FBS (10%, v/v), streptomycin (100 μg/ml) and penicillin (100 U/ml) and incubated in 5% CO humidified atmosphere at 37°C. To induce oxidative stress, menadione was freshly prepared from a stock solutions (10 mM), prior to each experiment. Menadione and baicalein were dissolved in a minimum amount of dimethyl Sulfoxide (DMSO) and then diluted with the culture medium to the desired concentration. The concentration of DMSO in the culture medium kept lower than 0.1% and the control cells were treated with the vehicle solution containing the same amount of DMSO.

2.3. Determination of Cell Viability

Cell viability was assessed by the 3-(4,5-dimethylthiazol-2-yl)-2,5-diphenyl tetrazolium bromide (MTT) reduction assay. Viable cells with active mitochondria reduce the yellow tetrazolium salt MTT giving dark blue water insoluble formazan crystals. To perform the assay for evaluation of the cytoprotective effects of baicalein and caspase inhibitor (z-VAD-fmk) on menadione-treated SK-N-MC cells, SK-N-MC cells were suspended in medium and seeded at a density of 5×10^4 cells/well in 96 well plates for a day. Cells were pretreated with various concentrations of baicalein (10, 20, 40, 50 μM) and pan-caspase inhibitor (50 μM) and then treated with menadione (35 μM) for additional 24 h at 37°C. MTT was dissolved at a concentration of 5 mg/ml in PBS and stored at 4°C, protected from light and tightly capped. After incubation, cells were treated with the 10 μl MTT solution for 4 h. Then, the medium was removed and 200 μl DMSO was added to each well. The formazan dye crystals were solubilized in 30 min, and absorbance was measured at 570 nm using an ELISA reader (Exert 96, Asys Hitch, Ec Austria). Results were expressed as the percentage of MTT reduction, assuming that the absorbance of the control cells was 100%.

2.4. Measurement of Intracellular ROS

Oxidation of 2',7'-dichlorofluorescein diacetate (DCFH-DA) to fluorescent DCF is taken as an index of overall oxidative stress in biological system according to LeBel method [24]. Cells were pre-treated with various concentrations of baicalein (10, 20, 40 μM) and 50 μM caspase inhibitor for 3 h followed by menadione treatment (35 μM) for 12 h at 37°C. Then the cells were incubated with 10 μM DCFH-DA for 1 h followed by washing twice with phosphate buffer saline and suspension in the same buffer. Finally, the fluorescent intensity was monitored using a varian-spectrofluorometer with excitation and emission wavelength of 485 and 530 nm, respectively.

2.5. Determination of Lipid Peroxidation

Malondialdehyde (MDA) levels were measured by the double heating method [25]. The method is based on spectrophotometric measurement of the purple color generated by the reaction of thiobarbituric acid (TBA) with MDA. Briefly, 0.5 ml of cell lysate was mixed with 2.5 ml of trichloroacetic acid (TCA, 10%, w/v) solution followed by boiling in a water bath at 95°C for 15 min. After cooling to room temperature, the samples were centrifuged at 3000 rpm for 10 min and 2 ml of each sample supernatant was transferred to a test tube containing 1 ml of TBA solution (0.67% w/v). Each tube was then placed in a boiling water bath for 15 min. After cooling to room temperature, the absorbance was measured at 532 nm with respect to the blank solution. The protein concentration was determined by Lowry's method [26]. The concentration of MDA was calculated based on the absorbance coefficient of the TBA-MDA complex ($\varepsilon = 1.56 \times 10^5\ cm^{-1} \cdot M^{-1}$) and it was expressed as nmol/mg of protein.

2.6. Determination of Protein Carbonyl Formation

The assessment of protein carbonyl content is a widely-used marker for oxidative protein modification. Protein carbonyls (PCOs) were measured using Reznick and Packer method [27]. Briefly, 1 ml of 10 mM DNPH in 2 M HCl was added to the cell lysates. Samples were incubated for 1 hr at room temperature and were vortexed every 15 min. Then, 1 ml of trichloroacetic acid (TCA 10% w/v) was added to each reaction mixture and centrifuged at 3000 rpm for 10 min. The pellets were washed twice with 2 ml of ethanol/ethyl acetate (1:1, v/v) and each dissolved in 1 ml of guanidine hydrochloride (6 M, pH 2.3) and incubated for 10 min at 37°C whilst mixing. The carbonyl content was calculated based on the molar extinction coefficient of DNPH ($\varepsilon = 2.2 \times 10^4\ cm^{-1} \cdot M^{-1}$).

2.7. Fluorescence Microscopy Evaluation of Apoptotic Cells

Acridine orange/ethidium bromide double staining was applied to observe the morphological changes among menadione-treated cell. Using this technique, cells can be distinguished as normal cells (uniformly stained green) and apoptotic cells that are stained orange because of cell membrane destruction and the intercalation of ethidium bromide between the nucleotide bases of DNA. After treatment, cells were washed twice with phosphate buffer saline and adjusted to a cell density of 1×10^4 cells/ml of phosphate solution (1:1 v/v). The nuclear morphology was evaluated by Axoscope 2 plus fluorescence microscope from Zeiss (Germany). The cells with condensed or fragmented nuclei were counted as apoptotic cells. All experiments were repeated three times, and the number of stained cells was counted in 10 randomly selected fields.

2.8. Evaluation of Intracellular Formation of Lipofuscin Pigments

Extraction of intracellular lipofuscin was achieved following lysis of each sample according to a published procedure with slight modification [28]. The cells were seeded in triplicate into 24-well plates for 24 h prior to pretreatments. After pretreatment with different doses of baicalein (10, 20, 40 μM) for 3 h, each cell sample was treated with 35 μM menadione for 24 h. The attached cells in each well were trypsinized with trypsin-EDTA solution followed by cell counting using a hemocytometer. Each plate was then centrifuged, the cell pellet was washed with PBS, and the cell content was lysed with lysis buffer containing 1% Triton X-100, 1 mM EDTA and 1 mM PMSF. Each cell lysate was harvested and its fluorescence intensity was monitored on a varian spectrofluometer, model Cary Eclipse, with an excitation wavelength of 310 nm and emission wavelength of 620 nm [29]. The fluorescence intensities of the samples were then normalized for equal cell numbers.

2.9. Measurement of Intracellular Iron Contents via Ferrozine-Based Colorimetric Assay

The assay was performed directly in 24-well plates. Cells were lysed by addition of 200 μl iron releasing reagent (a freshly mixed solution of equal volumes of 1.4 M HCl and 4.5% (w/v) $KMnO_4$ in H_2O_2) to each well. The plates were sealed with foil and incubated for 2 h at 60°C, after which 60 μl of the detection reagent (6.5 mM ferrozine, 6.5 mM EDTA, 2.5 M ammonium acetate and 1 M ascorbic acid dissolved in water) was added. After further incubation for 30 min at room temperature, 280 μl of the mixture was transferred to a well of a 96-well plate and its absorbance recorded at 550 nm and compared to the absorbance of the $FeCl_3$-treated standards under all equal experimental conditions. The determined intracellular iron concentration for each well was normalized against the protein content of replicate wells [30].

2.10. Western Blot Analysis

SK-N-MC cells were seeded at a density of 10^5 cells/ml in 12-well plates for 24 h. The cells were pretreated with baicalein (40 μM) and caspase inhibitor (50 μM). After 3 h, menadione (35 μM) was added to the cells and incubated at 37˚C for an additional 24 h. Then, the cells were harvested and lysed using lysis buffer containing 1% Triton X-100, 1% SDS, 10 mM Tris (pH 7.4), 100 mM NaCl, 1 mM EGTA, 1 mM EDTA, 20 mM sodium pyrophosphate, 2 mM Na_3VO_4, 1 mM NaF, 0.5% sodium deoxycholate, 10% glycerol, 1mM phenylmethylsulphonyl fluoride, 10 μg/ml leupeptin, 1 μg/ml pepstatin and 60 μg/ml aprotinin. Protein concentration of each sample was determined using Lowry' method (Lowry *et al.*, 1951). Equal quantities of protein (40 μg) were subjected to 12.5% SDS-polyacrylamide gel electrophoresis (PAGE) and were transferred to PVDF membranes. The blots were blocked with 5% (w/v) non-fat dry milk in Tris-buffered saline buffer containing 0.1% Tween-20 (TBS/T) for an overnight at 4˚C. The blocked blots were incubated with primary antibodies for 2 hr at room temperature using antibody dilutions as recommended by the manufacturer in Tris-buffered saline pH 7.4 containing 0.1% Tween-20. After 1-hr incubation with anti-rabbit or anti-mouse horseradish peroxidase (HRP)-conjugated secondary antibodies (Biosource), the proteins were detected by an enhanced chemiluminescence detection system (Amersham-Pharmacia, Piscataway, NJ, USA) according to the manufacturer's instructions. Blots were stripped at 50˚C for 30 min in 100 mM 2-mercaptoethanol, 2% SDS, 62.2 mM Tris-HCl pH 6.7 and reprobed for further investigations. For analysis of the western blotting data, densitometric analysis was performed using Image.J software, and the densities were normalized with respect to β-tubulin as the internal control.

2.11. Statistical Analysis

Data were expressed as percent of values of untreated control cells, and each value represents the mean ± SD (n = 3). The significant differences between the means of the treated and untreated cells were calculated by unpaired Student' *t*-test, and *p*-values < 0.05 were considered significant.

3. Results

3.1. Baicalein and Pan-Caspase Inhibitor (Z-VAD-Fmk) Shield SK-N-MC Cells against Menadione-Induced Cytotoxicity

Menadione is a quinone known to induce an oxidative stress generated primarily by superoxide radicals leading to cell death [31]. We found that menadione at 35 μM caused 55% cell death among SK-N-MC cells (**Figure 1**). In our previous study, we ascertained that no remarkable changes were seen among the cells in range of 10 - 50 μM of baicalein after 24 h [32]. Thus, cytoprotcetive effects of different doses of baicalein (10, 20, 40, 50 μM) on menadione (35 μM)-induced cytotoxicity in SK-N-C cells were investigated. The detrimental effects of menadione on SK-N-MC cells were considerably blocked by pretreatment with baicalein. The same result was observed for Z-VAD-fmk. As shown in **Figure 1**, the extent of survival was restored to 67%, 84%, 89% and 71% by pretreatment of cells with different doses of baicalein (10, 20, 40, 50 μM) for 3 h followed by treatment with 35 μM menadione for 24 h. Baicalein at a concentration of 40 μM, provided utmost protection against menadione insult producing a 44% increase in cell survival. Moreover, Z-VAD-fmk (50 μM) increased cell viability to 86% (**Figure 1**).

3.2. Baicalein but Not Z-VAD-Fmk, Mitigates Menadione-Induced Increase in Intracellular ROS Generation

Increase in ROS generation was measured as one of the indicators of menadione-induced oxidative stress in cells. As shown in **Figure 2**, generation of intracellular ROS (in term of DCF fluorescent intensity) in SK-N-MC cells increased by almost a factor of 6.2 after 12-h treatment

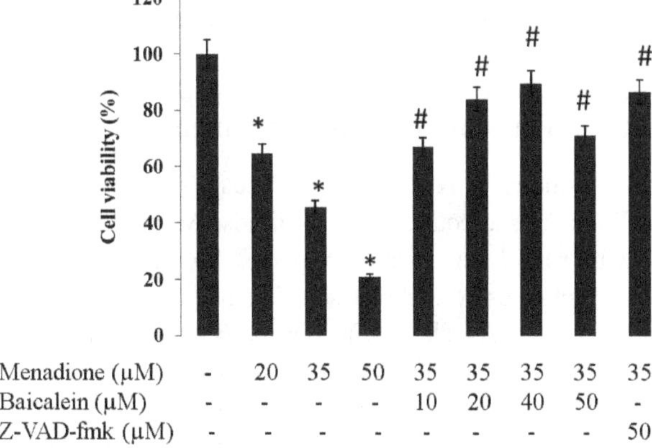

Figure 1. Cts of menadione, baicalein, and Z-VAD-fmk on viability of SK-N-MC cells. SK-N-MC cells were treated with different concentrations of menadione (20, 35, 50 μM) to find IC50 of menadione for further experiments (35 μM). Then, SK-N-MC cells were pretreated with different concentrations of baicalein (10, 20, 40, 50 μM) and Z-VAD-fmk (50 μM) for 3h and then incubated with menadione (35 μM) for 24 h. Cell viability was examined by MTT assay. Values correspond to means ± SD of three independent experiments. *significantly different from control cells (p < 0.05), #significantly different from menadione-treated cells (p < 0.05).

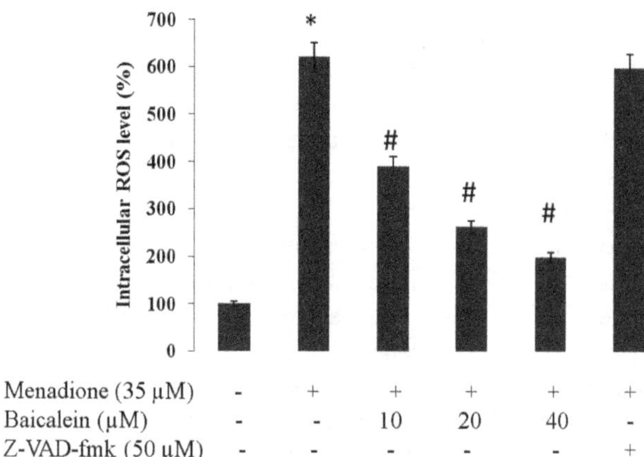

Figure 2. Effects of baicalein and Z-VAD-fmk on intracellular ROS level in menadione-treated SK-N-MC cells. SK-N-MC cells were pretreated with baicalein (10, 20, 40 μM) and Z-VAD-fmk (50 μM) for 3 h and then incubated with menadione for 12 h. ROS levels were monitored using 2', 7' dichlorofluorescein diacetate (DCFH-DA) staining. The fluorescence intensity was monitored on a varian-spectrofluorometer with excitation and emission wavelengths of 485 and 530 nm, respectively. Values correspond to means ± SD of three independent experiments. *significantly different from control cells ($p < 0.05$), #significantly different from menadione-treated cells ($p < 0.05$).

with menadione (35 μM) compared to ROS level of the untreated control cells. Pretreatment of the cells with different doses of baicalein (10, 20, 40 μM) attenuated ROS production in SK-N-MC cells by factors of 2.3, 3.6 and 4.3, respectively. However, pretreatment with Z-AD-mk (50 μM for 3 h) did not significantly change the ROS level in menadione-treated SK-N-MC cells.

3.3. Baicalein but Not Z-VAD-Fmk, Curbs Menadione-Induced Lipid Peroxidation

Menadione-induced oxidative stress causes oxidation of intracellular biomolecules such as lipids. MDA is produced while lipid peroxidation happens. So, MDA level measurement is used as a marker of menadione-induced oxidative stress. As shown in **Figure 3**, baicalein repressed lipid peroxidation in SK-N-MC cells. After 12 h of incubation with 35 μM menadione, MDA levels were significantly increased relative to the untreated control cells (0.41 nmol/mg protein in control cells versus 2.33 nmol/mg protein in menadione-treated cells). Pretreatment of cells with different doses of baicalein (10, 20, 40 μM) for 3 h followed by a 12 h treatment with menadione (35 μM) reduced MDA formation to 1.64, 1.01, and 0.66 nmol/mg protein, respectively, indicating that baicalein had quenched lipid peroxidation of the SK-N-MC cells. However, pretreatment with 50 μM Z-VAD-fmk did not significantly alter MDA contents in menadione-treated SK-N-MC cells.

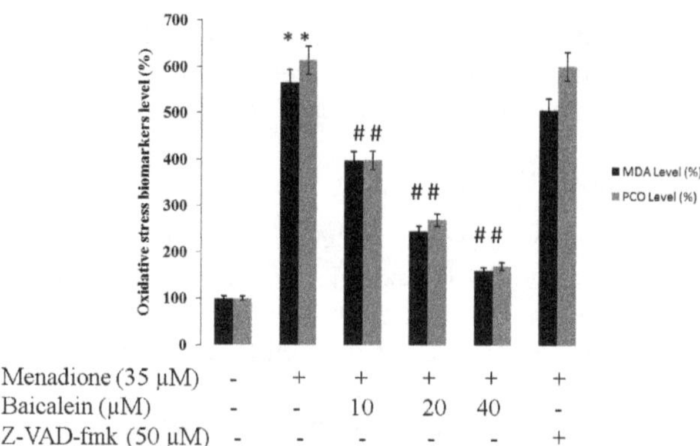

Figure 3. Effects of baicalein and Z-VAD-fmk on intracellular lipid peroxidation and protein carbonyl formation in menadione-treated SK-N-MC cells. SK-N-MC cells were pretreated with baicalein (10, 20, 40 μM) and Z-VAD-fmk (50 μM) for 3 h and then incubated with menadione for 12 h. lipid and protein oxidations were measured by analysis of MDA and PCO. Values correspond to means ± SD of three independent experiments. *significantly different from control cells ($p < 0.05$), #significantly different from menadione-treated cells ($p < 0.05$).

3.4. Baicalein but Not Z-VAD-Fmk, Diminishes Menadione-Induced Protein Carbonyl Formation

Protein carbonyl is a marker of protein oxidation in oxidative stress condition. We evaluated the effects of different doses of baicalein (10, 20, 40 μM) and Z-VAD-fmk (50 μM) on protein carbonyl formation in SK-N-MC cells. After treatment with menadione (35 μM), the amount of protein carbonyl increased to 4.03 nmol/mg protein compared to 0.65 nmol/mg protein of control cells. Pretreatment with baicalein (10, 20, 40 μM) reduced protein carbonyl formation to 2.6, 1.7 and 1.1 nmol/mg protein, respectively (**Figure 3**). However, pretreatment with 50 μM Z-VAD-fmk did not significantly alter PCO contents in menadione-treated SK-N-MC cells.

3.5. Baicalein and Z-VAD-Fmk Prevent Menadione-Induced Caspase-Dependent Apoptotic Cell Death

To study the protective effect of baicalein on SK-N-MC cells, acridine orange/ethidium bromide double staining technique was used to evaluate the occurrence of apoptosis in cells. As shown in **Figure 4**, the non-apoptotic control cells were stained green and the apoptotic cells had orange particles in their nuclei due to nuclear DNA fragmentation. The menadione treatment increased the extent of apoptosis relative to untreated control cells and pretreatment with baicalein (40 μM, 3 h) diminished apoptosis compared to menadione-treated cells (**Figure 4**). We also pretreated SK-N-MC cells with Z-VAD-fmk (50 μM) for 3 h followed by exposure to menadione (35

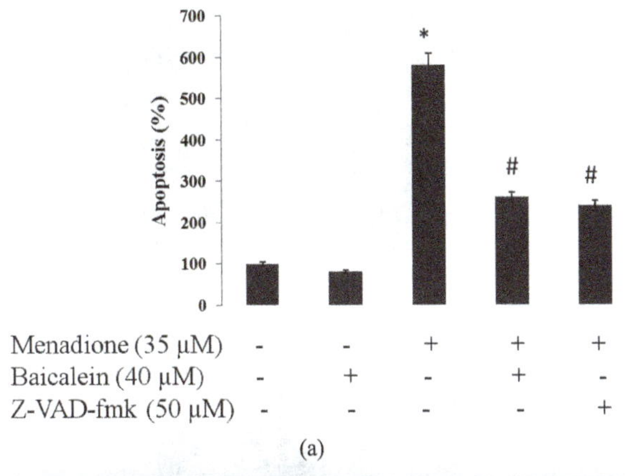

(a)

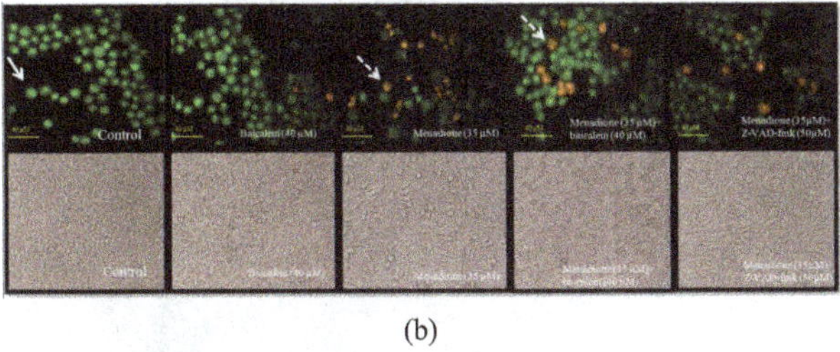

(b)

Figure 4. Effect of baicalein and Z-VAD-fmk treatments on menadione-induced apoptosis in SK-N-MC cells. (a) SK-N-MC cells were treated with baicalein (40 µM) and Z-VAD-fmk (50 µM) for 3 h followed by exposure to menadione (35 µM) for 24 h. cell pretreatment with baicalein and Z-VAD-fmk clearly decreased the number of apoptotic cells relative to cells treated only with menadione. Values correspond to means ± SD of three independent experiments. *significantly different from control cells ($p < 0.05$), #significantly different from menadione-treated cells ($p < 0.05$); (b) morphological analysis of SK-N-MC cells by double staining method. White arrow indicates live cells, dashed arrow shows apoptotic cells. Scale bar: 40 µM.

µM) for 24 h. As shown in **Figure 4**, Z-VAD-fmk reduced the extent of apoptosis relative to menadione-treated cells, confirming the caspase-dependent apoptosis of cells.

3.6. Effect of Baicalein on Menadione-Induced Lipofuscin Formation

Exposure of the cells to 35 µM menadione for 24 h caused 374% increase in the intracellular level of lipofuscin relative to menadione-untreated control cells. Pretreatment of the cells with baicalein (10, 20, 40 µM) diminished the formation of lipofuscin pigments by 155%, 192% and 214% after 24 h of exposure (**Figure 5**).

3.7. Baicalein Decreases Iron Accumulation in Menadione-Induced SK-N-MC Cells

Iron is important for electron transport in the respiratory chain and for various enzymatic reactions. When present in excess, however, iron can harm biological systems since in redox-active form it catalyzes the generation of highly reactive oxygen species [33]. Since both iron deficiency and overload impaired cellular functions, the quantitation of iron in cells and extracellular fluids is of considerable interest [34,35]. As shown in **Figure 6**, treatment of SK-N-MC cells with menadione elevated free iron contents compare to basal iron level in the control samples (2.17 nmol/mg proteins compare to 1.1 nmol/mg protein of control). However, pretreatments with different doses of baicalein (10, 20, 40 µM) diminished the iron contents to 1.75, 1.54 and 1.33, respectively.

3.8. Effects of Baicalein and Z-VAD-Fmk on Menadione-Induced Cell Death

Previous studies have shown that menadione-induced

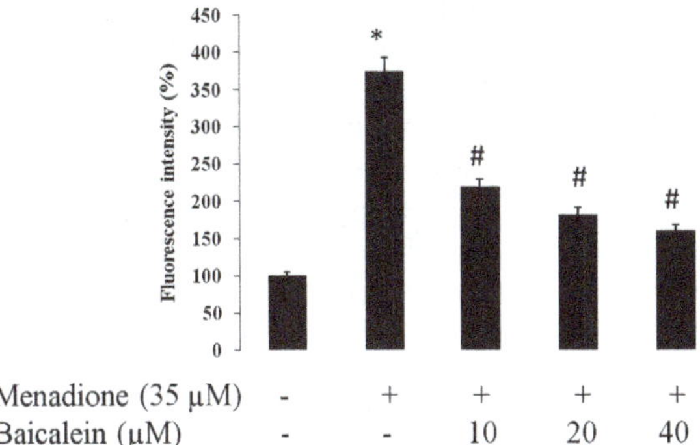

Figure 5. Inhibitory effect of baicalein on the menadione-treated accumulation of intracellular lipofuscin pigments. SK-N-MC cells were exposed to baicalein (10, 20, 40 µM) for 3 h followed by exposure to menadione (35 µM) for 24 h. Then, the extent of lipofuscin in cell lysates were evaluated using a varian spectrofluorometer, model Cary Eclipse, set at an excitation wavelength of 310 nm and an emission wavelength of 620 nm. *significantly different from control cells ($p < 0.05$), #significantly different from menadione-treated cells ($p < 0.05$).

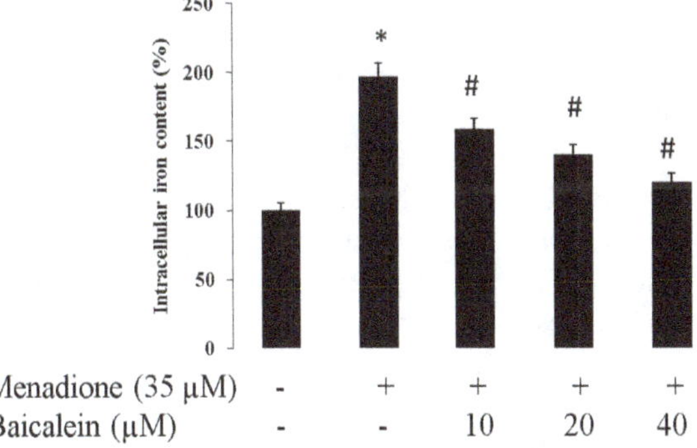

Figure 6. Effect of baicalein on intracellular iron contents in menadione-treated SK-N-MC cells. SK-N-MC cells were exposed to baicalein (10, 20, 40 µM) for 3 h followed by exposure to menadione (35 µM) for 24 h. Iron contents were evaluated by colorimetric ferrozine-based assay. *significantly different from control cells ($p < 0.05$), #significantly different from menadione-treated cells ($p < 0.05$).

apoptosis is associated with changes in apoptosis-related Bcl-2 family of regulatory proteins. Bax is a pro-apoptotic member of the Bcl-2 family which forms mitochondrial permeability pores for release of cytochrome c to the cytosol via binding to the anti-apoptotic Bcl-2 member. This event in turn will lead to cleavage of procaspase-9 and further activation of procaspase-3 and cell death through apoptosis [36]. Pretreatment of cells with baicalein prior to menadione treatment, reduced Bax/Bcl2 ratio and pretreatment of cells with baicalein and Z-VAD-fmk decreased cleaved caspase-9 in SK-N-MC cells which showed that baicalein inhibited caspase-dependent apoptosis in this cell line (**Figure 7**).

4. Discussion

One of the well-accepted theories for explicating the aging process is the free radical theory proposed by Denham Harman [5]. This theory illustrates that there is a causal relationship between oxidative stress and pathogenesis of age-related disorders [6]. Lipofuscin, a histological index of aging, is a highly oxidized cross-link aggregate consisting of oxidized proteins (30% - 58%) and lipids (19% - 51%) clusters accrues mostly in post-mitotic cells such as neurons, cardiac myocytes, skeletal muscle fibers and retinal pigments [7]. Since oxidative reactions are compulsory components of normal life processes, the incidence of reactive oxygen species with ensuing lipofuscin formation is an inexorable side effect of life [10]. Many studies have signified that many ROS-induced diseases such as neurodegenerative disorders are associated with high levels of lipofuscin within neuronal cells [37,38]. It has been widely reported that loosely bound iron in the cellular iron pool can react with endogenous hydrogen peroxide to produce the short-lived and highly reactive hydroxyl radicals through the Fenton reaction. These hydroxyl radicals, in turn, can oxidize nucleic acids, proteins or lipids leading to lipofuscin formation [23]. Oxidized proteins within lipofuscin are linked by intermolecular cross-links. Many of these cross-links are caused by non proteineous compounds including oxidized lipids such as Malondialdehyde (MDA) and 4-hydroxy-2-nonenal by means of reactions with lysine amino groups, cysteine sulfhydryl groups and histidine imidazole groups of proteins [39]. Thus, preventing biomolecules peroxidations and maintaining iron homeostasis play major roles in blocking lipofuscin formation.

Menadione (2-methyl-1,4 naphthoquinone) in the cells converts to menadione semiquinone radical via NADPH cytochrome c reductase activity. Then, semiquinone radical is recycled back to menadione through rapid reaction with molecular oxygen. This can result in the formation of superoxide radical which causes oxidative stress [40]. Although superoxide is chemically incapable of

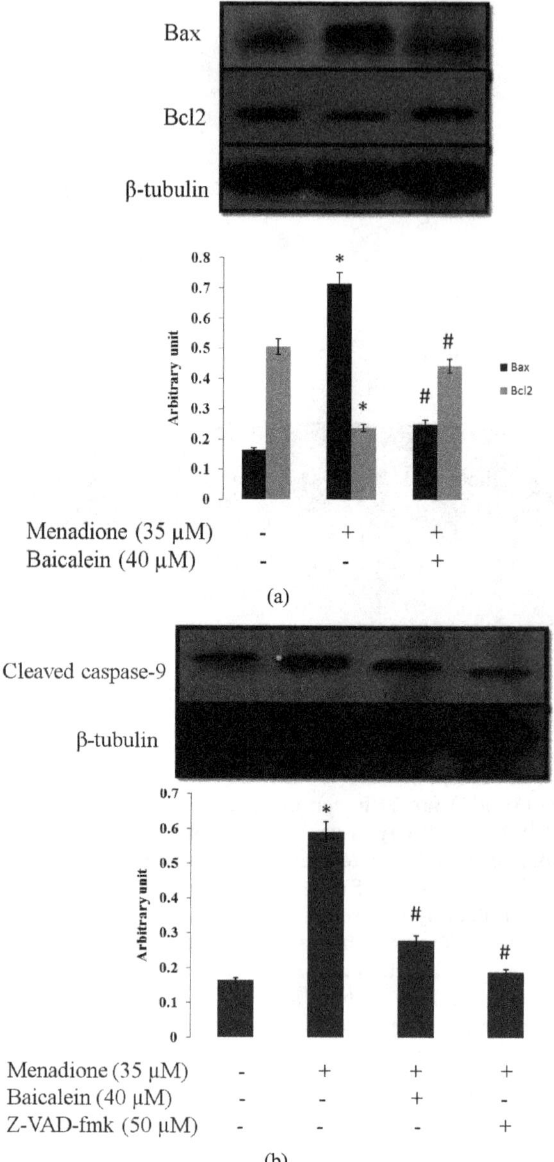

Figure 7. Analysis of Bcl-2, Bax and procaspase-9 activation in SK-N-MC cells treated with menadione, baicalein and Z-VAD-fmk. SK-N-MC cells were pretreated with baicalein (40 μM) and Z-VAD-fmk (50 μM) for 3 h and then incubated with menadione (35 μM) for 24 h. (a) bcl-2, Bax and the (b) procaspase-9 expression were estimated by immunoblots using relevant specific antibodies, and intensity of each band was estimated by densitometric analysis. Equal sample loadings were confirmed by tubulin band. Values correspond to means ± SD of three independent experiments. *significantly different from control cells ($p < 0.05$), #significantly different from menadione-treated cells ($p < 0.05$).

affecting biomolecules directly, it is assumed to do so indirectly by participating in the production of hydroxyl radicals through Fenton reaction. Superoxide radicals can provide free iron to catalyze peroxidation from two sources: release iron from ferritin and oxidizes the [4Fe -

4S] clusters of enzymes such as dehydratases, precipitating the release of one or more iron atoms [41]. Thus, menadione as a Fenton catalyst, assisted the production of free iron for production of hydroxyl radicals to ignite cross link of oxidized proteins and lipids in order to form lipofuscin.

There is an accumulating evidence denoting that lipofuscin can induce neurotoxicity via its capacity for binding metals such as iron, copper, zinc and calcium which stimulates generation of excessive ROS and decrease proteasomal and lysosomal degradation by inhibition of the proteasomal turnover [7]. Numerous studies have shown that intracellular iron accumulation contributes to the development of several common neurodegenerative diseases such as Alzheimer's disease (AD) and Parkinson's disease (PD) [33-35].

In order to restrain the destructive effects of ROS including superoxide radicals in neuronal cells, dietary flavonoids are shown to have potential anti-aging and brain-protective activities. Baicalein (5, 6, 7-trihydroxy-2-phenyl-4H-1-benzopyran-4-one), a naturally occurring flavonoid, is the major bioactive compounds found in traditional Chinese medicinal herb, Baikal Skullcap (*Scutellaria baicalensis* GEORGI) [22]. Baicalein produces promising results as a strong antioxidant. Its ability to cross blood brain barrier (BBB), hydrophobicity, presence of hydroxyl groups at C-5 and C-7, a double bond between C-2 and C-3, high trolox equivalent antioxidant capacity (TEAC) and DPPH free radical scavenging activity make baicalein a good ROS scavenger in neurons [20,21]. Presence of hydroxyl groups in baicalein structure results in scavenging of charged species such as superoxide radicals and hydroxyl radicals more efficiently compared to non-charged oxidant species [20]. On the other hand, baicalein can inhibit the production of endogenous hydroxyl radicals produced through the Fenton reaction by forming stable and inert complexes with iron [23]. Iron-binding motifs in some phenolic compounds can clarify the potential ability of them to modulate iron homeostasis in the body. Baicalein contains these motifs and thus expected to chelate iron. Some recent studies have shown that two hydroxyl groups at the 6 and 7 positions on the A ring seems to be the powerful metal binding site [20,23].

In support of what we have explained before, our studies showed that baicalein reduced the harmful effects of menadione by scavenging superoxide radicals which led to increased cell viability and decreased intracellular MDA and PCO. In addition, our results confirmed that baicalein has anti-Fenton properties since it decreased the free iron contents of SK-N-MC cells exposed to menadion treatment. We also observed that baicalein strongly inhibited lipofuscin formation in menadione-treated SK-N-MC cells and displays anti-aging features. Morphological analysis and western blot results implied that baicalein prevented apoptotic cell death through inhibition of Bax and procaspase-9 activations and induction of bcl2 expression which averted activation of further caspases and transcription factors, release of cytochrome c and resultant cell death. The results were confirmed by applying pan-caspase inhibitor (Z-VAD-fmk). Moreover, our experiments have shown that Z-VAD-fmk prevented cell death in SK-N-MC cells through inhibition of caspases and did not have any significant antioxidant characteristics.

Overall, flavonoid baicalein can be considered as a strong and auspicious antioxidant which could protect neuronal cells and hence, baicalein is a reliable option for antioxidant therapy in treatment of age-related and neurodegenerative disorders, pending further *in vivo* and clinical investigations.

5. Acknowledgements

The author appreciates the financial support of this investigation by the Research Council of University of Theran.

REFERENCES

[1] R. J. Sohal and R. Weindruch, "Oxidative Stress, Caloric Restriction and Aging," *Science*, Vol. 273, No. 5271, 1996, pp. 59-63. doi:10.1126/science.273.5271.59

[2] L. Rossi, S. Mazzitelli, M. Arciello, C. R. Capo and G. Rotilio, "Benefits from Dietary Polyphenols for Brain Aging and Alzheimer's Disease," *Neurochemical Research*, Vol. 33, No. 12, 2008, pp. 2390-2400. doi:10.1007/s11064-008-9696-7

[3] J. L. Martindale and N. J. Holbrook, "Cellular Response to Oxidative Stress: Signaling for Suicide and Survival," *Journal of Cell Physiology*, Vol. 192, No. 1, 2002, pp. 1-15. doi:10.1002/jcp.10119

[4] L. Moldovan and N. I. Moldovan, "Oxygen Free Radicals and Redox Biology of Organelles," *Histochemistry and Cell Biology*, Vol. 122, No. 4, 2004, pp. 395-412. doi:10.1007/s00418-004-0676-y

[5] S. Schmitt-Schillig, S. Schaffer, C. C. Weber, G. P. Eckert and W. E. Muller, "Flavonoids and the Aging Brain," *Journal of Physiology and Pharmacology*, Vol. 56, No. 1, 2005, pp. 23-36.

[6] D. Harman, "Free Radical Theory of Aging: Effect of Free Radical Reaction Inhibitors on the Mortality Rate Male LAF Mice," *Journal of Gerontology*, Vol. 23, No. 4, 1968, pp. 476-482. doi:10.1093/geronj/23.4.476

[7] T. Jung, N. Bader and T. Grune, "Lipofuscin: Formation, Distribution and Metabolic Consequences," *Annals of the New York Academy of Science*, Vol. 1119, 2007, pp. 97-111. doi:10.1196/annals.1404.008

[8] U. T. Brunk and A. Terman, "Lipofuscin: Mechanisms of Age-Related Accumulation and Influence on Cell Function," Free *Radical Biology and Medicine*, Vol. 33, No. 5,

2002, pp. 611-619. doi:10.1016/S0891-5849(02)00959-0

[9] T. Kurz, A. Terman and U. T. Brunk, "Autophagy, Aging and Apoptosis: The Role of Oxidative Stress and Lysosomal Iron," *Archieves of Biochemistry and Biophysics*, Vol. 462, No. 2, 2007, pp. 220-230. doi:10.1016/j.abb.2007.01.013

[10] A. Terman and U. T. Brunk, "Lipofuscin: Mechanisms of Formation and Increase with Age," *Acta Pathologica, Microbiologica et Immunologica Scandinavica*, Vol. 106, No. 2, 1998, pp. 265-276. doi:10.1111/j.1699-0463.1998.tb01346.x

[11] B. Halliwell, "Reactive Oxygen Species and the Central Nervous System," *Journal of Neurochemistry*, Vol. 59, No. 5, 1992, pp. 1609-1923. doi:10.1111/j.1471-4159.1992.tb10990.x

[12] B. Halliwell and J. M. Gutteridge, "Role of Free Radicals and Catalytic Metal Ions in Human Disease: An Overview," *Methos in Enzymology*, Vol. 186, 1990, pp. 1-85.

[13] M. Giorgio, M. Trinei, E. Migliaccio and P. G. Pelicci, "Hydrogen Peroxide: A Metabolic By-Product or a Common Mediator of Ageing Signals?" *Nature Reviews: Molecular Cell Biology*, Vol. 8, No. 9, 2007, pp. 722-728. doi:10.1038/nrm2240

[14] D. Vauzour, K. Vafeiadou, A. Rodriguez-Mateas, C. Rendeiro and J. P. E. Spencer, "The Neuroprotective Potential of Flavonoids: A Multiplicity of Effects," *Genes and Nutrition*, Vol. 3, No. 3-4, 2008, pp. 115-126. doi:10.1007/s12263-008-0091-4

[15] J. P. E. Spencer, "Flavonoids: Modulators of Brain Function?" *British Journal of Nutrition*, Vol. 99, No. 1, 2008, pp. ES60-77.

[16] D. Vauzour, "Dietary Polyphenols as Modulators of Brain Functions: Biological Actions and Molecular Mechanisms Underlying Their Beneficial Effects," *Oxidative Medicine and Cellular Longevity*, Vol. 2012, Article ID: 914273, pp. 1-16.

[17] T. Osawa, "Protective Role of Dietary Polyphenols in Oxidative Stress," *Mechanism of Ageing and Development*, Vol. 111, No. 2-3, 1999, pp. 133-139. doi:10.1016/S0047-6374(99)00069-X

[18] E. Middleton, C. Kandaswami and T. C. Theoharides, "The Effects of Plant Flavonoids on Mammalian Cells: Implications for Inflammation, Heart Disease and Cancer," *Pharmacological Reviews*, Vol. 52, No. 4, 2000, pp. 673-751.

[19] D. Atmani, N. Chaher, D. Atmani, M. Berboucha, N. Debbache and H. Boudaoud, "Flavonoids in Human Health: From Structure to Biological Activity," *Current Nutrition and Food Science*, Vol. 5, No. 4, 2009, pp. 225-237. doi:10.2174/157340109790218049

[20] Z. S. Markovic, M. Dimitric-Markovic, D. Milenkovic and N. Filipovic, "Structural and Electronic Features of Baicalein and Its Radicals," *Monatsh Chemistry*, Vol. 142, No. 2, 2011, pp. 145-152. doi:10.1007/s00706-010-0426-x

[21] O. Firuzi, A. Lacanna, R. Petrucci, G. Marrosu and L. Saso, "Evaluation of the Antioxidant Activity of Flavonoids by 'Ferric Reducing Antioxidant Power' Assay and Cyclic Voltammetry," *Biochemica et Biophysica Acta*, Vol. 1721, No. 1-3, 2005, pp. 174-184. doi:10.1016/j.bbagen.2004.11.001

[22] S. H. Zhang, J. Ye and G. Dong, "Neuroprotective Effects of Baicalein on Hydrogen Peroxide-Mediated Oxidative Stress and Mitochondrial Dysfunction in PC12 Cells," *Journal of Molecular Neuroscience*, Vol. 40, No. 3, 2010, pp. 311-320. doi:10.1007/s12031-009-9285-5

[23] C. A. Perez, Y. Wei and M. Guo, "Iron-Binding and Anti-Fenton Properties of Baicalein and Baicalein," *Journal of Inorganic Biochemistry*, Vol. 103, No. 3, 2009, pp. 326-332. doi:10.1016/j.jinorgbio.2008.11.003

[24] C. P. LeBel, H. Ischiropoulos and S. C. Bondy, "Evaluation of the Probe 2', 7'-Dichlorofluorescein as an Indicator of Reactive Oxygen Species Formation and Oxidative Stress," *Chemical Research in Toxicology*, Vo. 5, No. 2, 1992, pp. 227-231. doi:10.1021/tx00026a012

[25] H. H. Drapper and M. Hadley, "Malondialdehyde Determination as Index of Lipid Peroxiadation," *Methods in Enzymology*, Vol. 186, 1990, pp. 421-431. doi:10.1016/0076-6879(90)86135-I

[26] O. H. Lowry, N. J. Rosebrough, A. L. Farr and R. J. Randall, "Protein Measurement with the Folin Phenol Reagent," *Journal of Biological Chemistry*, Vol. 193, No. 1, 1951, pp. 265-275.

[27] A. Z. Reznick and L. Packer, "Oxidative Damage to Proteins: Spectrophotometric Method for Carbonyl Assay," *Methods in Enzymology*, Vol. 233, 1994, pp. 357-363. doi:10.1016/S0076-6879(94)33041-7

[28] S. Emig, D. Schmalz, M. Shakibaei and K. Buchner, "The Nuclear Pore Complex Protein p62 Is One of Several Sialic Acid-Containing Proteins of the Nuclear Envelope," *Journal of Biological Chemistry*, Vol. 270, No. 23, 1995, pp. 13787-13793. doi:10.1074/jbc.270.23.13787

[29] Y. Mochizuki, M. K. Park, T. Mori and S. Kawashima, "The Difference in Autofluorescence Features of Lipofuscin between Brain and Adrenal," *Zoological Science*, Vol. 12, No. 3, 1995, pp. 283-288. doi:10.2108/zsj.12.283

[30] J. Riemer, H. H. Hoepken, H. Czerwinska, S. R. Robinson and R. Dringen, "Colorimetric Ferrozine-Based Assay for the Quantitation of Iron in Cultured Cells," *Analytical Biochemistry*, Vol. 33, No. 2, 2004, pp. 370-375. doi:10.1016/j.ab.2004.03.049

[31] M. J. Czaja, H. Liu and Y. Wong, "Oxidant-Induced Hepatocytes Injury from Menadione is Regulated by ERK and AP-1 Signaling," *Hepatology*, Vol. 37, No. 6, 2003, pp. 1405-1413. doi:10.1053/jhep.2003.50233

[32] M. Moslehi, A. Meshkini and R. Yazdanparast, "Flavonoid Baicalein Modulates H_2O_2-Induced Mitogen-Activated Protein Kinases Activation and Cell Death in SK-N-MC Cells," *Cellular and Molecular Neurobiology*, Vol. 32, No. 4, 2012, pp. 549-560. doi:10.1007/s10571-011-9795-x

[33] R. R. Crichton, S. Wilmet, R. Legssyer and R. J. Ward, "Molecular and Cellular Mechanisms of Iron Homeostasis and Toxicity in Mammalian Cells," *Journal of Inorganic Biochemistry*, Vol. 91, No. 1, 2002, pp. 9-18. doi:10.1016/S0162-0134(02)00461-0

[34] J. L. Beard and J. R. Connor, "Iron Status and Neural Functioning," *Annual Review of Nutrition*, Vol. 23, 2003, pp. 41-58. doi:10.1146/annurev.nutr.23.020102.075739

[35] J. R. Burdo and J. R. Connor, "Brain Iron Uptake and Homeostatic Mechanisms: An Overview," *Biometals*, Vol. 16, No.1, 2003, pp. 63-75. doi:10.1023/A:1020718718550

[36] J. E. Chipuk and D. R. Green, "How Do Bcl2 Proteins Induce Mitochondrial Outer Membrane Permeabilization?" *Trends in Cell Biology*, Vol. 18, No. 4, 2008, pp. 157-164. doi:10.1016/j.tcb.2008.01.007

[37] M. R. D'Andrea, R. G. Nagele, N. A. Gumula, P. A. Reiser, D. A. Polkovitch, B. M. Hertzog and P. Andrade-Gordon, "Lipofuscin and Abeta42 Exhibit Distinct Distribution Patterns in Normal and Alzheimer's Disease Brain," *Neuroscience Letters*, Vol. 323, No. 1, 2002, pp. 45-49. doi:10.1016/S0304-3940(01)02444-2

[38] L. M. Drach, J. Bohl and H. H. Goebel, "The Lipofuscin Content of Nerve Cells of the Inferior Olivary Nucleus in Alzheimer's Disease," *Dementia*, Vol. 5, No. 5, 1994, pp. 234-239. doi:10.1073/pnas.97.2.611

[39] H. H. F. Refsgaard, L. Tsai and E. R. Stadtman, "Modifications of Proteins by Polyunsaturated Fatty Acid Peroxidation Products," *Proceedings of the National Academy of Sciences of the United States of America*, Vol. 97, No. 2, 2000, pp. 611-616.

[40] K. L. Seanor, J. V. Cross, S. M. Nguyen, M. Yan and D. J. Templeton, "Reactive Quinines Differentially Regulates SAPK/JNK and p38/mHOG Stress Kinases," *Antioxidants and Redox Signaling*, Vol. 5, No. 1, 2003, pp. 103-113.

[41] K. Keyer and J. A. Imlay, "Superoxide Accelerates DNA Damage by Elevating Free-Iron Levels," *Proceedings of the National Academy of Sciences of the United States of America*, Vol. 93, No. 24, 1996, pp. 13635-13640. doi:10.1073/pnas.93.24.13635

Abbreviation

MDA: Malondialdehyde;
MTT: 3-(4,5-dimethylthiazol-2-yl)-2,5-diphenyl tetrazolium bromide;
PBS: Phosphate buffer saline;
PCO: Protein carbonyl;
AD: Alzheimer's disease;
ROS: Reactive oxygen species.

18

The Cell Sorting Process of *Xenopus* Gastrula Cells Progresses in a Stepwise Fashion Involving Concentrification and Polarization

Ayano Harata, Takashi Matsuzaki, Koichi Ozaki, Setsunosuke Ihara*
Faculty of Life and Environmental Science, Department of Biological Science, Shimane University, Shimane, Japan
Email: *ihara@life.shimane-u.ac.jp

ABSTRACT

Animal pole cells (AC) and vegetal pole cells (VC) dissociated from early *Xenopus* gastrulae were intermingled, and the cell sorting process occurring within the aggregate was analyzed. The overall process of cell sorting was found to morphologically consist of two steps, "concentrification" and "polarization", as designated here. First, AC and VC clusters emerged at random positions in the aggregate, and the individual clusters gradually assembled themselves by 5 hours in culture (5 hC), forming a concentric arrangement, in which the AC cluster was enveloped by the VC cluster. This concentrification step is essentially consistent with the descriptions in earlier studies. As the next step, the AC and VC clusters moved up and down from 7.5 to 12 hC, resulting in the vertical polarization, namely, a serial array just like *in vivo*. Immunohistochemical analyses showed that AC expressed both C- and E-cadherins, while VC only expressed C-cadherin, as *in vivo*, suggesting the normal participation of cadherin system. On the other hand, the actin localization showed that the actin bundles accumulated at the edge of the AC cluster until the concentrification was completed, and gradually decreased during the polarization step. Another important finding was that AC cluster could generate cartilage tissues during the long-term (7 days) culture, evidence for a healthy inductive interaction between the AC and VC. Taken together, the present experimental system allows the AC and VC to be viable and grow into an embryo-like organization.

Keywords: Cell Sorting; *Xenopus laevis*; Concentrification; Polarization; Embryogenesis

1. Introduction

Upon dissociation, metazoan cells exert an intrinsic potential of sorting out by themselves. It has been repeatedly studied to date as a model to elucidate the principle of morphogenesis. For example, dissociated Hydra cells reassemble and restore the normal whole body [1,2]. Similarly, dissociated chick limb bud cells sort out [3] and form a proximodistal axis [4,5].

In amphibian early embryos, it is also widely known that dissociated cells derived from different germ layers first unite indiscriminately and reconstruct normal tissue structures under relatively simple culture conditions [6]. Although it has been clarified that cadherins as cell adhesion molecules play an essential role in cell sorting the in *Xenopus laevis* embryo [7,8], the dynamic aspect of the entire process of cell sorting is still poorly understood.

*Corresponding author.

We have now focused on observing in detail the behaveiors of individual germ layer cells when animal cells (AC) and vegetal cells (VC) obtained from early gastrula were intermingled. The two-cell stage embyos were microinjected with dextran rhodamine, if necessary for the precise identification of the cell sources. In *Xenopus laevis*, the regional identity starts to be created before midbrastula and the sorting behavior becomes conspicuously progressive as the development advances [9,10]. On the other hand, the germ layer specification is not very advanced at the early gastrula stage [11,12]. Therefore, we decided to use the gastrula stage, considering it fit for analysis.

In this paper, we have focused on the following two kinds of macromolecules: 1) C-cadherin and E-cadherin, both of which are major forms of Xenopus cell adhesion molecules at the embryonic stage [13-16], and 2) the cortex actin, cytoskeleton protein known to cooperate with cadherin during the regulation of the cell shape and cell motility [17]. Our objective was to examine how these macromolecules are expressed within the aggregates in the present *in vitro* system. We found that the dissociated AC and VC reassembled and concentrically arranged, and subsequently, the two clusters were mutually polarized up and down. Thus our results clearly showed that the cell sorting process of AC and VC could be divided into two steps. Furthermore, the AC cluster in the polarized aggregate differentiated into mesodermal tissues thereafter, implying that the present culture system starting from cell sorting allowed the so-called "Nieuwkoop's mesodermal induction" [18,19] between AC and VC. The expression patterns of the C- and E-cadherins were consistent with those in vivo, in terms of the specificity of expressing germ layers. The actin bundles localized at the edge of the AC cluster were reduced during cell sorting process. These findings suggest that the present aggregate culture system is useful for analyzing the germ layer interactions.

2. Materials and Methods

2.1. Animals and Emblyos

Sexually mature *Xenopus laevis* colonies were purchased from Sato Yosyoku (Chiba, Japan) and embryos were obtained by artificial fertilization. Oocytes were stripped from females injected 10 hours earlier with 800 units of human chorionic gonadotrophin (Gonatropin, Asuka Pharmaceutical Co., Tokyo, Japan), and fertilized with minced testis in De Boer's solution (110 mM NaCl, 1.3 mM KCl, 0.45 mM $CaCl_2$, 3 mM HEPES, pH 7.3 at 23˚C). The embryos were maintained in 10% Steinberg's solution at 23˚C (1× Steinberg's solution; 58 mM NaCl, 0.67 mM KCl, 0.34 mM $Ca(NO_3)_2 \cdot 4H_2O$, 0.83 mM $MgSO_4 \cdot 7H_2O$, 10 mM HEPES, pH 7.3 at 23˚C). The development stage was determined according to Nieuwkoop and Faber [20].

2.2. Cell Preparation and Cell Sorting Culture

At stage 10 (the early stage of gastrulation), the animal caps were collected from five dejellied and devitellined embryos using an eyebrow knife and tungsten needle. Likewise, parts of the vegetal hemisphere were carefully collected to be free of not only the animal caps but also the marginal zones. They were individually transferred into Ca^{2+}- and Mg^{2+}-free Modified Barth Saline (CMF-MBS: 88 mM NaCl, 1 mM KCl, 5 mM HEPES, 2.5 mM $NaHCO_3$, pH 7.8 at 23˚C) containing 50 μg/ml gentamicin statically for one hour. The outer layer of the animal cap was discarded because it was difficult to dissociate, and only the inner layer was dissociated by gentle pipetting. AC (animal cells) were mixed with an equal volume of VC (vegetal cells), then transferred to an agar-coated 4-well Nunc dish (Thermo Fisher Scientific, Roskilde, Denmark) filled with Ca^{2+}, Mg^{2+}-containing MBS. The mixtures were rotated at 70 rpm and 23˚C for one hour, and then they were subjected to a stationary culture at 23˚C. Aggregates were incubated for 2.5, 5, 7.5, 12, 24, 48, 72, 96, 120 and 148 h, observed by a streomicroscope, and fixed with MEMFA [21,22]. For the long culture, culture media were refreshed every 12 hours. They were dehydrated through an ethanol series and embedded in paraffin wax (Shandon Histoplast, Thermo Scientific, Cheshire, UK).

2.3. Cell Tracking by Dextran Rhodamine Labeling

At the two-cell stage, dejellied embryos were transferred into 4% Ficoll in 10% Steinberg's solution, and both blastomeres of them were injected with a total volume of 20nl containing 1% Dextran rhodamine (DR, D3312, Molecular Plobes, Eugene, USA) at their animal pole side. At different times in culture, the aggregates consisting of the DR-labeled and unlabeled cells were fixed with 4% paraformaldehyde (PFA) in amphibian CMF-PBS (6.4 g NaCl, 0.2 g KCl, 2.9 g $NaHPO_3 \cdot 12H_2O$, 0.2 g KH_2PO_4, and 200 ml water), then embedded in paraffin, and sectioned into 4 μm slices. The slices were deparaffinized, mounted using Fluoromount (Diagnostic Biosystems, Pleasantom, USA), and observed with a fluorescent microscope (OLYMPUS BX50, Tokyo, Japan). The nuclei had been stained with DAPI (4, 6-diamidino-2-phenylindole dihydrochloride, 0.5μg/ml).

2.4. Histological Examination

The paraffin sections were deparaffinized and stained with Alcian blue.

2.5. Immunohistochemistry

The paraffin sections were deparaffinized, rinsed with CMF-PBS, blocked with 1% normal horse serum in CMF-PBS for 20 minutes at room temperature, incubated with the primary antibody overnight at 4˚C, then extensively washed with CMF-PBS. The secondary antibody was then added and incubation was carried out for 2 hours at room temperature. The samples were stained for nuclei with DAPI, mounted using Fluoromount, and observed by fluorescent microscope. The following primary antibodies were used: the anti-E-cadherin monoclonal antibody (supernatant 5D3, Developmental Study Hybridoma Bank, Iowa, USA, 1/3 dilution), anti-C-cadherin monoclonal antibody (supernatant 6B6, Developmental Study Hybridoma Bank, 1/2 dilution), and anti-actin monoclonal antibody (MAB1501, Millipore, Billerica. USA, 1:100 dilution). An Alexa 488 conjugated goat anti mouse IgG_{2b} (A21141; Molecular Probes,

Eugene, USA, 1:500 dilution) and an Alexa 594 conjugated goat anti mouse IgG_1 (A21125; Molecular Probes, Eugene, USA, 1:1000 dilution) were used as the secondary antibodies. The fluorescence intensities specific to the actin filament at the edge of the AC clusters were determined using the software Image J (National Institutes of Health, USA).

2.6. Western Blot Analysis

Xenopus embryos at stages 10.5, 12.5, 15, 19, and 21 and aggregates were lysed in extraction buffer (1 × CMF-PBS, 1% triton X-100, 20 units/ml aprotinin, 1 mM EDTA (pH 8.0)) containing 1 mM diisopropyl fluorophosphates. The samples were quantified by the BCA protein assay kit (Thermo Fisher Scientific, Illiois, USA). They were combined with an equal amount of Leammli's 2× sample buffer containing 5% 2-mercaptoethanol, boiled for 2 minutes at 90°C, and separated in 8% SDS-PAGE gel by running for 60 minutes at 60 mA. Gels were transferred to an immobilon-P membrane (Millipore, Billerica, USA) using a semidry apparatus (Nihon Eido Co., Ltd., Tokyo, Japan) for 3 hours at 100 mA. All membranes were blocked with 1% NHS in PBT for 40 minutes at room temperature, and incubated with the primary antibody overnight at 4°C. The following primary antibodies were used: anti-E-cadherin mAb (supernatant 5D3, Developmental Study Hybridoma Bank, Iowa, 1/10 dilution), anti C-cadherin mAb (supernatant 6B6, Developmental Study Hybridoma Bank, Iowa, 1/10 dilution), and anti β-tubulin mAb (T4026; Sigma, St. Louis, USA, 1:1000 dilution). The HRP-conjugated mouse immunoglobulin (Dako, Glostrup, Denmark, 1:3000 dilution) was used as the secondary antibody for 2 hours at room temperature. The membranes conjugated with HRP were detected by DAB solutions (3 mg of 3-3' diaminobezidine (Dojindo Laboratories, Kumamoto, Japan), 10 ml of 50 mM Tris- HCl (pH 7.6), and 8 µl of 30% H_2O_2). The intensity of bands was quantitatively estimated by image J (National Institutes of Health, USA).

2.7. The Three-Dimensional Analysis of Aggregate

The thick paraffin sections (20 µm) were reacted with the anti-actin antibody, followed by incubation with Alexa 488 conjugated goat anti-mouse IgG_1 (A21121; Molecular Probes, 1:500 dilution). The samples were observed using a Leica TCS SP5 conforcal microscope (Leica Microsystem, Wetzlar, Germany) equipped with immersion lens. All images were taken with the size of 1024 × 1024 pixels, and individual 1.5 µm optical slices were assembled into stacks. Some of them ware axially tilted in the X and Y directions.

3. Results

3.1. Change in Spatial Arrangement Decision of Dissociated Embryonic Cells during Cell Sorting Culture

The phenomenon called "sorting out of cells", including those found in the pioneering studies by Townes and Holtfreter [6], have many intriguing issues to explore. We focused on the behaviors of dissociated embryonic cells during reaggregation in which AC and VC isolated from early *Xenopus* gastrulae had been intermingled.

To clarify whether or how the dissociated cells move to their own positions in the present aggregate culture system, we visualized the process of cell sorting using dextran rhodamine labeled cells (**Figure 1**). In control experiments with the mixture of labeled AC and unlabeled AC (**Figures 1 (A)-(C)**) and vice versa (**Figures 1(D)-(F)**), the cell arrangement of both the AC and VC aggregates remained at random, confirming that the labeling of cells had no effect on the sorting out, and the

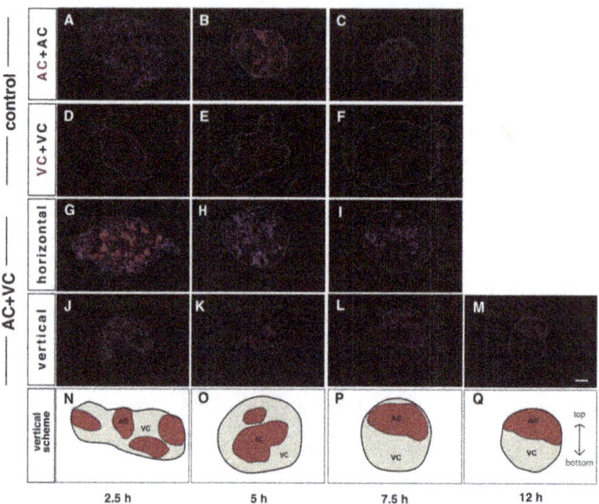

Figure 1. Overall cell rearrangement during cell sorting culture. Labeling with dextran rhodamine (DR) was performed at 2 cell stage. Reaggregatios were observed with a fluorescence microscope at 2.5 h ((A), (D), (G), (J)), 5 h ((B), (E), (H), (K)), 7.5 h ((C), (F), (I), (L)) and 12 h (M) after beginning of the stationary culture. Nuclei were counterstained with DAPI (blue). DR-labeled cells (DR(+), red) were discriminated from unlabeled cells (DR(-), dark). Control aggregates, DR(+)AC+DR(-)AC ((A)-(C)) and DR (+)VC+DR(-)VC(D-F), maintained random arrangement. On the other hand, the AC cluster of DR(+)AC+DR(-)VC aggregates became located at the center, enveloped by VC cluster, in horizontal sections ((G)-(I)), while the two clusters gradually rearranged up and down in vertical sections (J-M). White dotted line: the contour of the aggregate. Scale bar: 100 µm. The schemes of the rearrangement of clusters in vertical sections are shown ((N)-(P)). Red areas: AC clusters. White areas: VC clusters. Black dotted lines: boundaries between AC and VC clusters. Black solid line: the contour of the aggregate.

two kinds of cells could individually recognize themselves. The shape of the AC aggregates became nearly spherical with time (**Figures 1 (A)-(C)**), whereas the VC aggregates remained flat and rugged (**Figures 1(D)-(F)**). A temporal and spatial evaluation of the cell arrangement within the AC+VC aggregate was made by preparing the horizontal (**Figures 1(G)-(I)**) and vertical (**Figures 1(J)-(M)**) sections against the area facing the agar at different times during the stationary culture. Each of the AC and VC formed clusters and was arranged at random (**Figure 1(G)**) for the first 2.5 hours in culture (2.5 hC) and gradually self-assembled into large clusters, then they concentrically arranged, that is, the AC cluster was enveloped by the VC cluster (**Figures 1(G)** and **(I)**). This concentrification of clusters suggested that cell sorting progressed. Interestingly, as shown in the vertical sections, the AC and VC clusters started to be rearranged relatively up and down from about 7.5 hC (**Figures 1(J)** and **(L)**), and such rearrangement was completed by 12 hC (**Figure 1(M)**). These data suggest that the overall cell sorting process of the AC and VC is temporally divided into two steps, *i.e.*, concentrification and polarization, both of which are the original terms that we propose in this study. As the borders between AC and VC clusters in the actual fluorescence-stained sections are not necessarily discriminative, the schematic drawing of the vertical sectional views is shown in **Figures 1(N)-(Q)**.

3.2. The Expression Patterns of Cadherin within the Aggregates

The expression of the cell-cell adhesion molecule, cadherin, during the present *in vitro* culture was immunohistochemically examined (**Figure 2**). A maternal cadherin, C-cadherin, is known to be ubiquitously expressed in during the early development of *Xenopus laevis* [14, 15], and zygotic cadherin, E-cadherin, is activated in only the ectoderm from the gastrula stage [13]. In this study, C-cadherin showed a punctate distribution at the adjacent AC plasma membrane and E-cadherin was also weakly expressed there at 2.5 hC (**Figure 2(A)**). Both cadherins were concentrated similar to a line drawing along the AC plasma membrane after 5hC and particularly elevated from 7.5 hC to 12 hC (**Figures 2(B)-(D)**). Endoderm-derived VC always expressed C-cadherin very weakly as *in vivo*. The boundaries between the two clusters are illustrated in **Figure 2** (bottom panels). Next, we quantitatively assessed both cadherins in the aggregate by a western blot assay (**Figure 3**). The cadherin levels in the entire aggregate were sequentially compared, because it was difficult to separately analyze the AC and VC clusters. As a result, both the C- (**Figures 3(a)** and **(b)**) and E-cadherins (**Figures 3(c)** and **(d)**) showed almost a unidirectional increase during the test period. Indeed, this tendency of upregulation might explain the fact that the aggregates became more spherical with time. The persistence and up-regulation of cadherin further suggested that the expression of cadherin was also essential for maintaining the morphology of the aggregates during the polarization step, as will be described in Discussion.

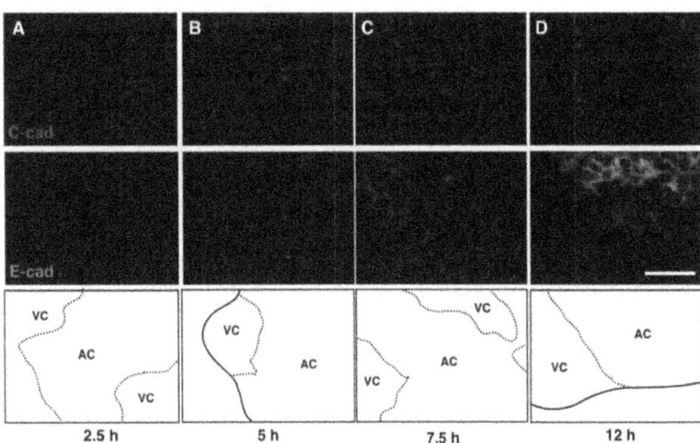

Figure 2. The spatial patterns of expression of C-cadherin and E-cadherin during cell sorting culture. Aggregates were double stained (upper, C-cadherin, red; middle, E-cadherin, green) at 2.5 h (A), 5 h (B), 7.5 h (C) and 12 h (D) after beginning stationary culture. The arrangements of AC and VC clusters in the same fields are illustrated (bottom). Both cadherins were positive in AC, while C-cadherin was faint in VC. At 2.5 h, C-cadherin displayed a punctate pattern and E-cadherin also weakly appeared (A). At and after 5 h (especially in 12 h), both cadherins increased as a continuous line along cell membrane ((B), (C), (D)). Nuclei were stained with DAPI (blue). Dotted lines: the boundaries between AC and VC clusters. Solid line: the contour of the aggregate. Scale bar: 50 μm.

3.3. The Expression of Actin Filaments in the Surface of Clusters

The process of sorting out is generally thought to result in an architecture with some concentric arrangements composed of different cell elements, as shown by the preceding theoretical studies (for example, [23]). On the contrary, the concentric arrangement of the AC and VC clusters was shifted to a polarized one during the later stage of the present stationary culture. How did the individual cells behave so that the clusters relocated? From the midblastula to neurula stages, *Xenopus laevis* embryos should be devoted to the germ layer specification progress, namely, a series of cooperative morphogenetic events involving such as convergent extension, invagination, cell migration, etc. [11,24]. Therefore, it is very likely that the AC and/or VC obtained from the gastrula stage may exert some active cell motility during the polarized rearrangement. Based on this presumption, we tried to verify the possibility that the production of lamellipodia from the AC and their polarity might be accompanied by rearrangement of the clusters (**Figure 4**). The expression of actin was positive in the AC mem-

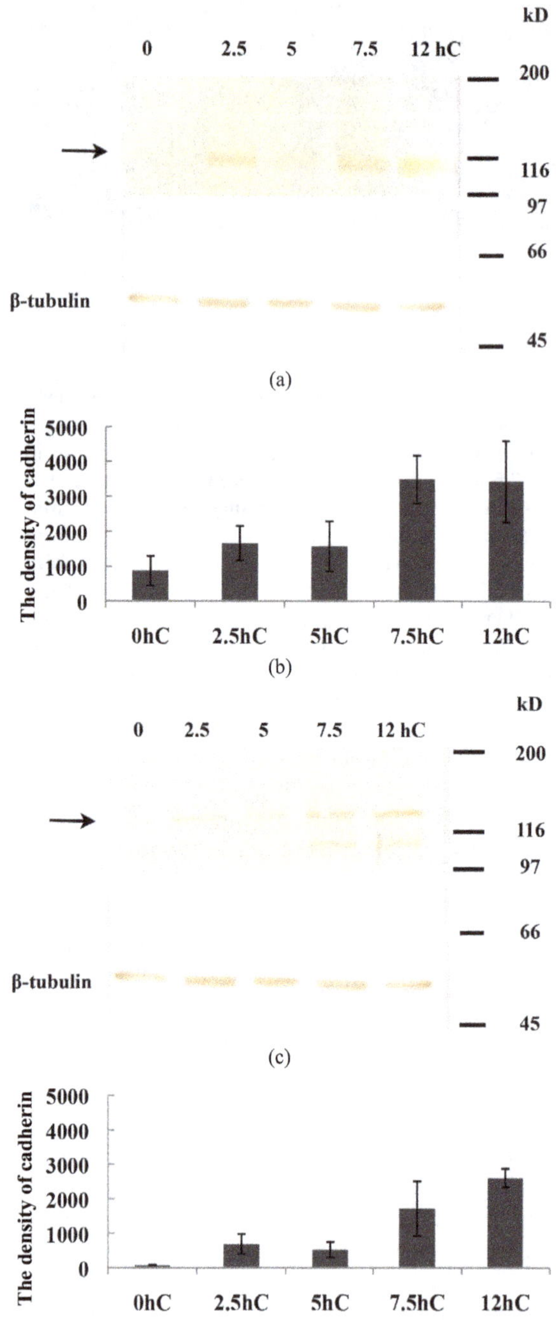

Figure 3. The quantitative change in cadherin during cell sorting culture. The expression levels of (c)-cadherin and E-cadherin were quantified by western blot analysis ((a) -(d)). (c)-cadherin ((a), arrow) and E-cadherin ((c), arrow) were seen at 120 kD and 140 kD, respectively, both of which showed an up-regulation with time ((b), (d)). Anti-β-tubulin (55 kD) was used as an internal standard. n = 4.

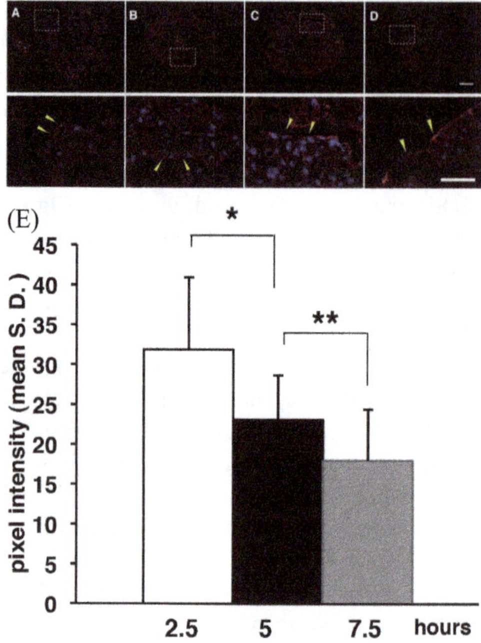

Figure 4. The expression patterns of actin filaments during cell sorting culture. Aggregates were stained with anti-actin antibody at 2.5 h (A), 5 h (B), 7.5 h (C) and 12 h (D) after the beginning stationary culture (the whole images of aggregate (upper) and high magnifications (lower). The pixel intensity of fluorescence for actin bundles at the edge of AC clusters was semiquantified by image J (E). Actin filaments were detectable along the membrane of AC, but not VC ((A)-(D)). Bundle-like expression was found along the edge of AC cluster at 2.5 h and 5 h (A, B, yellow arrowheads), and was intermittent at 7.5 h (C, yellow arrowheads). At 12 h, bundle-like expression was seen at the margin of the aggregates (D, white arrowheads). Actin (red), DAPI (blue). Scale bars: 100 μm (whole images); 50 μm (high images). Asterisks (*) indicate statistical significance (*, $p < 0.003$. **, $p < 0.03$. n = 3).

brane, but extremely low in the VC (**Figures 4(A)-(D)**). At 2.5 hC, actin bundles were observed at the edge of AC clusters (**Figure 4(A)**, yellow arrowheads) and gradually decreased from 5 hC (**Figure 4(B)**, yellow arrowheads) to 7.5 hC (**Figure 4(C)**, yellow arrowheads). At 12 hC at which the polarization was completed, cells closely contacted with each other, so that the space between the clusters was almost lost, and moreover, newly formed actin bundles were seen in the outermost margin of the aggregates (**Figure 4(D)**, yellow arrowheads). The results shown in **Figures 4(A)-(D)** are summarized as follows: The sign of the expected active movement of the cell population was not evident, but instead, actin bundles at the edge of the AC cluster seemed to be continuously down-regulated at least up to 7.5 hC. This possibility was supported by a semiquantitative estimation by image J (**Figure 4(E)**). It should be noted that it is not surprising that the actin network would be upregulated at the stage from 5 hC to 7.5 hC, because the initial stage of polarization might have then started, but that was not the case.

To assess how these actin bundles were localized in the AC clusters, we also analyzed the expression of the actin filament in thick sections (20 μm) using confocal laser microscopy (**Figure 5**) by which the structural

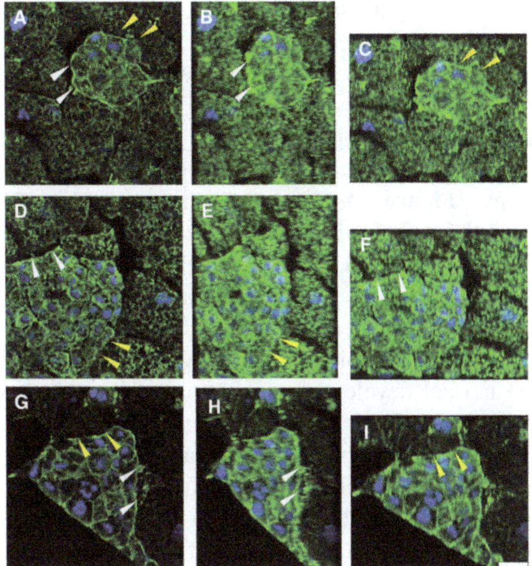

Figure 5. The accumulation of cortex actin in AC cluster revealed by conforcal images of the aggregates stained with anti-actin antibody. Samples were fixed and stained at 2.5 ((A)-(C)), 5 ((D)-(F)), 7.5 ((G)-(I)) after the beginning stationary culture. The optical slices were assembled into a stack without tilting ((A), (D), (G)), X axially tilted by 40 degrees ((B), (E), (H)), and Y axially tilted at 35 degrees ((C), (F), (I)). Such a tilting method clearly disclosed that the actin filaments, cortex actin (white arrowheads), preferentially accumulated at the margin of the AC cluster, and that they tended to gradually decrease during the stationary culture. The actin accumulation of the actin was not seen in the area where the cluster had come in contact with VC (yellow arrowheads). Fifteen to 20 optical slices (1.5 µm distance) were asembled into a stack. DAPI (blue). Scale bar: 25 µm.

artifacts after fixation would be minimized. When 15 to 20 optical slices (1.5 µm distance) were assembled into stacks (**Figures 5(A)**, **(D)** and **(G)**), and were X axially tilted at 40 degrees (**Figures 5(B)**, **(E)** and **(H)**), and Y axially tilted at 35 degrees (**Figures 5(C)**, **(F)** and **(I)**), the presence of the cortex actin, namely, the accumulation of the actin along the margin of the AC cluster was disclosed. At 2.5 hC, there were remarkable gaps between the AC and VC clusters (**Figure 5(A)**, white arrowheads), but the two clusters seemed to gradually contact with each other (**Figure 5**, yellow arrowheads). The cortex actin gradually decreased from 5 hC to 7.5 hC (**Figures 5(D)-(H)**, white arrowheads). These results indicated that the accumulation of actin along the margin of AC cluster was prominent during the concentrification step and declined by the beginning of the polarization step.

3.4. Aggregate Formation of Dissociated Embryonic Cells

The dissociated AC and VC were intermingled and underwent a long-term stationary culture. At first, they indiscriminately united and formed a flat aggregate. They then gradually rounded up (**Figures 5(A)-(D)**). After 12 hC, the pigmented AC began to occupy the upper part of the aggregate as if it reproduced the arrangement of the ectoderm and endoderm just like *in vivo* (**Figures 5(E)-(G)**). The AC and VC clusters were mutually apposed up and down, as shown by using the Dextran rhodamine labeled cells (**Figure 6(H)**). From 96 hC, opaque tissue arose from the AC cluster (**Figure 6(L)**, arrowhead) and gradually expanded (**Figure 6(J)**, arrowheads). The aggregates were fixed with MEMFA then processed and sectioned for staining with Alcian blue, revealing that the mesodermal differentiation into the mature cartilage fairly progressed in the AC cluster (**Figure 6(K)**, arrow, L).

4. Discussion

4.1. The Cell Sorting Process of AC and VC Is Considered to Comprise Two Steps, Concentrification and Polarization

Our experimental results (**Figure 1**) strongly suggest that

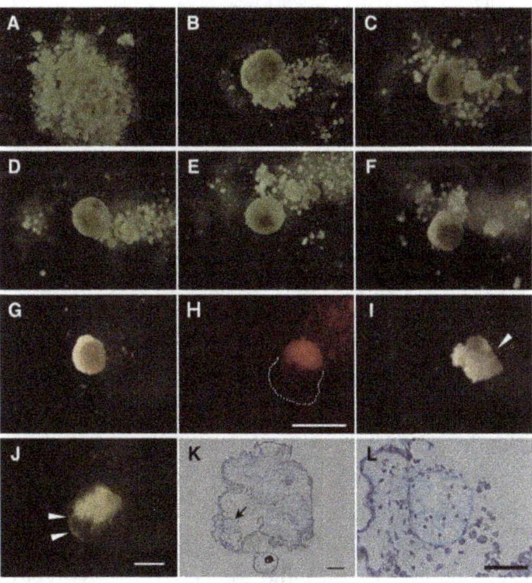

Figure 6. Long term culture of aggregates allowed cartilage formation, a typical mesodermal cell differentiation. AC + VC aggregates were incubated under the present stationary culture conditions for 0 h (A), 2 h (B), 4 h (C), 6 h (D), 12 h (E), 24 h (F), 48 h (G, H), 96 h (I) and 168 h (J). The dissociated cells reassembled and the pigmented AC appeared in the upper half after 12 h. H was a DR(+)AC+DR(-)VC aggregate. A close apposition of AC and VC clusters was noted in the aggregate (H). Around 96 h, opaque tissues emerged from AC and gradually expanded ((I), (J), arrowheads). Paraffin sections of aggregates cultured for 168h were stained with Alcian blue (K). (L) is the higher magnification of the area indicated by arrow in (K), showing that AC cluster had been differentiated into cartilage tissue. Scale bar: 500 µm ((H), (J)), 100 µm (K), and 50 µm (L).

the whole process of cell sorting of the AC and VC dissociated during the early gastrula stage can be roughly divided into two steps, *i.e.*, concentrification and polarization. The cells seem to be devoted to "cell sorting in a conventional meaning" for the initial several hours to form a concentric-layered structure, and then the concentrificated aggregates started to restore or mimic the in vivo animal-to-vegetal polarity by rearrangement of the AC and VC clusters. Our finding that these two steps sequentially proceeded but being temporally separated has never been argued by the past studies until now. It took about half a day for the aggregates to establish a polarized embryoid structure. The subsequent embryogenesis in the present *in vitro* culture system also proceeded very slowly, so that, for example, 96-hour cultured embryos (**Figure 6(H)**), which would otherwise reach swimming tadpole larvae, showed that their development was considerably delayed. This delay, however, did not directly result in death as individuals, because normal tissue differentiation, such as cartilage formation, was allowable even during the further long-term culture as will be described later. Anyway, it is noteworthy that embryonic cells have a latent adaptability to temporally detach the two tasks of sorting out and embryogenesis.

4.2. Upregulation of Cadherins Accompanied with the Rearrangement of the AC and VC Clusters

Homophilic binding molecule cadherins are responsible for cell sorting, which produces segregation of the cell populations and formation of a tissue boundary [25,26]. In *Xenopus laevis* embryos, the regional identity starts to be created before midblastula and the cell sorting behavior gradually becomes remarkable as the development advances [9,10]. Accordingly, cadherins are thought to play active roles in the specification of the germ layers. Indeed, C-cadherin is known to ubiquitously express as a maternal cadherin in *Xenopus laevis*, while E-cadherin starts to exclusively express in the ectoderm from the gastrula stage [13-16]. The ectoderm-derived AC would express both cadherins, and the endoderm-derived VC express only C-cadherin.

Although the homophilic binding dependence of the behavior of the cadherin-expressing cells has not been verified in this study, homotypic reassembly (AC-AC or VC-VC recombination) took place and the cell-type-specific expression patterns of these cadherins *in vivo* was reproduced in the present *in vitro* system (**Figure 2**). AC showed an intense staining for the C- and E-cadherins, so that AC was thought to be more cohesive than VC, resulting in segregation between the AC and VC clusters in the sorting out process. Our observation of the concentrification of the clusters, in which AC was enveloped by VC, was consistent with the differential adhesion hypothesis (DAH) that the selectivity of the inside and outside of the clusters in sorting out is decided by the relative strengths of the intercellular adhesiveness [27,28]. The expression of both cadherins lasted for a while after rearrangement, suggesting that in some way the cadherins were needed for the maintenance of the aggregate morphology (**Figure 2**). Moreover, the expression of cadherin in the AC cluster seemed to be increased after polarization as compared to the wrapped AC cluster during the cencentrification step. Quantitative analysis by western blot also showed that the C- and E-cadherins increased with the rearrangement of the clusters (**Figure 3**). In the normal development of intact embryos, however, the maternal C-cadherin showed no change in the expression level during and after the gastrula stage, and the expression of E-cadherin started from the gastrula stage, increasing with the progression of development (not shown, [29]). An upregulation of the C-cadherin, which is ordinarily considered as a maternal protein, might be a unique phenomenon occurring in the present culture system. However, we have realized one difficulty in answering the question of what is the cause or trigger for the polarization. Foty *et al.* [30] claimed that, if the total cell binding energy of the cell populations is equal between two cell populations, the clusters should appose. According to this theory, the AC adhesiveness should be downregulated as polarization of the clusters proceeded. Our result, a steadily increasing expression of the cadherins, was unexpected, and the occurrence of polarization cannot simply be explained by only the categories about cell adhesiveness that we know at present. Cell sorting can significantly occur between cells that express different cadherins and the heterophilic adhesion sometimes occurs as strongly as the homophilic one [8]. These results suggested that the specificity and strength of the cell adhesion depend not only on the interactions between the cadherins perse, but also other unknown parameters [31].

4.3. Is the Rearrangement of the AC and VC Clusters Modulated by the Change in the Surface Tension?

We have withheld the consideration of the role of cadherins, and tried to examine the expression and localization of the cytoskeletal actins, another major intracellular element in the cell motility, in order to find a clue to the question of what is the definitive change within the cells for the concentrified-to-polarized rearrangement of the cell clusters.

Morphogenetic movements linked to the active rearrangement of the cell populations progress during the gastrula stage in *Xenopus laevis* [32]. As we used gastrula embryonic cells as the starting materials, we simply expected that the active cell motility might be reutilized in the cell sorting process, especially in the polarization step. However, immunohistochemical analyses (**Figure 4**) did not show any signs directly supporting this possibility. The overall expression of the actin filaments continued to decrease throughout the entire period of cell sorting, including the polarization, although it was also plausible that no lamellipodia were detected in the aggregates due to the poor resolution of our optical system.

We also examined the possible participation of fibronectin (FN) in the polarization step, as this extracellular matrix component serves as a substrate for mesoderm migration during gastrulation [12]. FN was not localized at the border region between clusters until 7.5 hC (not shown), while it was positive at the periphery of the AC cluster only when they were in the middle of differentiation around 12 hC. Thus, FN does not contribute to the rearrangement of the clusters, but may emerge as tissue boundaries when the AC clusters differentiate [33].

Although the immunohistochmical analyses did not show any definitive involvement of the actin filaments and fibronectins in the motility of individual cells such as lamellipodal movement (though not yet experimentally ruled out), another finding to be considered is that a significant amount of actin bundles at the edge of the AC cluster was accumulated at least until completion of the concentrification (**Figure 4**). The morphology of these actin bundles resembled the appearance of the ectoderm with a high surface tension observed in the cell sorting assay using zebrafish germ layer progenitors [34,35], suggesting that such an ectodermal alignment was dependent on the surface tension. The actin bundles along the surface of the AC cluster in our system might also emerge through a similar surface-tension-dependent mechanism as in zebrafish. Holtfleter's "tissue affinity" of amphibian germ layers corresponds to the tissue surface tension, that is, the force minimizing the surface area of the cell population, and the cell population with a higher surface tension is surrounded by a lower tension one [36]. According to DAH, the tissue surface tension depends on the cell adhesion energy, which decides the outside-inside hierarchy [27,30]. Therefore, the AC cluster was assumed to have a higher tension than the VC cluster during the concentrification step.

We should now pay attention to the fact that the tangential accumulation of actin at the edge of the AC clusters showed a tendency to decrease as polarization of the clusters proceeded, as observed by an image analysis of these actin bundles (**Figure 4(E)**). Moreover, the three-dimensional analysis by confocal microscopy proved the existence of actin bundles, *i.e.*, cell cortex actin, at the edge of the AC clusters, and confirmed that such structures decreased with the polarization. Thus our experimental results were fairly consistent with the above-stated DAH, and led us to postulate that a reducetion in the surface tension of the AC cluster is a major cause for rearrangement into the apposition of the AC and VC clusters.

DAH explains that the tissue surface tension is simply proportional to the cell adhesion energy, while the differential interfacial tension hypothesis, DITH, was proposed in which the interfacial tension in individual cells is responsible for regulation of the tissue surface tension [37]. Some recent studies emphasized that the balance between the cell adhesion and cell cortex tension was a key factor for determination of the tissue surface tension [31,34,38]. Detailed experimental analyses on the relationship between the cell adhesiveness and surface tension as well as arguments toward integration of the relevant information are needed for fully understanding the process of cell sorting, including the polarization step in this study.

Apart from the polarization, we also observed a series of interesting phenomena with the entire aggregate. By the completion of the polarization, the aggregate itself became compacted, and the space between the AC and VC clusters was almost lost. More interestingly, the actin bundles at the outermost surface of the aggregate became remarkable (**Figure 4(D)**). Accumulation of the actin filament was probably considered as a sign of the elevating surface tension of the entire aggregate like that of the AC cluster.

4.4. The Cell Sorting Process Can Be Followed by Tissue Differentiation Via the Inductive Interactions between the AC and VC

The cell sorting properties of the dissociated cells from *Xenopus laevis* embryos at the blastula and gastrula stages have been studied to date, mainly on their involvement in determination of the germ layer identities in vivo and/or influence of the properties by exogenous inducing substances [9,39]. In this study, we employed the sorting out assay using the same staged *Xenopus* embryos as those in preceding studies. As a series of detailed temporal analyses of cell sorting, we tried a long term stationary culture up to 7 days. The AC clusters, which had established a polarized arrangement by 12 hC, were found to exhibit the differentiation of cartilage tissues though slowly but normally (**Figure 6**). This reminds us of an earlier study that dissociated cells obtained from early amphibian embryos could unite indiscriminately with any other and sort out by tissue affinities and they in turn reconstructed normal tissues distinctive of derived germ layers [6].

AC is derived from the blastocoelic roof (animal cap) possessing a multi-differentiatiation potencies from the blastula to early gastrula stage. The animal cap is known to be competent to respond to a member of TGF-β, activin, and differentiate into the mesoderm and endoderm in a dose-dependent manner [40]. It was reported that the treatment of animal caps in a mass with 100 ng/ml of activin resulted in differentiation into cartilage tissues [41,42]. It is no wonder that the present *in vitro* system using cells dissociated during the gastrula stage allows the occurrence of similar TGF-β-directed interactions between the AC and VC followed by a typical mesoderm induction such as cartilage differentiation.

In conclusion, our study showed that the cell sorting process of AC and VC from *Xenopus* early gastrulae could be divided into two steps, concentrification and polarization, and that, in this in vitro system, the expression of cadherin like *in vivo* and a reduction of actin accumulation at the edge of AC cluster occurred. The actin downregulation was considered to account for the rearrangement of the clusters. Moreover, by extending the period of the cell sorting culture, the AC cluster was differentiated into mature cartilage tissues probably via inductive interactions between the AC and VC. Thus, the aggregate culture system of *Xenopus* AC and VC allows progress of a certain number of principal events involved in the embryogenesis *in vivo*.

5. Acknowledgements

We would like to thank the members of our Morphogenesis Laboratories for their supports, including helpful discussions in weekly seminars, during this study.

REFERENCES

[1] A. Gierer, S. Berking, H. Bode, C. N. David, K. Flick, G. Hansmann, H. Schaller and E. Trenkner, "Regeneration of Hydra from Reaggregated Cells," *Nature New Biology*, Vol. 239, No. 91, 1972, pp. 98-101.

[2] Y. Takaku, T. Hariyama and T. Fujisawa, "Motility of Endodermal Epithelial Cells Plays a Major Role in Reorganizing the Two Epithelial Layers in Hydra," *Mechanisms of Development*, Vol. 122, No. 1, 2005, pp. 109-122. doi:10.1016/j.mod.2004.08.004

[3] H. Ide, N. Wada and K. Uchiyama, "Sorting out of Cells from Different Parts and Stages of the Chick Limb Bud," *Developmental Biolog*, Vol. 162, No. 1, 1994, pp. 71-76. doi:10.1006/dbio.1994.1067

[4] H. Yajima, S. Yoneitamura, N. Watanabe, K. Tamura and H. Ide, "Role of N-Cadherin in the Sorting-Out of Mesenchymal Cells and in the Positional Identity along the Proximodistal Axis of the Chick Limb Bud," *Developmental Dynamics*, Vol. 216, No. 3, 1999, pp. 274-284.

[5] H. Yajima, K. Hara, H. Ide and K. Tamura, "Cell Adhesiveness and Affinity for Limb Pattern Formation," *The International Journal of Developmental Biology*, Vol. 46, No. 7, 2002, pp. 897-904.

[6] P. L. Townes and J. Holtfreter, "Directed Movements and Selective Adhesion of Embryonic Amphibian Cells," *Journal of Experimental Zoology*, Vol. 128, No. 1, 1955, pp. 53-120. doi:10.1002/jez.1401280105

[7] M. Marsden and D. W. DeSimone, "Integrin-ECM Interactions Regulate Cadherin-Dependent Cell Adhesion and Are Required for Convergent Extension in *Xenopus*," *Current Biology*, Vol. 13, No. 14, 2003, pp. 1182-1191. doi:10.1016/S0960-9822(03)00433-0

[8] C. M. Niessen and B. M. Gumbiner, "Cadherin-Mediated Cell Sorting Not Determined by Binding or Adhesion Specificity," *Journal of Cell Biology*, Vol. 156, No. 2, 2002, pp. 389-399. doi:10.1083/jcb.200108040

[9] A. Turner, A. M. Snape, C. C. Wylie and J. Heasman, "Regional Identity Is Established before Gastrulation in the *Xenopus* Embryo," *Journal of Experimental Zoology*, Vol. 251, No. 2, 1989, pp. 245-252. doi:10.1002/jez.1402510212

[10] C. C. Wylie, A. Snape, J. Heasman and J. C. Smith, "Vegetal Pole Cells and Commitment to Form Endoderm in *Xenopus laevis*," *Developmental Biology*, Vol. 119, No. 2, 1987, pp. 496-502. doi:10.1016/0012-1606(87)90052-2

[11] R. Keller, L. Davidson, A. Edlund, T. Elul, M. Ezin, D. Shook and P. Skoglund, "Mechanisms of Convergence and Extension by Cell Intercalation," *Philosophical Transactions of the Royal Society B: Biological Sciences*, Vol. 355, No. 1399, 2000, pp. 897-922. doi:10.1098/rstb.2000.0626

[12] R. Winklbauer and R. E. Keller, "Fibronectin, Mesoderm Migration, and Gastrulation in *Xenopus*," *Developmental Biology*, Vol. 177, No. 2, 1996, pp. 413-426. doi:10.1006/dbio.1996.0174

[13] Y. S. Choi and B. Gumbiner, "Expression of Cell Adhesion Molecule E-Cadherin in *Xenopus* Embryos Begins at Gastrulation and Predominates in the Ectoderm," *Journal of Cell Biology*, Vol. 108, No. 6, 1989, pp. 2449-2458. doi:10.1083/jcb.108.6.2449

[14] Y. S. Choi, R. Sehgal, P. McCrea and B. Gumbiner, "A Cadherin-Like Protein in Eggs and Cleaving Embryos of *Xenopus laevis* Is Expressed in Oocytes in Response to Progesterone," *Journal of Cell Biology*, Vol. 110, No. 5, 1990, pp. 1575-1582. doi:10.1083/jcb.110.5.1575

[15] G. Levi, D. Ginsberg, J. M. Girault, I. Sabanay, J. P. Thiery and B. Geiger, "EP-Cadherin in Muscles and Epithelia of *Xenopus laevis* Embryos," *Development*, Vol. 113, No. 4, 1991, pp. 1335-1344.

[16] G. Levi, B. Gumbiner and J. P. Thiery, "The Distribution of E-Cadherin during *Xenopus laevis* Development," *Development*, Vol. 111, No. 1, 1991, pp. 159-169.

[17] Q. Tao, S. Nandadasa, P. D. McCrea, J. Heasman and C. Wylie, "G-Protein-Coupled Signals Control Cortical Actin Assembly by Controlling Cadherin Expression in the early *Xenopus* Embryo," *Development*, Vol. 134, No. 14, 2007, pp. 2551-2561. doi:10.1242/dev.002824

[18] E. Agius, M. Oelgeschlager, O. Wessely, C. Kemp and E. M. De Robertis, "Endodermal Nodal-Related Signals and Mesodermal Induction in *Xenopus*," *Development*, Vol. 127, No. 6, 2000, pp. 1173-1183.

[19] P. D. Nieuwkoop, "The Formation of the Mesoderm in *Urodelean* Amphibians. I. Induction by the Endoderm,"

Wilhelm Roux' Archiv für Entwicklungsmechanik der Organismen, Vol. 162, No. 4, 1969, pp. 334-373. doi:10.1007/BF00578701

[20] P. D. Nieuwkoop and J. Faber, "Normal Table of *Xenopus laevis*," Daudin, North-Holland Publishing Company, Amsterdam, 1956.

[21] R. M. Harland, "*In Situ* Hybridization: An Improved Whole Mount Method for *Xenopus* Embryos," *Methods in Cell Biology*, Vol. 36, 1991, pp. 685-695. doi:10.1016/S0091-679X(08)60307-6

[22] H. L. Sive, R. M. Grainger and R. M. Harland, "Early Development of *Xenopus laevis*: A laboratory Manual," Cold Spring Harbor Laboratory Press, New York, 1998.

[23] R. A. Foty, C. M. Pfleger, G. Forgacs and M. S. Steinberg, "Surface Tensions of Embryonic Tissues Predict Their Mutual Envelopment Behavior," *Development*, Vol. 122, No. 5, 1996, pp. 1611-1620.

[24] J. L. Stubbs, L. Davidson, R. Keller and C. Kintner, "Radial Intercalation of Ciliated Cells during *Xenopus* Skin Development," *Development*, Vol. 133, No. 13, 2006, pp. 2507-2515. doi:10.1242/dev.02417

[25] A. Nose, A. Nagafuchi and M. Takeichi, "Expressed Recombinant Cadherins Mediate Cell Sorting in Model Systems," *Cell*, Vol. 54, No. 7, 1988, pp. 933-1001. doi:10.1016/0092-8674(88)90114-6

[26] M. Takeichi, "The Cadherins: Cell-Cell Adhesion Molecules Controlling Animal Morphogenesis," *Development*, Vol. 102, No. 4, 1988, pp. 639-655.

[27] R. A. Foty and M. S. Steinberg, "The Differential Adhesion Hypothesis: A Direct Evaluation," *Developmental Biology*, Vol. 278, No. 1, 2005, pp. 255-263. doi:10.1016/j.ydbio.2004.11.012

[28] M. S. Steinberg, "Differential Adhesion in Morphogenesis: A Modern View," *Current Opinion in Genetics & Development*, Vol. 17, No. 4, 2007, pp. 281-286. doi:10.1016/j.gde.2007.05.002

[29] S. Nandadasa, Q. Tao, N. R. Menon, J. Heasman and C. Wylie, "N- and E-Cadherins in *Xenopus* Are Specifically Required in the Neural and Non-Neural Ectoderm, Respectively, for F-Actin Assembly and Morphogenetic Movements," *Development*, Vol. 136, No. 8, 2009, pp. 1327-1338. doi:10.1242/dev.031203

[30] R. A. Foty and M. S. Steinberg, "Cadherin-Mediated Cell-Cell Adhesion and Tissue Segregation in Relation to Malignancy," *The International Journal of Developmental Biology*, Vol. 48, No. 5-6, 2000, pp. 397-409. doi:10.1387/ijdb.041810rf

[31] T. Lecuit and P. F. Lenne, "Cell Surface Mechanics and the Control of Cell Shape, Tissue Patterns and Morphogenesis," *Nature Reviews Molecular Cell Biology*, Vol. 8, No. 8, 2007, pp. 633-644. doi:10.1038/nrm2222

[32] R. Winklbauer, M. Nagel, A. Selchow and S. Wacker, "Mesoderm Migration in the *Xenopus* Gastrula," *International Journal of Developmental Biology*, Vol. 40, No. 1, 1996, pp. 305-311.

[33] L. A. Davidson, R. Keller and D. W. DeSimone, "Assembly and Remodeling of the Fibrillar Fibronectin Extracellular Matrix during Gastrulation and Neurulation in *Xenopus laevis*," *Developmental Dynamics*, Vol. 231, No. 4, 2004, pp. 888-895. doi:10.1002/dvdy.20217

[34] M. Krieg, Y. Arboleda-Estudillo, P. H. Puech, J. Kafer, F. Graner, D. J. Muller and C. P. Heisenberg, "Tensile Forces Govern Germ-Layer Organization in Zebrafish," *Nature Cell Biology*, Vol. 10, No. 4, 2008, pp. 429-436. doi:10.1038/ncb1705

[35] E. M. Schotz, R. D. Burdine,F. Julicher, M. S. Steinberg, C. P. Heisenberg and R. A. Foty, "Quantitative Differences in Tissue Surface Tension Influence Zebrafish Germ Layer Positioning," *HFSP Journal*, Vol. 2, No. 1, 2008, pp. 42-56. doi:10.2976/1.2834817

[36] G. S. Davis, H. M. Phillips and M. S. Steinberg, "Germ-Layer Surface Tensions and 'Tissue Affinities' in *Rana pipiens* Gastrulae: Quantitative Measurements," *Developmental Biology*, Vol. 192, No. 2, 1997, pp. 630-644. doi:10.1006/dbio.1997.8741

[37] A. K. Harris, "Is Cell Sorting Caused by Differences in the Work of Intercellular Adhesion? A Critique of the Steinberg Hypothesis," *Journal of Theoretical Biology*, Vol. 61, No. 2, 1976, pp. 267-285. doi:10.1016/0022-5193(76)90019-9

[38] M. L. Manning, R. A. Foty, M. S. Steinberg and E. M. Schoetz, "Coaction of Intercellular Adhesion and Cortical Tension Specifies Tissue Surface Tension," *Proceedings of the National Academy of Sciences of USA*, Vol. 107, No. 23, 2010, pp. 2517-12522. doi:10.1073/pnas.1003743107

[39] H, Kuroda, H. Sakumoto, K. Kinoshita and M. Asashima, "Changes in the Adhesive Properties of Dissociated and reaggregated *Xenopus laevis* Embryo Cells," *Current Opinion in Genetics & Development*, Vol. 41, No. 3, 1999, pp. 283-291. doi:10.1046/j.1440-169X.1999.413428.x

[40] K. Okabayashi and M. Asashima, "Tissue Generation from Amphibian Animal Caps," *Current Opinion in Genetics & Development*, Vol. 13, No. 5, 2003, pp. 502-507. doi:10.1016/S0959-437X(03)00111-4

[41] Y. Fukui, M. Furue, Y. Myoishi, J. D. Sato, T. Okamoto and M. Asashima, "Long-Term Culture of *Xenopus* Presumptive Ectoderm in A Nutrient-Supplemented Culture Medium," *Current Opinion in Genetics & Development*, Vol. 45, No. 5-6, 2003, pp. 499-506. doi:10.1111/j.1440-169X.2003.00717.x

[42] M. Furue, Y. Myoishi, Y. Fukui, T. Ariizumi, T, Okamoto and M. Asashima, "Activin A induces Craniofacial Cartilage from Undifferentiated *Xenopus* Ectoderm *in Vitro*," *Proceedings of the National Academy of Sciences of USA*, Vol. 99, No. 24, 2002, pp. 15474-15479. doi:10.1073/pnas.242597399

19

Generation of Unfolded DNA in Human Neutrophils Following Hypothermal Treatment

Jin Kawata[1,2], Makoto Kikuchi[2], Hisato Saitoh[1*]

[1]Department of New Frontier Sciences, Graduate School of Science and Technology, Kumamoto University, Kumamoto, Japan
[2]Kumamoto Health Science University, Kumamoto, Japan
Email: *hisa@kumamoto-u.ac.jp

ABSTRACT

By visualizing DNA with diamidino phenylindole (DAPI), we found that hypothermal incubation followed by rewarming of human neutrophils resulted in an increased number of DAPI-positive objects representative of extensive DNA unfolding seemingly similar to neutrophil extracellular traps (NETs). In contrast to canonical NET formation, diphenylene iodonium (DPI), an NADPH oxidase inhibitor, exhibited negligible effects on formation of the DAPI-positive objects. Moreover, multiple instances of DNA damage were detected in the objects, but not in canonical NETs. Our results thus suggest the potential of hypothermia for triggering DNA structural alteration in neutrophils, which is similar to but distinct from NET formation.

Keywords: Hypothermal Treatment; DNA Unfolding; Neutrophil Extracellular Trap (NET)

1. Introduction

Low-temperature conditions, referred to as hypothermia, are generally used for the storage of cells, tissues, organs and bodies for both scientific and clinical purposes. Hypothermia is an important means of slowing down cellular metabolism during storage, thus inhibiting injurious processes caused by the deficiency of oxygen and substrate supply. However, hypothermia can give rise to cell injury, including cell death [1,2].

Neutrophils are a main type of effector cell in the innate immune system [3,4], which circulate in the blood and engulf invading microorganisms such as bacteria and fungi by phagocytosis. In addition to such activities, Brinkmann *et al.* have reported that, following activation by microorganisms, neutrophils can undergo morphological changes detectable by microscopic observations [5]. These changes include loss of the lobular-shaped nucleus followed by disintegration of the nuclear envelope, which allow nuclear, cytoplasmic and granular components to mix together and subsequently rupture the cell membrane to release the DNA/chromatin into the extracellular environment [5]. The result is that the unfolded DNA/chromatin fibers with attached bactericidal proteins can function as neutrophil extracellular traps (NETs) for microorganisms. NETs appear to be the result of a unique form of cell death. Therefore, as opposed to apoptosis and necrosis, Steinberg and Grinstein coined the term "NETosis" for neutrophil cell death, which leads to the formation of NETs [6].

In addition to microorganism infection, several physiological inducers of NETs have been reported [7 and references herein]. For instance, platelets activated via Toll-like receptor 4 rapidly induce NET formation [8]. Antibodies [9], antibody-antigen complexes [10,11], human immunodeficiency virus (HIV-1) [12], and microbial components such as lipopolysaccharide [13,14] are also known to induce the formation of NETs. Although the intracellular signaling pathway(s) that transmit these physiological stimuli remain largely unknown, reactive oxygen species (ROS) generation was demonstrated to be an absolute requirement for NET formation [15,16]. Thus, one of the most widely-used agents to induce NETs in *in vitro* experiments is phorbol-12-myristate-13-acetate (PMA), which directly stimulates protein kinase C (PKC) leading to potent activation of nicotinamide adenine dinucleotide phosphate (NADPH) oxidase, which in turn generates superoxide [5,7,17]. Therefore, it is reasonable to use diphenylene iodonium (DPI), a NADPH oxidase inhibitor [18], to block the formation of PMA-stimulated NETs [15,17,19].

In this study, we found that hypothermal incubation of

*Corresponding author.

human neutrophils at 4°C for up to 1 h followed by incubation at 37°C resulted in an increased number of DAPI-stainable objects similar to global DNA unfolding observed in PMA-stimulated NETs. However, our additional experimental data revealed that hypothermia/rewarming-induced DNA unfolding was regulated in a manner similar to, but biochemically and pharmacologically distinct from, canonical NETs. Although the molecular mechanism of this phenomenon is not fully understood, we inferred, based on our experimental data, the possible role of ROS, which were generated during hypothermia/rewarming-treatment in a manner independent of NADPH oxidase activity in the formation of the DAPI-positive, NET-like objects.

2. Materials and Methods

2.1. Peripheral Blood Preparation and Culture

Human peripheral blood preparations (from two normal male donors, collected in compliance with Kumamoto Health Science University and approved by the University Oversight Committee) were enriched for neutrophils by density gradient centrifugation with HISTOPPAQUE 117 (Sigma-Aldrich) and Lymphocyte Separation Solution 1.119 (Nakarai Tesque) according to the procedures described by the supplier. Washed enriched neutrophilic fractions were counted and examined for purity using Wright Giemsa staining (Sigma-Aldrich).

2.2. Drug and Hypothermal Treatments

Cells were incubated in culture dishes containing an immersed coverslip in RPMI 1640 (Sigma-Aldrich) supplemented with 5% fetal bovine serum (FBS), 1% penicillin/streptomycin and 0.1% gentamaycin in a humidified atmosphere containing 5% CO_2. To induce NETs, PMA (Wako Pure Chemical Industries) was added to the culture medium at a concentration of 50 nM and incubated for 4 h at 37°C. To inhibit NADPH oxidase activity, DPI (Cayman Chemical) was added to the culture medium at a concentration of 20 μM. Hypothermal treatment and rewarming of cells were performed by incubation in a humidified atmosphere containing 5% CO_2. After drug and/or hypothermal/rewarming treatment, the coverslips were removed from the cultures and subjected to appropriate assays. DNA was visualized by staining with DAPI.

2.3. Antibodies and Immunostaining

Cells were washed once for 5 min with ice-cold PBS and then fixed with 4% paraformaldehyde in PBS for 5 min at room temperature. After fixation, the cells were rinsed once with PBS and subjected to indirect-immunofluorescence analysis using anti-neutrophil elastase (Calbiochem), anti-histone H3 (Santa Cruz Biotechnology), and anti-histone H3 citrulline R26 (Abcam) antibodies. The secondary antibodies were obtained from Santa Cruz Biotechnology and Sigma-Aldrich.

2.4. Bacteria Trapping Assay

Escherichia coli BL21 (*DE*3) were transformed with pET28-EGFP, a plasmid for expression of green fluorescent protein (GFP), and cultured in Luria-Bertani (LB) medium containing kanamycin at 37°C for 16 h. 10^7 *E. coli* cells were incubated with a coverslip containing hypothermia/rewarming-induced DAPI-positive objects in RPMI 1640 supplemented with 5% FBS at 37°C. After 20 min at room temperature, the coverslips were washed three times with PBS followed by incubation with 4% paraformaldehyde. DNA fibers were stained with DAPI. Because the *E. coli* expressed GFP, bacteria trapped by DNA fibers could be detected by fluorescence microscopy. For DNase I treatment, the coverslips were treated with PBS containing 100 U/ml DNase I (Takara) at 37°C for 1 hr. The numbers of *E. coli* with GFP signals on the coverslips were counted by fluorescent microscopy.

2.5. TUNEL Assay

Terminal deoxynucleotidyl transferase dUTP nick end labeling (TUNEL) assays were performed using the MEBSTAIN Apoptosis TUNEL Kit II (MBL) according to the manufacturer's instructions. The TUNEL-positive cells were counted under a microscope. The percentage of TUNEL-positive cells was defined by the number of positive cells among the total number of cells in each sample. For one experiment, cells were counted in at least three different microscopic fields of view.

2.6. Intracellular ROS Detection Assay

Intracellular ROS production was monitored using the cell permeable fluorescent dye, CellROX Deep Red Reagent (Invitrogen). This agent can readily react with ROS to form a fluorescent product proportional to the amount of ROS generated in the cells. The cells were incubated with 5 μM CellROX Deep Red for 30 min and then harvested. The fluorescence intensity of the cells was measured using a FACSVerse flow cytometer (BD Biosciences).

2.7. MitoTracker Analysis

After fixation with 4% paraformaldehyde, cultured human neutrophils were stained with the mitochondrion-specific dye, MitoTracker Red CMXRRos (Invitrogen), according to the manufacturer's instruction. The cells were immediately analyzed using a FACSVerse flow cytometer.

2.8. Statistical Analysis

Unless otherwise stated, all data are presented as the mean ± SD. Within individual experiments, data points were based on a minimum of triplicate representative samples and experiments were repeated at least three times.

3. Results and Discussion

3.1. Effect of Hypothermia on Human Neutrophils in Culture

After isolating human neutrophils from peripheral blood preparations (see Materials and Methods), the cells (4.5 × 10^6 cells) were incubated at 4°C for 1 h followed by incubation at 37°C for 5 h in the 6-cm culture dish supplemented with 2 ml of the culture medium, in which a coverslip was immersed (**Figure 1(a)**). We found that most of the cells were present as non-adherent forms, and were thus floating in the culture medium. These non-adherent cells appeared morphologically intact as shown by Wright–Giemsa staining (**Figure 1(b)**, left panel). In contrast, we unexpectedly found that small, but substantial, numbers of Wright-Giemsa-stainable materials, which looked different from intact neutrophils, were present on the coverslip (**Figure 1(b)**, right panel).

When the materials on the coverslip were stained with DAPI without any fixative treatment, we observed bright fluorescent signals under fluorescent microscopy, many of which appeared to consist of multiple DAPI-positive strings (**Figure 1(c)**). Because DAPI is a fluorescent dye that intercalates into double-stranded DNA, and that living neutrophils are less permeable to the dye than dead neutrophils, we thought it probable that these bright DAPI-stained signals represented global DNA unfolding of dead neutrophils, which somehow adhered to the coverslip. It should be mentioned that there were few neutrophils with normal morphology on the coverslip per view field, implying that most of neutrophils floated in the culture medium under the standard culture conditions.

(a)

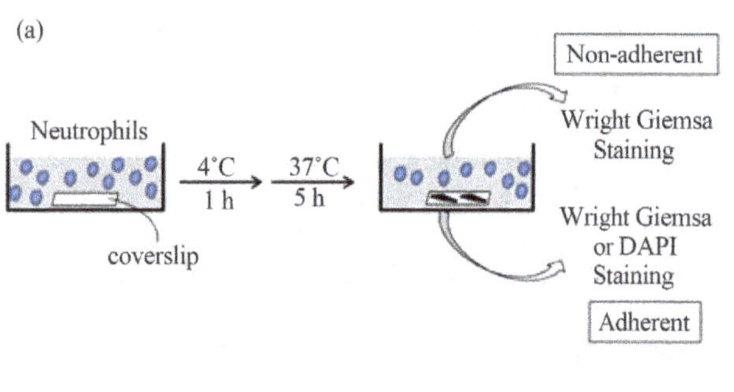

(b)

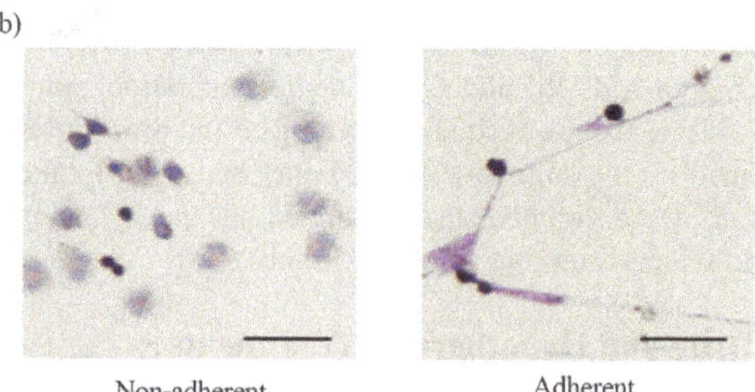

(c)

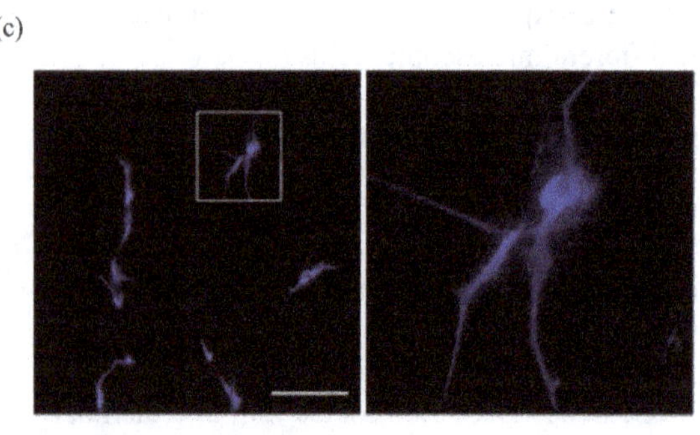

(d)

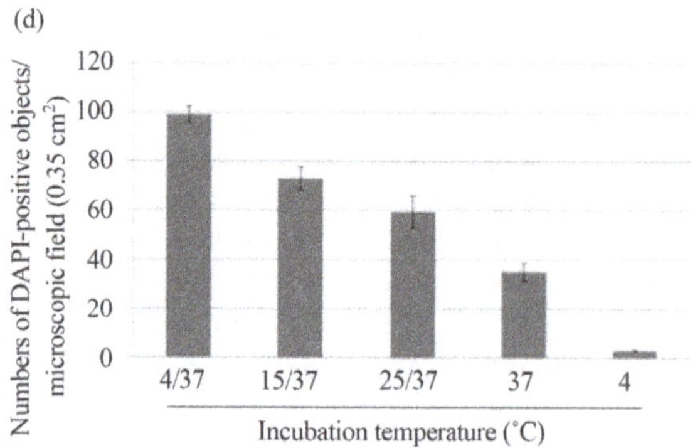

Figure 1. DAPI-stained objects in the hypothermia/rewarming-treated human neutrophils. (a) Schematic representation of procedure for detecting the hypothermia/rewarming-induced DAPI-positive objects. Human neutrophils from peripheral blood preparations were cultured in dishes containing a coverslip at 4°C for 1 h followed by rewarming at 37°C for 5 h. Non-adherent and adherent materials in the culture medium stained with Wright Giemsa; (b) Non-adherent and adherent materials in the culture were stained with Wright Giemsa (left and right panels). Bar indicates 50 μm; (c) The morphologies of the DAPI-stained objects adherent to the coverslip were detected by fluorescence microscopy (left panel). Bar indicates 50 μm. The panel on the right shows a higher magnification of the region indicated in the left panel; (d) Human neutrophils from peripheral blood preparations (1 × 10^6 cells) were cultured in dishes containing a coverslip for 6 h at 4°C (4), at 4°C for 1 h followed by rewarming at 37°C for 5 h (4/37), at 15°C for 1 h followed by rewarming at 37°C for 5 h (15/37), at 25°C for 1 h followed by rewarming at 37°C for 5 h (25/37), and at 37°C for 6 h (37). After incubation under the conditions as indicated, the numbers of DAPI-positive objects on the coverslips in the microscopic field (0.35 cm^2) were counted. The values shown represent means ± SE of three independent experiments.

We then investigated whether the requirement for DAPI-positive object production was simple exposure to hypothermia or rather the combination of hypothermia/rewarming. When human neutrophils were maintained at a constant temperature of either 4°C or 37°C, significantly less DAPI-positive signals were detected as compared with cells cultured either at 4°C, 15°C, or 25°C for 1 h followed by incubation at 37°C for 5 h (**Figure 1(d)**). These results suggest that the appearance of the DAPI-positive objects was associated with incubation of neutrophils under hypothermal conditions followed by rewarming.

3.2. Comparison of the Biochemical and Immunohistochemical Properties of Hypothermia/Rewarming-Induced DAPI-Positive Objects and PMA-Stimulated NETs

When we observed the DAPI-positive objects in hypothermia/rewarming-treated human neutrophils, we noticed that morphological similarities between the objects and DAPI-stained PMA-stimulated NETs, leading us to suspect that the DAPI-positive objects *per se* might represent NETs (**Figure 2(a)**). To investigate this possibility, we first asked whether the that DAPI-positive objects possessed the ability to bind bacteria. Given NETs are defined as extracellular DNA-proteinaceous structures that exhibit the ability to associate with a wide variety of Gram-positive and Gram-negative pathogens [7], we expected that the DAPI-positive structures might also show similar properties. As shown in **Figure 2(b)**, when GFP-expressing *E. coli* was incubated with the coverslip containing DAPI-positive objects, we found multiple GFP signals present together with the DAPI-signals. Their ability to trap bacteria appeared equivalent to that of PMA-stimulated NETs, because the number and distribution of GFP-signals associated with the DAPI-positive objects were very similar to those associated with PMA-stimulated NETs, suggesting that the objects possessed the ability to trap bacteria. It should be mentioned that the number of bacteria trapped to the DAPI-positive objects was reduced when the coverslips were treated with DNase I (**Figure 2(c)**). Similar results were obtained when PMA-stimulated NETs were treated with DNase I. These results imply that both structures are equally susceptible to DNase I treatment with respect to bacterial trap.

To further evaluate the similarities between the DAPI-positive objects and canonical NETs, we performed indirect-immunofluorescence analysis using antibodies that recognize marker proteins for NETs: anti-neutrophil

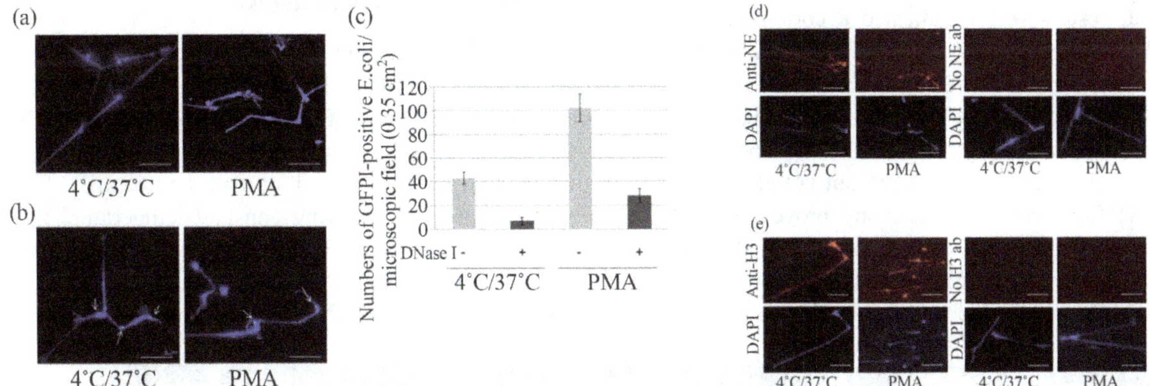

Figure 2. Hypothermia/rewarming-induced DAPI-positive objects exhibited several features similar to PMA-stimulated NETs. (a) Human neutrophils were incubated in a culture dish containing a coverslip at 4°C for 1 h followed by incubation at 37°C for 5 h. The coverslip was removed and fixed in PBS containing 4% paraformaldehyde and then stained with DAPI (left). For the control, PMA-stimulated neutrophils, which exhibit canonical NETs, were fixed with 4% paraformaldehyde and subjected to DAPI-staining (right). Bar indicates 50 μm; (b) Human neutrophils were incubated in a culture dish containing a coverslip at 4°C for 1 h followed by incubation at 37°C for 5 h. The coverslip was removed and then transferred to culture medium containing *E. coli* expressing recombinant GFP, followed by incubation for 15 min at 37°C (left). For the control, PMA-stimulated neutrophils were treated in the same way (right). The arrows indicate GFP-signals that represent *E. coli.* Bar indicates 50 μm; (c) After hypothermia/rewarming-(left) or PMA-incubation (right), the coverslips were treated with DNase buffer alone (gray bars) or DNase buffer containing 100 U/ml DNase I (black bars) at 37°C for 1 hr. The numbers of *E. coli* with GFP signals on the coverslips in the microscopic field (0.35 cm^2) were counted. The values shown represent means ± SE of three independent experiments; (d) The hypothermia/rewarming-induced DAPI-positive objects and PMA-stimulated NETs were immunostained with (upper panels in left and middle-right columns) or without (upper panels in middle-right and right columns) anti-NE antibodies. DAPI-stained images of each treatment are shown at the bottom. Bar indicates 50 μm; (e) The hypothermia/rewarming-induced DAPI-positive objects and PMA-stimulated NETs were immunostained with (upper panels in left and middle-right columns) or without (upper panels in middle-right and right columns) anti-histone H3 antibodies. DAPI-stained images of each treatment are shown at the bottom. Bar indicates 50 μm.

elastase (NE) and anti-histone H3 antibodies [7]. As shown in **Figures 2(d)** and **(e)**, in the presence of these anti-bodies, the signals were detected not only on PMA-stimulated NETs but also on the DAPI-positive objects. In contrast, in the absence of the antibodies, these signals were barely detected, suggesting existence of NE and histone H3 on both the DAPI-positive objects and canonical NETs. Taken together, at least with regard to their ability to trap bacteria and the existence of two marker proteins for NETs, our results indicated that the DAPI-positive objects possessed biochemical similarities to canonical NETs.

It should be noted, however, the experiments described above are not sufficient to conclude that the DAPI-positive objects have anti-bacterial activity. We are therefore investigating whether NE and histones on the DAPI-positive objects can indeed inactivate bacterial toxic proteins, called "virulence factors," and inhibit bacterial growth. In addition, we wish to test whether the DAPI-positive objects can capture microorganisms besides Gram-negative bacteria (*E. coli*), such as fungi and parasites.

3.3. The DAPI-Positive Objects Exhibited Several Biochemical Properties Different from Those of PMA-Stimulated NETs

Although our results so far indicated a correlation between DAPI-positive objects and NETs, several differences were also revealed. For instance, when indirect immunofluorescence analysis was conducted using anti-histone H3 citrulline R26 (anti-H3cit) antibody, we found that the antibody stained many, but not all, PMA-stimulated NETs, whereas the antibody proved poor at detecting the DAPI-positive objects (**Figure 3(a)**). Because it has been reported that peptidylarginine deiminase 4 (PAD4/PADI4), which catalyzes histone hy percitrullination, mediates chromatin decondensation and is vital to NET formation [13,20,21], our observation of different staining patterns with anti-H3cit antibody between the DAPI-positive objects and PMA-stimulated NETs implies activation of PAD4 in PMA-stimulated cells, while the enzyme might not be fully activated in the hypothermia/rewarming-treated cells.

We also detected differences between the objects and NETs in TUNEL assays. As shown in **Figure 3(b)**, TUNEL-positive signals were negligible in PMA-treated cells, confirming the previous report that no TUNEL-positive DNA damage is detectable in canonical NETs [15]. In contrast, TUNEL assay visualized more than 90% of the hypothermia/rewarming-induced DAPI-positive objects, indicating the existence of multiple sites of TUNEL-positive DNA damage on extensively unfolded DNAs in the objects. Because TUNEL-positive signals are frequently associated with apoptotic cells, these data indicated that hypothermia/rewarming-incubation might trigger, to some extent, activation of apoptosis-related DNA cleavage enzyme(s) in neutrophils, suggesting the possible contribution of apoptotic signaling pathways, at least in part, to DNA structural alterations during the formation of the DAPI-positive, NET-like structures.

3.4. Involvement of ROS Elevation to Generate the DAPI-Positive Objects

Given that ROS generation is an absolute requirement for the formation of NETs [15,16], we next assessed whether hypothermia/rewarming of neutrophils coincided with the generation of ROS. Thus, we measured ROS in hypothermia/rewarming-treated human neutrophils. As shown in **Figure 4(a)**, during constant temperature incubation at either 4˚C or 37˚C, no increase/decrease in ROS was detected. In contrast, we found a significant elevation of ROS when cells were kept at 4˚C followed

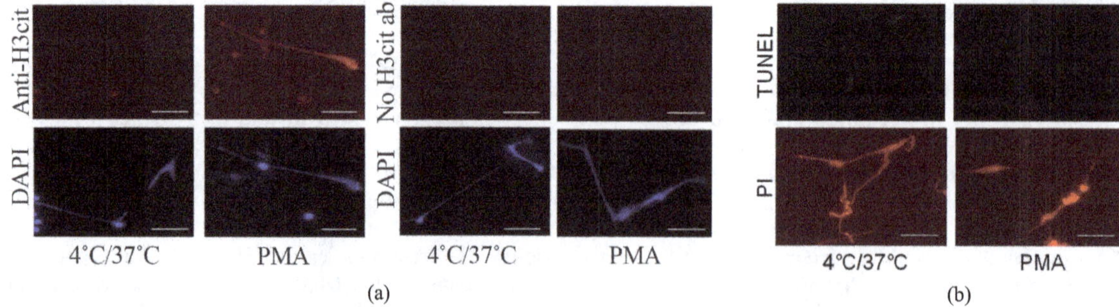

Figure 3. Hypothermia/rewarming-induced DAPI-positive, NET-like structures exhibited several features distinct from PMA-stimulated NETs. (a) The hypothermia/rewarming-induced DAPI-positive objects and PMA-stimulated NETs were immunostained with (upper panels in left and middle-right columns) or without (upper panels in middle-right and right columns) anti-H3cit antibodies. DAPI-stained images of each treatment are shown at the bottom. Bar indicates 50 μm; (b) Hypothermia/rewarming-induced DAPI-positive, NET-like structures (left column) and PMA-stimulated NETs (right column) were subjected to TUNEL assay (upper panel). DNA was visualized with propidium iodide (PI; lower panel). Bar indicates 50 μm.

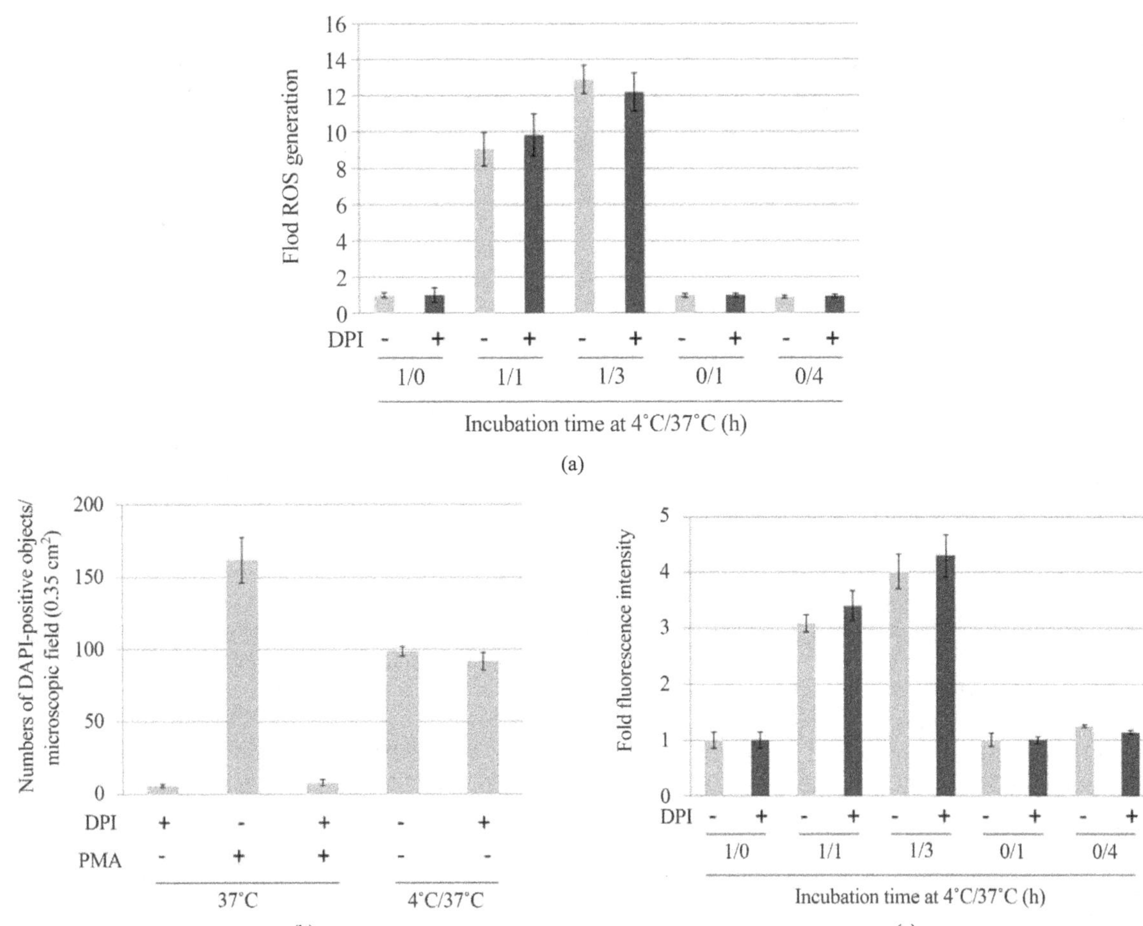

Figure 4. Hypothermia/rewarming-induced DAPI-positive, NET-like structures are biochemically and pharmacologically distinct from PMA-stimulated NETs. (a) Human neutrophils were incubated at 4˚C for 1 h followed by incubation at 37˚C for 1 h (indicated as "1/1") or 3 h (indicated as "1/3") in culture medium with (+; black bars) or without (−; gray bars) 20 μM DPI. ROS formation by cells incubated at 4˚C and 37˚C was quantified using CellROX Deep Red Reagent and fold ROS generation (ROS formation at 37˚C over that at 4˚C) was calculated. For the control, ROS formation was quantified during cell culture at 4˚C for 1 h (indicated as "1/0") and 37˚C for 1 h or 4 h (indicated as "0/1" or "0/4", respectively) in the presence or absence of DPI, and fold ROS generation during each culture period (ROS formation at the start over that at the end of culture) was calculated; (b) Human neutrophils (1 × 10^6 cells) were incubated at 4˚C for 1 h followed by incubation at 37˚C for 5 h in culture medium with (black bars) or without (gray bars) 20 μM DPI. The numbers of DAPI-positive objects on the coverslips in the microscopic field (0.35 cm^2), which were in proportion to total numbers of the DAPI-positive objects in the cultures, were counted. For the control, human neutrophils were incubated at 37˚C in the presence (+) or absence (−) of 20 μM DPI or 50 nM PMA and the numbers of NETs were counted; (c) Human neutrophils were incubated at 4˚C for 1 h followed by incubation at 37˚C for 1 h (indicated as "1/1") or 3 h (indicated as "1/3") in culture medium with (+; black bars) or without (−; gray bars) 20 μM DPI. Cells were subjected to MitoTracker Red-staining and fold changes in fluorescent signals (signal at 37˚C over that at 4˚C) were calculated. For the control, fluorescent signals were quantified during cell culture at 4˚C for 1 h (indicated as "1/0") and 37˚C for 1 h or 4 h (indicated as "0/1" or "0/4", respectively) in the presence or absence of DPI, and fold fluorescent signal changes during each culture period (signal at the start over that at the end of culture) were calculated.

by incubation at 37˚C for 1 h or 3 h; an approximate 10-fold increase in ROS was apparent in the cells after incubation at 37˚C.

We next examined whether NADPH oxidase contributed to ROS production in hypothermia/rewarming-treated cells. Given that PMA-induced NET formation is effectively inhibited by DPI, an inhibitor of NADPH oxidase activity [15,17,18 and see **Figure 4(b)**], we tested the effect of this drug. Intriguingly, neither ROS generation nor the formation of DAPI-positive objects was significantly affected by administration of DPI (**Figures 4(a)** and **(b)**), indicating that the mechanism of ROS generation in hypothermia/rewarming-treated neutrophils could be pharmacologically distinguished from that in PMA-stimulated cells with respect to the involvement of NADPH oxidase in ROS generation.

Although where and how ROS are produced in the hypothermia/rewarming-treated cells remains unclear, it is noteworthy that the mitochondrion-specific dye, MitoTracker Red, detected structural and/or functional alterations in mitochondria in the hypothermia/rewarming-treated cells (**Figure 4(c)**). Because mitochondria are known to produce ROS in aerobic organisms, including humans [22], these data suggest scenario that this organelle may generate ROS during hypothermia/rewarming, leading to the formation of the DAPI-positive, NET-like structures. However, DPI is also known as a potent inhibitor of mitochondrial reactive oxygen species production [23]. If DPI inhibits generation of ROS from both mitochondria and the NADPH oxidase pathway during hypothermia/rewarming of neutrophils, we need to consider the possibility of an alternative pathway to generate ROS, besides the NADPH oxidase pathway or the mitochondria pathway.

4. Conclusion

DAPI-positive objects with extensive DNA unfolding were observed in human neutrophils cultured in hypothermic conditions followed by rewarming. Our experimental data indicated that such DNA structural alterations in neutrophils may be related to NET formation, but can be biochemically and pharmacologically discriminated from NET formation. We also considered that the objects might not represent apoptotic cells, given that apoptotic cells contain condensed DNA enclosed in membrane, which is not observed in the objects. We thus suggest that the hypothermia/rewarming-induced DNA unfolding is regulated in a manner distinct from either canonical NETosis or canonical apoptosis, arguing the existence of a previously unappreciated signaling pathway that alters global genomic DNA structures in eukaryotic cells. Further, the results indicate that cold-treatment followed by warming may affect NET formation, which is an important consideration because many researchers use hypothermal conditions during the isolation and culture of neutrophils.

5. Acknowledgements

We thank all the members of the Saitoh Laboratory for helpful discussion. This work was supported by research grant to H. S. from Astellas Foundation for Research on Metabolic Disorders, and by intramural founding in Kumamoto Health Science University to J. K.

REFERENCES

[1] J. Kruuv, D. Glofcheski, K. H. Cheng, S. D. Campbell, H. M. Al-Qysi, W. T. Nolan and J. R. Lepock, "Factors Influencing Survival and Growth of Mammalian Cells Exposed to Hypothermia. I. Effects of Temperature and Membrane Lipid Perturbers," *Journal of Cellular Physiology*, Vol. 115, No. 2, 1983, pp. 179-185. doi:10.1002/jcp.1041150212

[2] P. W. Hochachka, "Defense Strategies against Hypoxia and Hypothermia," *Science*, Vol. 231, No. 4735, 1986, pp. 234-241. doi:10.1126/science.2417316

[3] N. Borregaard, "Neutrophils, From Marrow to Microbes," *Immunity*, Vol. 33, No. 5, 2010, pp. 657-670. doi:10.1016/j.immuni.2010.11.011

[4] B. Amulic, C. Cazalet, G. L. Hayes, K. D. Metzler and A. Zychlinsky, "Neutrophil Function: From Mechanisms to Disease," *Annual Reviews*: *Annual Review of Immunology*, Vol. 30, 2012, pp. 459-489. doi:10.1146/annurev-immunol-020711-074942

[5] V. Brinkmann, U. Reichard, C. Goosmann, B. Fauler, Y. Uhlemann, D. S. Weiss, Y. Weinrauch and A. Zychlinsky, "Neutrophil Extracellular Traps Kill Bacteria," *Science*, Vol. 303, No. 5663, 2004, pp. 1532-1535. doi:10.1126/science.1092385

[6] B. E. Steinberg and S. Grinstein, "Unconventional Roles of the NADPH Oxidase: Signaling, Ion Homeostasis, and Cell Death," *Science's STKE*, Vol. 2007, No. 379, 2007, p. 11. doi:10.1126/stke.3792007pe11

[7] V. Brinkmann and A. Zychlinsky, "Neutrophil Extracellular Traps: Is Immunity the Second Function of Chromatin?" *Journal of Cellular Physiology*, Vol. 198, No. 5, 2012, pp. 773-783. doi:10.1083/jcb.201203170

[8] S. R. Clark, A. C. Ma, S. A. Tavener, B. McDonald, Z. Goodarzi, M. M. Kelly, K. D. Patel, S. Chakrabarti, E. McAvoy, G. D. Sinclair, E. M. Keys, E. Allen-Vercoe, R. Devinney, C. J. Doig, F. H. Green and P. Kubes, "Platelet TLR4 Activates Neutrophil Extracellular Traps to Ensnare Bacteria in Septic Blood," *Nature Medicine*, Vo. 13, No. 4, 2007, pp. 463-469. doi:10.1038/nm1565

[9] K. Kessenbrock, M. Krumbholz, U. Schönermarck, W. Back, W. L. Gross, Z. Werb, H. J. Gröne, V. Brinkmann and D. E. Jenne, "Netting Neutrophils in Autoimmune Small-Vessel Vasculitis," *Nature Medicine*, Vol. 15, No. 6, 2009, pp. 623-625. doi:10.1038/nm.1959

[10] G. S. Garcia-Romo, S. Caielli, B. Vega, J. Connolly, F. Allantaz, Z. Xu, M. Punaro, J. Baisch, C. Guiducci, R. L. Coffman, F. J. Barrat, J. Banchereau and V. Pascual, "Netting Neutrophils Are Major Inducers of Type I IFN Production in Pediatric Systemic Lupus Erythematosus," *Science Translational Medicine*, Vol. 3, No. 73, 2011, p. 73ra20. doi:10.1126/scitranslmed.3001201

[11] R. Lande, D. Ganguly, V. Facchinetti, L. Frasca, C. Conrad, J. Gregorio, S. Meller, G. Chamilos, R. Sebasigari, V. Riccieri, R. Bassett, H. Amuro, S. Fukuhara, T. Ito, Y. J. Liu and M. Gilliet, "Neutrophils Activate Plasmacytoid Dendritic Cells by Releasing Self-DNA-Peptide Complexes in Systemic Lupus Erythematosus," *Science Translational Medicine*, Vol. 3, No. 73, 2011, p. 73ra19. doi:10.1126/scitranslmed.3001180

[12] T. Saitoh, J. Komano, Y. Saitoh, T. Misawa, M. Takahama, T. Kozaki, T. Uehata, H. Iwasaki, H. Omori, S. Yamaoka, N. Yamamoto and S. Akira, "Neutrophil Extracellular Traps Mediate a Host Defense Response to

Human Immunodeficiency Virus-1," *Cell Host Microbe*, Vol. 12, No. 1, 2012, pp. 109-116. doi:10.1016/j.chom.2012.05.015

[13] I. Neeli, N. Dwivedi, S. Khan and M. Radic, "Regulation of Extracellular Chromatin Release from Neutrophils," *Journal of Innate Immunity*, Vol. 1, No. 3, 2009, pp. 194-201. doi:10.1159/000206974

[14] M. B. Lim, J. W. Kuiper, A. Katchky, H. Goldberg and M. Glogauer, "Rac2 is Required for the Formation of Neutrophil Extracellular Traps," *Journal of Leukocyte Biology*, Vol. 90, No. 4, 2011, pp. 771-776. doi:10.1189/jlb.1010549

[15] T. A. Fuchs, U. Abed, C. Goosmann, R. Hurwitz, I. Schulze, V. Wahn, Y. Weinrauch, V. Brinkmann and A. Zychlinsky, "Novel Cell Death Program Leads to Neutrophil Extracellular Traps," *The Journal of Cell Biology*, Vol. 176, No. 2, 2007, pp. 231-241. doi:10.1083/jcb.200606027

[16] M. Bianchi, A. Hakkim, V. Brinkmann, U. Siler, R. A. Seger, A. Zychlinsky and J. Reichenbach, "Restoration of NET Formation by Gene Therapy in CGD Controls Aspergillosis," *Blood*, Vol. 114, No. 13, 2009, pp. 2619-2622. doi:10.1182/blood-2009-05-221606

[17] A. Hakkim, T. A. Fuchs, N. E. Martinez, S. Hess, H. Prinz, A. Zychlinsky and H. Waldmann, "Activation of the Raf-MEK-ERK Pathway Is Required for Neutrophil Extracellular Trap Formation," *Nature Chemical Biology*, Vol. 7, No. 2, 2011, pp. 75-77. doi:10.1038/nchembio.496

[18] B. V. O'Donnell, D. G. Tew, O. T. Jones and P. J. England, "Studies on the Inhibitory Mechanism of Iodonium Compounds with Special Reference to Neutrophil NADPH Oxidase," *Biochemical Journal*, Vol. 290, No. 1, 1993, pp. 41-49.

[19] Q. Remijsen, T. Vanden Berghe, E. Wirawan, B. Asselbergh, E. Parthoens, R. De Rycke, S. Noppen, M. Delforge, J. Willems and P. Vandenabeele, "Neutrophil Extracellular Trap Cell Death Requires Both Autophagy and Superoxide Generation," *Cell Research*, Vol. 21, No. 2, 2011, pp. 290-304. doi:10.1038/cr.2010.150

[20] I. Neeli, S. N. Khan and M. Radic, "Histone Deimination as a Response to Inflammatory Stimuli in Neutrophils," *The Journal of Immunology*, Vol. 180, No. 3, 2008, pp. 1895-1902.

[21] Y. Wang, M. Li, S. Stadler, S. Correll, P. Li, D. Wang, R. Hayama, L. Leonelli, H. Han, S. A. Grigoryev, C. D. Allis and S. A. Coonrod, "Histone Hypercitrullination Mediates Chromatin Decondensation and Neutrophil Extracellular Trap Formation," *The Journal of Cell Biology*, Vol. 184, No. 2, 2009, pp. 205-213. doi:10.1083/jcb.200806072

[22] M. Maryanovich and A. Gross, "A ROS Rheostat for Cell Fate Regulation," *Trends in Cell Biology*, Vol. 23, No. 3, 2013, pp. 129-134. doi:10.1016/j.tcb.2012.09.007

[23] L. Yunbo and A. Michael, "Diphenyleneiodonium, an NAD(P)H Oxidase Inhibitor, Also Potently Inhibits Mitochondrial Reactive Oxygen Species Production," *Biochemical and Biophysical Research Communications*, Vol. 253, No. 2, 1998, pp. 295-299. doi:10.1006/bbrc.1998.9729

20

The Expression of *TIMP*1, *TIMP*2, *VCAN*, *SPARC*, *CLEC*3*B* and *E*2*F*1 in Subcutaneous Adipose Tissue of Obese Males and Glucose Intolerance

Dmytro Minchenko[1,2], Oksana Ratushna[1], Yulia Bashta[1], Ruslana Herasymenko[1], Oleksandr Minchenko[1*]

[1]Department of Molecular Biology, Palladin Institute of Biochemistry, National Academy of Sciences of Ukraine, Kyiv, Ukraine
[2]Department of Pediatrics, National Bogomolets Medical University, Kyiv, Ukraine
Email: ominchenko@yahoo.com

ABSTRACT

We investigated the expression of *TIMP*1, *TIMP*2, *SPARC*, *VCAN*, and *CLEC*3*B* genes, encoded matricellular proteins with pleiotropic functions, and glucose intolerance in obese male subjects with normal and impaired glucose tolerance. The purpose of this study was to examine the association between the gene expressions and glucose intolerance in obesity. The results indicate that obesity leads to significant increase of *TIMP*1, *TIMP*2, *E*2*F*1 and *CLEC*3*B* gene expressions in subcutaneous adipose tissue, especially *TIMP*2 gene. However, more significant increase of the expression of *TIMP*1 and *TIMP*2 was found in adipose tissue of obese patients with glucose intolerance. No significant changes were found in the expression of *VCAN* and *SPARC* genes in adipose tissue of obese subjects with normal glucose tolerance but increased in the group of obese subjects with glucose intolerance. At the same time, the *E*2*F*1 and *CLEC*3*B* gene expressions were decreased in adipose tissue of obese patients with glucose intolerance. Results of this study provide evidence that changes in the expression of genes encoded *TIMP*1, *TIMP*2, *VCAN*, *SPARC*, *E*2*F*1 and *CLEC*3*B* in subcutaneous adipose tissue of obese individuals associate with glucose intolerance.

Keywords: Obesity; Glucose Intolerance; Men; Gene Expression; *TIMP*1; *TIMP*2; *VCAN*; *SPARC*; *CLEC*3*B*; *E*2*F*1

1. Introduction

Obesity is one of the most profound public health problems. Much has been learned regarding the regulation of body weight and adiposity [1,2]. Like many other medical conditions, obesity, metabolic syndrome and type 2 diabetes are the results of interaction between environmental and genetic factors, including biological rhythms which control metabolism and are intrinsically and inter dependently linked [3,4].

Circadian rhythms govern or contribute to regulation of a large array of metabolic and physiological functions, including insulin sensitivity, energy homeostasis, satiety signaling, cellular proliferation, and cardiovascular function [5-7]. Recent studies have demonstrated relationships between the dysfunction of circadian clock system and the development of metabolic abnormalities, including adiposity and excessive levels of circulating lipids [8-11].

Adipose tissue growth is tightly associated with cell proliferation processes and angiogenesis which is an important component of different proliferative processes. In particular, fat tissue growth is regulated by different tightly interconnected factors [12,13]. Special interest deserve the matrix proteins with pleiotropic roles, such as the tissue inhibitor of matrix metalloproteinase (*TIMP*1 and *TIMP*2) [14,15], versican (*VCAN*) [16] and *SPARC* (secreted protein acidic and rich in cysteine, also known as osteonectin) [17]. These proteins are linked to human obesity and diabetes complications as they to reveal anti-angiogenic properties. Moreover, they seem to regulate cell growth and apoptosis, particularly through affecting the vascular endothelial growth factor (*VEGF*), at resulting in *VEGF* resistance in diabetes mellitus despite the presence of the functionally active VEGF receptor 1 [18, 19].

*TIMP*s are multifunctional and can act either directly through cell surface receptors or indirectly through modulation of protease activity to direct cell fate [20]. Thus,

*Corresponding author.

*TIMP*1 is able to promote cell proliferation in a wide range of cell types, and may also have an anti-apoptotic function [14]. However, at the same time, *TIMP*1 is a potent inhibitor of tumor growth and angiogenesis and is involved in cell adhesion and migration [14]. TIMP1 is a negative regulator of adipogenesis both in mice and in humans [15]. *TIMP*2 thought to be a metastasis suppressor and has a unique role among *TIMP* family members in its ability to directly suppress the proliferation of endothelial cells [14]. Nevertheless, *TIMP*-2 has been shown to induce gene expression, to promote G(1) cell cycle arrest, and to inhibit cell migration [14]. It was shown that TIMP-2 involved in binding to the receptor integrin $\alpha3\beta1$, and mediates angio-inhibitory and tumor suppressor activity [20].

It was also observed that enhanced neovascularization correlated with down regulation of anti-angiogenic connective tissue growth factor (CTGF). The latter binds VEGFA and its angiogenic activity are inhibited in the VEGFA-CTGF complex form [12]. But, stability of this complex as well as the angiogenic activity of VEGF depends from the matrix metalloproteinases and its inhibitors [12,21].

SPARC (secreted protein acidic and rich in cysteine), also known as osteonectin or BM-40, is a widely expressed profibrotic protein with pleiotropic roles, is linked to human obesity, insulin resistance, diabetes mellitus and diabetes complications such as diabetic retinopathy and nephropathy because it has anti-angiogenic properties and appears to regulate cell growth [17]. Elevated plasma levels of this protein were observed in patients with newly diagnosed type 2 diabetes mellitus [22]. Furthermore, evidence suggests that adipose tissue becomes increasingly fibrotic in obesity and SPARC as a regulator of the extracellular matrix also contributes to this adipose-tissue fibrosis [17,23]. Fibrosis of subcutaneous adipose tissue may restrict accumulation of triglycerides in this type of tissue. These are, therefore, deposited as ectopic lipids in other non-adipose tissues such as the liver or as intramyocellular lipids in skeletal muscle and predisposes to insulin resistance [17].

Versican (*VCAN*), also known as chondroitin sulfate proteoglycan 2 (*CSPG*2), is one of the main components of the extracellular matrix and is considered to be crucial to several key cellular processes involved in development and disease; *i.e.* cellular adhesion, proliferation, differentiation, migration, and angiogenesis, as well as, in intercellular signaling and connecting cells with the extracellular matrix [16,24,25]. Recently, the Wnt/β-catenin/TCF response elements in the human versican promoter, which is essential for activation of versican transcription, was identified [25]. Moreover, the versican was identified among six proteins at the center of the angiogenesis-associated network [26]. Hence, the angiogenesis phenomena are governed by complex signaling networks.

The transcription factor *E2F*1 regulates the expression of genes involved in a wide range of cellular processes; it is needed for regulation of nuclear receptor networks by retinoblastoma and via the CDK4-pRB-E2F1 regulatory pathway is involved in glucose homeostasis, defining a new link between cell proliferation and metabolism [27].

The endoplasmic reticulum stress is recognized as an important determinant of type 2 diabetes and contributes to the expression profile of many regulatory genes resulting in peripheral insulin resistance and diabetic complications, acting by inhibiting insulin receptor signaling [28,29]. However, detailed molecular mechanisms of the involvement of the endoplasmic reticulum stress in the development of obesity and its complications with different degree of dysglycemia are not yet clear.

Because the function of some matricellular proteins with pleiotropic roles is closely linked to metabolic homeostasis and these proteins can modulate the intracellular signaling network and lead to development of obesity and its complications (insulin resistance and glucose intolerance), the main goal of this work was to study the role of gene expressions encoding for the matricellular proteins with pleiotropic roles, in subcutaneous adipose tissue of obese individuals for evaluation of its possible significance to development of human obesity and glucose intolerance.

2. Materials and Methods

2.1. Patients

All participants gave written informed consent and the studies were approved by the local research ethics committees of Institute of Experimental Endocrinology Slovak Academy of Sciences. Patients clinical characteristics are shown in **Table 1**. The 18 male subjects participate in the study. They were divided into three groups (6 men in each group): 1) clinically healthy lean individuals, 2) patients with obesity and normal glucose tolerance (NGT) and 3) obesity and impaired glucose tolerance (IGT).

The lean (control) participant groups were individuals with mean age 44 ± 3 years and mean body mass index (BMI) 23 ± 0.6 kg/m^2. The obese participant groups were individuals with mean age (45 ± 3 and 44 ± 3 years, respectively) and mean BMI (32 ± 0.6 and 34 ± 0.6 kg/m^2, respectively). Thus, BMI in these groups of patients was significantly higher as compared to lean individuals (**Table 1**).

In the group of obese patients with impaired glucose tolerance 2 h glucose level in the blood was 7.8 mmol/l during an OGTT as compared with 5.3 mmol/l in obese participants (NGT) (**Table 1**). Moreover, in this group of obese patients with glucose intolerance is decreased insulin sensitivity index T (2.70 ± 0.19 vs 5.09 ± 0.67 mg/kg/min in an obese patients) and increased the level of fasting triglycerides (2.17 ± 0.44 vs 1.36 ± 0.2 mmol/l in an obese patients) as well as fasting insulin (15.2 ± 2.3 vs 9.37 ± 1.6 μIU/ml in an obese patients (NGT), as shown in **Table 1**.

2.2. RNA Isolation

RNasy Lipid Tissue Mini Kit (QIAGEN, Germany) was used for RNA extraction from subcutaneous adipose tissue.

2.3. Reverse Transcription and Quantitative Real-Time PCR Analysis

The expression levels of genes encoded *TIMP* metallopeptidase inhibitor 1 (*TIMP*1), *TIMP*2, *SPARC* (secreted protein, acidic, cysteine-rich) and versican (*VCAN*) were measured in subcutaneous fat tissue by real-time quantitative polymerase chain reaction of complementary DNA (cDNA). QuaniTect Reverse Transcription Kit (QIAGEN, Germany) was used for cDNA synthesis. The 7900 HT Fast Real-Time PCR System (Applied Biosystems), Absolute QPCR SYBR Green Mix (Thermo Scientific, UK) and pair of primers specific for each studied gene (Sigma, USA) were used for quantitative real-time polymerase chain reaction (qPCR). The primers sequences are shown in **Table 2**. The expression of beta-actin mRNA was used as control of analyzed RNA quantity.

An analysis of quantitative real-time polymerase chain reaction was performed using special computer program "Differential expression calculator" and statistical analysis—in Excel program. The values of different gene expressions were normalized to the expression of beta-actin mRNA and represent as percent of control (lean individuals, 100%) and are the means ± SEM for up to six different samples.

3. Results

In this study, we analyzed the expression levels of genes

Table 1. Characteristics of the study participants.

Variable	Lean, NGT	Obese, NGT	Obese, IGT
Age at visit (years) (*n*)	44 ± 3.4 (6)	45 ± 3 (6)	44 ± 3.2 (6)
Body mass index (BMI) (kg/m²) (*n*)	23 ± 0.6 (6)	32 ± 0.6* (6)	34 ± 0.6* (6)
Fasting glucose (mmol/l) (*n*)	4.5 ± 0.09 (6)	5.0 ± 0.22* (6)	5.5 ± 0.26 (6)
2 h oral glucose tolerance test (OGTT) glucose (mmol/l) (*n*)	5.08 ± 0.64 (5)	5.31 ± 0.88 (6)	7.83 ± 0.36*^ (6)
Insulin sensitivity index (T; mg/kg/min) (*n*)	7.9 ± 0.58 (6)	5.1 ± 0.67* (6)	2.7 ± 0.19*^ (5)
Fasting triglycerides (mmol/l) (*n*)	1.0 ± 0.19 (6)	1.36 ± 0.2 (6)	2.17 ± 0.44*^ (6)
Fasting insulin (μIU/ml) (*n*)	8.0 ± 2.8 (3)	9.37 ± 1.6 (3)	15.2± 2.3* (4)

Data are means ± SEM. NGT: normal glucose tolerance; IGT: impaired glucose tolerance; *P < 0.05 vs control (lean group); ^P < 0.05 vs obese (NGT) group.

Table 2. Characteristics of the primers used for quantitative real-time polymerase chain reaction.

Gene symbol	Gene name	Primer's sequence	Nucleotide numbers in sequence	Gen Bank accession number
*TIMP*1	tissue inhibitor of matrix metalloproteinase 1	F; 5'-ATTCCGACCTCGTCATCAGR: 3'-GCAGTTTTCCAGCAATGAG	301 - 320 530 - 511	NM_003254
*TIMP*2	tissue inhibitor of matrix metalloproteinase 1	F; 5'-GATGCACATCACCCTCTGTGR: 3'-GTCGAGAAACTCCTGCTTGG	665 - 684 950 - 931	NM_003255
SPARC	Secreted protein acidic and rich in cysteine; osteonectin	F; 5'-TGCCTGATGAGACAGAGGTG R: 3'-AAGTGGCAGGAAGAGTCGAA	176 - 195 479 - 460	NM_003118
VCAN (*CSPG2*)	versican;chondroitin sulfate proteoglycan 2	F; 5'- CCTTTCTGGGGAAGAACTCC R: 3'-GGTCACATAGGAAGCGTGGT	102 - 121 272 - 253	NM_004385
CLEC3B	C-type lectin domain family 3, member B, tetranectin	F; 5'-ACCCAGAAGCCCAAGAAGATR: 3'-GAAGGTCTTCGTCTGGGTGA	169 - 188 369 - 350	NM_003278
*E2F*1	E2F transcription factor 1	F; 5'-GGGCTCTAACTGCACTTTCGR: 3'-AGGGAGTTGGGGTATCAACC	1830 - 1849 2096 - 2077	NM_005225
ACTB	beta-actin	F; 5'-GGACTTCGAGCAAGAGATGG R: 3'-AGCACTGTGTTGGCGTACAG	747 - 766 980 - 961	NM_001101

encoding tissue inhibitor of matrix metalloproteinase 1 and 2, *SPARC*, versican, *CLEC3B* and *E2F*1 in subcutaneous adipose tissue from three groups of participants: lean (control), obese with normal glucose tolerance and obese with impaired glucose tolerance. As shown in **Figure 1**, the expression level of *TIMP*1 and *TIMP*2 mRNA in subcutaneous adipose tissue of obese individuals with normal glucose tolerance increases as compared to the control (lean) subjects: +56% and +183%, correspondingly. Moreover, in obese patients with impaired glucose tolerance (IGT) the expression of these genes was significantly higher as compared both to the lean subjects and the obese individuals with normal glucose tolerance.

As shown in **Figure 2**, the expression level of *SPARC* and *VCAN* mRNA did not change significantly in subcutaneous adipose tissue of the obese individuals, compared with lean participants. At the same time, the expression level of *SPARC* and *VCAN* mRNA is increased in subcutaneous adipose tissue of the obese individuals with impaired glucose tolerance, when compared to the groups of obese subjects with normal glucose tolerance or lean participants, being much more intense for the *SPARC* gene (**Figure 2**).

As shown in **Figure 3**, the expression level of *CLEC3B* and *E21F*1 mRNA significantly increased in subcutaneous adipose tissue of the obese individuals with NGT, compared with lean participants, being much more intense for the *CLEC3B* gene. At the same time, the expression level both of these genes in adipose tissue of the obese individuals with impaired glucose tolerance was decreased when compared to the groups of obese subjects with normal glucose tolerance (**Figure 3**).

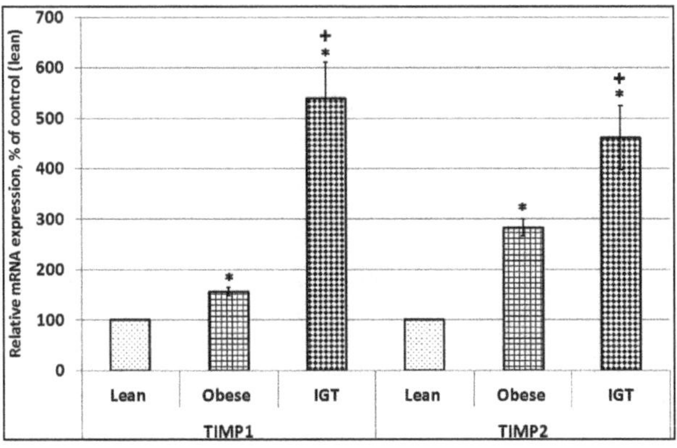

Figure 1. The expression level of TIMP metallopeptidase inhibitor 1 and 2 (*TIMP*1 and *TIMP*2) mRNA in subcutaneous adipose tissue of lean (Lean) and obese individuals with normal glucose tolerance (Obese) as well as in obese patients with glucose intolerance (IGT). The values of *TIMP*1 and *TIMP*2 mRNA expressions were normalized to the expression of beta-actin mRNA and are represented as a percent of control (Lean, 100%). Data is expressed as mean ± SEM of values from each group; *P < 0.05 vs group of lean individuals; +P < 0.05 vs group with obesity and normal glucose tolerance test (NGT).

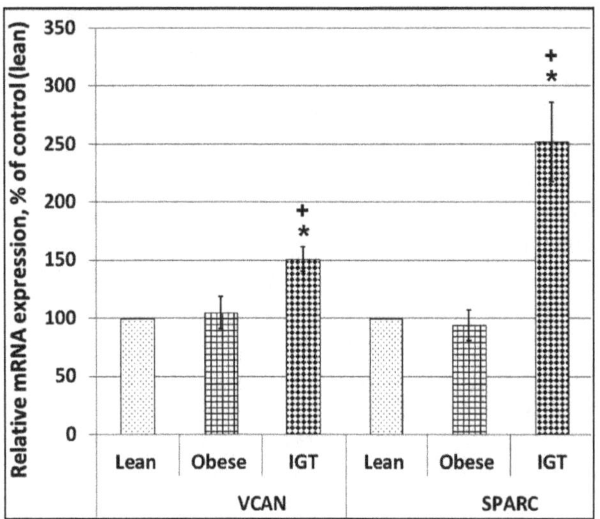

Figure 2. Versican (*VCAN*) and secreted protein acidic and rich in cysteine (*SPARC*) mRNA expression level in subcutaneous adipose tissue of lean and obese individuals with normal glucose tolerance (Obese) as well as in obese patients with impaired glucose tolerance (IGT). The values of *VCAN* and *SPARC* mRNA expressions were normalized to the expression of beta-actin mRNA and are represented as a percent of control (Lean, 100%). Data is expressed as mean ± SEM of values from each group; *P < 0.05 vs lean group; +P < 0.05 vs obese group.

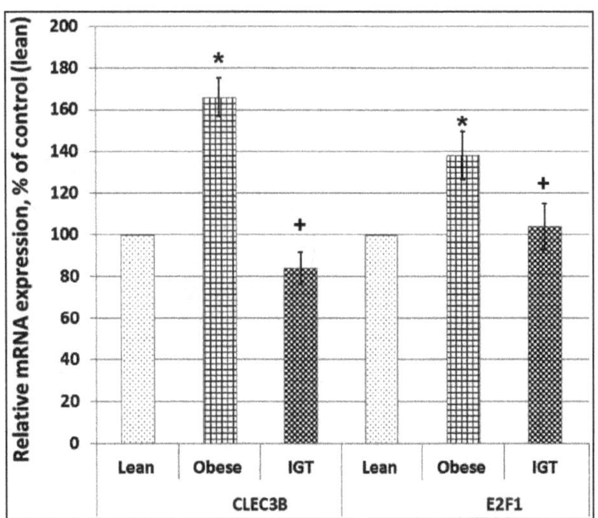

Figure 3. C-type lectin domain family 3, member B (CLEC3B) and transcription factor E2F1 mRNA expression level in subcutaneous adipose tissue of lean and obese individuals with normal glucose tolerance (Obese) as well as in obese patients with impaired glucose tolerance (IGT). The values of *VCAN* and *SPARC* mRNA expressions were normalized to the expression of beta-actin mRNA and are represented as a percent of control (Lean, 100%). Data is expressed as mean ± SEM of values from each group; *P < 0.05 vs lean group; +P < 0.05 vs obese group.

We also analyzed the correlation between different clinical characteristics and gene expressions. There is positive correlation between increased BMI and *TIMP*1 (r = 0.93), *TIMP*2 (r = 0.95), *E2F*1 (r = 0.89) and *CLEC3B* (r = 0.92) gene expressions in the obese men with normal glucose tolerance versus lean subjects (**Figure 4**). Moreover, as shown in **Table 1**, obese individuals with glucose intolerance versus obese subjects with normal glucose tolerance have significantly higher the 2 h oral glucose tolerance test (OGTT), fasting triglycerides and fasting insulin levels, but decreased insulin sensitivity index. We observed negative correlation between this insulin sensitivity index and *TIMP*1 (r = −0.70), *TIMP*2 (r = −0.64), *SPARC* (r = −0.64), *VCAN* (r = −0.66), and *CLEC3B* (r = 0.75) gene expressions in the obese individuals with glucose intolerance versus obese men with normal glucose tolerance (**Figure 5**).

4. Discussion

In this study, we have demonstrated that obesity is associated with dysregulation of the *TIMP*1, *TIMP*2, *VCAN*, *CLEC3B*, *E2F*1 and *SPARC* genes, which encode the important regulatory factors with pleiotropic functions, in particularly, control angiogenesis and growth processes [14,15,17,25]. Moreover, these factors are involved in the development of the obesity, glucose tolerance and type 2 diabetes, the most profound public health problems [15,17,27]. Angiogenesis is an important component of different proliferative processes, in particular, fat tissue growth; however, angiogenesis can also contribute to the development of complications associated with obesity, insulin resistance and glucose intolerance. In this study we have shown that obesity leads to a significant increase of *TIMP*1 and *TIMP*2 gene expressions in men subcutaneous adipose tissue, besides that, the most significant increase was shown for *TIMP*2 gene. It is possible that proteins encoded by these genes are involved the development of obesity, because both the *TIMP*1 and *TIMP*2 genes are implicated in direct regulation of cell growth, apoptosis and angiogenesis [14].

However, the tissue inhibitors of metalloproteinases

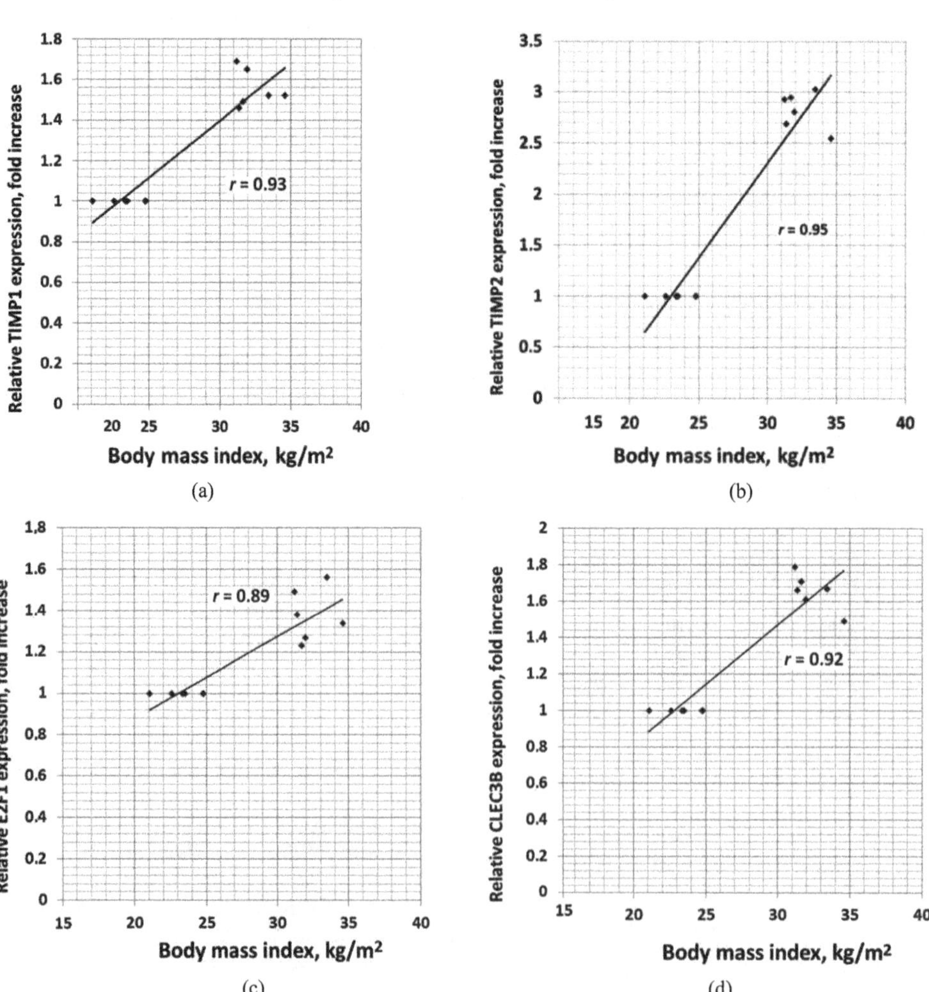

Figure 4. The correlation between body mass index (BMI) and *TIMP*1 (a); *TIMP*2 (b); *E2F*1 (c) and *CLEC3B* (d) gene expressions in the obese men with normal glucose tolerance versus lean subjects.

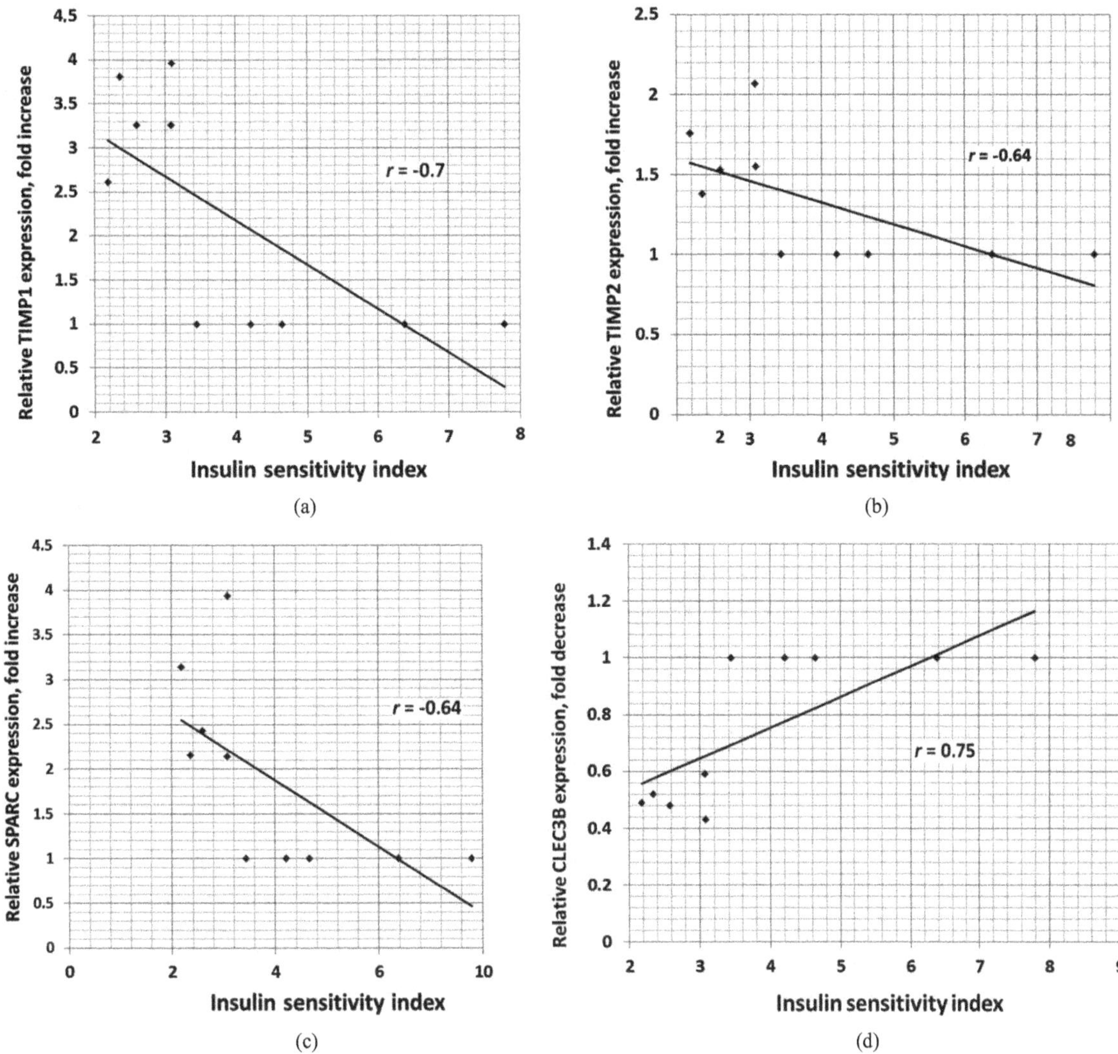

Figure 5. The correlation between insulin sensitivity index and*TIMP1* (a); *TIMP2* (b); *SPARC* (c) and *CLEC3B* (d) gene expressions in the obese individuals with glucose intolerance versus obese men with normal glucose tolerance.

(*TIMPs*) are multifunctional and can both promote and inhibit of cells growth in cell specific manner via different signaling pathways [14]. *TIMPs* can act either directly through cell surface receptors or indirectly through modulation of protease activity to direct cell fate. Thus, Meissburger *et al.* [15] have shown that tissue inhibitor of matrix metalloproteinase 1 controls adipogenesis in obesity in humans. At the same time, *TIMP1* can inhibit cell growth and apoptosis via binding to CD63 [14]. Tissue inhibitors of metalloproteinases suppress matrix metalloproteinase activity critical for extracellular matrix turnover associated with both physiologic and pathologic tissue remodeling; however, anti-proliferative as well as anti-angiogenic effects of *TIMP2* are independent of metalloproteinase inhibition [20]. It was also shown that these effects of *TIMP2* require alpha 3 beta 1 integrin-mediated binding of this tissue inhibitor of metalloproteinases to endothelial cells. Further, *TIMP2* induces a decrease in total protein tyrosine phosphatase activity associated with beta1 integrin subunits as well as dissociation of the phosphatase SHP-1 from beta 1; however, *TIMP2* also results in a concomitant increase in protein tyrosine phosphatase activity associated with tyrosine kinase receptors FGFR-1 and KDR [20]. Furthermore, *TIMP2* has been shown to induce gene expression, to promote cell cycle arrest and to inhibit cell migration [14]. Thus, these findings establish an unexpected, metalloproteinase-independent mechanism for *TIMP2* inhibition of endothelial cell proliferation and reveal an important component of the anti-angiogenic effect of this metalloproteinase inhibitor [20].

Moreover, it is possible that these genes are also involved the development of obesity associated complications, like glucose intolerance, because we have shown significant up-regulation of both *TIMP1* and *TIMP2* as well as *SPARC* and *VCAN* gene expressions in adipose tissue of subjects with impaired glucose intolerance versus the obese individuals with normal glucose tolerance. This observation agrees with data which recently has been shown that transcription factor retinoid-related orphan receptor gamma as a negative regulator of adipocyte differentiation modulates insulin sensitivity in obesity through expression of metalloproteinase-3 and its inhibitors [30]. It is also known that *SPARC* expression is predominant in subcutaneous adipose tissue, its expression contributing to metabolic dysregulation in obesity and is associated with insulin resistance and glucose intolerance [17,22]. The augmented expression of *SPARC* gene in the obese participants with glucose intolerance is possibly associated with hyperinsulinemia, because insulin as well as leptin can induce its expression [23].

It is possible that changes in *TIMP*1, *TIMP*2, *SPARC, CLEC*3*B*, *E*2*F*1, and *VCAN* gene expressions in obese individuals with impaired glucose tolerance are probably a result, at least partly, of endoplasmic reticulum stress mediated by the insulin resistance and elevated level of glucose. Recently, the endoplasmic reticulum stress is recognized as an important determinant of obesity and type 2 diabetes as well as a central feature of peripheral insulin resistance and glucose homeostasis [28,31,32].

Furthermore, SPARC as a regulator of the extra cellular matrix may contributes to metabolic dysregulation in obesity, insulin resistance and diabetes complications such as diabetic retinopathy and nephropathy, conditions that are ameliorated in the *SPARC*-knockout mouse model, via adipose-tissue fibrosis [17]. Hence, our results agree with this data that the *SPARC* protein is associated with diabetes complications, because *SPARC* protein plays an important role in angiogenesis [33]. At the same time, the primary function of *SPARC* in angiogenesis remains unclear, because SPARC activity in some circumstances promotes angiogenesis, while in others is more consistent with an anti-angiogenic activity [34]. Undoubtedly, the mercurial nature of *SPARC* belies a redundancy of functional proteins in angiogenesis. Moreover, fibrosis of subcutaneous adipose tissue may restrict accumulation of triglycerides in this type of tissue which are, therefore, diverted and deposited as ectopic lipids in other tissues such as the liver or as intramyocellular lipids in skeletal muscle, which predisposes to insulin resistance [17].

Results of this study have also shown enhanced level of *VCAN* expression in subcutaneous adipose tissue of obese individuals with glucose intolerance, which possibly contributes to proliferative processes, angiogenesis, metabolic dysregulation, and insulin resistance, because this chondroitin sulfate proteoglycan as an one of the main components of the extracellular matrix participates in cell adhesion, proliferation, migration, apoptosis and angiogenesis [16,25]. Most of these effects of versican are possibly mediated through the Wnt/β-catenin/TCF response elements in the *VCAN* promoter [35].

We also observed significant reduction of *E*2*F*1 gene expression in adipose tissue of obese patients with glucose intolerance. This data is consistent with results of Blanchet *et al.* [36] concerning functional activity of transcription factor E2F1 which regulate both proliferative and metabolic genes and coordinates cellular responses by acting as a regulatory switch between cell proliferation and metabolism. This reduction of the expression of *E*2*F*1 gene is possibly associated with activation of pro-proliferative and suppression of anti-proliferative gene expressions. Thus, tumor suppressor kinase LKB 1 is turning out to be a key regulator of the body's metabolic activities, including its handling of glucose body's level via activation of AMPK [37]. Moreover, knockout of AMPK related gene NUAK2 leads to developing of obesity, metabolic syndrome as well as cancer [38].

It is interesting to note that angiogenesis like many other biological processes is regulated by complex network of different factors which are tightly interconnected [12,19,21,39]. Thus, thrombospondin-1, a matrix-bound adhesive glycoprotein, has been shown to modulate tumor progression and up-regulates tissue inhibitor of metalloproteinase-1 production in human tumor cells and also up-regulates matrix metalloproteinases MMP-2 and MMP-9 [39]. This data suggested that the balance between matrix metalloproteinases and tissue inhibitors of metalloproteinases is a key determinant in different biological effects of THBS1, including tumor cell invasion, and may provide an explanation for the divergent activeties reported for thrombospondin-1 in tumor progression. Thus, the THBS1 is involved in influencing the critical balance between MMPs and their inhibitors, maintaining the controlled degradation of the extracellular matrix needed to support metastasis and possibly in obesity as well as in obesity with different degree of dysglycemia. Moreover, different cells express a lot of pro-angiogenic and anti-angiogenic genes (e.g., *TIMP*1, *TIMP*2); however, detailed analysis of melanoma cells from different patients do not show a significantly higher median number of expressed pro-angiogenic or anti-angiogenic genes, but 97% of these cell samples aberrantly express at least one of the angiogenic factors [40]. Thus, theangiogenesis is regulated by complex network of pro-angiogenic and anti-angiogenic factors and aberrant expression at least one of these factors can modify an angiogenesis.

It is possible that *TIMP*1, *TIMP*2, *VCAN*, *CLEC*3*B*, *E*2*F*1 and *SPARC* are also included in this network and the balance between different regulatory factors which participate in the control of angiogenesis, including vascular endothelial growth factor, a key pro-angiogenic factor, insulin, insulin-like growth factors, leptin and others, really determinate an angiogenesis both in obesity and in obesity with glucose intolerance.

Results of this study also provide strong evidence that expression of *TIMP*1, *TIMP*2, *VCAN* and *SPARC* genes encoded the key regulatory factors with pleiotropic functions in subcutaneous adipose tissue of the obese individuals with glucose intolerance is deregulated and that these changes can be determined both by insulin resistance and endoplasmic reticulum stress. Collectively, the results of this study underscore the crucial role of *TIMP*1, *TIMP*2, *SPARC* and *VCAM* regulatory factors with pleiotropic function in the developing the obesity and its complications, especially glucose intolerance via participation in the intracellular signaling network, responsible for regulation of angiogenesis and glucose metabolism.

5. Acknowledgements

We thank Prof. Iwar Klimes (Institute of Experimental Endocrinology, Slovakia) for support and interest in this work as well as the gift of RNA from men's adipose tissue and clinical characteristics of patients.

REFERENCES

[1] M. S. Bray and M. E. Young, "Circadian Rhythms in the Development of Obesity: Potential Role for the Circadian Clock within the Adipocyte," *Obesity Reviews*, Vol. 8, No. 2, 2007, pp. 169-181.

[2] M. S. Bray and M. E. Young, "The Role of Cell-Specific Circadian Clocks in Metabolism and Disease," *Obesity Reviews*, Vol. 10, No. 2, 2009, pp. 6-13. doi:10.1111/j.1467-789X.2009.00684.x

[3] J. Kovac, J. Husse and H. Oster, "A Time to Fast, a Time to Feast: The Crosstalk between Metabolism and the Circadian Clock," *Molecules and Cells*, Vol. 282, No. 2, 2009, pp. 75-80. doi:10.1007/s10059-009-0113-0

[4] E. M. Scott, A. M. Carter and P. J. Grant, "Association between Polymorphisms in the Clock Gene, Obesity and the Metabolic Syndrome in Man Clock Polymorphisms and Obesity," *International Journal of Obesity*, Vol. 32, 2008, pp. 658-662. doi:10.1038/sj.ijo.0803778

[5] C. B. Green, J. S. Takahashi and J. Bass, "The Meter of Metabolism," *Cell*, Vol. 134, No. 5, 2008, pp. 728-742. doi:10.1016/j.cell.2008.08.022

[6] K. M. Ramsey, B. Marcheva, A. Kohsaka and J. Bass, "The Clock Work of Metabolism," *Annual Review of Nutrition*, Vol. 27, 2007, pp. 219-240. doi:10.1146/annurev.nutr.27.061406.093546

[7] M. S. Bray and M. E. Young, "Regulation of Fatty Acid Metabolism by Cell Autonomous Circadian Clocks: Time to Fatten Up on Information?" *The Journal of Biological Chemistry*, Vol. 286, 2011, pp. 11883-11889. doi:10.1074/jbc.R110.214643

[8] H. Ando, T. Takamura, N. Matsuzawa-Nagata, K. R. Shima, T. Eto, H. Misu, M. Shiramoto, T. Tsuru, S. Irie, A. Fujimura and S. Kaneko, "Clock Gene Expression in Peripheral Leucocytes of Patients with Type 2 Diabetes," *Diabetologia*, Vol. 52, No. 2, 2009, pp. 329-335. doi:10.1007/s00125-008-1194-6

[9] H. Ando, M. Kumazaki, Y. Motosugi, K. Ushijima, T. Maekawa, E. Ishikawa and A. Fujimura, "Impairment of Peripheral Circadian Clocks Precedes Metabolic Abnormalities in ob/ob mice," *Endocrinology*, Vol. 152, No. 4, 2011, pp. 1347-1354. doi:10.1210/en.2010-1068

[10] W. Huang, K. M. Ramsey, B. Marcheva and J. Bass, "Circadian Rhythms, Sleep, and Metabolism," *Journal of Clinical Investigation*, Vol. 121, No. 6, 2011, pp. 2133-2141. doi:10.1172/JCI46043

[11] S. Shimba, T. Ogawa, S. Hitosugi, Y. Ichihashi, Y. Nakadaira, M. Kobayashi, M. Tezuka, Y. Kosuge, K. Ishige, Y. Ito, K. Komiyama, Y. Okamatsu-Ogura, K. Kimura and M. Saito, "Deficient of a Clock Gene, Brain and Muscle Arnt-Like Protein-1 (BMAL1), Induces Dyslipidemia and Ectopic Fat Formation," *PLoS One*, Vol. 6, 2011, e25231. doi:10.1371/journal.pone.0025231

[12] G. Hashimoto, I. Inoki, FujiiY, T. Aoki, E. Ikeda and Y. Okada, "Matrix Metalloproteinases Cleave Connective Tissue Growth Factor and Reactivate Angiogenic Activity of Vascular Endothelial Growth Factor 165," *The Journal of Biological Chemistry*, Vol. 277, 2002, pp. 36288-36295. doi:10.1074/jbc.M201674200

[13] I. Inoki, T. Shiomi, G. Hashimoto, H. Enomoto, H. Nakamura, K. Makino, E. Ikeda, S. Takata, K. Kobayashi and Y. Okada, "Connective Tissue Growth Factor Binds Vascular Endothelial Growth Factor (VEGF) and Inhibits VEGF-Induced Angiogenesis," *FASEB Journal*, Vol. 16, 2002, pp. 219-221. doi:10.1096/fj.01-0332fje

[14] W. G. Stetler-Stevenson, "Tissue Inhibitors of Metalloproteinases in Cell Signaling: Metalloproteinase-Independent Biological Activities," *Science Signal*, Vol. 1, No. 27, 2008. doi:10.1126/scisignal.127re6

[15] B. Meissburger, L. Stachorski, E. Roder, G. Rudofsky and C. Wolfrum, "Tissue Inhibitor of Matrix Metalloproteinase 1 (TIMP1) Controls Adipogenesis in Obesity in Mice and in Humans," *Diabetologia*, Vol. 54, No. 6, 2011, pp. 1468-1479. doi:10.1007/s00125-011-2093-9

[16] W. W. Du, B. B. Yang, B. L. Yang, Z. Deng, L. Fang, S. W. Shan, Z. Jeyapalan, Y. Zhang, A. Seth and A. J. Yee, "Versican G3 Domain Modulates Breast Cancer Cell Apoptosis: A Mechanism for Breast Cancer Cell Response to Chemotherapy and EGFR Therapy," *PLoS One*, Vol. 6, 2011, Article ID:E26396. doi:10.1371/journal.pone.0026396

[17] K. Kos and J. P. Wilding, "SPARC: A Key Player in the Pathologies Associated with Obesity and Diabetes," *Nature Reviews Endocrinology*, Vol. 6, 2010, pp. 225-235. doi:10.1038/nrendo.2010.18

[18] V. Tchaikovski, S. Olieslagers, F. D. Böhmer and J. Waltenberger, "Diabetes Mellitus Activates Signal Transduction Pathways Resulting in Vascular Endothelial Growth Factor Resistance of Human Monocytes," *Circulation*, Vol. 120, 2009, pp. 150-159. doi:10.1161/CIRCULATIONAHA.108.817528

[19] J. Waltenberger, "VEGF Resistance as a Molecular Basis to Explain the Angiogenesis Paradox in Diabetes Melli-

tus," *Biochemical Society Transactions*, Vol. 37, 2009, pp. 1167-1170. doi:10.1042/BST0371167

[20] D. W. Seo, H. Li, L. Guedez, P. T. Wingfield, T. Diaz, R. Salloum, B. Y. Wei and W. G. Stetler-Stevenson, "TIMP-2 Mediated Inhibition of Angiogenesis: An MMP-Independent Mechanism," *Cell*, Vol. 114, No. 2, 200, pp. 171-180. doi:10.1016/S0092-8674(03)00551-8

[21] M. Dews, A. Homayouni, D. Yu, D. Murphy, C. Sevignani, E. Wentzel, E. E. Furth, W. M. Lee, G. H. Enders, J. T. Mendell and A. Thomas-Tikhonenko, "Augmentation of Tumor Angiogenesis by a Myc-Activated MicroRNA Cluster," *Nature Genetics*, Vol. 38, 2006, pp. 1060-1065. doi:10.1038/ng1855

[22] D. Wu, L. Li, M Yang, H. Liu and G. Yang, "Elevated Plasma Levels of SPARC in Patients with Newly Diagnosed Type 2 Diabetes Mellitus," *European Journal of Endocrinology*, Vol. 165, 2011, pp. 597-601. doi:10.1530/EJE-11-0131

[23] K. Kos, S. Wong, B. Tan, A. Gummesson, M. Jernas, N. Franck, D. Kerrigan, F. H. Nystrom, L. M. Carlsson, H. S. Randeva, J. H. Pinkney and J. P. Wilding, "Regulation of the Fibrosis and Angiogenesis Promoter SPARC/Osteonectin in Human Adipose Tissue by Weight Change, Leptin, Insulin, and Glucose," *Diabetes*, Vol. 58, No. 8, 2009, pp. 1780-1788. doi:10.2337/db09-0211

[24] P. S. Zheng, J. Wen, L. C. Ang, W. Sheng, A. Viloria-Petit, Y. Wang, Y. Wu, R. S. Kerbel and B. B. Yang, "Versican/PG-M G3 Domain Promotes Tumor Growth and Angiogenesis," *FASEB Journal*, Vol. 18, 2004, pp. 754-756,. doi:10.1096/fj.03-0545fje

[25] M. Rahmani, B. W. Wong, L. Ang, C. C. Cheung, J. M. Carthy, H. Walinski and B. M. McManus, "Versican: Signaling to Transcriptional Control Pathways," *Canadian Journal of Physiology and Pharmacology*, Vol. 84, 2006, pp. 77-92. doi:10.1139/y05-154

[26] C. G. Rivera, J. S. Bader and A. S. Popel, "Angiogenesis-Associated Crosstalk between Collagens, CXC Chemokines, and Thrombospondin Domain-Containing Proteins," *Annals of Biomedical Engineering*, Vol. 39, No. 8, 2011, pp. 2213-2222. doi:10.1007/s10439-011-0325-2

[27] J. S. Annicotte, E. Blanchet, C. Chavey, I. Iankova, S. Costes, S. Assou, J. Teyssier, S. Dalle, C. Sardet and L. Fajas, "The CDK4-pRB-E2F1 Pathway Controls Insulin Secretion," *NatCellBiol*, Vol. 11, 2009, pp. 1017-1023. doi:10.1038/ncb1915

[28] A. Lombardi, L. Ulianich, A. S. Treglia, C. Nigro, L. Parrillo, D. D. Lofrumento, G. Nicolardi, C. Garbi, F. Beguinot, C. Miele and B. Di Jeso, "Increased Hexosamine Biosynthetic Pathway Flux Dedifferentiates INS-1E Cells and Murine Islets by an Extracellular Signal-Regulated Kinase (ERK)1/2-Mediated Signal Transmission Pathway," *Diabetologia*, Vol. 55, No. 1, 2012, pp. 141-153. doi:10.1007/s00125-011-2315-1

[29] U. Ozcan, Q. Cao, E. Yilmaz, A. H. Lee, N. N. Iwakoshi, E. Ozdelen, G. Tuncman, C. Gorgun, L. H. Glimcher and G. S. Hotamisligil, "Endoplasmic Reticulum Stress Links Obesity, Insulin Action, and Type 2 Diabetes," *Science*, Vol. 306, No. 5695, 2004, pp. 457-461. doi:10.1126/science.1103160

[30] B. Meissburger, J. Ukropec, E. Roeder, N. Beaton, M. Geiger, D. Teupser, B. Civan, W. Langhans, P. P. Nawroth, D. Gasperikova, G. Rudofsky and C. Wolfrum, "Adipogenesis and Insulin Sensitivity in Obesity Are Regulated by Retinoid-Related Orphan Receptor Gamma," *EMBO Molecular Medicine*, Vol. 3, 2011, pp. 637-651. doi:10.1002/emmm.201100172

[31] S. W. Park, Y. Zhou, J. Lee, J. Lee and U. Ozcan, "Sarco(endo)plasmic Reticulum Ca^{2+}-ATPase 2b Is a Major Regulator of Endoplasmic Reticulum Stress and Glucose Homeostasis in Obesity," *Proceedings of the National Academy of Sciences of USA*, Vol. 107, 2010, pp. 19320-19325. doi:10.1073/pnas.1012044107

[32] Y. Zhou, J. Lee, C. M. Reno, C. Sun, S. W. Park, J. Chung, J. Lee, S. J. Fisher, M. F. White and S. B. Biddinger, U. Ozcan, "Regulation of Glucose Homeostasis through a XBP-1-FoxO1 Interaction," *Nature Medicine*, Vol. 17, 2011, pp. 356-365. doi:10.1038/nm.2293

[33] G. P. Nagaraju and D. Sharma, "Anti-Cancer Role of SPARC, an Inhibitor of Adipogenesis," *Cancer Treatment Reviews*, Vol. 37, No. 7, 2011, pp. 559-566. doi:10.1016/j.ctrv.2010.12.001

[34] A. D. Bradshaw, "Diverse Biological Functions of the SPARC Family of Proteins," *The International Journal of Biochemistry & Cell Biology*, Vol. 44, No. 3, 2012, pp. 480-488. doi:10.1016/j.biocel.2011.12.021

[35] M. Rahmani, J. M. Carthy and B M. McManus, "Mapping of the Wnt/β-Catenin/TCF Response Elements in the Human Versican Promoter," *Methods in Molecular Biology*, Vol. 836, 2012, pp. 35-52. doi:10.1007/978-1-61779-498-8_3

[36] E. Blanchet, J. S. Annicotte, S. Lagarrigue, V. Aguilar, C. Clapé, C. Chavey, V. Fritz, F. Casas, F. Apparailly, J. Auwerx and L. Fajas, "E2F Transcription Factor-1 Regulates Oxidative Metabolism," *Nature Cell Biology*, Vol. 13, 2011, pp. 1146-1152. doi:10.1038/ncb2309

[37] R. J. Shaw, "LKB1 and AMP-Activated Protein Kinase Control of mTOR Signalling and Growth," *Acta Physiologica*, Vol. 196, 2009, pp. 65-80. doi:10.1111/j.1748-1716.2009.01972.x

[38] K. Tsuchihara, T. Ogura, R. Fujioka, S. Fujii, W. Kuga, M. Saito, T. Ochiya, A. Ochiai and H. Esumi, "Susceptibility of Snark-Deficient Mice to Azoxymethane Induced Colorectal Tumorigenesis and the Formation of Aberrant Crypt Foci," *Cancer Science*, Vol. 99, No. 4, 2008, pp. 677-682. doi:10.1111/j.1349-7006.2008.00734.x

[39] A. S. John, X. Hu, V. L. Rothman and G. P. Tuszynski, "Thrombospondin-1 (TSP-1) Up-Regulates Tissue Inhibitor of Metalloproteinase-1 (TIMP-1) Production in Human Tumor Cells: Exploring the Functional Significance in Tumor Cell Invasion," *Experimental and Molecular Pathology*, Vol. 87, No. 3, 2009, pp. 184-188. doi:10.1016/j.yexmp.2009.09.002

[40] D. Hose, J. Moreaux, T. Meissner, A. Seckinger, H. Goldschmidt, A. Benner, K. Mahtouk, J. Hillengass, T. Rème, J. De Vos, M. Hundemer, M. Condomines, U. Bertsch, J. F. Rossi, A. Jauch, B. Klein and T. Möhler, "Induction of Angiogenesis by Normal and Malignant Plasma Cells," *Blood*, Vol. 114, No. 1, 2009, pp. 128-143. doi:10.1182/blood-2008-10-184226

Alteration of Major Insulin Signaling Molecules by Chronic Ethanol in Hypertensive Vascular Smooth Muscle Cells of Rats

Sparkle D. Williams, Benny Washington
Department of Biological Sciences, Tennessee State University, Nashville, USA
Email: bwashington@tnstate.edu

ABSTRACT

Insulin resistance is an important risk factor in the development of cardiovascular diseases such as hypertension and atherosclerosis. However, despite its importance, the specific role of insulin resistance in the etiology of these diseases is poorly understood. At the same time, ethanol (ETOH) is a potent vasoconstrictor that primarily induces down regulation of mitogen activated protein kinases (MAPKs) which could exacerbate insulin resistance and possibly lead to cardiovascular diseases. This article describes how chronic ETOH exposure interferes with insulin signaling in hypertensive vascular smooth muscle cells (HVSMCs) which leads to the alteration of MAPKs, the major signaling molecules. Elevated (50 - 800 mM) chronic exposure (24 hr) of HVSMCS to ETOH prior to insulin stimulation decreased insulin-induced ERK 1/2 (MAPKs) and AKT expression. Similar experiments were conducted using normotensive cells from rat. These cells showed reductions in insulin-induced ERK 1/2 phosphorylation as well, but only at higher concentrations of ETOH (400 - 800 mM). These alterations in insulin signaling could provide an alternative molecular mechanism that may increase the risk of insulin resistance, thus increasing the possibility of cardiovascular diseases.

Keywords: HVSMCs; Ethanol; Insulin; ERK 1/2; AKT

1. Introduction

Insulin resistance can lead to the development of cardiovascular diseases such as hypertension and atherosclerosis [1-3]. However, there is little information known about the role of insulin resistance in cardiovascular diseases. As a result of chronic alcohol consumption, insulin resistance can occur. Insulin release and activation is important in the body to alleviate the outcome of diabetes and in some cases, hypertension [4,5]. Chronic ethanol (ETOH) consumption can also affect insulin's ability to bind to its receptor, therefore, leading to insulin resistance [4,5]. However, there are not many studies to prove this effect, especially in a hypertensive phenotype. Studies have shown that moderate consumption of alcohol is not a precursor to coronary artery disease, but chronic consumption of alcohol can be detrimental [5].

Insulin signaling is important in the cell for the release of insulin to control glucose intake and lipid metabolism [5]. Insulin has an important role in the biological process because it binds to its receptor (insulin receptor) to initiate glucose metabolism [5]. The insulin receptor is a transmembrane domain spanning tyrosine kinase receptor composed of alpha and beta subunits that mediate the actions of insulin [6-8]. Insulin is closely associated with hypertension, non-insulin dependent diabetes, atherosclerosis, and dyslipidemia [6,7]. As a result, insulin activation of PI3K-AKT (Phosphoinositide 3-Kinase), which mediates neuronal survival, motility, energy metabolism, and plasticity is impaired [6,8,9].

The extracellular signal-regulated kinases (ERKs), also known as extracellular mitogen activated protein kinases (MAPKs), have been found to be altered by ethanol treatment of vascular smooth muscle cells from the aorta of a rat [5,9-11]. This effect of ETOH has been shown to be manifested via several pathways by the use of signaling inhibitors, such as PKC, leading to a cascading effect in treated cells [11,12]. Several studies have shown that insulin activates a complex set of intracellular responses, including the activation of mitogen-activated protein kinases ERK 1/2 [13] and AKT [5].

The normal role of AKT in the cell is to propagate insulin receptor signaling to downstream effectors [14]. AKT is downstream of PI3K, both of which are a part of one of the major pathways in insulin signaling. The other major pathway is the mitogen activated protein kinase pathway. It is also known that AKT has a role in activating insulin response [10]. Alteration of AKT expression is exhibited by ETOH impaired insulin signaling in the body and a decrease in glucose transport of rat cardiac muscle cells [10]. Although insulin response can be altered with chronic ETOH, there are other factors that can alter this response such as genetics and the person's environment [5].

Chronic ETOH consumption in experimental animal models has been shown to alter insulin signaling events via the mitogen activated protein kinases producing insulin resistance in the liver. These adverse effects of ethanol have been shown to be the result of the inhibition of insulin or insulin-like growth factor which alters mRNA and DNA synthesis and the activation of proapoptotic signals through PI3K and AKT [15-17]. Results from previous studies suggest that ETOH impairment of insulin action is likely to be downstream from PI3K, however, the mechanisms underlying the effects of ETOH on insulin resistance and the effect of insulin resistance on the development of cardiovascular diseases remain to be determined [5,18]. This paper describes how chronic ETOH exposure can alter insulin signaling in HVSMCs using mitogen protein kinases as indicators. Chronic ETOH exposure's effect on insulin signaling has not been analyzed before in hypertensive cells.

2. Materials and Methods

2.1. Reagents and Antibodies

ERK 1/2 (p44/p42) and AKT antibodies were purchased from Cell Signaling (Beverly, MA); anti-rabbit IgG antibodies (horseradish peroxidase linked) from Amersham Bioscience (Piscataway, NJ); and ECL detection system was obtained from Pierce Biotechnology (Rockford, IL). Other supplies include Dulbecco's Modified Eagle's Medium (Amersham), fetal bovine calf serum, penicillin and streptomycin were purchased from Sigma/Aldrich (St. Louis, MO).

2.2. Cell Culture

Vascular smooth muscle cells (VSMCs) were received from Vanderbilt University. Cells were cultured in DMEM containing 10% fetal bovine serum, 2% penicillin and streptomycin. Subcultured passages were between 3 and 12. Cells were maintained at a pH of 7.1 in 75 cm^2 flasks under a humidified atmosphere of 5% CO_2, 95% O_2 at 37°C and plated in 6-well falcon plates.

2.3. Insulin Treatment

For dose response experiments, cells were stimulated with 1, 2, 4, 8, and 16 μM insulin for 30 min. Insulin was aspirated from each well. For time course experiments, cells were stimulated with 8 μM of insulin for 1, 5 10, 20, and 40 min. Cells were lysed with 300 - 500 μl Laemmli Sample Buffer (2% SDS, 25% glycerol, 0.01% bromophenol blue, and 62.5 mM Tris-HCl pH 6.8). Cells were then scraped from the monolayer surface and collected in microcentrifuge tubes.

2.4. Ethanol Treatment

Hypertensive and normal rat cells were induced with 50, 100, 200, 400, and 800 mM ETOH. Control cells were induced with a DMEM (-) solution containing no serum that aids in cell proliferation. After a 24 hr incubation period, ETOH was aspirated from all wells and cells (except controls) were stimulated with 8 μM of insulin. Insulin was aspirated from each well and the cells were lysed with 300 - 500 μl Laemmli Sample Buffer (2% SDS, 25% glycerol, 0.01% bromophenol blue, and 62.5 mM Tris-HCl pH 6.8). Cells were then scraped from the monolayer surface and collected in microcentrifuge tubes.

2.5. Western Blot Analysis

Whole cell lysates were collected and diluted with sample buffer to equal concentrations of 40 μg/20 μl. Lowry protein assay was conducted to determine standard protein concentration. Protein samples were then separated along with rainbow markers to measure the molecular weight of proteins on a 10% SDS-polyacrylamide gel from Bio-Rad Laboratories (Hercules, CA) at 200 volts for approximately 50 min. Proteins were then transferred to a nitrocellulose membrane from Amersham Biosciences (Piscataway, NJ) using a semi-dry transfer apparatus at 10 volts for 90 min. Blots were blocked with 2% non-fat dry milk in TBS (Tris-Buffered Saline) for at least one hour. After which blots were incubated with primary antibodies ERK 1/2 or AKT overnight followed by anti-rabbit secondary antibody for 1 hour. Blots were immersed in chemiluminscent solution and developed in a dark room.

2.6. Statistical Analysis

All experiments were performed in triplicate and expressed as means ± SE of the density using arbitrary units from three individual experiments. Statistical significance was determined with paired or unpaired one-tailed Student's t-test, with $P < 0.05$ considered significant.

3. Results

3.1. Insulin Induction Increases ERK 1/2 and AKT Expression in Hypertensive VSMCs

Before we could determine the effect of ETOH on insulin signaling, we first had to determine the maximum concentration of insulin it would take to stimulate phosphorylation of ERK 1/2 and AKT via Western Blotting analysis. HVSMCs were stimulated with a concentration range of 1 - 16 μM of insulin for 30 min. **Figure 1** denotes that stimulating HVSMCs with 1 - 16 μM of insulin, increased ERK 1/2 phosphorylation by approximately 23%. In addition, this insulin stimulation increased in AKT expression 33% above basal with a maximum expression detected with 8 μM of insulin (**Figure 2**). This data suggests that AKT activation in HVSMCs is more sensitive than ERK 1/2 to insulin signaling.

3.2. Insulin Induction Increases ERK 1/2 and AKT Expression in Normal VSMCs

In order to determine the effect of insulin signaling in normal cells, VSMCs were stimulated with a concentration range of 1 - 16 μM of insulin. Maximal expression of phosphorylated ERK 1/2 occurred with 8 μM of insulin, which was significantly different from basal (**Figure 2**). No significant increases in ERK 1/2 were observed between 1 - 4 mM (**Figure 2**). Similar experiments showed that AKT expression was significantly increased throughout all concentrations except 1 μM with maximal expression at 8 μM. Insulin seemingly, also induced increases in AKT with 1 - 8 μM in normal VSMCs (**Figure 2**). This increase in expression of AKT was not observed when cells were stimulated with 16 μM insulin which resulted in complete inhibition.

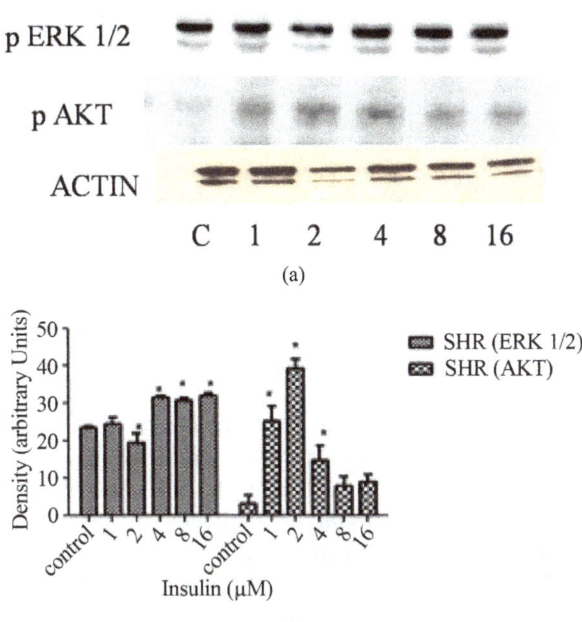

Figure 1. Dose response curve for insulin on ERK 1/2 and AKT expression in HVSMCs. Cells were stimulated with 1, 2, 4, 8, and 16 μM insulin for 30 min and lysate harvested. (a) Western Blot Analysis of SHR (Spontaneously Hypertensive) lysate probed with antibodies for ERK 1/2 and AKT expression; (b) Graphical representation of data by densitometry analysis software was taken from a mean of three experiments $p < 0.05$ compared to control.

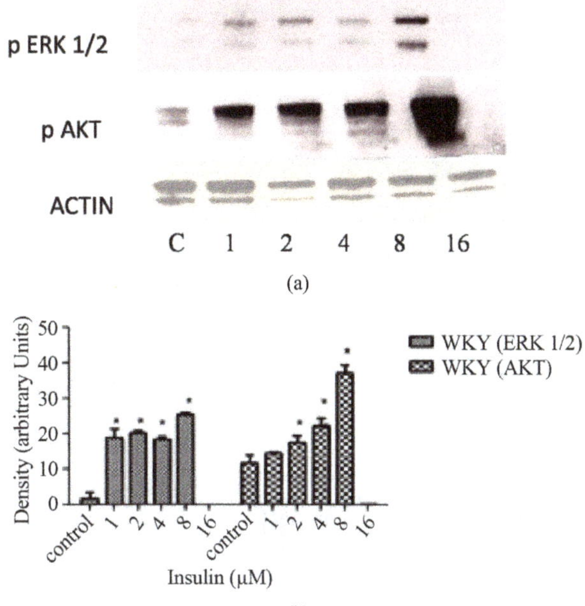

Figure 2. Dose response curve for insulin on ERK 1/2 and AKT expression in normal VSMCs. Cells were stimulated with 1, 2, 4, 8, and 16 μM insulin for 30 min and lysate harvested. (a) Western Blot Analysis of WKY (Wistar Kyoto) lysate probed with antibodies for ERK 1/2 and AKT expression; (b) Graphical representation of data by densitometry analysis software was taken from a mean of three experiments $p < 0.05$ compared to control.

3.3. Chronic ETOH Impairs Insulin Signaling in Hypertensive VSMCs

In order to investigate whether ETOH alters insulin signaling in HVSMCs, cells were treated chronically (24 hrs) with 50 - 800 mM of ETOH. After ETOH treatment HVSMCs were stimulated with 8 μM insulin for 30 min and Western Blotting was performed. As a result, ERK 1/2 expression significantly decreased with 50 - 800 mM ETOH treatment by 10% compared to insulin stimulation only (**Figure 3**). Using the same treatment range (50 - 800 mM) AKT expression was evaluated by chronic ETOH treatment. After 24 hours ETOH treatment, AKT expression HVSMCs significantly decreased (**Figure 3**). This decrease was gradual as the concentration increased.

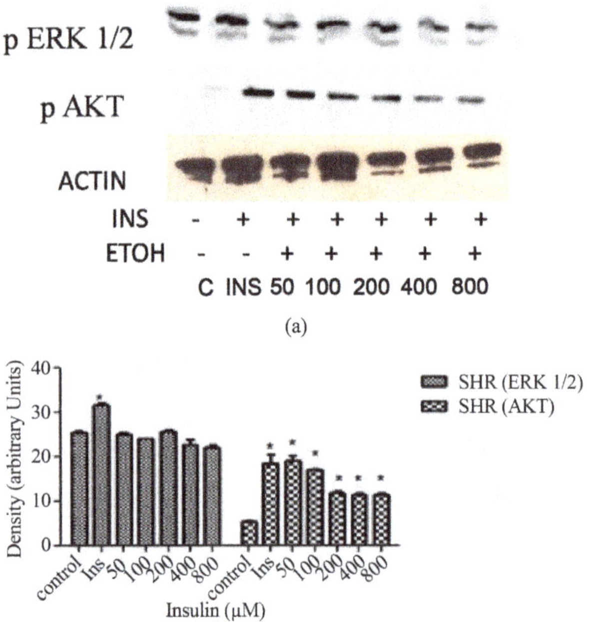

Figure 3. Twenty-four hour ETOH treatment of HVSMCs. Cells were treated with specified concentrations of ETOH (50 - 800 mM) and stimulated with 8 μM insulin for 30 min and lysate collected. (a) Western Blot Analysis of SHR (Spontaneously Hypertensive) lysate probed with antibodies for ERK 1/2 and AKT expression; (b) Graphical representation of data by densitometry analysis software was taken from a mean of three experiments $p < 0.05$ compared to control.

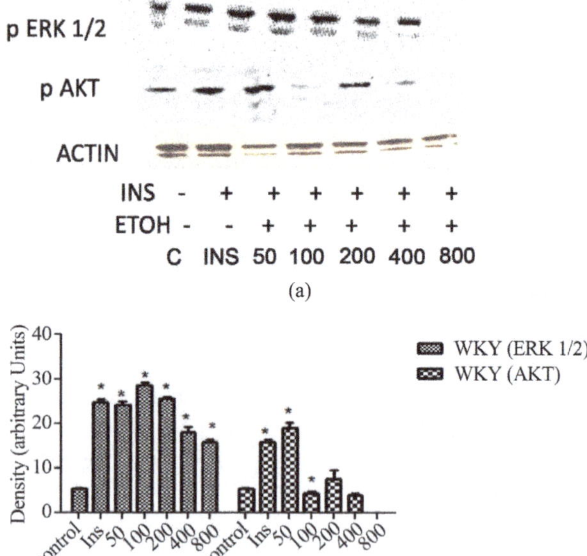

Figure 4. Twenty-four hour ETOH treatment of normal VSMCs. Cells were treated with specified concentrations of ETOH (50 - 800 mM) and stimulated with 8 μM insulin for 30 min and lysate collected. (a) Western Blot Analysis of WKY (Wistar Kyoto) lysate probed with antibodies for ERK 1/2 and AKT expression; (b) Graphical representation of data by densitometry analysis software was taken from a mean of three experiments $p < 0.05$ compared to control.

3.4. Chronic ETOH Impairs Insulin Signaling in Normal VSMCs

In order to investigate whether ETOH alters insulin signaling in normal VSMCs, we evaluated chronic ETOH treatment in them as well. Cells were treated chronically with 50 - 800 mM of ETOH for 24 hr. After ETOH treatment, VSMCs were stimulated with 8 μM insulin for 30 min. Fifty and 100 mM of ETOH had no significant effect on insulin-induced AKT expression when compared to insulin stimulation alone. However, 200 - 800 mM reduced insulin-induced AKT expression approximately 20%. Chronic ETOH treatment of normal VSMCs seems to cause an increase in signal at 50 mM and a decrease at higher concentrations when compared to the insulin induced increase alone (**Figure 4**).

4. Discussion

The research findings in this paper provide evidence for changes in mitogen-activated protein kinases possibly contributing to the onset of cardiovascular disease. The focus of this study is to show the association of chronic ethanol-induced changes in ERK 1/2 and AKT expression in HVSMCs. Recent reports show that hypertensive persons are predisposed to the development of diabetes. [19]. Seventy-five percent of cardiovascular diseases are attributed to diabetes, hypertension, and alcohol, indicating that this is a contributing factor to the onset of hypertension.

In **Figure 1**, ERK 1/2 phosphorylation was increased in hypertensive cells with increasing concentrations of insulin. This confirms that cells are able to survive with higher concentrations of insulin indicated by ERK 1/2. On the other hand, when AKT stimulation occurs, it has been reported that cellular expression decreases. This could be due to the alteration of insulin signaling. In **Figure 2**, normal VSMCs with ERK 1/2 and AKT decreased with 1 - 8 μM insulin. With 16 μM of insulin, both ERK 1/2 and AKT expression was inhibited. In **Figure 3**, hypertensive cells treated with ethanol ERK 1/2 and AKT both decrease in expression as the concentrations of ETOH increase which could mean that ethanol is affecting insulin signaling. **Figure 4** depicts a biphasic effect of ETOH on insulin signaling in normal cells which suggest that there may be different effect in normal cells.

Cardiovascular diseases include atherosclerosis and hypertension. Risk factors for such diseases of the cardiovascular system include family history, diabetes, obesity, smoking, excessive alcohol intake, and a diet high in salt and/or low in antioxidant nutrients. Individuals with hypertension are at increased risk for atherosclerotic diseases such as stroke, heart, and kidney disease which can be exacerbated by diabetes and alcohol [20]. The adverse effects of long-term excessive use of alcohol are similar to those seen with other sedative-hypnotics drugs (apart from organ toxicity which is much more problematic with alcohol).

Though the underlying mechanisms remain undefined, accumulating evidence strongly suggests that ETOH interferes with insulin's action by altering mitogen activated protein kinases. The major signaling molecules, MAPKs, implicated in the biological actions of insulin, and the expression of the insulin receptor may be major factors leading to cardiovascular diseases.

5. Acknowledgements

The authors would like to acknowledge The Research and Engineering Training Program (REAP) and Title III Program for providing support for this work.

REFERENCES

[1] P. De Meyts, "The Structural Basis of Insulin and Insulin-Like Growth Factor-I Receptor Binding and Negative Cooperativity, and Its Relevance to Mitogenic versus Metabolic Signaling," *Diabetologia*, Vol. 37, Suppl. 2, 1994, pp. S135-S148. http://dx.doi.org/10.1007/BF00400837

[2] E. Conti, F. Andreotti, A. Sciahbasi, P. Riccardi, G. Marra, E. Menini, *et al.*, "Markedly Reduced Insulin-Like Growth Factor-1 in the Acute Phase of Myocardial Infarction," *Journal of the American College of Cardiology*, Vol. 38, No. 1, 2001, pp. 26-32. http://dx.doi.org/10.1016/S0735-1097(01)01367-5

[3] A. Juul, T. Scheike, M. Davidsen, J. Gyllenborg and T. Jorgenson, "Low Serum Insulin-Like Growth Factor 1 Is Associated with Increased Risk of Ischemic Heart Disease: A Population-Based Case Control Study," *Circulation*, Vol. 106, No. 8, 2002, pp. 939-944. http://dx.doi.org/10.1161/01.CIR.0000027563.44593.CC

[4] L. Macho, S. Zorad, Z. Radikova, P. Patterson-Buckedahl and R. Kvetnansky, "Ethanol Consumption Affects Stress Response and Insulin Binding Tissue of Rats Endocrine Regulations," *Endocrine Regulation*, Vol. 37, No. 4, 2003, pp. 195-202.

[5] A. Natali, S. Vichi, P. Landi, S. Severi, A. L'Abbate and E. Ferrannini, "Coronary Atherosclerosis in Type II Diabetes: Angiographic Findings and Clinical Outcome," *Diabetologia*, Vol. 43, No. 5, 2000, pp. 632-641. http://dx.doi.org/10.1007/s001250051352

[6] J. E. P. Brown, D. Onyango, and S. J. Dunmore, "Resistin Down-Regulates Insulin Receptor Expression, and Modulates Cell Viability in Rodent Pancreatic Beta-Cells," *FEBS Letters*, Vol. 581, No. 17, 2007, pp. 3273-3276. http://dx.doi.org/10.1016/j.febslet.2007.06.031

[7] L. He, J. C. Marecki, G. Serrero, F. A. Simmen, M. Ronis and T. Badger, "Dose-Dependent Effects of Alcohol on Insulin Signaling: Partial Explanation for Biphasic Alcohol Impact on Human Health," *Molecular Endocrinology*, Vol. 21, No. 21, 2007, pp. 2541-2550. http://dx.doi.org/10.1210/me.2007-0036

[8] T. Limin, X. Hou, J. Liu, X. Zhang, N. Sun, L. Gao, *et al.*, "Chronic Ethanol Consumption Resulting in the Down-regulation of Insulin Receptor-Subunit, Insulin Receptor Substrate-1, and Glucose Transporter 4 Expression in Rat Cardiac Muscles," *Alcohol*, Vol. 43, 2009, pp. 51-58. http://dx.doi.org/10.1016/j.alcohol.2008.11.001

[9] B. Washington, C. Mtshali, S. Williams, H. Smith, J. D. Li, B. Shaw, *et al.*, "Ethanol-Induced Mitogen Activated Protein Kinase Activity Mediated through Protein Kinase C," *Cell and Molecular Biology*, Vol. 49, No. 8, 2003, pp. 1351-1356.

[10] A. L. Johnson, G. D. Goode, C. Mtshali, E. L. Myles and B. Washington, "Protein Kinase C-Alpha/BetaII, Delta, and Zeta/Lambda Involvement in Ethanol-Induced MAPK Expression in Vascular Smooth Muscle Cells," *Cell and Molecular Biology*, Vol. 53, No. 4, 2007, pp. 38-44.

[11] H. Wang, S. Doronin and C. C. Malbon, "Insulin Activation of Mitogen-Activated Protein Kinases Erk1,2 Is Amplified via Beta-Adrenergic Receptor Expression and Requires the Integrity of the Tyr350 of the Receptor," *The Journal of Biological Chemistry*, Vol. 275, No. 46, 2000, pp. 36086-26093. http://dx.doi.org/10.1074/jbc.M004404200

[12] S. Vasdev, V. Gill and P. Singal, "Beneficial Effect of Low Ethanol Intake on the Cardiovascular System: Possible Biochemical Mechanisms," *Journal of Vascular Health and Risk Management*, Vol. 2, No. 3, 2006, pp. 263-276. http://dx.doi.org/10.2147/vhrm.2006.2.3.263

[13] S. Gitlow, "Substance Use Disorders: A Practical Guide," 2nd Edition, Lippincott Williams and Wilkins, Philadelphia, 2006, pp. 52, 103-121.

[14] B. Taylor, J. Rehm and G. Gmel, "Moderate Alcohol Consumption and the Gastrointestinal Tract," *Digital Distribution*, Vol. 23, No. 3-4, 2005, pp. 170-176.

[15] A. E. Seiler, B. N. Ross and R. Rubin, "Inhibition of Insulin-Like Growth Factor-1 Receptor and IRS Signaling by Ethanol in SH-SY5Y Neuroblastoma Cells," *Journal of Neurochemistry*, Vol. 76, No. 2, 2001, pp. 573-581. http://dx.doi.org/10.1046/j.1471-4159.2001.00025.x

[16] D. T. Furuya, R. Binsack and U. F. Machado, "Low Ethanol Consumption Increases Insulin Sensitivity in Wistar Rats," *Brazilian Journal of Medical and Biological Research*, Vol. 36, No. 1, 2003, pp. 125-130. http://dx.doi.org/10.1590/S0100-879X2003000100017

[17] J. R. Sowers, M. Epstein, E. D. Frohlich and N. R. Campbell, "Management of Hypertension for People with Diabetes," *Clinical Summary*, 2011, pp. 1-52.

[18] L. Hansen, Y. Ikeda, G. Olsen, A. Busch and L. Mosthaf, "Insulin Signaling Is Inhibited by Micromolar Concentrations of H_2O_2," *The Journal of Biological Chemistry*, Vol. 274, No. 35, 1999, pp. 25078-25084. http://dx.doi.org/10.1074/jbc.274.35.25078

[19] E. Motley, K. Eguchi, C. Gardner, A. Hicks, C. Reynolds,

G. Frank, *et al.*, "Insulin-Induced Akt Activation Is Inhibited by Angiotensin II in the Vasculature through Protein Kinase C," *Hypertension*, Vol. 41, No. 3, 2002, pp. 775-780. http://dx.doi.org/10.1161/01.HYP.0000051891.90321.12

[20] Y. Sasaki, N. Hayashi, T. Ito, H. Fusamoto, T. Kamada and J. R. Wands, "Influence of Ethanol on Insulin Receptor Substrate-1-Mediated Signal Transduction during Rat Liver Regeneration," *Alcoholism Supplements*, Vol. 29, No. 1, 1994, pp. 99-106.

22

Immunohistochemical Analysis of the Acid Secretion Potency in Gastric Parietal Cells

Rie Irie-Maezono[1], Shinichiro Tsuyama[2]
[1]Department of Gene Therapy and Regenerative Medicine, Kagoshima University, Graduate School of Medical and Dental Sciences, Kagoshima, Japan
[2]Laboratory for Neuroanatomy, Kagoshima University, Graduate School of Medical and Dental Sciences, Kagoshima, Japan
Email: maezono@m.kufm.kagoshima-u.ac.jp

ABSTRACT

Gastric parietal cells are important in acid secretion, but it is unclear which cells throughout the gastric gland have the highest secretion potency. Here, we used immunohistochemical methods with anti-H^+, K^+-ATPase, phosphoryl ezrin and CD44 antibodies to study the distribution of gastric acid secretion activity. Stomach tissues from freely fed and starved rats were cryofixed for light microscopy or fixed by high-pressure freezing for electron microscopy. Parietal cells from freely fed animals corresponded to the active secretion phase and to the inactive resting phase from starved rats. Anti-H^+, K^+-ATPase and anti-phosphoryl ezrin labeling were observed on the membrane of the intracellular canaliculi and the tubulovesicle from freely fed rats, while cells from starved animals showed weak labeling with anti-phosphoryl ezrin antibody staining. Morphometrical analysis at the electron microscopic level was performed on active and inactive acid secretory phases between the upper and base regions of the gland. H^+, K^+-ATPase and CD44 were distributed on both sites of the microvillous and tubulovesicle membrane in the same cells, but phosphoryl ezrin localized predominantly on the microvillous membrane in active cells of the glandular neck and upper base. Therefore, the highest secreting potency appeared to be in cells of the glandular neck and upper base.

Keywords: Gastric Parietal Cells; Secretory Potency; Phosphoryl Ezrin; Histochemical Morphometry

1. Introduction

Gastric parietal cells play a major role in acid secretion and are widely distributed from the pit to the base of rat gastric glands. They show characteristic aspects of intracellular canaliculi (IC) with numerous microvilli and tubulovesicles (TV) in the cytoplasm, which are thought to be interconvertable structures. Although the conversion mechanism for these structures is unclear, various hypothesizes have been proposed. During the active acid-secreting phase of parietal cells, the IC is markedly expanded, but the cells undergo a morphological transformation during their inactive resting phase when the IC reduces in width and the TV mass increases [1,2]. The secretion activity alternates according to the physiological phases of feeding or starving.

Proton potassium ATPase (H^+, K^+-ATPase; "the proton pump") is an important enzyme for gastric acid secretion and exists as an integral membrane-protein along the IC and TV throughout the parietal cell membrane. We previously used high-pressure freezing followed by freeze-substitution to investigate the histochemistry of gastric gland cells and the ultrastructural alterations that occur in both fed and starved phases [3,4]. Cryofixation using rapid freezing (especially high pressure rapid freezing for capable freezing depth) is believed to be superior to conventional chemical fixation with regard to morphological preservation and retention of soluble components. Using antibodies against the proton potassium ATPase α- and β-subunits, we also showed that the enzyme localized on both IC and TV membranes in almost all parietal cells throughout the length of the gland [3-5].

Parietal cells contain more actin than other glandular cells. Transformation between IC and TV occurs with redistribution of actin in the cell. Filamentous actins are anchored to the plasma membrane via phosphoryl ezrin, and most actin molecules are thought to form a globular structure in the inactive resting state, which molecules polymerize rapidly to form a filamentous structure upon active acid secretion [2,6-11]. Ezrin is a member of the ERM (ezrin/radixin/moesin) family of proteins that is implicated in linking functional activities of the plasma membrane to the actin cytoskeleton. In addition, actin binds to intramembranous CD44 via phosphoryl ezrin in the plasma membrane [12-14]. It has previously been suggested that the above-mentioned morphological changes are induced and triggered by this cytoskeletal reorganization of β-actin [15-17].

The purpose of the present study was to investigate which parietal cells are more active than others in terms of acid secretion, based on the distribution of phosphoryl ezrin and CD44 throughout the gland using immunohistochemical techniques.

2. Materials and Methods

2.1. Tissue Preparation

Ten male Wistar rats were used in the experiments and divided into two groups of five animals each. One group was fed freely and the other was starved for 48 h with free access to water. The rats were anesthetized with an intraperitoneal injection of sodium pentobarbital, and the pH of the luminal gastric juice was determined. The stomach tissues were cut into small pieces and cryofixed using a rapid freezing device (RF-6, Eiko, Japan) using liquid propane for cryofixing and liquid nitrogen for light microscopy. The specimens were then freeze-substituted with acetone containing 0.2% glutaraldehyde at −79°C for 72 h and embedded in paraffin [17].

For electron microscopy, the specimens underwent high-pressure freezing under a 21×10^5 hPa atmosphere (HPM 010, BAL-TEC, Liechtenstein). The frozen specimens were freeze-substituted with acetone containing 1% osmium tetroxide or 0.2% glutaraldehyde at −79°C for 72 h and were embedded in Epon812 or Lowicryl K4M resin, respectively [3].

2.2. Primary and Secondary Antibodies

A rabbit antibody against the H^+, K^+-ATPase (proton pump) α-subunit (Immunogen; C-terminal synthetic peptide based on the porcine H^+, K^+-ATPase α-subunit sequence) was purchased from Calbiochem-Novabiochem (San Diego, CA). The antibody was used at a dilution of 1:100 in phosphate buffered saline (PBS). A mouse monoclonal antibody against the H^+, K^+-ATPase β-subunit (Immunogen; purified 34-kDa core peptide from deglycosylated hog gastric microsomes) was purchased from Abcam Ltd. (Cambridge, UK) and diluted 1:200 in PBS. A rabbit polyclonal antibody against phosphoryl ezrin (Immunogen: KLH-conjugated, synthetic phosphopeptide corresponding to residues surrounding Thr567 of human ezrin) and an anti-CD44 antibody were purchased from CHEMICON International, Inc. (Temecula, CA). These were used at a dilution of 1: 10 - 20 in PBS. A mouse monoclonal antibody against actin (pan Ab-5; Clone ACTN05) was purchased from LAB VISION Co. (Fremont, CA). This antibody reacts with all six known isoforms of vertebrate actin (MW-42 kD) and also with two highly homologous cytoplasmic actins (β, γ). This antibody was diluted 1: 10 - 80 in PBS. The antibodies were confirmed to show cross-reactivity against the rat. The following were used as secondary antibodies: biotinylated goat anti-mouse immunoglobulin ($F(ab')_2$) or biotinylated swine anti-rabbit immunoglobulin (DAKO Cytomation, Glostrup, Denmark) diluted 1: 100 - 200 in PBS; horseradish peroxidase (HRP)-conjugated streptavidin (DAKO Cytomation) (1:200 in Tris-buffered-saline; TBS); colloidal gold (CG)-conjugated streptavidin (1:1 in PBS) from British BioCell International (Cardiff, UK).

2.3. Immunohistochemical Staining for Light Microscopy

Specimens embedded in paraffin were cut into 4 μm-thick slices with a sliding Jung-model microtome, mounted on silconized glass slides and air-dried. Sections were deparaffinized, rehydrated and immersed in PBS. After blocking endogenous peroxidase activity with 0.3% hydrogen peroxide in methanol, specimens were incubated with primary antibodies against the H^+, K^+-ATPase α- and β-subunits. They were then labeled with biotinylated anti-rabbit or anti-mouse IgG antibodies overnight followed by HRP-conjugated streptavidin for 1 h. Visualization was performed using 3, 3'-diaminobenzidine tetrahydrochloride (DAB; DAKO Cytomation) for 10 min. Finally, sections were rinsed in distilled water, counterstained with Mayer's hematoxylin, dehydrated in a graded ethanol series, cleared in xylene and mounted with Eukitt (O. Kindler, Germany).

2.4. Electron Microscopy

Ultrathin sections from specimens fixed with 1% osmium tetroxide and embedded in epoxy resin were cut with a Reichert Ultracat-N ultramicrotome and stained with uranyl acetate and Reynolds' lead citrate. They were observed using a HITACHI H-7000 electron microscope at an acceleration voltage of 80 kV.

Specimens fixed with 0.2% glutaraldehyde and embedded in Lowicryl K4M resin were stained by the immunogold method (particle size of colloidal gold (CG): 15 nm and 10 nm). Thin sections were incubated with unlabeled-streptavidin (Southern Biotech, Birmingham, AL) for 30 min at room temperature to block endogenous biotin. Immunogold staining was performed as described previously [4,18]. Briefly, the sections were incubated with anti-H$^+$, K$^+$-ATPase α- and β-subunits, anti-phosphoryl ezrin and anti-CD44 antibodies followed by biotinylated anti-rabbit or anti-mouse IgG antibodies and labeled with streptavidin-colloidal gold. Finally, the sections were counterstained with uranyl acetate and Millonig's lead acetate.

2.5. Morphometrical Analysis of Labeling Density with Phosphoryl Ezrin Immunogold Staining

The parietal cell labeling density with the anti-phosphoryl ezrin antibody was analyzed using Image-J NIH software. Thus, the labeling number of gold particles (on IC containing multiple microvilli) was counted on electron microscopic photographs taken at 15,000 × magnification of the neck or base region (n = 20 each) of active phase glands (fed animals) and inactive resting phase glands (starved animals). The labeling density was estimated as the number of gold particles per unit area (μm^2) of IC. For the assessment of the phosphoryl ezrin labeling-density, the results were statistically analyzed by *t*-test using Microsoft Excel software. Statistical comparisons were made between the neck and the base area of the gland, and between active versus inactive glands from starved animals. The differences between sites or feeding/starving were evaluated by *t*-test. $P < 0.01$ or 0.05 was considered significant. The results are expressed as the arithmetic mean ± SE.

3. Results

3.1. Morphological Observation

The average pH in the fed rats was 2.0, compared with 6.4 in the starved rats. We therefore hypothesized that the former corresponds to the active phase and the latter to the inactive resting phase of gastric juice secretion.

The parietal cells showed excellent ultrastructural preservation at the electron microscopic level. The ultrastructure of IC, TV, and other organelles was well preserved for each active and inactive phase of glands when specimens were processed successfully by means of HPF-followed by FS.

3.2. Immunohistochemical Observation with Anti-H$^+$, K$^+$-ATPase Antibody

The parietal cells were labeled intensely and clearly by immunohistochemical staining with the H$^+$, K$^+$-ATPase anti-α- and -β-subunit antibodies. Cells in the neck and upper base were labeled particularly strongly in active phase animals. Staining was evenly distributed from the deep pit to the glandular base of the cells (**Figures 1(A)** and **(B)**), and the staining pattern was similar between the anti-β-subunit antibody and α-subunit antibody (data not shown). In the active phase, the microvillous membrane and apical cell membrane of the IC were labeled with the anti-α- and -β-H+, K+-ATPase subunit antibodies and with the anti-phosphorylated-ezrin antibody (corresponding to residues surrounding Thr566 and 567), while TV membranes were hardly stained. In the inactive phase, IC microvilli were labeled weakly with this antibody. The anti-CD44 antibody staining pattern was similar to that of anti-α- (and -β-) H$^+$, K$^+$-ATPase subunit antibodies in inactive phase animals (**Figures 2**, **3(A)** and **(B)**). A regional labeling difference was evident from the neck to the upper base and lower base.

3.3. Morphometric Analysis with Anti-Phosphoryl Ezrin Antibody and Immunogold Labeling

Immunogold staining was performed to examine the intracellular distribution and the labeling density of the

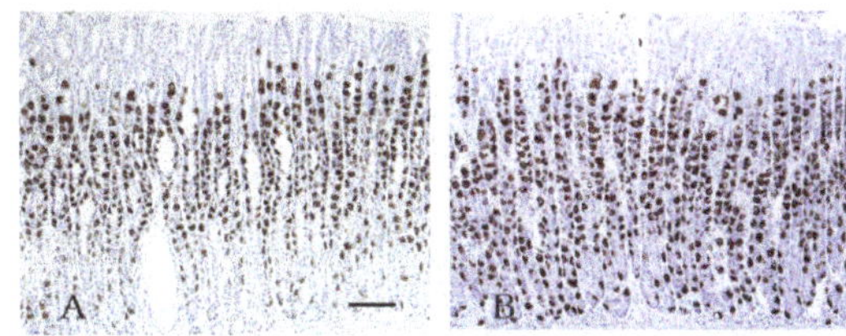

Figure 1. Active phase rat gastric gland immunostained with the anti-H$^+$, K$^+$-ATPase α-subunit antibody. (A) Parietal cells throughout the gland are stained strongly. Cells are large and plump and become smaller and more slender as they migrate downwards. Reaction products are thread-like in shape; (B) Inactive resting phase gland with similar staining. Parietal cells scattered throughout the gland are also stained positively and reaction products were observed diffusely in the cytoplasm. Scale bar = 100 μm.

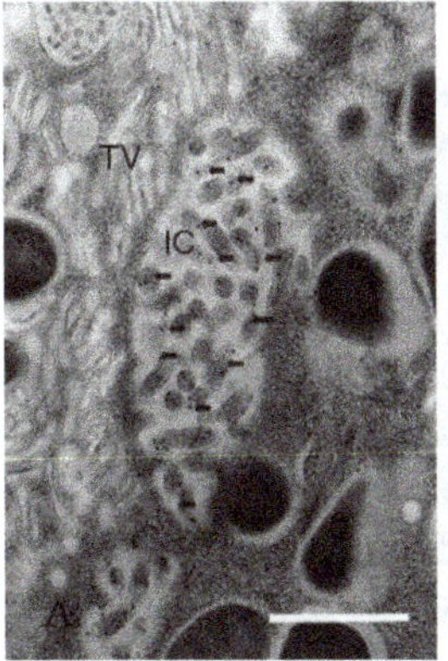

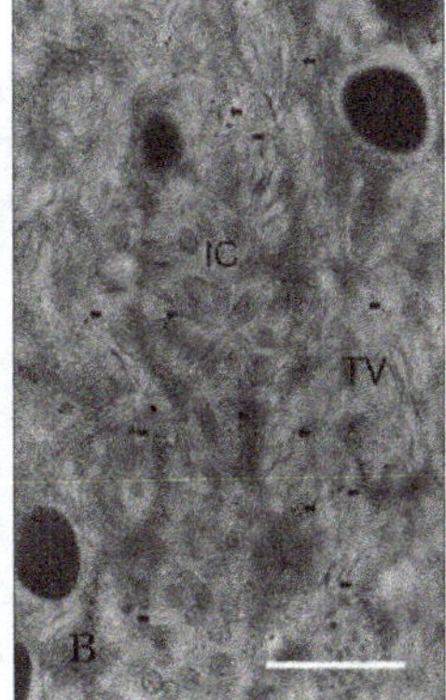

Figure 2. Section of parietal cells stained with immunogold method. (A) Anti-phosphoryl ezrin antibody staining. The IC membrane and its microvilli were stained with anti-phosphoryl ezrin antibody, but little staining was visible on the TV membrane; (B) Anti-CD44 ant-ibody staining. The two organelles (IC and TV) were labeled with this antibody. Scale bar = 1 μm.

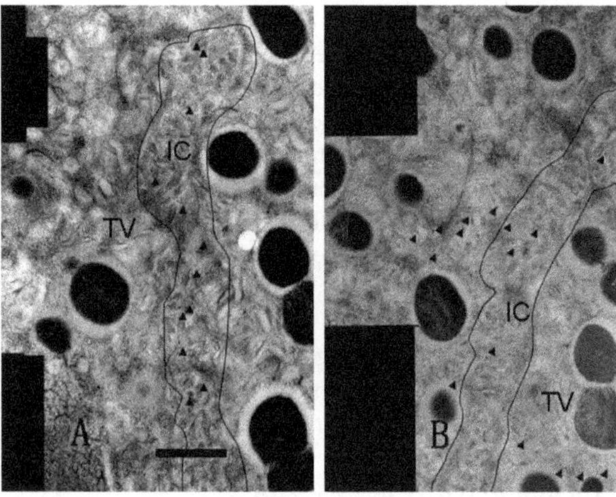

Figure 3. Adjacent serial sections immunostaining. Each photograph is composed of several distinct pictures (montaged pictures) and serially sectioned. Labeled gold particles are shown with arrowheads. The anti-phosphoryl ezrin antibody bound only to the IC microvilli (A), while the anti-CD44 antibody labeled both the IC and TV (B). Scale bar = 1 μm.

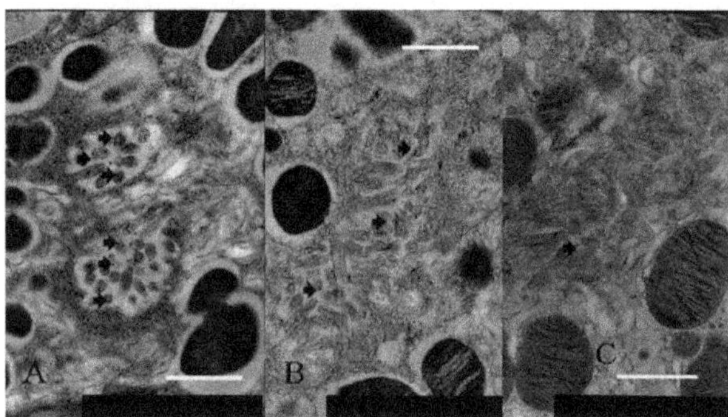

Figure 4. Labeling differences in each level of gland of parietal cell stained with immunogold using anti-phosphoryl ezrin antibody. (A) Neck region, adjoining mucous neck cell in active animals; (B) Lower region of base, neighboring chief cell; (C) Neck region from starved rats. Anti-phosphoryl ezrin antibody labeling is strong on IC membrane of parietal cell from fed (active secreting) rats (A), and moderate to weak in lower half of base (B) and through gland from starved (inactive resting or inactive secreting) rats (C). Scale bar = 1 μm.

anti-phosphoryl ezrin antibody in each cell between the gland segments. Parietal cells adjacent to mucous neck cells or chief cells were deemed to be in the neck or base region, respectively. The gold particle numbers were compared between the neck and the base region of the glands, and between active and inactive resting glands (**Figure 4**). The number of labeled gold particles was divided by the IC area to give the labeling density (per μm^2). Labeling density zonation was clear from the neck to the base, with a significantly higher density in parietal cells located in the isthmus to neck region (mean ± SE; 25.501 ± 3.736 μm^2) compared with the glandular lower base (17.082 ± 7.275 μm^2) or from inactive starved rats (1.926 ± 0.465 μm^2) (**Figures 5(A)** and **(B)**).

Statistical analysis using IMAGE-J revealed that phosphoryl ezrin expression in the neck and upper base was significantly higher than that in lower base (25.501 ± 3.736 vs 17.082 ± 7.275, $p < 0.05$) and than that in starved gland (25.501 ± 3.736 vs 1.926 ± 0.465, $p < 0.01$). These findings suggest that the phosphoryl ezrin assemble in the membrane of active parietal microvilli at neck to upper base.

4. Discussion

The component cells of gastric glands include pit mucous cells, progenitor cells, parietal cells, mucous neck cells, chief cells and endocrine cells and have previously been studied in rodents. These cells undergo mitosis in the isthmus, from where they migrate and differentiate along the longitudinal axis of the gland in an upward or downward direction [19-21]. The parietal cells migrate upwards and downwards then mature, while chief cells derive from the mucous neck cell through an intermediate cell type to the mature chief cell in a downwards migration [17]. Parietal cells adjoining the mucous neck cell are considered to be in the neck region and those next to the chief cell are in the base region of the gland.

Gastric juice is very acidic, with a pH of around 1.5. The average pH value measured in this study was pH 2.0 in fed rats and pH 6.4 in starved animals, indicating that the parietal cells of fed rats correspond to cells in the active secretory-phase, while those of starved animals correspond to cells in the inactive resting phase. The pa-

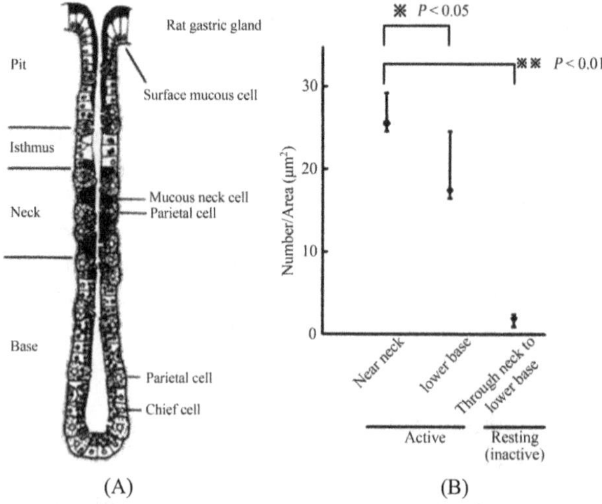

Figure 5. (A) Schematic drawing of rat gastric gland. After proliferation in the glandular isthmus, the parietal cell migrates and differentiate through neck to base of gland; (B) Statistical analysis of phosphoryl-ezrin plotting. The labeling of the cell in the near neck site was significantly higher than that in the lower base, or whole site of the gland in the resting (the inactive).

rietal cells are distributed broadly through the gland from the isthmus to the base. The cells are large and plump from the isthmus to the upper base, although they become smaller and more slender in the lower base as they migrate downwards [19-22].

Routine light microscopy revealed parietal cells of an acidophilic nature with numerous mitochondria in the cytoplasm. The cells were stained with immunolabeling using an anti-H^+, K^+-ATPase antibody, and the reaction products had a thread-like appearance, especially in the isthmus to the neck region [5,18]. The expression pattern followed the contour line of the stained apical-cell membrane. In previous studies, we used high pressure-frozen, freeze-substituted and resin-embedded tissue samples to demonstrate the intracellular localization of H^+, K^+-ATPase [3,23]. This antibody clearly stained the IC containing many microvilli in the active phase, while TV was stained in the inactive resting phase. Diffuse cytoplasmic staining was observed in cells scattered widely from the isthmus to the glandular base. This aspect is thought to derive from TV membrane staining at the light microscopy level.

Yao *et al.* [15] revealed that most actin in the gastric gland is present in parietal cells, which are largely globular during the inactive resting secretion phase. However, during the active phase, actin forms a filamentous structure. Jöns *et al.* [24] reported that the binding of actin to the plasma membrane is dependent on phosphoryl ezrin molecules, and the initiation of ultrastructural changes in parietal cells is thought to be induced by the polymerization of cytoskeletal actin [2,7,15] followed by binding of the actin molecule to the C-terminal of phosphoryl ezrin and intramembranous CD44 to the N-terminus [8-10, 16,25,26]. Non-phosphoryl ezrin molecules are distributed throughout the cytoplasm without binding actin in the inactive resting phase [16,29-29], while localization of the ezrin-actin linkage in the cytoplasm occurs in the active phase during the formation of phosphorylated ezrin molecules and filamentous actin-molecule binding. This indicates that phosphoryl ezrin and actin are related to the cell surface distribution of the H^+, K^+-ATPase proton pump.

In the present study, immunostaining with the phosphoryl ezrin antibody resulted in heavy labeling in the neck to the upper base region (**Figures 3** and **4**), revealing that the labeled cells are in the active phase [5]. A statistical morphometric data comparison of neck and base cells showed that phosphoryl ezrin and H^+, K^+-ATPase were present in similar sites within the same cell [18]. The apical membrane of IC containing microvilli in the active phase was labeled by immunogold staining with the anti-phosphoryl ezrin antibody. On the other hand, labeling of these structures was weak in the inactive resting phase (**Figures 4** and **5**).

These findings suggest that the parietal cells in the neck and upper base are more active than those in the lower base. Karam *et al.* [20] described the intracellular mitochondrial distribution of parietal cells at various levels of the mouse gastric gland and revealed a high number in the isthmus and neck. Moreover, the integral H^+, K^+-ATPase is exposed on the luminal surface when IC is enlarged, and the enzyme is activated by a successive K^+ ion supply. On the other hand, the increasing TV volume results in a decreased surface area of the canaliculi membrane as the IC and TV derive from similar membrane systems [12]. The TV is transformed into the apical plasma membrane during acid secretion [2]. H^+, K^+-ATPase then becomes an intravesicular membrane protein and acid secretion is halted because of an interruption in the K^+ ion supply. This transformation is limited at sites with small amounts of phosphoryl ezrin, such as the middle to the bottom region of the gland [18,22] where parietal cells are small and slender. Here, it is thought that the acid secretory capacity is limited and unresponsive to physiological or feeding conditions.

The purpose of this work was to decide at which level parietal cells are most active. Acid secretion activity in the gland appears to be higher in the segment from the isthmus to the upper glandular base, which agrees with the findings of Fykse *et al.* [30] who reported that parietal cells in the second part of four segments in the gland were activated by histamine treatment. Jiang *et al.* [5] reported that the acid-secreting potency of individual parietal cells was higher in the upper third of the gland (containing the superficial part). Further experiments are required to clarify the secretion system of gastric acid in parietal cells.

5. Acknowledgements

We would like to thank Mr. S. Nonaka for technical assistance and Professor K. Kosai and Professor emeritus S. Nakagawa for helpful comments. This work was supported in part by a Grant-in-Aid for Scientific Research (No. 15590164 to S. Tsuyama) from the Ministry of Education, Culture, Sports, Science and Technology of Japan and the Kodama Memorial Fund for Medical Research.

REFERENCES

[1] N. Sugai, S. Ito, A. Ichikawa and M. Ichikawa, "The Fine Structure of the Tubulovesiclular System in Mouse Gastric Parietal Cell Processed by Cryofixation Method," *Journal of Electron Microscopy*, Vol. 34, No. 2, 1985, pp. 113-122.

[2] B. J. Agnew, J. G. Duman, C. Watson, D. E. Coling and J. G. Forte, "Cytological Transformations Associated with Parietal Cell Stimulation: Critical Steps in the Activation Cascade," *Journal of Cell Science*, Vol. 112, No. 16, 1999,

pp. 2639-2646.

[3] S. Tsuyama, S. Matsushita, T. Takatsuka, S. Nonaka, K. Hasui and F. Murata, "Cytochemical Investigation of Gastric Gland Component Cells with High-Pressure Freezing Followed by Freeze-Substitution and Hydrophilic Resin Embedding," *Anatomical Science International*, Vol. 77, No. 1, 2002, pp. 74-83. http://dx.doi.org/10.1046/j.0022-7722.2002.00011.x

[4] D. Wakamatsu, S. Tsuyama, R. Maezono, K. Katoh, S. Ogata, S. Takao, S. Natsugoe, T. Aikoh and F. Murata, "Immunohistochemical Detection of the Cytoskeletal Components in Gastric Parietal Cells," *Acta Histochemica et Cytochemica*, Vol. 38, No. 5, 2005, pp. 331-337. http://dx.doi.org/10.1267/ahc.38.331

[5] X. Jiang, E. Suzaki and K. Kataoka, "Immunofluorescence Detection of Gastric H(+)/K(+)-ATPase and Its Alterations as Related to Acid Secretion," *Histochemistry and Cell Biology*, Vol. 117, No. 1, 2002, pp. 21-27. http://dx.doi.org/10.1007/s00418-001-0369-8

[6] C. Andréoli, M. Martin, R. R. Le Borgne and P. Mangeat, "Ezrin Has Properties to Self-Associate at the Plasma Membrane," *Journal of Cell Science*, Vol. 107, No. 9, 1994, pp. 2509-2521.

[7] X. Yao, L. Chen and J. G. Forte, "Biochemical Characterization of Ezrin-Actin Interaction," *The Journal of Biological Chemistry*, Vol. 271, No. 12, 1996, pp. 7224-7729. http://dx.doi.org/10.1074/jbc.271.12.7224

[8] D. T. Dransfield, A. Bradford, J. Smith, M. Martin, C. Roy, P. H. Mangeat and J. R. Goldenring, "Ezrin Is a Cyclic AMP-Dependent Protein Kinase Anchoring Protein," *The EMBO Journal*, Vol. 16, No. 1, 1997, pp. 35-43. http://dx.doi.org/10.1093/emboj/16.1.35

[9] X. Cao, X. Ding, Z. Guo, R. Zhou, F. Wang, F. Long, F. Wu, F. Bi, Q. Wang, D. Fan, J. G. Forte, M. Teng and X. Yao, "PALS1 Specifies the Localization of Ezrin to the Apical Membrane of Gastric Parietal Cells," *The Journal of Biological Chemistry*, Vol. 280, No. 14, 2005, pp. 13584-13592. http://dx.doi.org/10.1074/jbc.M411941200

[10] X. Ding, H. Deng, D. Wang, J. Zhou, Y. Huang, X. Zhao, X. Yu, M. Wang, F. Wang, T. Ward, F. Aikhionbare and X. Yao, "Phspho-Regulated ACAP4-Ezrin Interaction Is Essential for Histamine-Estimated Parietal Cell Secretion," *The Journal of Biological Chemistry*, Vol. 285, No. 24, 2010, pp. 18769-18780. http://dx.doi.org/10.1074/jbc.M110.129007

[11] M. Nishi, F. Aoyama, F. Kisa, H. Zhu, M. Sun, P. Lin, H. Ohta, B. Van, S. Yamamoto, S. Kakizawa, H. Sakai, J. Ma, A. Sawaguchi and H. Takeshima, "TRIM50 Regulates Vesicular Trafficking for Acid Secretion in Gastric Parietal Cell," *The Journal of Biological Chemistry*, Vol. 287, No. 40, 2012, pp. 33523-33532. http://dx.doi.org/10.1074/jbc.M112.370551

[12] P. V. Jensen and L.-I. Larson, "Actin Microdomains on Endothelial Cells: Association with CD44, ERM Proteins, and Signaling Molecules during Quiescence and Wound Healing," *Histochemistry and Cell Biology*, Vol. 121, No. 5, 2004, pp. 361-369. http://dx.doi.org/10.1007/s00418-004-0648-2

[13] K. L. Brown, D. Birkenhead, J.-C. Y. Lai, L. Li, R. Li and R. Johnson, "Regulation of Hyaluronan Binding by F-Actin and Colocalization of CD44 and Phosphorylated Ezrin/Radixin/Moesin (ERM) Proteins in Myeloid Cells," *Experimental Cell Research*, Vol. 303, No. 2, 2005, pp. 400-414. http://dx.doi.org/10.1016/j.yexcr.2004.10.002

[14] V. Orian-Rousseau, H. Morrison, A. Matzke, T. Kastilan, G. Pace, P. Herrliich and H. Ponta, "Hepatocyte Growth Factor-Induced Ras Activation Requires ERM Proteins Linked to Both CD44v6 and F-Actin," *Molecular Biology of the Cell*, Vol. 18, No. 1, 2007, pp. 76-83. http://dx.doi.org/10.1091/mbc.E06-08-0674

[15] X. Yao, C. Chaponnier, G. Gabbiani and J. G. Forte, "Polarized Distribution of Actin Isoforms in Gastric Parietal Cells," *Molecular Biology of the Cell*, Vol. 6, No. 5, 1995, pp. 541-557. http://dx.doi.org/10.1091/mbc.6.5.541

[16] L. Zhu, J. Crothers Jr., R. Zhou and J. G. Forte, "A Possible Mechanism for Ezrin to Establish Epithelial Cell Polarity," *American Journal of physiology Cell Physiology*, Vol. 299, No. 2, 2010, pp. C431-C443. http://dx.doi.org/10.1152/ajpcell.00090.2010

[17] K. Ihida, T. Suganuma, S. Tsuyama and F. Murata, "Glycoconjugate Histochemistry of the Rat Fundic Gland Using Griffonia Simplicifolia Agglutinin-II during the Development," *American Journal of Anatomy*, Vol. 182, No. 3, 1988, pp. 250-256.

[18] S. Tsuyama, S. Matsushita, S. Nonaka, S. Yonezawa and F. Murata, "Cytochemical of Gastric Parietal Cells with High-Pressure Freezing Followed by Freeze-Substitution," *The Journal of Electron Microscopy*, Vol. 52, No. 2, 2003, pp. 145-151. http://dx.doi.org/10.1093/jmicro/52.2.145

[19] S. M. Karam and C.-P. Leblond, "Dynamics of Epithelial Cells in Corpus of the Mouse Stomach. IV Bidirectional Migration of Parietal Cells Ending in Gradual Degeneration and Loss," *The Anatomical Record*, Vol. 236, No. 2, 1993, pp. 314-332. http://dx.doi.org/10.1002/ar.1092360205

[20] S. M. Karam, X. Yao and J. G. Forte, "Functional Heterogeneity of Parietal Cells along the Pit-Gland Axis," *American Journal of Physiology*, Vol. 272, No. 1, 1997, pp. G161-G171.

[21] S. M. Karam, T. Staiton, W. M. Hassan and C.-P. Leblond, "Defining Epithelial Cell Progenitors the Human Oxyntic Mucosa," *Stem Cell*, Vol. 21, No. 3, 2003, pp. 322-336. http://dx.doi.org/10.1634/stemcells.21-3-322

[22] D.-H. Yang, S. Tsuyama, Y.-B. Ge, D. Wakamats, J. Ohmori and F. Murata, "Proliferation and Migration Kinetics of Stem Cells in the Rat Fundic Gland," *Histology and Histopathology*, Vol. 12, No. 3, 1997, pp. 719-727.

[23] K. Tyagarajan, D. Chow, A. Smolka and J. G. Forte, "Structutral Interactions between α- and β-Subunits of the Gastric H,K-ATPase," *Biochimica et Biophysca Acta*, Vol. 1236, No. 1, 1995, pp. 105-113. http://dx.doi.org/10.1016/0005-2736(95)00044-4

[24] T. Jöns, H. Heim, U. Kistner and G. Ahnert-Hilger, "SAP97 Is a Potential Candidate for Basolateral Fixation of Ezrin in Parietal Cell," *Histochemistry and Cell Biology*, Vol. 111, No. 4, 1999, pp. 313-318. http://dx.doi.org/10.1007/s004180050362

[25] D. Liu, L. Ge, F. Wang, H. Takahashi, D. Wang, Z. Guo, S.-H. Yoshimura, T. Ward, X. Ding, K. Takeyasu and X. Yao, "Single-Molecule Detection of Phosphorylation-Induced Plasticity Changes during Ezrin Activation," *FEBS Letters*, Vol. 581, No. 18, 2007, pp. 3563-3571. http://dx.doi.org/10.1016/j.febslet.2007.06.071

[26] L. Zhu, R. Zhu, S. Mettler, T. Wu, A. Abbas, J. Delaney and J. G. Forte, "High Turnover Ezrin T567 Phophorylation: Comformation, Activity, and Cellular Function," *American Journal of Physiology Cell Physiology*, Vol. 293, No. 3, 2007, pp. C874-C884. http://dx.doi.org/10.1152/ajpcell.00111.2007

[27] D. Hanzel, H. Reggio, A. Bretscher, J. G. Forte and P. Mangeat, "The Secretion-Stimulated 80K Phophoprotein of Parietal Cells Is Ezrin, and Has Properties of a Membrane Cytoskeletal Linker in the Induced Apical Membrane," *The EMBO Journal*, Vol. 10, No. 9, 1991, pp. 2363-2373.

[28] R. Zhou, X. Cao, C. Watson, Y. Miao, Z. Guo, J. G. Forte and X. Yao, "Characterization of Protein Kinase A-Mediated Phosphorylation of Ezrin in Gastric Parietal Cell Activation," *The Journal of Biological Chemistry*, Vol. 278, No. 37, 2003, pp. 35651-35659. http://dx.doi.org/10.1074/jbc.M303416200

[29] R. Zhou, L. Zhu, A. Kodani, P. Hauser, X. Yao and J. G. Forte, "Phosphorylation of Ezrin on Threonine 567 Produces a Change in Secretory Phenotype and Repolarizes the Gastric Parietal Cell," *Journal of Cell Science*, Vol. 118, No. 19, 2005, pp. 4381-4391. http://dx.doi.org/10.1242/jcs.02559

[30] V. Fykse, E. Solligård, MØ. Bendheim, D. Chen, J. E. Grønbech, A. K. Sandvik and H. L. Waldum, "ECL Cell Histamine Mobilization and Parietal Cell Stimulation in the Rat Stomach by Microdialysis and Electron Microscopy," *Acta Physiologica*, Vol. 186, No. 1, 2006, pp. 37-43. http://dx.doi.org/10.1111/j.1748-1716.2005.01504.x

23

Screening of Total Organophosphate Pesticides in Agricultural Products with a Cellular Biosensor

Kelly Lokka[1], Panagiotis Skandamis[2], Spiridon Kintzios[1*]
[1]Laboratory of Enzyme Technology, Faculty of Biotechnology, Agricultural University of Athens, Athens, Greece
[2]Laboratory of Food Quality Control and Hygiene, Faculty of Food Science & Technology, Agricultural University of Athens, Athens, Greece
Email: [*]skin@aua.gr

ABSTRACT

Organophosphates belong to the most important pesticides used in agricultural practice worldwide. Although their analytical determinations are quite feasible with various conventional methods, there is a lack of efficient screening methods, which will facilitate the rapid, high-throughput detection of organophosphates in different food commodities. This study presents the construction of a rapid and sensitive cellular biosensor test based on the measurement of changes of the cell membrane potential of immobilized cells, according to the working principle of the Bioelectric Recognition Assay (BERA). Two different cell types were used, derived either by animal (neuroblastoma) or plant cells (tobacco protoplasts). The sensor was applied for the detection of a mixture of two organophosphate pesticides, diazinon and chlorpyrifos in two different substrates (tomato, orange). The pesticides in the samples inhibited the activity of cell membrane-bound acetylcholinesterase (AChE), thus causing a measurable membrane depolarization in the presence of achetylcholine (Ach). Based on the observed patterns of response, we demonstrate that the sensor can be used for the qualitative and, in some concentrations, quantitative detection of organophosphates in different substrates with satisfactory reproducibility and sensitivity, with a limit of detection at least equal to the official Limit of Detection (LOQ). The assay is rapid with a total duration of 3 min at a competitive cost. The sensitivity of the biosensor can be further increased either by incorporating more AChE-bearing cells per test reaction unit or by using cells engineered with more potent AChE isoforms. Standardization of cultured cell parameters, such as age of the cells and subculture history prior to cell immobilization, combined with use of planar electrodes, can further increase the reproducibility of the novel test.

Keywords: Bioelectric Recognition Assay (BERA); Matrix Effects; N2a Cells; Organophosphates; Tobacco Protoplasts

1. Introduction

Organophosphate insecticides have been used widely in agriculture and in household applications as pesticides due to their high insecticidal activity and relatively low persistence [1]. Their mechanism of action is the irreversible inhibition of acetylcholinesterase (AchE), a key enzyme in the recycling of the neurotransmitter acetylcholine (Ach) [2]. Organophosphates phosphorylate the serine hydroxyl group at the site of action of acetylcholine. They bind irreversibly, deactivating the esterase, resulting in accumulation of acetylcholine at the endplate. Decrease in plasma cholinesterase results in a decrease of cholinesterase activity in the central, parasympathetic, and sympathetic nervous systems. Accumulation of acetylcholine at the neuromuscular junction causes persistent depolarization of skeletal muscle, while neural transmission in the central nervous system is disrupted [3]. Long-term exposure to organophosphates has been associated with irritability, fatigue, headache, difficulties with memory and concentration and other neurophysiological abnormalities [4-6].

The conventional analysis of pesticide residues in food commodities is a labor intensive procedure, since it is necessary to cover a wide range of different chemicals, using a single procedure. Standard analysis methods include gas chromatography and high performance liquid chromatography to achieve the necessary selectivity and sensitivity for the different classes of compounds under detection [7]. As a consequence, current methods of analysis provide a limited sample analysis capacity, on a day/instrument basis [8]. While the analytical determina-

[*]Corresponding author.

tion of the pesticide residues in an unknown sample will always be carried out using sophisticated methods based on conventional technology, rapid screening is the only solution for assuring food control by means of high-throughput detection of organophosphates in different food commodities. Providing novel solutions for food quality monitoring is also in accordance with the new EU and international regulations for minimal residue concentration in marketed food and agricultural products.

Based on the inhibition of AchE and choline oxidase, many biosensors have been developed for the detection of organophosphorous and carbamate pesticides such as those described by Andres *et al.* [9], Choi *et al.* [10] and Andreou *et al.* [11]. It has been recently demonstrated that a successful pesticide assay could be based on a BERA platform, comprising electrically active cells (*i.e.* with an active transport of ions through the cell membrane) interfaced with microelectrodes which allow the capture of extracellular spikes or impedance changes associated with cellular response against the pesticide under detection [12]. More specifically, the inhibition of AchE by organophosphates and carbamate residues caused an increase of available Ach in the assay solution, which in turn caused the depolarization of the membranes of immobilized neuronal cells in a concentration-dependnet manner. In other words, the presence of the pesticides was detected by the degree of inhibition of cellular AChE, which is inversely associated with ACh concentration. Inhibition of AChE leads in increased excitatory ACh transmission and depolarization of the cell membrane, which was measured as a change of the sensor's potential, due to changes in the concentration of electrolytes in the immediate vicinity of the working electrode. This novel type of biosensor was further included in a validation test against E.U. proficiency test samples [13].

The aim of the present study was to further develop this novel biosensor principle by incorporating, for the first time, plant (tobacco) protoplasts as organophosphate biorecognition elements. Their reliability as sensor components was compared to those of neuronal cells. In addition, both biosensor versions were applied for the detection of a mixture of organophosphate pesticides in different substrates, thus evaluating, also for the first time, possible matrix effects on the sensor's performance.

2. Materials and Methods

2.1. Materials

Diazinon (Diethoxy-[(2-isopropyl-6-methyl-4-pyramidinyl)oxy]-thioxophosphorane; CAS [333-41-5]; Mw = 304.35) and chlorpyrifos (*O,O*-diethyl *O*-3,5,6-trichloro-2-pyridyl phosphorothioate; CAS [2921-88-2]; Mw = 350.59) (both purchased from Chem Device, West Chester, PA, USA) were used as standard organophosphate insecticides. Pesticide mixtures which contained 10 μM of each pesticide, were prepared daily in acetone solution. All other reagents were purchased from Fluka (Switzerland).

2.2. Cell Culture and Sensor Fabrication from Plant Protoplast and Neuroblastoma N2a Cells

Plant protoplasts were isolated from tobacco (*Nicotiana tabacum*) leaves by preplasmolysing 0.5 g of them in 20 ml of CPW solution [14] supplemented with 0.7 M mannitol for one hour and then incubating them in 20 ml solution of the same composition and additionally supplemented with 3 mg pectinase (8.5 units/mg, from *Aspergilus niger*) and 2 mg cellulase (9.5 units/mg, from *Trichoderma viridae*) for 20 hours. One ml of protoplast solution contained 3.5×10^6 cells/ml. Mouse neuroblastoma (N2a) cell cultures were originally provided from LGC promochem (UK). Cells were cultured in Dulbecco's medium with 10% heat-inactivated foetal calf serum (FCS), 10% antibiotics (streptomycin) and 10% l-glutamine. For incorporation into the biosensor, cells were detached from the culture vessel by adding trypsine/ EDTA for 10 min at 37°C and further concentrated by centrifugation (6 min, 1200 rpm, 25°C). For manufacturing consumable biosensors, 1 ml of plant or animal cells (at a density of 2.5×10^6/ml) were mixed with 2 ml of 4% (w/v) sodium alginate solution and then the mixture was added drop wise, by means of a 22G syringe, in 0.8 M $CaCl_2$. Each of the resulting calcium alginate beads had an approximate diameter of 2 mm and contained approximately 5×10^4 cells. The sensors were storable at room temperature in culture medium, under normal atmospheric conditions (*i.e.* non-CO_2 enriched) for at least three weeks without any loss of performance.

2.3. Sample Preparation

Organically-grown, pesticide-free tomato and orange fruits were used for the preparation of the assayed samples. Two groups of samples were prepared: The first group (*fortified samples*) comprised of blended fruits spiked with mixtures of diazinon and chlorpyrifos at various concentrations. Subsequently, 15 g of the blended fruit-pesticide mixture were homogenized in 30 ml acetone using an Ultra-Turrex homogeniser (model Silent Crusher, Heidolph) at 6000 rpm for one min. In following, 30 ml dichloromethane, then 30 ml petroleum ether were added and the mixture was stirred for one min. Then the mixture was centrifuged at 4000 rpm for 4 min at 20°C. After removing 25 ml of the supernatant, the rest was left to dry in a waterbath, at 65°C - 70°C. Finally, the dry samples were redissolved in 5 ml methanol (10% v/v). The second group (*matrix standards*) was created by spiking final dry samples, dissolved in methanol, with the pesticide mixtures.

The concentration of each pesticide in the samples was calculated so that each pesticide was present either at the Limit of Quantification (LOQ), the Minimum Residue Level (MRL) or one-tenth thereof (MRL/10), according to the concentrations specified by the European Union Directive 839/2008 (**Table 1**). Control samples were created as the analytical sample but omitting spiking with pesticide mixtures. Dried control samples were also re-dissolved in methanol.

2.4. Assay Principle

According to the working principle of the method, the presence of organophosphate compounds is detected by the degree of inhibition of cellular AChE, which is inversely associated with ACh concentration. ACh is an excitatory neurotransmitter (Hoang *et al.*, 2007), the activity of which is regulated by AChE. Therefore, inhibition of AChE can lead in increased excitatory ACh transmission, which can be measured by the depolarization of the cell membrane. In other words, inhibition of AChE by pesticide residues in the sample will result to excessive stimulation of N2a cells by ACh, which will further lead to membrane depolarization above a pre-determined threshold. Membrane depolarization events and associated electrolyte influx/efflux will reflect themselves on the sensor's response as a change of the sensor's potential, due to changes in the concentration of electrolytes in the immediate vicinity of the working electrode [12,13].

2.5. Assay Procedure

Each cell-bearing bead (cell sensor) was manually connected to a working electrode (the electrode was inserted through the entire length of the sensor without extruding from the opposite end) made from pure silver, electrochemically coated with an Ag/AgCl layer and having a diameter of 0.75 mm. The distance between working and reference electrode was 3 cm. Electrodes were connected to the recording device, which comprised the PMD-1608FS A/D card (Measurement Computing, Middleboro, MA) (**Figures 1(a)** and **(b)**). The software responsible for recording the signal and data processing was InstaCal (Measurement Computing).

For each assay, the sensor system, comprising of the bead attached to the working electrode and a reference electrode, was immersed into each sample solution (200 μl). The sample solution comprised 150 μl of 50 mM Tris-aminomethane (Tris) buffer (pH 8), containing 0.5 mM acetylcholine iodide (Ach) and 50 μl of pesticide sample. The response of each sensor was estimated by recording the average change of the sensor potential for a period of 180 sec after sample application.

Table 1. LOQ and MRL points laid down by the European Union Directive 839/2008.

	Tomato		Orange	
	LOQ (ppm)	MRL (ppm)	LOQ (ppm)	MRL (ppm)
Diazinon	0.02	0.5	0.024	1
Chlorpyrifos	0.03	0.5	0.03	0.3

2.6. Data Analysis and Experimental Design

Experiments were set up in a completely randomized design and each experiment was repeated three times. In each application, a set of six biosensors was tested against each individual sample. Correlations between the sensor's response and pesticide concentrations were done using MS-Excel. Data means among different days were compared using Duncan's multiple range test (with significance at $p < 0.05$).

The effect of the extraction procedure and the concentration of the pesticides on the screening efficiency (SE%) in each sample was calculated individually for each sample, according to the following equation:

$$SE\% = 100 \times \frac{\text{average response to the fortified sample}}{\text{average response to the matrix standard}}$$

3. Results and Discussion

3.1. Response of the Sensor to Organophosphate Pesticides in Tomato Samples

The response of the cell-based biosensor against different concentrations of a mixture of diazinon and chlorpyrifos in tomato samples is presented in **Figure 2**. The results

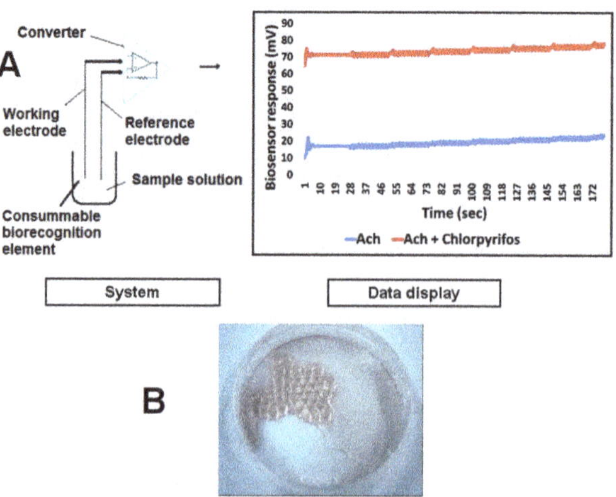

Figure 1. (A) Schematic outline of the biosensor system. The graph shows the considerable increase of the response of the cellular biorecognition element to Ach after the addition of an organophosphate (chlorpyriphos) (*red line*). The insert (B) shows a Petri dish with consumable biorecognition elements (gel beads with immobilized cells).

of the assay using neuroblastoma as the sensor's biorecognition elements are shown in **Figure 2(a)**, while the results of the assay using plant protoplasts are shown in **Figure 1(b)**. In the case of neuroblastoma N2a cells, when no pesticide was present in the sample (control sample), a sensor response of −142 ± 2 mV was observed. The sensors responded to increasing pesticide concentrations by considerable positive increase of the sensor's potential (**Figure 2(a)**). This observation is in accordance with the assay principle, where inhibition of AChE can lead in increased excitatory ACh transmission, which can be measured by the depolarization of the cell membrane (hence the shift of the sensor measurements to more positive values). The response of the sensor against matrix standards was quite reproducible (average variation = 12.7%) with the exception of the response against the highest concentration (MRL). Considering matrix standards only, a satisfactory correlation was observed between sensor response and total pesticide concentration ($r^2 = 0.9624$, $y = 0.042x - 0.1765$). On the contrary, the response against the fortified samples was less reproducible, although a concentration-dependent pattern was again observed. The better reproducibility of the response of matrix standards vs. fortified samples could be due to the more homogenous distribution of the pesticides under detection in the matrix standards compared to the fortified samples.

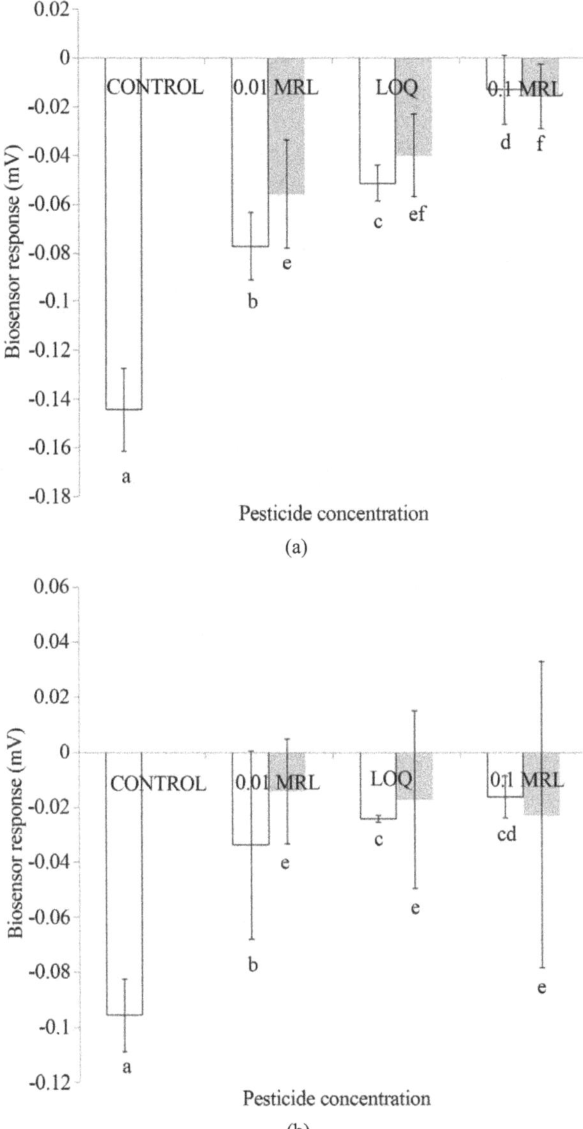

Figure 2. Response of the cell-based biosensor against different concentrations of a mixture of diazinon and chlorpyrifos in tomato samples. The biosensor was based either on neuroblastoma cells (a) or tobacco protoplasts (b). Pesticide concentrations are expressed as the corresponding LOQ, MRL and MRL/10 values, according to Table 1. Sensor response is expressed as a change in the membrane potential of immobilized cells. (n = 6 replications (different sensors) for each sample and error bars represent standard errors of the average value of all replications with each sample). The white columns represent matrix standards and the grey columns fortified samples, as described analytically in the Materials section. Columns sharing a common letter are not statistically different ($p \geq 0.05$).

When plant protoplasts were used, the response to control solutions was less negative (−95 ± 1 mV) than with animal cells, indicating a lower state of membrane hyperpolarization of the immobilized cells. In this case, a correlation between pesticide concentration and biosensor response was observed only against matrix standards, not fortified samples (**Figure 2(b)**). In addition, a considerable variation of the sensor response was observed, much higher than for the N2a-based sensor. Depending on the concentration of the pesticides, the screening efficiency ranged from 42% to 141%. Thus, spiked pesticide concentrations lower than or equal to the LOQ were underestimated, while higher concentrations (0.1 MRL) were overestimated.

3.2. Response of the Sensor to Organophosphate Pesticides in Orange Samples

The response of the cell-based biosensor against different concentrations of a mixture of diazinon and chlorpyrifos in orange samples is presented in **Figure 3**. The results of the assay using neuroblastoma as the sensor's biorecognition elements are shown in **Figure 3(a)**, while the results of the assay using plant protoplasts are shown in **Figure 3(b)**. When no pesticide was present in the sample (control sample), a sensor response of −161 ± 3 mV (animal cells) or −119 ± 7 mV (plant protoplasts) was observed, *i.e.* steady-state membrane hyperpolarization was higher than for tomato controls. This was probably due to the different matrix effect: Mizayawa *et al.* [15] have previously reported that constituents of the essential oils of *Citrus* sp. considerably affected the activity of

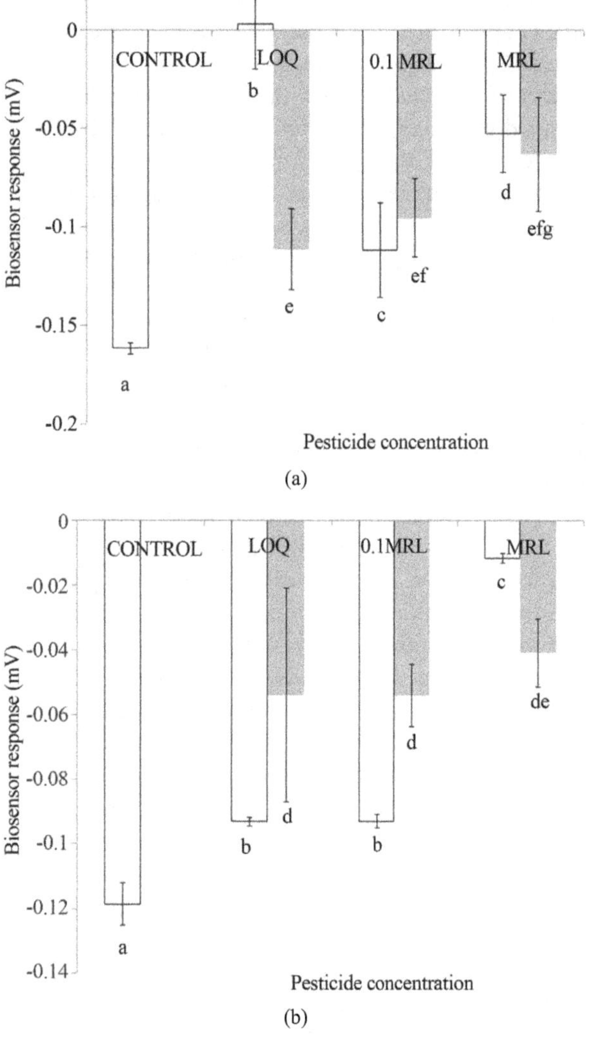

Figure 3. Response of the cell-based biosensor against different concentrations of a mixture of diazinon and chlorpyrifos in orange samples. The biosensor was based either on neuroblastoma cells (a) or tobacco protoplasts (b). Pesticide concentrations are expressed as the corresponding LOQ, MRL and MRL/10 values, according to Table 1. Sensor response is expressed as a change in the membrane potential of immobilized cells. (n = 6 replications (different sensors) for each sample and error bars represent standard errors of the average value of all replications with each sample). The white columns represent matrix standards and the grey columns fortified samples, as described analytically in the Materials section. Columns sharing a common letter are not statistically different ($p \geq 0.05$).

membrane-bound AChE, thus affecting the potential difference along the membrane of the immobilized cells. Similarly to tomato samples, the response of the sensor dependent on the total pesticide concentration, although this effect was less pronounced using the plant cell-based version (**Figure 3(b)**). In addition, the response against orange samples was more reproducible than for tomato samples, especially against analytical orange samples. This lack of reproducibility (*i.e.* a considerable variation of response) was even more obvious when protoplast-based sensors were used to assay fortified samples, especially at the LOQ (**Figure 3(b)**). Depending on the concentration of the pesticides, the screening efficiency ranged from 38% to 350%, *i.e.* pesticide concentrations were either under- or overestimated, depending on the spiked concentration and the type of the biosensor, similarly to the assay of the tomato samples. On the other hand, and considering only matrix standards, a satisfactory correlation was observed between sensor response and total pesticide concentration for both animal cell-based (r^2 = 0,8947, y = 0.0018x – 0,1454) and plant protoplast-based biosensors (r^2 = 0.7923, y = 0.0321x – 0.1593).

However, as indicated by the concentration-dependent screening efficiency for both types of biosensors, it was not possible to obtain a reliable quantitative response. By testing more pesticide concentrations in the future, it might be possible to identify a range of concentrations for each matrix X biosensor combination where the screening efficiency will be close to 100%, thus allowing for quantitative determination as well.

Differences in the sensor responses between tomato and orange samples pesticides may be due to differences of the matrix effect. For example, tomato is rich in carotenoids, such as lycopene. It has been previously shown that carotenoids and their oxidation products promote gap junctional communication [16], therefore possibly affecting the electrophysiological behaviour of neuroblastoma cells.

Matrix effects aside, differences in biosensor response to different batches of the same sample could be due to factors related to the cellular biorecognition element itself, such as the age of the cells (days elapsed between the detachment of cells from culture, subsequent immobilization and use of the biosensor) as well as subculture history prior to cell immobilization. It is strongly recommended to standardize these factors in order to minimize variability in biosensor response.

As expected, biosensors based on neuroblastoma cells were more sensitive than protoplast-based ones, as demonstrated by the greater difference in response (cell membrane depolarization) compared to control samples. This is due to their membrane-bound AChE [17]. The occurrence of this enzyme on the membranes of tobacco protoplasts is also a fact, as first reported by Madhavan *et al.* [18] (though in guard, not mesophyll cells). However, the concentration of the enzyme on the plant cell membrane is considerably lower than on neuroblastoma cells, as corroborated by colorimetric assays by our group (unpublished data). Yet the sensitivity of the sensor system presented in this study is high enough to allow for the measurement of the pesticide-AChE interaction, even at the LOQ.

The AChE-based higher level of response of neuroblastoma cells could be further exploited in order to increase the sensitivity of the novel assay, either by incorporating more AChE-bearing cells per test reaction unit, or by using cells engineered with more potent AChE isoforms. The first approach is currently under investigation in our laboratory with promising, though preliminary, results.

The reproducibility of the system could be improved considerably by redesigning the cell-electrode interface. Quite recent experiments in our lab has shown that using screen-printed electrodes reduced variation in response against organophosphate pesticides in solution, though no matrix effects were yet assessed [19]. We plan to repeat the experiments described in the present study using planar electrodes.

The novel assay has also been used for the detection of the avermectin abamectin and the pyrethroid α-cypermethrin [20], whereas various operational parameters, including assay temperature and electrode material were validated. A variation of the experimental approach described in the present study involved the seeding of N2a neuroblastoma cells on top of PEDOT electrodes treated with Nafion and Polylysine [21].

4. Conclusion

There are several areas of future work that would improve the utility of cell-based biosensors. Numerous applications would benefit from the development of parallel systems that allow for simultaneous measurements on multiple cell lines, thus improving both the breadth of sensitivity and the ability to discriminate or classify different groups of analytes. As demonstrated in the present study, the biosensor system can be used only for screening purposes, since it was not possible to achieve satisfactory quantitative determination. In addition, due to the vast differentiation of food commodities, a much wider range of samples, both in the context of matrix and residue composition should be tested before the novel system can be practically employed. Despite this, the novel biosensor for screening organophosphate pesticides offers a number of advantages over other conventional biosensor techniques, such as sensitivity and low cost, as well as the ability to monitor, in real-time, the presence of pesticide residues in food products. A particular advantage is the high speed of the assay (analysis time = 3 min), although this does not include the time required for sample extraction. However, we should mention that, for the practical application of the novel assay as a rapid screening tool, a single-or two-step extraction of the sample in an organic solvent would suffice. It must be emphasized that it is not meant to replace elaborate analytical methods but rather to assist the scaling up of food quality control, primarily designed to screen rapidly large amounts of agricultural products and food commodities for the presence of pesticide residues at the site of production (field), packaging, processing and/or sale. The application scope of the novel assay principle has already been extended to include complex organic contaminants [22] and mycotoxins [23]. Therefore, it represents a totally new generation of analytical instruments, enabling the implementation of food safety analysis by even minor users, such as small agricultural unions or food companies and can be potentially used by all parties involved in the chain of food production, processing and distribution, a market with a volume of 3.6 billion € in the EU-27 region only.

5. Acknowledgements

The research project was partially funded by the EMBIO Project of the Cypriot Ministry of Industry, Commerce and Tourism. The authors express their thanks to Dr. Georgia Moschopoulou for her comments and assistance in the compilation of the manuscript.

REFERENCES

[1] FAOSTAT, "Database on Pesticide Consumption. Rome: Food and Agriculture Organization of the United Nations," Statistical Analysis Service, 2009.

[2] M. Pohanka, K. Musilek and K. Kuca, "Progress of Biosensors Based on Cholinesterase Inhibition," *Current Medicinal Chemistry*, Vol. 16, No. 14, 2009, pp. 1790-1798. doi:10.2174/092986709788186129

[3] R. A. Maselli and B. C. Soliven, "Analysis of the Organophosphate-Induced Electromyographic Response to Repetitive Nerve Stimulation: Paradoxical Response to Edrophonium and D-Tubocurarine," *Muscle & Nerve*, Vol. 14, No. 12, 1991, pp. 1182-1188. doi:10.1002/mus.880141207

[4] B. Eskenazi and N. A. Maizlish, "Effects of Occupational Exposure to Chemicals on Neurobehavioral Functioning," In: R. E. Tarter, D. H. V. Thiel and K. L. Edwards, Eds., *Medical Neuropsychology: The Impact of Disease on Behavior*, New York, 1988, pp. 409-419. doi:10.1007/978-1-4757-1165-3_9

[5] L. Rosenstock, M. Keifer and W. E. Daniell, "Chronic Central Nervous System Effects of Acute Organophosphate Pesticide Intoxication. The Pesticide Health Effects Study Group," *Lancet*, Vol. 338, No. 8761, 1991, pp. 223-227. doi:10.1016/0140-6736(91)90356-T

[6] M. Eddieston, N. A. Buckley, P. Eyer and A. H. Dawson, "Medical Management of Acute Organophosphorus Pesticide Self-Poisoning," *Lancet*, Vol. 371, No. 9612, 2008, pp. 597-607. doi:10.1016/S0140-6736(07)61202-1

[7] J. Haib, I. Hofer and J. M. Renaud, "Analysis of Multiple Pesticide Residues in Tobacco Using Pressurized Liquid

Extraction, Automated Solid-Phase Extraction Clean-Up and Gas Chromatography-Tandem Mass Spectrometry," *Journal of Chromatography A*, Vol. 1020, No. 2, 2003, pp. 173-187. doi:10.1016/j.chroma.2003.08.049

[8] J. Sherma, "Review of Advances in the Thin Layer Chromatography of Pesticides: 2004-2006," *Journal of Environmental Science & Health B*, Vol. 42, No. 4, 2007, pp. 429-440. doi:10.1080/03601230701316440

[9] R. T. Andres and R. Narayanaswamy, "Fibre-Optic Pesticide Biosensor Based on Covalently Immobilized Acetylcholinesterase and Thymol Blue," *Talanta*, Vol. 44, No. 8, 1997, pp. 1335-1352. doi:10.1016/S0039-9140(96)02071-1

[10] J. W. Choi, Y. K. Kim, I. H. Lee, J. Min and W. H. Lee, "Optical Organophosphorus Biosensor Consisting Acetyl Cholinesterase/Viologen Hetero Langmuir-Blodjett Film," *Biosensors and Bioelectronics*, Vol. 16, No. 9-12, 2001, pp. 937-943. doi:10.1016/S0956-5663(01)00213-5

[11] V. G. Andreou and Y. D. Clonis, "A Portable Fiber-Optic Pesticide Biosensor Based on Immobilized Cholinesterase and Sol-Gel Entrapped Bromcresol Purple for In-Field Use," *Biosensors and Bioelectronics*, Vol. 17, No. 1-2, 2002, pp. 61-69. doi:10.1016/S0956-5663(01)00261-5

[12] S. Mavrikou, K. Flampouri, G. Moschopoulou, O. Mangana, A. Michaelides and S. Kintzios, "Assessment of Organophosphate and Carbamate Pesticide Residues in Cigarette Tobacco with a Novel Cell Biosensor," *Sensors*, Vol. 8, No. 4, 2008, pp. 2818-2832. doi:10.3390/s8042818

[13] E. Flampouri, S. Mavrikou, S. Kintzios, G. Miliadis and P. Aplada-Sarli, "Development and Validation of a Cellular Biosensor Detecting Pesticide Residues in Tomatoes," *Talanta*, Vol. 80, No. 5, 2009, pp. 1799-1804. doi:10.1016/j.talanta.2009.10.026

[14] J. Reinert and M. Yeoman, "Plant Cell Tissue Culture," Springer-Verlag, Berlin, 1982. doi:10.1007/978-3-642-81784-7

[15] M. Mizayawa, H. Tougo and M. Ishihara, "Inhibition of Acetylcholinesterase Activity by Essential Oil from *Citrus paradise*," *Natural Products Letters*, Vol. 15, No. 3, 2001, pp. 205-210. doi:10.1080/10575630108041281

[16] O. Aust, N. Ale-Agha, L. Zhang, H. Wollersen, H. Sies and W. Stahl, "Lycopene Oxidation Product Enhances Gap Junctional Communication," *Food Chemistry and Toxicology*, Vol. 41, No. 10, 2003, pp. 1399-1407. doi:10.1016/S0278-6915(03)00148-0

[17] J. Flaskos, W. G. McLean, M. J. Fowler and A. J. Hargreaves, "Tricresyl Phosphate Inhibits the Formation of Axon-Like Processes and Disrupts Neurofilaments in Cultured Mouse N2a and Rat PC12 Cells," *Neuroscience Letters*, Vol. 242, No. 2, 1998, pp. 101-104. doi:10.1016/S0304-3940(98)00054-8

[18] S. Madhavan, G. Sarath, B. H. Lee and R. S. Pegden, "Guard Cell Protoplasts Contain Acetylcholinesterase Activity," *Plant Science*, Vol. 109, No. 2, 1991, pp. 119-127. doi:10.1016/0168-9452(95)04164-P

[19] D. Ferentinos, C. P. Yialouris, P. Blouchos, G. Moschopoulou, V. Tsourou and S. Kintzios, "The Use of Artificial Neural Networks as a Component of a Cell-Based Biosensor Device for the Detection of Pesticides," *Procedia Engineering*, Vol. 47, 2012, pp. 989-992. doi:10.1016/j.proeng.2012.09.313

[20] E. Voumvouraki and S. Kintzios, "Differential Screening of the Neurotoxicity of Insecticides by Means of a Novel Electrophysiological Biosensor," *Procedia Engineering*, Vol. 25, 2011, 964-967. doi:10.1016/j.proeng.2011.12.237

[21] E. Flampouri and S. Kintzios, "Nafion and Polylysine Treated PEDOT Mammalian Cell Biosensor," *Procedia Engineering*, Vol. 25, 2011, pp. 964-967.

[22] V. Varelas, N. Sanvicens, M. P. Marco and S. Kintzios, "Development of a Cellular Biosensor for the Detection of 2, 4, 6-Trichloroanisole (TCA)," *Talanta*, Vol. 84, No. 3, 2010, pp. 936-940. doi:10.1016/j.talanta.2011.02.029

[23] E. Larou, I. Yiakoumettis, G. Kaltsas, A. Petropoulos, P. Skandamis and S. Kintzios, "High Throughput Cellular Biosensor for the Ultra-Sensitive, Ultra-Rapid Detection of Aflatoxin M1," *Food Control*, Vol. 29, No. 1, 2012, pp. 208-212. doi:10.1016/j.foodcont.2012.06.012

Changes in the Nuclei of Infected Cells at Early Stages of Infection with EMCV

Zaven A. Karalyan*, Hranush R. Avagyan, Hovakim S. Zakaryan, Liana O. Abroyan, Lina H. Hakobyan, Aida S. Avetisyan, Elena M. Karalova
Institute of Molecular Biology of the National Academy of Sciences of the Republic of Armenia, Laboratory of Cell Biology, Yerevan, Armenia
Email: *zkaralyan@yahoo.com

ABSTRACT

By the methods of quantitative cytophotometry, we have identified the changes in the nucleus and of some intranuclear compartments in the early stages of infection with encephalomyocarditis virus (EMCV). They can be characterized as early 1 - 2 hours post infection (hpi) and temporary increase (duration about 1 hour) in the content of the acidic proteins of the nucleolus, changing their decline to the control values. Then (after 1 - 2 hours) follows an increase in RNA content of nucleoli to 4 hours post infection (the process takes about 2 hours). The increase in RNA content in nucleoli in approximately the same time (slightly behind) with the activation of PML bodies (2 - 4 hpi). Then, the RNA content in nucleoli decreased to the control values, while simultaneously decreasing activity of PML bodies (ranging from 5 - 6 hpi). The early stages of infection EMCV is also characterized by the tendency to increase in the size of the nuclei of infected cells, and preserves at a later time. Then there is an increase in RNA content in the nucleus, roughly coinciding with the increased content of RNA in the nucleoli.

Keywords: EMCV; Acidic Proteins; RNA; Nucleolus; Nucleus

1. Introduction

The early stages of the replication of picornaviruses represent an extreme interest, this is due to a relatively short period of time that the viruses which they will start their own replication, but also establish control over a number of important cellular metabolic processes. The result of this phenomenon is a radical restructuring of the entire physiology of the infected cell. Part of the mechanism of this transformation has been clear for a long time, while others have only recently been clarified, but much of it is still unknown.

It is well known that picornaviruses are able to block cap-dependent translation. Since picornaviral translation is cap-independent by virtue of the 5 IRES, many of these viruses have evolved potent mechanisms to inhibit cellular cap-dependent translation during infection, thereby thwarting detrimental antiviral responses. The enteroviruses and aphthoviruses, for example, encode secondary proteases at their 2A and L positions respectively, which target eIF4G [1]. Cardioviruses do not have secondary proteases. Their L and 2A proteins have essential host shut-off roles, but use non-proteolytic mechanisms to achieve them. The EMCV L (67 aa) contributes to the inhibition of cap-dependent translation by triggering disruption of nucleocytoplasmic trafficking during infection. For the disruption of cap-dependent translation of the host cells with EMCV, 2A protein is responsible. As shown by [1], 2A NLS sequence is required for virus shutoff of cap-dependent host protein synthesis.

It has been shown that proteins of EMCV-the 2A, 3B (VPg), 3C (pro) and 3D (pol) can also be found within the nucleoli. The localization of these proteins occurs in the first 2 - 4 hours following infection of cells [2,3].

Therefore, it is important to study the influence of the picornaviruses on the nuclear structure of infected cells. As it is well known, the replication of picornaviruses occurs within the cytoplasm. However, at the earliest stages of viral infection many viral proteins are observed in the nucleus of affected cells. The research reported in this paper was directed towards determining the DNA, RNA and acidic proteins (non histone) as well as PML bodies in nuclei of EMCV infected cells.

*Corresponding author.

2. Materials and Methods

2.1. Cells

SK-N-MC human neuroblastoma cells were cultured in Dulbecco's Modified Eagle Medium (DMEM) (Sigma) supplemented with 10% heat-inactivated fetal bovine serum (FBS) at 37˚C in 5% CO_2.

2.2. Virus

EMCV (Columbia-SK strain) was used at multiplicity of infection 105 TCD50/ml on SK-N-MC. Viral titers were calculated by the method of Kärber. As a control the parallel conducted passages of noninfected cultures were used.

2.3. Image Cytophotometry

In order to quantitative DNA analyze of the received data, the cells preparations were fixed in 96% ethyl alcohol for 30 minutes and painted in fresh Shiffs reactive, by Feulgen (hydrolyze 5N HCL 60 minutes at 22˚C). The content of DNA in a nucleus and nucleolus was defined by computer-equipped microscope-photometer SMP 05 (OPTON). The image cytophotometry of DNA was performed on 575 nm wave [4]. Unstimulated human lymphocytes were used as diploid standards.

For quantification of RNA was used gallocyanin chromalum stain. To obtain reproducible staining results with these large sections, the method of Einarson was adapted to quantitative [5] and image analytical requirements. The image cytophotometry was performed on 610 nm wave. In each case controls were evaluated as 100%.

Fast green FCF staining (for acidic proteins) was used in Deitch modification [6,7]. The image cytophotometry was performed on 434 nm wave. In each case controls were evaluated as 100%.

2.4. Determination of the PML

Cells grown on glass cover slips were fixed in 4% paraformaldehyde/PBS (pH 7.5) for 5 minutes at room temperature, permeabilized in 0.5% Triton X-100 in PBS for 5 minutes at room temperature. PML protein was visualized with the monoclonal antibody. The determination of the PML was performed using monoclonal antibodies "PML PG-M3, Santa Cruz Biotechnology Inc"-catalogue no. sc-966 FITC [8].

2.5. Statistics

All experiments were conducted in triplicate. The significance of virus-induced changes was evaluated by two-tailed Student's t-test. p values < 0.05 were considered significant. SPSS version 15.0 software package (SPSS Inc., Chicago, IL, USA) was used for statistical analyses.

3. Results

The lytic EMCV infection was received by virus introduction on 48 h confluences of SK-N-MC culture. At 8 hours post infection (hpi) the virus titer reached 3.0 lg TCD50/ml, at 12-4.0, at 24-4.5, at 24-72-6.5 (**Figure 1**). The time-course of a single cycle of EMCV reproduction in SK-N-MC cells take place about 8-12 h.

The average content of the nucleolus to a nucleus does not change during the whole period of the experiment with the exception of some minuscule increases of the content of nucleolus in a nucleus (2.5-control, 2.9-at 6 hpi). The cytometry of the nucleolus values revealed the following results: in the whole there are not definite changes, but there may an emphasized tendency to increase the given value up to the 6 hpi. The total area of the nucleoli increases almost by 35%, but due to large variations in individual performance the difference is not reliable.

Thus, when the infection with picornaviruses occurs the activation of the nucleolar values is detected. It has a temporary nature—first of all increases in the synthesis of the acidic proteins and after that an increase in the content of the RNA and the area of the nucleoli occurs.

The quantitative indicators of DNA of the nucleus do not differ from the control values throughout the experiment, nor in the DNA content and neither in the distribution of DNA in ploidy classes.

The size of the nucleus tends to increase in comparison with the control ($p < 0.1$), by 1 hpi and remain so until the end of the observation period (6 hpi) (**Figure 2(a)**). In addition the increase in the size of the nucleus is small and varies from 8% - 11% from the baseline. It is important to note that the increase in the size of the nuclei of infected cells is not accompanied by an increase in the RNA or acidic proteins of the nucleus. However, if the content of acidic proteins of the nucleus remained in a relatively stable index throughout the study, the RNA has a tendency ($t = 1.8$, $p < 0.1$) increased by 5hpi, and coin-

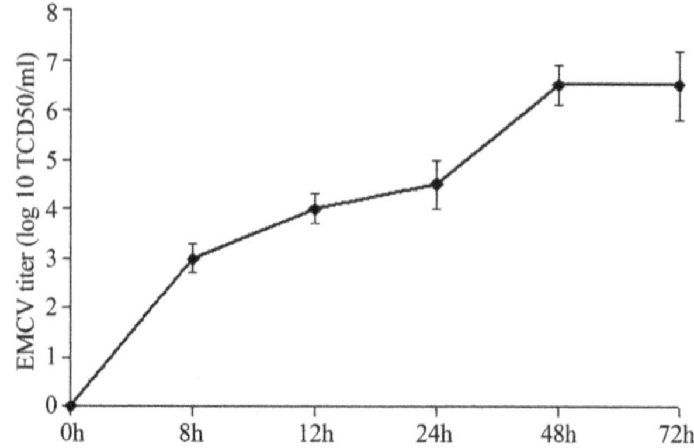

Figure 1. The titer of EMCV calculated on SK-B-MC cells.

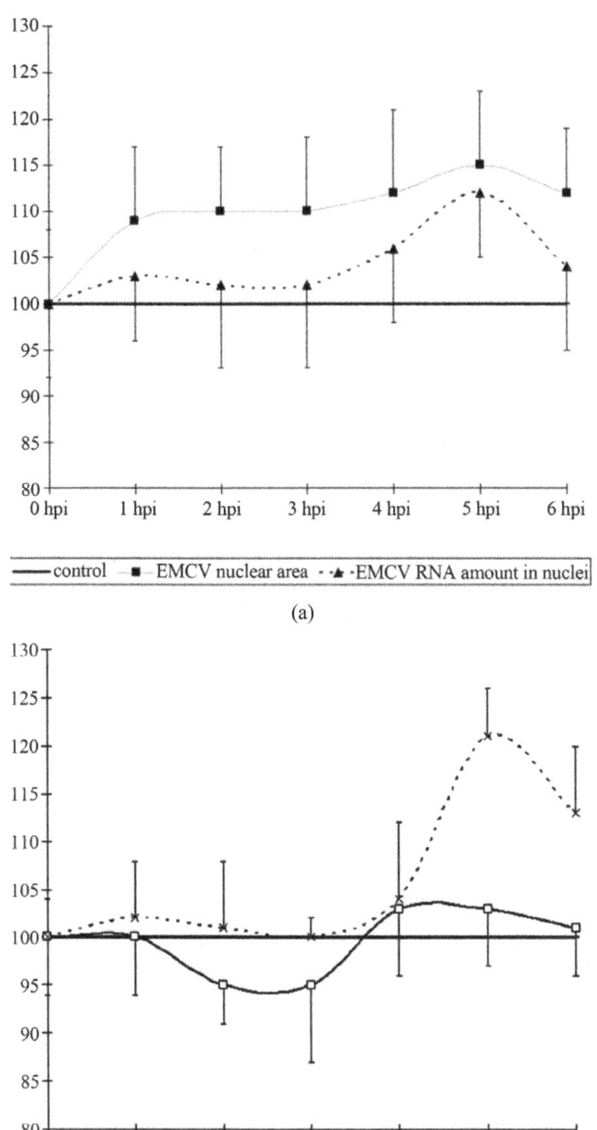

Figure 2. Dynamics of nuclear (a) and nucleolar (b) indices under the influence of EMCV infection. Data show percentage of control levels (average of 300 cells = 100%).

cides with an increase in RNA in the nucleoli (**Figure 2(b)**).

As it follows from the **Figure 3**, in the early stages of the infection, there is an increase in the content (amount, concentration) of the nucleolar acidic proteins (with a 1-2 hpi) ($t = 3.07$, $p < 0.05$). The elevated level of acidic proteins has a short-period character, and after 1 hour (to 3 hpi) its levels in the nucleolus does not differ from the benchmarks. The increased levels of the acidic proteins vary between 20% - 25% of the initial content.

The RNA in the nucleoli of the infected and control cells do not differ from each other, up to 4 hpi, when an

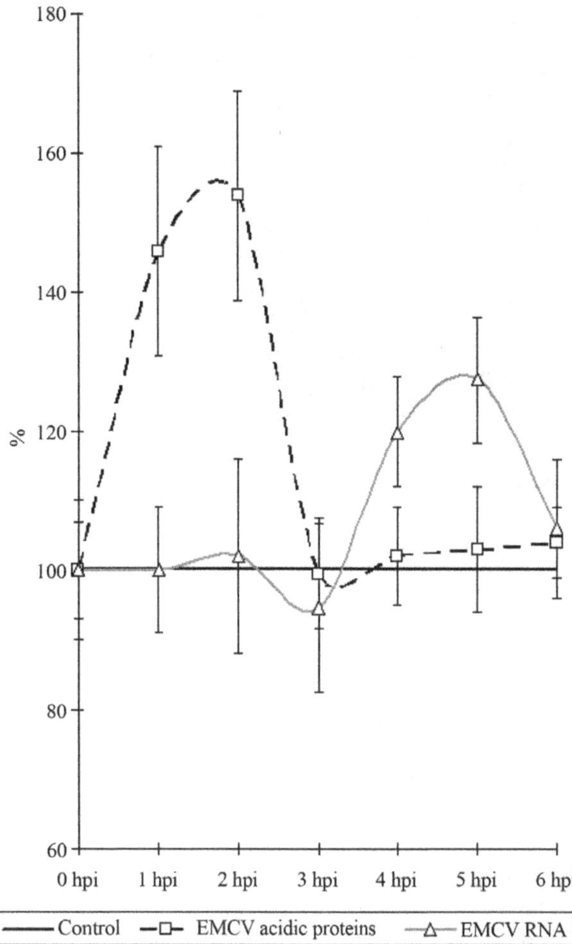

Figure 3. Dynamics of nucleolar RNA and acidic proteins under the influence of EMCV infection. Data show percentage of control levels (average of 300 cells = 100%).

increase in the RNA content in the nucleoli of infected cells begins to be observed (**Figure 3**) ($t = 2.17$, $p < 0.05$). The increase of the RNA in the nucleolus reaches 30% - 35% in comparison with the control values. The increase of the RNA as long with the proteins has a short period character and finishes after 5 hpi, decreasing down to the levels of the control values (6 hpi).

The intensity of the luminescence PML bodies, in the nuclei of the infected cells does not differ from the background values in the range from 1 to 2 hpi. Then there is a sharp increase in the intensity of luminescence, which indicates about the increased activity of PML bodies by 3 hpi. Increased activity of PML bodies completely fading away to a 4 HPI, after which it does not differ from the control values (**Figure 4**).

We have identified the changes in the nucleus and of some intranuclear compartments in the early stages of infection EMCV.

They can be characterized as early (1 - 2 hpi) and temporary increase (duration about 1 hour) in the content of the acidic proteins of the nucleolus, changing their

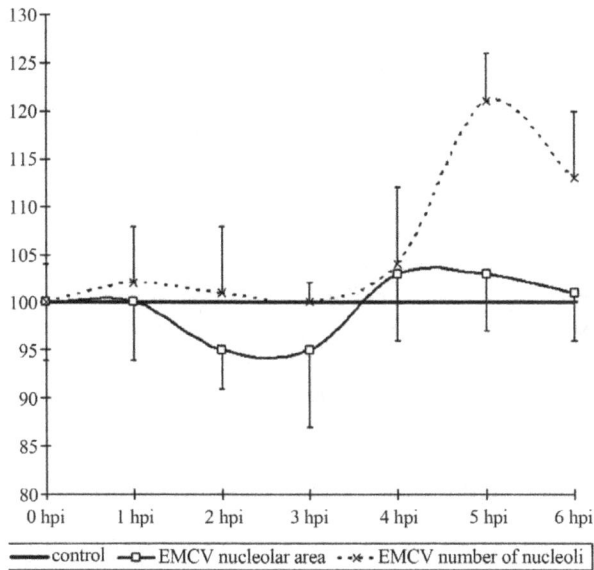

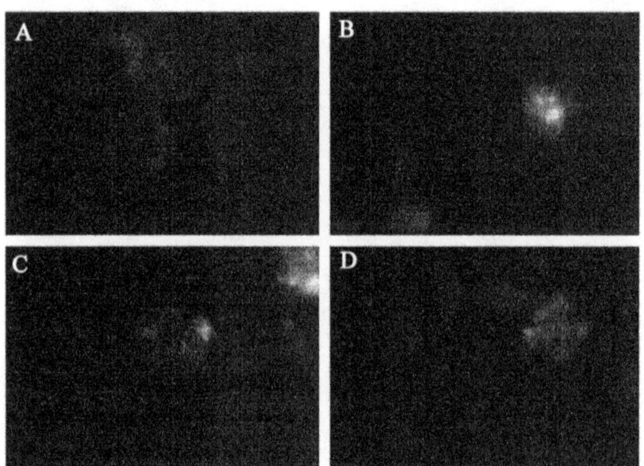

Figure 4. PML bodies in EMCV infected SK-N-MC cells. (A) PML bodies in control nuclei of SK-N-MC cells; (B) PML bodies in infected nuclei of SK-N-MC cells (3 h.p.i). (C) PML bodies in infected nuclei of SK-N-MC cells (6 h.p.i. EMCV); (D) PML bodies in infected nuclei of SK-N-MC cells (9 h.p.i. EMCV) (×400).

decline to the control values. Then (after 1 - 2 hours) follows an increase in RNA content of nucleoli to 4 hpi (the process takes about 2 hours). The increase in RNA content in nucleoli in approximately the same time (slightly behind) with the activation of PML bodies (2 - 4 hpi). Then, the RNA content in nucleoli decreased to the control values, while simultaneously decreasing activity of PML bodies (ranging from 5 - 6 hpi). The early stages of infection EMCV is also characterized by the tendency to increase in the size of the nuclei of infected cells, and preserves at a later time. Then there is an increase in RNA content in the nucleus, roughly coinciding with the increased content of RNA in the nucleoli.

The total content of acidic proteins of the nucleus is a relatively stable index, changing a little both in the control and in the early stages of infection EMCV.

4. Discussion

For EMCV is found the involvement of the viral proteins in the nuclear processes. Already in the early stages of infection (2 - 3 hpi) cardiovirus protein 2A, as well as proteins 3BVpg, 3Cpro, 3Dpolas a single precursor 3BCD are located in the nucleoli of the cells, where the synthesis of rRNA and ribosome assembly take place (Aminev *et al.*, 2003a, Aminev *et al.*, 2003b). The function of the cardiovirus proteins in the nucleus of infected cells have not been studied sufficiently. Protease 3Spro presumably responsible for the inhibition of synthesis of cellular mRNA, whereas the work of RNA polymerases I and III in cardiovirus infection are not inhibited [3]. The function performed by the protein 2A of cardiovirus in the nucleoli of the cells is not completely clear. Amines and colleagues suggest that the protein 2A in the nucleoli embedded in the mature ribosomal subunits, resulting in formation of modified ribosomes engaged mainly in the cytoplasm of virus-specific protein synthesis [2]. As shown by us, at this stage is a series of important structural and functional changes in the nucleoli of infected cells.

With regard of the above said the increased content of nucleolar acidic proteins in early stages of infection EMCV could be explained by two reasons. The first is the accumulation of the viral proteins and particularly proteins 2A, 3BVpg, 3Cpro, 3Dpol and its predecessor 3BCD. The second - the accumulation of nucleolar proteins with enzymatic activity (discussed below).We have identified the data which allow us to assert that in the early stages of infection (2 - 4 hpi) significant changes occur in the nucleoli, coinciding with the localization of viral proteins in them. These effects disappear in the later stages of infection (6 hpi and later). This suggests a short-term increase in the functional activity of the nucleolus [9] under the influence of infection EMCV. In addition to changes in the nucleoli, this process is accompanied by an increase in the PML bodies. PML bodies-small spherical domains are present in the nucleus of cells, they undergo morphological changes during the cell cycle. Their number is intensely variable, depending on the physiological state of the cells, cell cycle stage in viral infections and so on. The PML bodies are destroyed during some viral infections.

Incubation of the cells with IFN induces the synthesis of the PML-protein and inhibits the multiplication of the viruses. These cells are destroyed during a viral infection and for the replication of adenovirus DNA their destructtion is a necessary step that underlines the possible involvement of the corpuscles to provide antiviral activity of cells. Incubation of the cells with interferon induces the synthesis of PML-protein and inhibits proliferation of some viruses [10]. Upon infection of cells with the virus EMCV, it is shown a decrease in the number of the PML bodies, under the influence of the viral 3C protease. As the mechanism of reduction of PML bodies, the authors consider a possible direct degradation of the PML bodies 3C protease EMCV [11]. Thus, the sharp decrease in emission intensity of monoclonal antibodies to PML, to 5.6 HPI, is explained by the influence of non-structural viral proteins (3Cprotease), and aimed at suppressing IFN-mediated protection of the infected cell.

We have also investigated the acidic proteins of chromatin and the nucleolus in a control experiment and under the influence of the virus. Acidic proteins of the chromatin play an important role in the regulation of the genetic activity. It was shown that in proliferating tissues contain more acidic proteins than that of resting, and these proteins are richer with euchromatin than with heterochromatin. The acidic proteins restore the DNA-dependent RNA synthesis, inhibited by histone, which enhance the transcription of chromatin in vitro and the activation of genes. Acidic proteins, in contrast to the histones are tissue specific and bind to DNA only within the tissue from which they are allocated [12,13]. Taking into account of our data it should be concluded that the activation of the transcription processes are in the range of 2 - 4 HPI.

Nucleolar acidic proteins—are a large group of proteins associated with the functional activity of the nucleolar proteins and the equivalent argentophilic proteins [14]. The most important argentophilic proteins: RNA polymerase I, transcription factor UBF, nucleolin (C-23), nucleophosmin (nyumatrin or B-23), etc. In contrast to the widespread methods of silver protein, the quantitative cytochemical determination of acid proteins, allows us more accurate identify variations in protein content, and hence the functional activity of the nucleolus [15-17]. The role of these proteins in the replication of EMCV is unquestionable, since it is next to nucleophosmin (B-23) are localized EMCV proteins-the 2A, 3B (VPg), 3C (pro) and 3D (pol).

It should be noted that the EMCV does not need nuclear structures sensitive cells, as it is able to replicate in a cell-free medium containing the individual components of the damaged cells in the lysates [18]. Consequently, for successful replication the virus does not require modification of nucleolar apparatus.

We can therefore conclude that any changes in the nucleoli are the result of the reaction of cells to viral infection or virus modifies the activity of the nucleolus to the successful suppression of cellular antiviral mechanisms.

In our experiments, an elevated level of the acidic proteins in the nucleoli was preceded by the increased content of RNA. The time gaps were approximately 1 - 2 hours. RNA synthesis in the nucleoli is directly related to the formation of the nucleoli. It should be noted that in the temporal boundaries the biosynthesis of the nucleoli fits into a space equivalent to about 1 hour. 7 - 10 minutes is the transcription of the rDNA, the synthesis of the small subunit of the ribosome—15 - 30 min, the synthesis of the large subunit of the ribosome—20 - 40 minutes [19]. It is possible that identified by us the time difference in the content of the acidic proteins and RNA in the nucleolus is associated with the time of the biosynthesis of ribosomes.

REFERENCES

[1] R. Groppo, B. A. Brown and A. C. Palmenberg, "Mutational Analysis of the EMCV 2A Protein Identifies a Nuclear Localization Signal and an eIF4E Binding Site," *Virology*, Vol. 410, No. 1, 2011, pp. 257-267. doi:10.1016/j.virol.2010.11.002

[2] A. G. Aminev, S. P. Amineva and A. C. Palmenberg, "Encephalomyocarditis Viral Protein 2A Localizes to Nucleoli and Inhibits Cap-Dependent mRNA Translation," *Virus Research*, Vol. 95, No. 1-2, 2003, pp. 45-57. doi:10.1016/S0168-1702(03)00162-X

[3] A. G. Aminev, S. P. Amineva and A. C. Palmenberg, "Encephalomyocarditis Virus (EMCV) Proteins 2A and 3BCD Localize to Nuclei and Inhibit Cellular mRNA Transcription but Not rRNA Transcription," *Virus Research*, Vol. 95, No. 1-2, 2003, pp. 59-73. doi:10.1016/S0168-1702(03)00163-1

[4] A. D. Deich, "Introduction to Quantitative Cytochemistry," Academic Press New York/London, 1966, pp. 65-67.

[5] W. Sandritter, G. Kifer and W. Rik, "Gallocyanin-chromalum Stain," In: G. L. Wied, Ed., *Introduction to Quantitative Cytochemistry*, Academic Press, New York and London, 1966, pp. 153-170.

[6] A. C. Dhar and C. K. Shah, "Cytochemical Method to Localize Acidic Nuclear Proteins," *Stain Technology*, Vol. 57, No. 3, 1982, pp. 151-155.

[7] B. R. Zirkin, "A cytochemical Study of the Nonhistone Protein Content of Condensed and Extended Chromatin," *Experimental Cell Research*, Vol. 78, No. 2, 1973, pp. 394-398. doi:10.1016/0014-4827(73)90084-0

[8] C. C. Yin, A. B. Glassman, P. Lin, J. R. Valbuena, D. Jones, R. Luthra and L. J. Medeiros, "Morphologic, Cytogenetic, and Molecular Abnormalities in Therapy-Related Acute Promyelocytic Leukemia," *American Journal of Clinical Pathology*, Vol. 123, No. 6, 2005, pp. 840-848. doi:10.1309/TJFFK819RPCLFKJ0

[9] C. C. Morton, J. A. Brown, W. M. Holmes, W. E. Nance and B. Wolf, "Stain Intensity of Human Nucleolus Organizer Region Reflects Incorporation of Uridine in to Mature Ribosomal RNA," *Experimental Cell Research*, Vol. 145, No. 2, 1983, pp. 405-413. doi:10.1016/0014-4827(83)90019-8

[10] H. Zakaryan and T. Stamminger, "Nuclear Remodelling during Viral Infections," *Cellular Microbiology*, Vol. 13, No. 6, 2011, pp. 806-813. doi:10.1111/j.1462-5822.2011.01596.x

[11] B. El McHichi, T. Regad, M. A. Maroui, M. S. Rodriguez, A. Aminev, S. Gerbaud, N. Escriou, L. Dianoux and M. K. Chelbi-Alix, "SUMOylation Promotes PML Degradation during Encephalomyocarditis Virus Infection," *Journal of Virology*, Vol. 84, No. 22, 2010, pp. 11634-11645. doi:10.1128/JVI.01321-10

[12] R. Baserga and G. Stein, "Nuclear Acidic Proteins and Cell Proliferation," *Federation Proceedings*, Vol. 30, No. 6, 1971, pp. 1752-1759.

[13] G. Stein and R. Baserga, "Cytoplasmic Synthesis of Acidic Chromosomal Proteins," *Biochemical and Biophysical Research Communications*, Vol. 44, No. 1, 1971, pp. 218-223.

[14] F. Thiebaut, J. P. Rigaut and A. Reith, "Improvement in the Specificity of the Silver Staining Technique for AgNOR-Associated Acidic Proteins in Paraffin Sections," *Stain Technology*, Vol. 59, No. 3, 1984, pp. 181-188.

[15] M. Okuwaki, "The Structure and Functions of NPM1/ Nucleophsmin/B23, a Multifunctional Nucleolar Acidic

Protein," *The Journal of Biochemistry*, Vol. 143, No. 4, 2008, pp. 441-448. doi:10.1093/jb/mvm222

[16] E. A. Sorokina, J. A. Wesson and J. G. Kleinman, "An Acidic Peptide Sequence of Nucleolin-Related Protein Can Mediate the Attachment of Calcium Oxalate to Renal Tubule Cells," *Journal of the American Society of Nephrology*, Vol. 15, No. 8, 2004, pp. 2057-2065. doi:10.1097/01.ASN.0000133024.83256.C8

[17] R. Voit, A. Kuhn, E. E. Sander and I. Grummt, "Activation of Mammalian Ribosomal Gene Transcription Requires Phosphorylation of the Nucleolar Transcription Factor UBF," *Nucleic Acids Research*, Vol. 23, No. 14, 1995, pp. 2593-2599. doi:10.1093/nar/23.14.2593

[18] Y. V. Svitkin and N. Sonenberg, "Cell-Free Synthesis of Encephalomyocarditis Virus," *Journal of Virology*, Vol. 77, No. 11, 2003, pp. 6551-6555. doi:10.1128/JVI.77.11.6551-6555.2003

[19] A. S. Stoykova, M. D. Dabeva, R. N. Dimova and A. A. Hadjiolov, "Ribosome Biogenesis and Nucleolar Ultrastructure in Neuronal and Oligodendroglial Rat Brain Cells," *Journal of Neurochemistry*, Vol. 45, No. 6, 1985, pp. 1667-1676. doi:10.1111/j.1471-4159.1985.tb10521.x

Wound Closure on the Neonatal Rat Skin I. The Modulation of the Thickness of Epidermis at the Closing Incisional Wounds

Mary Arai, Takashi Matsuzaki, Setsunosuke Ihara*

Department of Biological Science, Faculty of Life and Environmental Science, Shimane University, Matsue, Japan

Email: *ihara@life.shimane-u.ac.jp

ABSTRACT

Full-thickness incisional wounds were made on the dorsal skin of 1-day-old rats to elucidate the mechanism of the fluctuation of the epidermal thickness after the wound closure. The thickness of the epidermis covering the wound reached a peak around 96 h post-wounding (PW), and became thinner thereafter. The analyses of the cell proliferation and apoptosis at the epidermal wound regions revealed that the rate of TUNEL-positive cells that displays the cells undergoing apoptosis increased as the epidermis became thinner around 120 h PW. Next, immunohistochemical analyses using antibodies against keratinocyte differentiation marker proteins indicated that the delay or interruption of the spinous to granular transition from 96 to 120 h PW might result in the epidermal thickening in the wound region. Third, the region undyed with anti-caspase-14 antibody extended downward in the thickened epidermis by 96 h PW, and in turn, it became intensely and widely stained with this antibody in the thinning epidermis by 120 h PW. Taken together, it is likely that the delay and acceleration of the terminal differentiation, including cornification of the epidermal keratinocytes may coordinately cause the fluctuation of the thickness of the epidermis at the wound site in rat neonates.

Keywords: Wound Healing; Reepithelialization; Rat; Neonate; Epidermis; Terminal Differentiation; Cornification

1. Introduction

Reepithelialization, a part of the process of the wound healing, is essential for organisms to survive. The patterns of reepithelialization are known to differ during the developmental stages [1-6]. First, the wounds in mammalian adult skin are sealed with fibrin clot and then the wound closure is driven by epidermis crawling from the wound margin [7]. Meanwhile, the wounds in the neonatal rat skin are covered rapidly with the extended wound-surrounding epidermis, and the abnormal expression patterns of keratins and cadherins at the wound site have been revealed by immunohistochemical analyses [6,8].

However, the late stages of reepithelialization after the wound closure have not yet been well investigated. In this study, we focused on the phenomenon that, once full-thickness incisional wounds made on the skin of neonatal rats were closed, the thickness of the covering epidermis increased and temporarily overran and, in turn, restored its normal level.

First, we examined the possible fluctuation of the epidermal cell number at the wound site, and calculated the rate of the cell proliferation as well as that of the cell death detected by anti-BrdU immunostaining and TUNEL, respectively. Second, we analyzed the terminal differentiation of the epidermal keratinocytes. In order to identify the differentiation stage of the wound epidermis after the wound closure, the immunohistochemistry was performed using antibodies of involucrin, profilaggrin/filaggrin, and loricrin as the epidermal differentiation markers [9-12]. Finally, we examined the distribution of CCAAT/enhancer-binding protein-α (C/EBP-α) and caspase-14. These two proteins are known to be endogenous factors regulating the terminal differentiation of the epidermal keratinocytes at the early or late stages, respectively [12-20].

The findings in the present study are summarized as follows. The delay or derangement in the terminal differentiation of keratinocyte occurs within the areas from the upper spinous to the granular layer in the thickening epidermis at the wound region. The thinning of the

*Corresponding author.

wound epidermis may also be associated with the cornification probably accompanied by an enhancement of the cell death and the action of caspase-14.

2. Materials and Methods

2.1. Wounding

Sprague-Dawley rats were anesthetized at 1 day after birth. Two incisional wounds 3 mm in length were made in the back skin on either side of the dorsal midline with a disposable scalpel (FEATHER). The wound sites were immediately covered with the cover agent Opsite (Smith & Nephew), and the wounded individuals were returned to the mother rats. The Opsite treatment had no effects on the wound healing except that it avoided unnecessary retardation of healing due to the infection that otherwise occasionally occurred (Koizumi *et al*., 2004).

2.2. BrdU Injection

Two hours before sampling, 10 μl/gram body-weight 10 mM 5-bromo-2'-deoxyuridine (BrdU) were injected into the abdominal cavity of rats.

2.3. Histology and Immunohistochemistry

The skin surrounding the wound was cut out at 0, 48, 72, 96, 120, or 144 h PW. Pieces of the skin were fixed with 4% paraformaldehyde in Ca^{2+}, Mg^{2+}-free phosphate-buffered saline [PBS(−)] for 1 h at room temperature, and embedded in paraffin. Sections (4 μm thickness) were stained with hematoxylin-eosin (HE) or analyzed immunohistochemically. For immunohistochemistry, deparaffinized sections were pretreated with citric acid buffer (pH 6.0) at 95˚C, and rinsed with PBS(−). Alternatively, the sections were pretreated with 2N HCl, followed by digestion with 0.05% trypsin at 37˚C. The pretreated sections were incubated with 1% normal horse serum in PBS(−) for 20 min, and then reacted with primary antibodies at 4˚C overnight. They were washed with PBS(−), then incubated with secondary antibodies for 2 h. Samples were washed with PBS(−), counterstained with 4',6-diamino-2-phenylindole dihydrochloride (DAPI, Polysciences) at 0.1 μg/ml, and mounted with Fluoromount (Japan Tanner Corporation). The following primary antibodies were used: mouse anti-cytokeratin 14 monoclonal antibody (mAb) (Chemicon, 1:100); mouse anti-cytokeretin 10/13 mAb (Lab Vision, 1:100); mouse anti-cytokeratin 6 mAb (Progen, 1:50); rabbit anti-mouse involucrin poriclonal antibody (pAb) (Berkley antibody company, 1:500); rabbit anti-mouse loricrin pAb (abcam, 1:500); rabbit anti-rat profilaggrin/filaggrin pAb (466) (generously supplied by Dr. R. B. Presland, 1:250); and mouse anti-BrdU mAb (Dako Cytomation, 1:100). Alexa fluor 594 goat anti-mouse IgG_1 (Molecular Probes, 1:1500); Alexa fluor 488 goat anti-mouse IgG_{2a} (Molecular Probes, 1:500); Alexa fluor 488 goat anti-mouse IgG_1 (Molecular Probes, 1:500); and Alexa fluor 488 goat anti-rabbit IgG (Molecular Probes, 1:500) were used as secondary antibodies.

The double staining of loricrin with involucrin and involucrin with filaggrin, and the staining of C/EBP-*α* and caspase 14 were performed using immunoenzyme technique. Deparaffinized sections were treated with citric acid buffer (pH 6.0) at 95˚C or 10% TritonX-100 in PBS(−), and endogenous peroxidase was inactivated by H_2O_2 before blocking process. After incubation with HRP-conjugated anti-rabbit immunogrobulins (DakoCytomation, 1:100) as the secondary antibody, the samples were reacted with DAB for color reaction. The samples for the double staining were subjected to staining with the next primary antibodies, followed by staining the alkaline phosphatase-conjugated anti-rabbit immunogrobulins (sigma-aldrich, 1:100) as the next secondary antibody. The immunostained sections were finally counterstained with hematoxylin. The following primary antibodies were used: anti-C/EBP-*α* rabbit pAb (Santa Cruz, 1:100); and anti-caspase 14 rabbit pAb (IMGENEX, 1:1000).

2.4. TUNEL

The TUNEL reaction was performed using *in situ* cell death detection kit and TMR red (Roche Diagnotics) according to the manufacture's instructions.

2.5. Numerical Applications of the Rate of BrdU and TUNEL Positive Cells

The ratios of the numbers of BrdU- or TUNEL-positive cells to those of DAPI positive cells in the wound areas or the intact areas of epidermis were calculated. The epidermal regions between both wound edges and the normal interfollicular epidermal regions more than 5 mm away from the wound edges were defined as the wound areas and the intact areas of epidermis, respectively, in this study.

3. Results

3.1. Morphological and Immunohistochemical Changes in the Incisional Wound

The standard technique using a scalpel No. 11 (FEATHER) in this study usually generated the full-thickness incisional wounds 1 mm in depth, as shown in **Figure 1(A)** (0 h PW). By 48 h PW, wound closure was completed by the epidermal sheets that crawled from the wound edge and contacted with each other at the wound center (**Figure 1(B)**). From 72 to 120 h PW, the epidermis covering the wound was thickened downward the dermis as compared with the normal area of epidermis, and the thick-

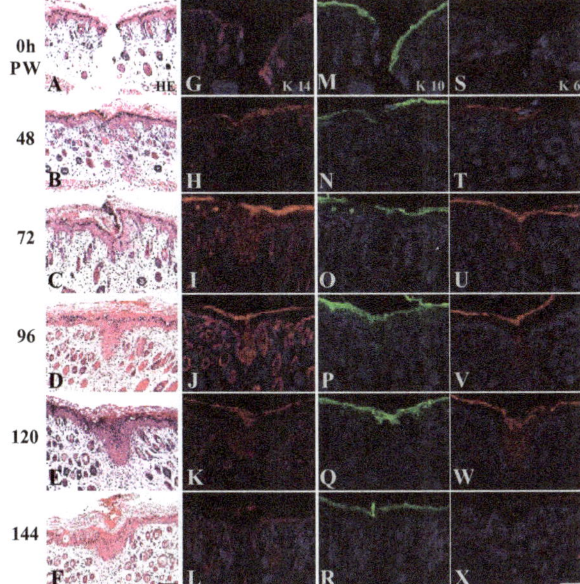

Figure 1. Morphological and immunohistochemical changes in neonatal rat skin wounds during reepithelialization. Serial cross sections were prepared and stained with hematoxylin-eosin ((A)-(F)), and anti-K14 ((G)-(L)), K10 ((M)-(R)), or K6 ((S)-(X)) antibodies. At 0 h PW, K14 and K10 keratins were localized at the basal layer (G), and the suprabasal layer (M), respectively. In contrast, K6 keratin was undetectable at 0 h PW (S). Around 48 h PW, wound closure was completed (B). By that time, the distribution of K14-positive cells were expanded vertically throughout the suprabasal layers and stretched horizontally from the wound edge (H); K10-positive cells were absent from the wound-covering areas (N); and K6 keratin became detectable in the suprabasal layers through the wound-covering and the wound-surrounding areas (T). By 72 h PW when the wound had been closed, the epidermis was thickened downward the dermis (C). The thickening lasted up to 96 h PW (D). These extending patterns of K 14 and K 6 were gradually shifted to the normal patterns from the distal region of the wounds from 72 to 120 h PW ((I)-(K), (U)-(W)). Conversely K10-positive cells were gradually increased with time in the wounded areas ((O)-(Q)). At 120 h PW, the thickened epidermis was beginning to be thinner ((E), (F)). The normal localization of these keratins was almost recovered in the wounded areas at 144 h PW ((L), (R), (X)). *Scale bars*, 100 μm; and *Counterstained*, with DAPI.

ening peaked around 96 h PW (**Figures 1(C)-(E)**). At 144 h PW, the thickened epidermis tended to become thinner (**Figure 1(F)**), and finally, the wound-covering epidermis almost restored the normal thickness 192 h PW (data not shown).

In addition, the immunohistochemical analyses using the epidermal differentiation markers showed that K14 keratin was localized at the basal layer, and the localization of K10 keratin was observed at the suprabasal layer, whereas K6 keratin was not detectable, at 0 h PW (**Figures 1(G), (M), (S)**). The localizations of these keratins-positive cells in the wound-covering and the wound-surrounding epidermis fell into disorder within 48 hours after wounding (**Figures 1(H), (N), (T)**), and gradually recovered the normal patterns from the distal region of wound during reepithelialization (**Figures 1(I)-(L), (O)-(R), (U)-(X)**).

3.2. The Analyses of the Process in Which the Temporarily Thickened Epidermis Becomes Thinner

From the above-mentioned histological observations, we noticed that the epidermis was thickened just after wound closure, and thereafter became thinner. Then, we became interested in what was the cause of the fluctuation of the epidermal thickness at the wound areas. To begin with, we speculated that the number of the epidermal cells would change at the wound. Exactly the living cells in the thickened epidermis outnumbered those in the control epidermis, as long as the total numbers of DAPI-positive nuclei were compared (**Figure 2(Q)**). Therefore, we further analyzed cell proliferation and apoptosis in the wound and intact areas of epidermis to find the answer to the fluctuation of the epidermal cell number.

BrdU-positive cells as the proliferating cells were detected in the basal layer at 0 h PW and in the wound-covering epidermis from 48 h PW onward, too (**Figures 2(A)-(F)**). Here, the absolute number of BrdU-positive cells in the epidermal wound regions suggested that the cell proliferation took place during the thickening of the epidermis in the wound region. However, when the ratios of the number of the proliferating cells to the total number of living cells were estimated, the rate in the wound areas of epidermis were almost constant throughout the period tested and, moreover, there was no great difference between these rates in the wound and the intact epidermis (**Figures 2(G)** and **(H)**). Consequently, it was likely that the down-regulation of cell proliferation, if any, did not so much contribute to the thinning of the wound-covering epidermis.

Next, we did TUNEL staining to test the possibility that apoptosis is induced in the thickened epidermis as the epidermis became thinner and reduced in cell number. From 0 h to 96 h PW, the rates of TUNEL-positive cells in both the wound and intact epidermis were generally similar, though more or less variable (**Figures 2(I)-(L), (O), (P)**). At 120 h PW, however, the rate of TUNEL-positive cells in the epidermal wound site significantly increased, as compared with that in the intact or 0 h PW (**Figures 2(M), (O), (P)**). The wound region at 144 h PW had still a high rate of TUNEL-positive cells (**Figures 2(N), (O), (P)**). The epidermis in the wound area reached a rate equivalent to that of the normal epidermis at 192 h PW (data not shown). It should be noted that TUNEL-positive cells were observed only in the vicinity of the nucleated upper granular layer in both

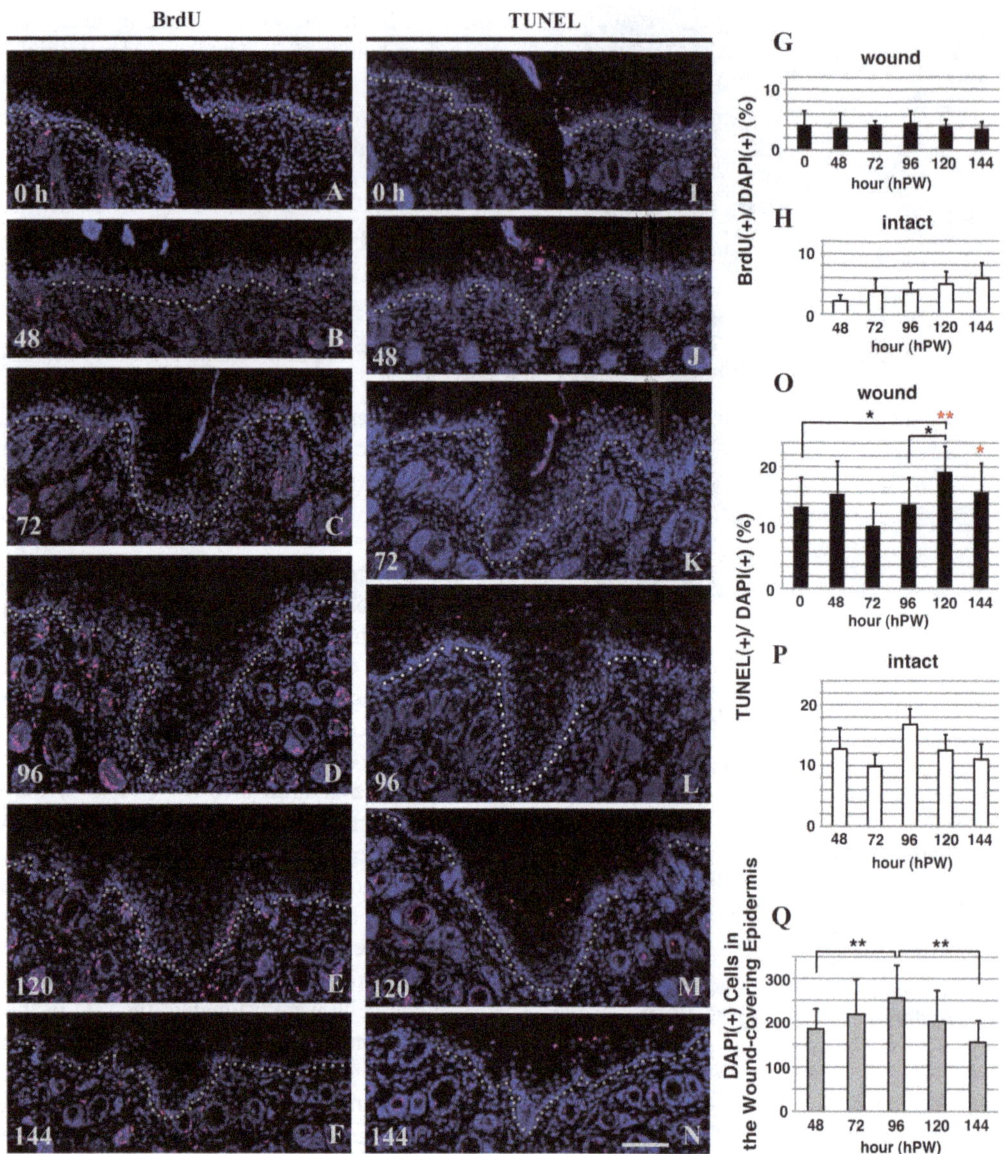

Figure 2. Cell proliferation and apoptosis in the wound and intact areas of epidermis. The rate of the number of "BrdU-positive cells (stained red)", which represents the proliferating cells, to the total number of living cells in the wound areas of epidermis were almost constant throughout the period tested (0 - 144 h PW, (A)-(G)). In contrast, the rate of "TUNEL-positive cells (stained red)" that displays the cells undergoing apoptosis increased as the epidermis became thinner (around 120 h PW), ((I)-(O)). The total numbers of DAPI-positive nuclei in the wound areas of epidermis at 48 - 144 h PW were counted as the living cells (Q). ***Scale bar*, 100 μm; *counterstain*, DAPI; *dashed lines*, the border between the epidermis and dermis; *, $p < 0.05$; **, $p < 0.01$; and *Red asterisks*, comparisons of intact values.**

wound and intact epidermis.

These results suggested that cornification accompanying cell death is significantly elevated during the thinning of the epidermis at the wound region.

3.3. The Rates of the Terminal Differentiation May Be Perturbed in the Epidermal Wound Areas

The epidermal keratinocytes generate diverse differentiation marker proteins during the terminal differentiation. We examined the localizations of these markers. First, the double immunostaining of loricrin and involucrin was executed during the process of wound healing. In both the wound and normal areas of the neonatal skin at 0 h PW, loricrin (brown) was expressed at and above the granular layer, whereas involucrin (blue) at and above the upper spinous layer (**Figure 3(A)**): *i.e.*, there was a clear spatial (and possibly temporal) difference in the start of expression between loricrin and involucrin. At 48 h PW when epidermis from the wound margin completed wound closure, such normal expression pattern remained

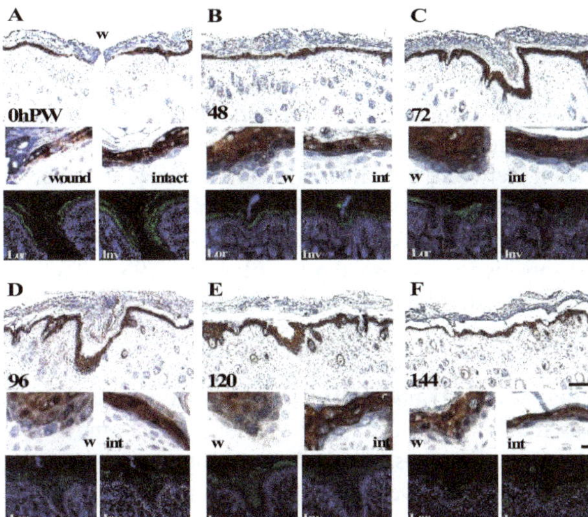

Figure 3. The loricrin-positive lowermost epidermal layer became identical to the involucrin-positive one from the middle of reepithelialization. The upper and middle panels (the magnifications of wound (left) and intact (right) areas) show the expression of anti-loricrin (brown) and involucrin (blue) detected by double staining using the enzyme-labeled antibodies, and the lower panels show these proteins single-stained with fluorescence-labeled antibodies (green) in the adjacent sections. Loricrin (brown) was expressed at and above the granular layer, and involucrin (blue) appeared from the upper spinous layer in the neonatal skin at 0 h PW (A). It should be noted that these two proteins were apt to be stained from the same layer at 96 and 120 h PW in the epidermal thickening regions ((D) and (E) wound). *Scale bars*, 100 μm, 10 μm, and 100 μm from above; *counterstain*, hematoxylin and DAPI; and *dashed lines*, the border between the epidermis and dermis.

at the wound-covering epidermis (**Figure 3(B)**). On the contrary, at 96 and 120 h PW when the epidermis at the wound areas was thickened, these two proteins were expressed from the identical cell layer in the epidermal thickening regions (**Figures 3(D)** and **(E)**), although it remained unclear which protein sifted. Later, the epidermal wound area getting thinner inchmeal recovered the normal expression pattern (144 h PW, **Figure 3(F)**).

Next, we carried out the double immunostaining of the epidermis of the neonates using the anti-involucrin and anti-profilaggrin/filaggrin antibodies. In both the wound and normal areas at 0 h PW, the expression of profilaggrin/filaggrin (blue) was observed at the zone lower than involucrin (brown), concretely, at and above the middle spinous cell layer (**Figure 4(A)**). In contrast, unlike the relationship of expression of loricrin and involucrin, the spatial difference in the expression of involucrin and profilaggrin/filaggrin were maintained in the same way as in the normal areas in the same thickening epidermis specimens. Such an expression pattern remained unchanged by the thinning phase of the wound-covering epidermis (**Figures 4(B)-(F)**).

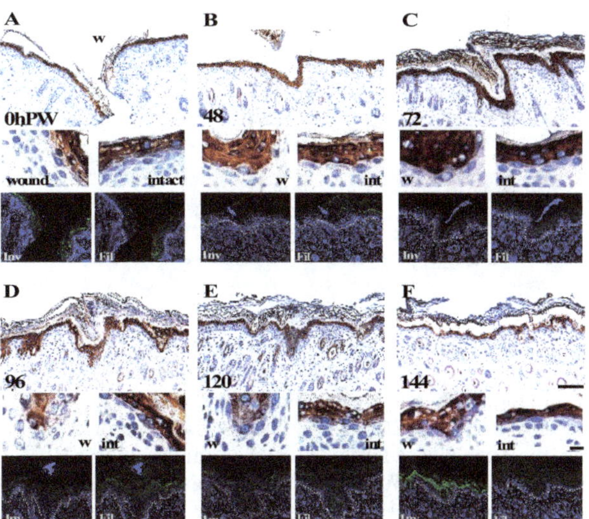

Figure 4. The expression patterns of filaggrin remained unchanged during reepithelialization. The results of double staining of involucrin (brown) and filaggrin (blue) and single staining in the adjacent sections (green) are arranged in the same way as in Figure 3. Unlike the expression of involucrin (brown) that started from upper spinous layer, filaggrin (blue) was expressed at and above the middle spinous layer in the 0 h PW (A), and maintained this hierarchical relation until 144 h PW ((B)-(F)). *Scale bars*, 100 μm, 10 μm, and 100 μm from above; *counterstain*, hematoxylin and DAPI; and *dashed lines*, the border between the epidermis and dermis.

3.4. The Localizations of Expression of Molecules Participating in the Epidermal Terminal Differentiation Fluctuated during Reepithelialization

We immunohistochemically examined the expression of C/EBP-α and caspase-14, known to participate in the terminal differentiation of the epidermal keratinocytes at the early and late stages, respectively. At 0 h PW, C/EBP-α was intensely stained at both the nuclei and the cytoplasm of the cells in the suprabasal layer, while caspase-14 at the cytoplasm in the granular layer (**Figures 5(A)** and **(H)**). The staining intensities of both molecules at the leading edge of the epidermis were obviously fallen by 24 h PW (**Figures 5(B)** and **(I)**). At 48 h PW, these immunostaining intensities were restored to the normal levels at the wound-covering epidermis. The epidermis at 72 h PW showed a similar tendency (**Figures 5(C), (D), (J), (K)**). By 96 h PW when the thickness of the wound-covering epidermis reached the peak, the region intensely stained with anti-caspase-14 was confined to the upper layer at the thickening epidermis, and the unstained region remarkably extended (**Figure 5(L)**). Subsequently, however, the region intensely stained with anti-caspase-14 in turn enlarged downward by 120 h PW in the midst of the thinning phase (**Figure 5(M)**). From 96 to 120 h PW, the localization of C/EBP-α-positive cells was

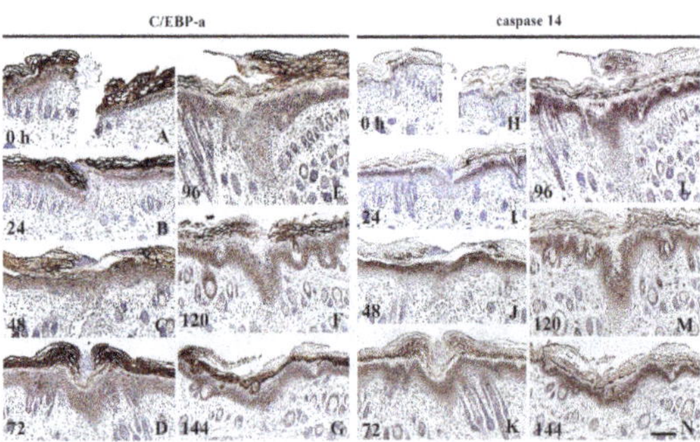

Figure 5. The expression of C/EBP-α and caspase-14 in the wound areas. The localization of C/EBP-α and caspase-14 was immunohistochemically examined. In the neonatal epidermis at 0 h PW, C/EBP-α and caspase-14 were distributed in the suprabasal layer (A) and granular layer (H), respectively. At 24 h PW, both of these molecules were diminished at the wound edge ((B), (I)). After wound closure was completed, these immunostaining intensities were restored to the normal level at the wound-covering epidermis by 48 h PW ((C), (J)). After that, the localization of C/EBP-α-positive cells was hardly changed, although the thickness of wound epidermis was varied ((D)-(G)). In contrast, the region unstained with anti-caspase-14 extensively extended downward in the thickened epidermal region by 96 h PW (L). Subsequently, the region intensely stained with anti-caspase-14, in turn, became enlarged more downward in the thinning epidermis at 120 h PW (M). ***Scale bar*, 100 µm; and *counterstained*, hematoxylin.**

hardly changed (**Figures 5(E)** and **(F)**). The epidermis at 144 h PW showed that both of these two molecules tended to restore the normal level of expression (**Figures 5(G)** and **(N)**).

These results suggested that the modulation of the expression of caspase-14 might be responsible for the epidermal thickness.

4. Discussion

HE and immunostaining experiments in this study revealed the healing process of the incisional wounds on neonatal rat skin to be outlined as follows (**Figure 1**). The opened wounds were first closed by the epidermal cells that, keeping the cell-to-cell contact, rapidly extended from the wound margin. Simultaneously, the wound-covering epidermis became thickened. Subsequently, the epidermis at the wound area became thinner up to the normal thickness, accompanying an elevation of the levels of the basal cells and the basement membrane. So, we have further continued morphological analyses of the incisional wounds, with the aim of pursuing the cause of above-stated fluctuation in the epidermal thickness.

4.1. The Phase of Thinning of Epidermis at the Wound Area and That of the Cell Death in the Upper Epidermis Overlap Each Other

We estimated the rates of cell proliferation and cell death using anti-BrdU immunostaining and TUNEL staining, respectively, to verify the possible causal relationship between the epidermal cell numbers and the fluctuation of the epidermal thickness in the wound area. The analyses of cell proliferation revealed that there is no clear correlation between the fluctuation of thickness at the wound-covering epidermis and the rate of the proliferating cells (**Figures 2(A)-(H)**). By contrast, the rate of TUNEL-positive cells rose shortly after peak (96 h PW) of the thickness (and the cell number) of the epidermis in the wound area, and reached the peak at 120 h PW, at which the thinning phase of epidermis was considered to start (**Figures 2(M)** and **(O)**). The peak value was 1.4 times as much as that at 96 h PW. It should be emphasized here that the rates of cell death in the wound areas at 120 and 144 h PW were significantly higher than those in the intact areas at 120 and 144 h PW, respectively (**Figures 2(O)** and **(P)**). Thus, the fluctuation of the rate of cell death and that of the thickness of epidermis seemed to coincide with each other. In other words, an elevation of the apoptosis in some cell populations in the upper epidermis may partially contribute to the reduction in the epidermal thickness.

A previous study showed that TUNEL-positive cells were often observed at the granular layer just below the horny layer of epidermis [21]. However, it has recently been reported that apoptosis and the terminal differentiation of keratinocytes are two different independent pathways [22,23]. Meanwhile, there is an article claiming that keratinocytes in the granular layer that express active caspase-14 display TUNEL-positive, being indicative of the physiological role of caspase-14 in relation to the DNA degradation [24]. Namely, an elevation of cell death (TUNEL-positive cells) at the upper part of the granular layer in the midst of thinning of the epidermis may result in an augmentation of cornification.

4.2. The Derangement Is Generated in the Terminal Differentiation on the Thickness-Fluctuating Epidermis at the Wound Area

One of the two daughter cells that generates from an epidermal stem cell via the asymmetric cell division is known to have the developmental fate toward the terminal differentiation through which, following several cycles of proliferation, the cell differentiates into spinous, granular and horny cells in order [25]. Such a terminal differentiation process involves not only the shift of keratin expression from K5 and K14 (in the basal layer) to K1 and K10 (in the suprabasal layers) but also a sequential switch-on of various differentiation markers such as involucrin, filaggrin, and loricrin in the suprabasal zone [26]. We made the simultaneous comparison of the intraepidermal localization of these differentiation marker proteins, using a series of immunostaining experiments, by which we could detect a one-cell-layer difference, if any, between the lowermost localizations of these proteins in the normal neonatal rat skin (**Figure 3**). Our experimental results showed that, at 96 and 120 h PW, loricrin and involucrin were expressed above the same cell layer in the thickened epidermis (**Figures 3(D)** and **(E)**). On the other hand, profilaggrin/filaggrin was found to be distributed from a cell layer lower than involucrin. The difference in staining regions lasted until 144 h PW (**Figure 4**). The results obtained from the two sets of analyses of the differentiation markers support the possibility that the stasis or the alteration of the rates in the epidermal terminal differentiation, which occurs probably at the zone from the upper spinous layer to the granular layer, may contribute to the fluctuation of the epidermal thickness at the wound areas.

Taken together, we speculated the post-wound-closure process of the epidermal wound healing, as follows (**Figure 6**): First, the epidermis at the wound site after the wound closure is thickened by an increase in the number of keratinocytes by an elevation of their proliferation and/or by a delay of differentiation within the

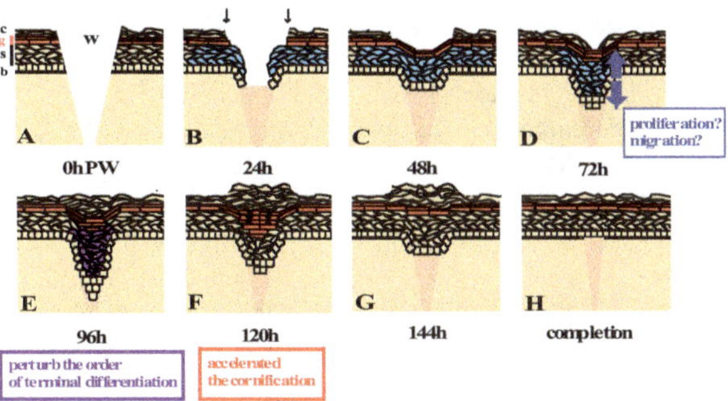

Figure 6. The scheme of re-epithelialization process in the wound of neonatal rat skin. The epidermal stratification at 0 h PW (A) is arrayed in the same order as that of the normal skin. The granular layer is colored red throughout this figure. The epidermis of the wound edge, which possibly comprises the basal cells and "immature spinous cells" (K14- and K6-positive, but the other differentiation makers-negative; blue) (Koizumi *et al.* 2004, 2005), migrates into the wound bed by 24 h PW (B). Wound closure is completed around 48 h PW (C). The wound-covering epidermis begins to thicken by 72 h PW, though the cause is unclear (blue arrows in (D)). Around 96 h PW, the epidermal thickening region in the wound area becomes thicker, and the terminal differentiation in the suprabasal layers is deranged ((E); purple). At 120 h PW, wound-induced thickened epidermis starts to be thinner by some mechanisms possibly involving the accelerated cornification ((F); increase granular cells, red; TUNEL-positive cells, green nuclei). Then, the wound-covering epidermis restores the normal thickness from 144 h PW onward ((G), (H)). *b*, stratum basale; *s*, stratum spinosum; *g*, stratum granulosum; *c*, stratum corneum; *w*, wound; and *black arrows*, the initial position of the wound edge.

spinous and granular layers. Next, the epidermal keratinocytes are rapidly cornified by a sudden acceleration of the differentiation to restore the steady state of epidermis. Although more evidence is necessary to establish this line of scheme, another noteworthy fact is the presence of the thick-multilayered horny layers on the thickening epidermis at the wound. This is an indication of the occurrence of transient and active cornification that may couple with the thinning of epidermis at the wound region.

4.3. A Spatial Expansion of Caspase-14 May Correlates with the Thinning of the Epidermis at the Wound Area

As the last analysis in this study, we examined the expression of C/EBP-α and caspase-14 that are known to closely participate in the terminal differentiation of keratinocytes. C/EBP-α is a transcription factor, acting on the early phase of the terminal differentiation [14,15], whereas caspase-14 a proteolytic enzyme involved in the cornification step [12,17]. During the early stage when the epidermal sheets were migrating for the wound closure, C/EBP-α and caspase-14 were absent from the leading edges (**Figure 5**), fairly consistent with the decrease of K10 expression and the spatial expansion of K14 staining (**Figure 1**). Following that, at 96 h PW, the pachychromatic region for caspase-14 was confined to the upper layers at the wound epidermis, and the region undyed was widely observed in the lower layers than that (**Figure 5(L)**). We inferred from these results that, at the thickening epidermis in the wound site, the cornification might be delayed by some down-regulation of the expression of caspase-14. The proportion of the granular layer to the total number of the epidermal cells at wounds might otherwise be reduced by a delay in the terminal differentiation process or activation of cell proliferation. Around the middle of the thinning phase at 120 h PW, the areas intensely stained with anti-caspase-14 at the wound epidermis were extended more downward (**Figure 5(M)**). Accordingly, the breakthrough of the terminal differentiation that had been arrested until then may be related to the expansion of expression of caspase-14 or the rapid increase of granular cells there.

5. Conclusion

In conclusion, one of the most plausible causes for the fluctuation of the epidermal thickness at the wound areas is considered to be a delay and acceleration of the terminal differentiation, including cornification, in the epidermal keratinocytes. Further analyses from the other viewpoints of cellular migration, intercellular and intracellular signaling, and regeneration of some specific cells, etc. are needed for thorough elucidation of how and why the thickness of the wound-covering epidermis fluctuates.

6. Acknowledgements

We thank the members of our Morphogenesis Laboratories for their support and helpful discussions in weekly seminars for furthering this study.

REFERENCES

[1] J. D. Burrington, "Wound Healing in the Fetal Lamb," *Journal of Pediatric Surgery*, Vol. 6, No. 5, 1971, pp. 523-528. http://dx.doi.org/10.1016/0022-3468(71)90373-3

[2] J. B. Dixon, "Inflammation in the Foetal and Neonatal Rat: the Local Reactions to Skin Burns," *The Journal of Pathology and Bacteriology*, Vol. 80, No. 1, 1960, pp. 73-82. http://dx.doi.org/10.1002/path.1700800109

[3] S. Ihara, Y. Motobayashi, E. Nagao and A. Kistler, "Ontogenetic Transition of Wound Healing Pattern in Rat Skin Occurring at the Fetal Stage," *Development*, Vol. 110, No. 3, 1990, pp. 671-680.

[4] M. T. Longaker, D. J. Whitby, N. S. Adzick, T. M. Crombleholme, J. C. Langer, B. W. Duncan, S. M. Bradley, R. Stern, M. W. Ferguson and M. R. Harrison, "Studies in Fetal Wound Healing, VI. Second and Early Third Trimester Fetal Wounds Demonstrate Rapid Collagen Deposition without Scar Formation," *Journal of Pediatric Surgery*, Vol. 25, No. 1, 1990, pp. 63-68, Discussion 8-9.

[5] P. Martin, "Mechanisms of Wound Healing in the Embryo and Fetus," *Current Topics in Developmental Biology*, Vol. 32, 1996, pp. 175-203. http://dx.doi.org/10.1016/S0070-2153(08)60428-7

[6] M. Koizumi, T. Matsuzaki and S. Ihara, "The Subsets of Keratinocytes Responsible for Covering Open Wounds in Neonatal Rat Skin," *Cell and Tissue Research*, Vol. 315, No. 2, 2004, pp. 187-195. http://dx.doi.org/10.1007/s00441-003-0823-0

[7] P. Martin and S. M. Parkhurst, "Parallels between Tissue Repair and Embryo Morphogenesis," *Development*, Vol. 131, No. 13, 2004, pp. 3021-3034. http://dx.doi.org/10.1242/dev.01253

[8] M. Koizumi, T. Matsuzaki and S. Ihara, "Expression of P-Cadherin Distinct from That of E-Cadherin in Re-Epithelialization in Neonatal Rat Skin," *Development, Growth & Differentiation*, Vol. 47, No. 2, 2005, pp. 75-85. http://dx.doi.org/10.1111/j.1440-169x.2004.00784.x

[9] C. Byrne, M. Tainsky and E. Fuchs, "Programming Gene Expression in Developing Epidermis," *Development*, Vol. 120, No. 9, 1994, pp. 2369-2383.

[10] E. R. Li, D. M. Owens, P. Djian and F. M. Watt, "Expression of Involucrin in Normal, Hyperproliferative and Neoplastic Mouse Keratinocytes," *Experimental Dermatology*, Vol. 9, No. 6, 2000, pp. 431-438. http://dx.doi.org/10.1034/j.1600-0625.2000.009006431.x

[11] R. B. Presland, M. K. Kuechle, S. P. Lewis, P. Fleckman and B. A. Dale, "Regulated Expression of Human Filaggrin in Keratinocytes Results in Cytoskeletal Disruption, Loss of Cell-Cell Adhesion, and Cell Cycle Arrest," *Experimental Cell Research*, Vol. 270, No. 2, 2001, pp. 199-213. http://dx.doi.org/10.1006/excr.2001.5348

[12] G. Denecker, P. Ovaere, P. Vandenabeele and W. Declercq, "Caspase-14 Reveals Its Secrets," *The Journal of Cell Biology*, Vol. 180, No. 3, 2008, pp. 451-458. http://dx.doi.org/10.1083/jcb.200709098

[13] P. F. Jhonson, "Molecular Stop Signs: Regulation of Cell-Cycle Arrest by C/EBP Transcription Factors," *Journal of Cell Science*, Vol. 118, 2005, pp. 2545-2555. http://dx.doi.org/10.1242/jcs.02459

[14] E. V. Maytin, J. C. Lin, R. Krishnamurthy, N. Batchvarova, D. Ron, P. J. Mitchell and J. F. Habener, "Keratin 10 Gene Expression during Differentiation of Mouse Epidermis Requires Transcription Factors C/EBP and AP-2," *Developmental Biology*, Vol. 216, No. 1, 1999, pp. 164-181. http://dx.doi.org/10.1006/dbio.1999.9460

[15] H. S. Oh and R. C. Smart, "Expression of CCAAT/Enhancer Binding Proteins (C/EBP) Is Associated with Squamous Differentiation in Epidermis and Isolated Primary Keratinocytes and Is Altered in Skin Neoplasms," *The Journal of Investigative Dermatology*, Vol. 110, No. 6, 1998, pp. 939-945. http://dx.doi.org/10.1046/j.1523-1747.1998.00199.x

[16] M. Van de Craen, G. Van Loo, S. Pype, W. Van Criekinge, I. Van den brande, F. Molemans, W. Fiers, W. Declercq and P. Vandenabeele, "Identification of a New Caspase Homologue: Caspase-14," *Cell Death and Differentiation*, Vol. 5, No. 10, 1998, pp. 838-846. http://dx.doi.org/10.1038/sj.cdd.4400444

[17] S. Lippens, M. Kockx, M. Knaapen, L. Mortier, R. Polakowska, A. Verheyen, M. Garmyn, A. Zwijsen, P. Formstecher, D. Huylebroeck, P. Vandenabeele and W. Declercq, "Epidermal Differentiation Does Not Involve the Pro-Apoptotic Executioner Caspases, but Is Associated with Caspase-14 Induction and Processing," *Cell Death and Differentiation*, Vol. 7, No. 12, 2000, pp. 1218-1224. http://dx.doi.org/10.1038/sj.cdd.4400785

[18] L. Alibardi, E. Tschachler and L. Eckhart, "Distribution of Caspase-14 in Epidermis and Hair Follicles Is Evolutionarily Conserved among Mammals," *The Anatomical Record Part A: Discoveries in Molecular, Cellular, and Evolutionary Biology*, Vol. 286, No. 2, 2005, pp. 962-973. http://dx.doi.org/10.1002/ar.a.20234

[19] E. Hoste, P. Kemperman, M. Devos, G. Denecker, S. Kezic, N. Yau, B. Gilbert, S. Lippens, P. De Groote, R. Roelandt, P. Van Damme, K. Gevaert, R. B. Presland, H. Takahara, G. Puppels, P. Caspers, P. Vandenabeele and W. Declercq, "Caspase-14 Is Required for Filaggrin Degradation to Natural Moisturizing Factors in the Skin," *The Journal of Investigative Dermatology*, Vol. 131, No. 11, 2011, pp. 2233-2241. http://dx.doi.org/10.1038/jid.2011.153

[20] L. Eckhart and E. Tschachler, "Cuts by Caspase-14 Control the Proteolysis of Filaggrin," *The Journal of Investigative Dermatology*, Vol. 131, No. 11, 2011, pp. 2173-2175. http://dx.doi.org/10.1038/jid.2011.282

[21] C. A. McCall and J. J. Cohen, "Programmed Cell Death

in Terminally Differentiating Keratinocytes: Role of Endogenous Endonuclease," *The Journal of Investigative Dermatology*, Vol. 97, No. 1, 1991, pp. 111-114. http://dx.doi.org/10.1111/1523-1747.ep12478519

[22] M. Rendl, J. Ban, P. Mrass, C. Mayer, B. Lengauer, L. Eckhart, W. Declercq and E. Tschachler, "Caspase-14 Expression by Epidermal Keratinocytes is Regulated by Retinoids in a Differentiation-Associated Manner," *The Journal of Investigative Dermatology*, Vol. 119, No. 5, 2002, pp. 1150-1155. http://dx.doi.org/10.1046/j.1523-1747.2002.19532.x

[23] S. Lippense, G. Denecker, P. Ovaere, P. Vandenabeele and W. Declercq, "Death Penalty for Keratinocytes: Apoptosis versus Cornification," *Cell Death and Differentiation*, Vol. 12, Suppl. 2, 2005, pp. 1497-1508. http://dx.doi.org/10.1038/sj.cdd.4401722

[24] T. Hibino, E. Fujita, Y. Tsuji, J. Nakanishi, H. Iwaki, C. Katagiri and T. Momoi, "Purification and Characterization of Active Caspase-14 from Human Epidermis and Development of the Cleavage Site-Directed Antibody," *Journal of Cellular Biochemistry*, Vol. 109, No. 3, 2010, pp. 487-497.

[25] E. Fuchs, "Skin Stem Cells: Rising to the Surface," *The Journal Cell Biology*, Vol. 180, No. 2, 2008, pp. 273-284. http://dx.doi.org/10.1083/jcb.200708185

[26] H. Tseng and H. Green, "Association of Basonuclin with Ability of Keratinocytes to Multiply and with Absence of Terminal Differentiation," *The Journal Cell Biology*, Vol. 126, No. 2, 1994, pp. 495-506. http://dx.doi.org/10.1083/jcb.126.2.495

26

The Ortholog of LYVE-1 Is Required for Thoracic Duct Formation in Zebrafish*

Wen-Han Chen[1#], Wen-Fang Tseng[2#], Gen-Hwa Lin[3#], Andrew Schreiner[1], Hsiao-Rong Chen[4], Mark M. Voigt[5], Chiou-Hwa Yuh[2], Jen-Leih Wu[3], Shuan Shian Huang[6], Jung San Huang[1†]

[1]Department of Biochemistry and Molecular Biology, Saint Louis University School of Medicine, Doisy Research Center, St. Louis, USA

[2]Division of Molecular and Genomic Medicine, National Health Research Institutes, Chunan, Taiwan

[3]Institute of Cellular and Organismic Biology, Academia Sinica, Taipei, Taiwan

[4]Institute of Systems Biology and Bioinformatics, National Central University, Jhongli, Taiwan

[5]Department of Pharmacology and Physiological Science, Saint Louis University School of Medicine, St. Louis, USA

[6]Auxagen Inc., St. Louis, USA

Email: †huangjs@slu.edu

ABSTRACT

LYVE-1 (also termed CRSBP-1), a 120-kDa disulfide-linked dimeric type I membrane glycoprotein, is a specific marker for lymphatic endothelial cells (LECs) and exhibits multiple ligand (hyaluronic acid and growth factors/cytokines) binding activity in mammals. Recent studies indicate that LYVE-1/CRSBP-1 ligands (VEGF-A^{165}, PDGF-BB, oligopeptides containing the cell-surface retention sequence (CRS) motifs of VEGF-A^{165} and PDGF-BB) induce opening of lymphatic intercellular junctions *in vitro* and *in vivo*. To determine the function of the ortholog of mammalian LYVE-1 in zebrafish, we cloned it (zLyve-1). The cloned cDNA (*zlyve1*) encodes a 328-amino-acid type I membrane glycoprotein. The protein and genomic structure evidence supports the notion that the cloned zLyve-1 is the ortholog of LYVE-1 in zebrafish. zLyve-1 expressed in cultured cells by transfection exhibits hyaluronic acid binding activity but lacks the growth factor binding activity seen in mammalian homologs. Knockdown of zLyve-1 levels by embryo microinjection with a specific antisense morpholino oligonucleotide (MO2) in wild-type and *Tg(fli1:EGFP)*-transgenic zebrafish causes defects in thoracic duct (TD) formation. Such zebrafish injected with MO2 also exhibit impaired TD flow (as determined by intramuscular injection of FITC-dextran). The phenotypes in these zebrafish injected with MO2 are reversed by co-injection with *zlyve1*cDNA. *In situ* hybridization reveals that zLyve-1 is expressed in the posterior cardinal vein (PCV). Expression of zLyve-1 at the highest level in the PCV occurs at 3 dpf which coincides with the time for TD formation in zebrafish development. These results suggest that zLyve-1 is required for TD formation. They also suggest that zLyve-1 is distinct from mammalian LYVE-1 in its role in lymphatic function.

Keywords: zLyve-1; Morpholino Knockdown; Thoracic Duct; Posterior Cardinal Vein Endothelial Cells

1. Introduction

Cell-surface retention sequence (CRS)-binding protein-1 (CRSBP-1) was identified in cultured transformed cells and plasma membranes purified from liver tissue in 1995 by its ability to bind PDGF-BB, VEGF-A^{165} and synthetic oligopeptides containing the CRS motifs of PDGF-BB and VEGF-A^{165} [1,2]. The CRS motifs contain a cluster of basic amino acid residues (Arg, Lys, His). cDNA cloning, sequencing and expression of bovine CRSBP-1 in 2003 [3] revealed that CRSBP-1 is identical to lymphatic vessel endothelial hyaluronic acid (HA) receptor-1 (LYVE-1), which was identified by CD44 homology cloning in 1999 [4]. LYVE-1 has been used as a specific marker for lym-

*Conflict of Interest: Jung San Huang and Shuan Shian Huang had equity positions in Auxagen, Inc. when the research was performed. Part of the research was supported by NIH grants awarded to Auxagen, Inc. (AR052578).

This work was presented at the Meeting "Lymphatic Circulation in Health and Disease", May 3-4, 2013, New Haven, CT, USA.

#Contributed equally

†Corresponding author.

phatic endothelial cells (LECs) and lymphatic vessels in mammals [5]. CRSBP-1/LYVE-1 expressed in cultured cells is a 120-kDa disulfide-linked dimeric type I membrane glycoprotein, and exhibits multiple ligand (hyaluronic acid and growth factors/cytokines) binding activity in mammals [3]. Recent studies indicate that CRSBP-1/LYVE-1 ligands (such as hyaluronic acid, PDGF-BB, VEGF-A^{165} and synthetic oligopeptides containing the CRS motifs of VEGF-A^{165} and PDGF-BB) induce opening of lymphatic intercellular junctions *in vitro* and *in vivo*, resulting in an increase in the transit of fluid, large molecules and cells from the interstitial space into lymphatic vessel lumens (termed interstitial-lymphatic transit) in mice [6-8].

We hypothesized that CRSBP-1/LYVE-1 plays an import role in lymphatic function in all vertebrate animals, including zebrafish. To test this hypothesis, our studies aimed at cloning the zebrafish ortholog (zLyve-1) of mammalian LYVE-1 and defining its role in lymphatic vessel formation and function. The lymphatic vasculature has been characterized in zebrafish [9-12]. Zebrafish possess a lymphatic vascular system that has morphological, molecular and functional characteristics like those found in other vertebrates [9-12]. It is an excellent animal system in which to investigate how lymphatic vessel formation and function are determined by proteins specifically expressed in LECs. Gene expression knockdown by embryo microinjection with specific antisense morpholino oligonucleotides (MOs) is a simple and effective strategy to define the roles of target genes in the development of lymphatic vessels in zebrafish in a relatively short time (days) [13,14]. Compared to other vertebrates, the zebrafish offers several other advantages including small size, optical clarity (fluorescent blood and lymphatic vessel structures are easy to see in transgenic *Tg*(*fli*1:*EGFP*) zebrafish) and rapid development (days) of embryos [15].

In this communication, we show the cloning of *zlyve1* cDNA from zebrafish RNA at the 2 - 5 dpf stages by RT-PCR. The analyses of genomic and putative protein structures of the identified zLyve-1 suggest that zLyve-1 is the ortholog of mammalian LYVE-1 in zebrafish. We also show that zLyve-1 expressed in cultured cells exhibits its hyaluronic acid-binding activity but lacks the growth factor-binding activity seen in mammalian LYVE-1, as determined by cetylpyridinium chloride precipitation and ^{125}I-labeled VEGF (vascular endothelial cell growth factor) peptide-affinity labeling [3], respectively. This is consistent with the presence of Link module and absence of the putative growth factor binding domain (the acidic-amino-acid-rich region) in zLyve-1 molecule. In addition, we show that zebrafish derived from both wild-type and fluorescent transgenic *Tg*(*fli*1:*EGFP*) embryos microinjected with a specific antisense morpholino oligonucleotide (MO2) exhibit defects in thoracic duct formation and impaired thoracic duct flow (as determined by intramuscular injection of FITC-dextran). These phenotypes can be reversed by co-injection with *zlyve1* cDNA. Furthermore, we show that zLyve-1 is expressed in the posterior cardinal vein (PCV), as evidenced by *in situ* hybridization. Since PCV endothelial cells are known to be precursor cells for LECs forming the thoracic duct (TD), this suggests that zLyve-1 expressed in PCV endothelial cells may play an important role in lymphatic development in zebrafish.

2. Materials and Methods

2.1. Materials

Na^{125}I (100 mCi/mL), was purchased from MP Biochemicals (Solon, OH). VEGF peptide (a 25-mer peptide containing the amino acid sequence KKSVRGKGKGQ-KRKRKKSRYKSWSV) was synthesized by C S Bio Co. (Menlo Park, CA). Cetylpyridinium chloride (CPC), 1-ethyl-3-(3-dimethylaminopropyl) carbodiimide HCl (EDAC), Triton X-100, hyaluronic acid (M.W. ~2,000,000) from human umbilical cord (~98% purity), chloramine T and FITC-dextran were obtained from Sigma (St. Louis, MO). Anti-hemagglutinin (HA) epitope antibody was obtained from Cell Signaling (Danvers, MA). pCMV-HA was obtained from Clontech (Mountain View, CA). H1299/bLYVE-1 and H1299/vector cells, which were stably transfected with pCEP-bLYVE-1 cDNA and pCEP vector only, respectively, were prepared as previously described [3]. H1299 cells (human lung carcinoma cells) were transiently transfected with *zlyve1* cDNA in the pCMV-HA vector (pCMV-*zlyve1*-HA) and termed H1299/*zlyve1* cells which expressed HA-tagged zLyve-1.

2.2. *zlyve1* and *zlyve1l* cDNA Construction

zlyve1 cDNA was amplified from zebrafish RNA at the 2 - 5 dpf stages by RT-PCR using the high capacity RNA-to-cDNA Kit^{TM} (Applied Biosystems, Foster City, CA), and KOD FX DNA polymerase (Toyobo Biochemicals, Osaka, Japan). The primer sequences used for PCR were 5'-ATGACACGAGTCTGCATGGG-3' (forward), and
5'-TCATTGTTCTGCAGAGCTACTGTCG-3' (reverse). The amplified cDNA was inserted into the yT & A vector (Yeastern, Taipei, Taiwan) via T-A cloning. The KpnI and XbaI fragment was then sub-cloned into pCMV-HA at KpnI and NotI fill-in sites and termed pCMV-*zlyve1*-HA. The cDNA (*zlyve1l*) of zebrafish Lyve-1-like protein was also cloned from zebrafish RNA at the 2 - 5 dpf stages using RT-PCR in a similar manner. The CMV promoter is a universal promoter and functional in the zebrafish system.

2.3. Multiple Alignments, Phylogenetic Analysis and Genomic Structural Analysis of *zlyve1* (zLyve-1), *zlyve1l* (zLyve-1-Like) and Other LYVE-1 Vertebrate Homologs

After cDNA cloning of zLyve-1 and *zlyve1*-like protein, multiple alignments of these cDNAs with those of other vertebrate orthologs (human, chimpanzee, dog, cow, mouse, rat, chicken and salmon) were performed using the NCBI multiple alignment program. In addition to the multiple alignments, a phylogenetic tree was created using the results from the multiple alignment section of the same program. Genomic structure analysis was also performed using the NCBI sequence viewer for *zlyve1*, *zlyve1l* and other vertebrate orthologs.

2.4. Transfection

H1299 cells were transiently transfected with pCMV-*zlyve1*-HA and pCMV-HA, using lipofectamine as described [3] and termed H1299/*zlyve1* and H1299/vector cells, respectively.

2.5. Hyaluronic Acid Binding Activity Assay

The 0.1% Triton X-100 extracts (100 μg protein) of H1299 cells transiently transfected with pCMV-*zlyve1*-HA (H1299/*zlyve1* cells) were incubated with several concentrations of hyaluronic acid (average M.W. ~2,000,000) at room temperature for 1 h. The reaction mixtures were then subjected to CPC precipitation and analyzed by 7.5% SDS-PAGE, followed by Western blot analysis using anti-HA antibody (to immunoprecipitate HA-tagged zLyve-1 expressed in H1299/*zlyve1* cells). The antigens on the blots were visualized using horseradish peroxidase-conjugated anti-rabbit IgG antibody and the ECL system, as described [3]. The relative intensities of antigen (HA-tagged zLyve-1 which formed complexes with hyaluronic acid) bands on X-ray films were quantified by densitometry.

2.6. ^{125}I-VEGF Peptide Binding Assay

^{125}I-VEGF peptide was prepared using Na^{125}I and chloramine T as described [1-3,7]. The ^{125}I-VEGF peptide binding assay in cultured cells was carried out according to our published procedure [1,3]. Briefly, H1299/vector, H1299/bovine LYVE-1 (H1299/bLYVE-1) [3] and H1299/*zlyve1* cells seeded at the same cell density were incubated with 6 μM ^{125}I-VEGF peptide with and without 100-fold excess of unlabeled VEGF peptide. After 2.5 h at 4°C, the specific binding of ^{125}I-VEGF peptide was estimated by subtracting nonspecific binding, which was determined in the presence of 100-fold excess of unlabeled VEGF peptide, from total binding.

2.7. ^{125}I-VEGF Peptide-Affinity Labeling

The ^{125}I-VEGF peptide-affinity labeling in cultured cells was carried out as described [1-3,7]. H1299/vector, H1299/bLYVE-1, and H1299/*zlyve1* cells were incubated with 6 μM ^{125}I-VEGF peptide with and without 100-fold excess of unlabeled VEGF peptide. After 2.5 h at 4°C, ^{125}I-VEGF peptide-affinity labeling was carried out by adding EDAC into cells and incubating at 4°C for an additional 15 min. ^{125}I-VEGF peptide-affinity labeled cells were subjected to Triton X-100 lysis followed by 7.5% SDS-PAGE and autoradiography.

2.8. Husbandry and Maintenance of Zebrafish

We performed zebrafish (*Danio rerio*) studies according to the US National Institute of Child Health and Human Development Animal Care and Use Committee guidelines for the use of laboratory animals (St. Louis University IACUC office) and those of Academia Sinica, Taiwan, and National Health Research Institutes, Taiwan. Wild-type zebrafish (AB strain) were obtained from natural crosses and subsequent collection of embryos. Embryos were collected by natural spawning, raised in 0.3× Danieau's buffer and then maintained in fish water at 28.5°C, with feeding twice daily until experiments. The wild-type zebrafish and fluorescent transgenic *Tg* (*fli1*: *EGFP*) zebrafish were also maintained at the facilities (St. Louis University School of Medicine and the Institute of Cellular and Organismic Biology, Academia Sinica, Taipei, Taiwan and Division of Molecular and Genomic Medicine, National Health Research Institutes, Chunan, Taiwan) in a controlled environment with a 14/10-h light-dark cycle at 28°C. The fish spawned soon after light onset, and fertilized eggs were collected at the 1-cell stage. The embryos were collected using natural-mating and cultured at 28°C in Ringer's solution.

2.9. Antisense Morpholino Oligonucleotide (MO) Microinjection

MO1, MO2, MO3, mis-paired MO2, mis-paired MO3 and control MO were obtained from Gene Tools, LLC (Philomath, OR). Six MOs (MO1, MO2, mis-paired MO2, MO3, mis-paired MO3 and control MO) included: splice site blockers that targeted the 5'UTR (MO1: 5'-TCATCCCC-ATGCAGACTCGTGTCAT-3'); the boundary between exon1 and intron 1 (MO2: 5'-TCGAGGCAGATCTTA-CCTTTACAA-3'; mis-paired MO2: 5'-TCGAGcCAc-ATCTTAgCTTTAAgAA), and the boundary between intron 1 and exon 2 (MO3: 5'-TGCACTGTTCATCA-GAAGAAAGCAG-3'; mis-paired MO3: 5'-TGCAgTc-TTCATCAcAAcAAAcCAG-3'), and a control MO (5'-TCGAGCCACATCTTAGCTTTAAGAA-3') (MO). MOs were dissolved in doubly distilled H_2O to make a stock solution with a concentration of ~8.4 ng/nl or 1 mM, stored at RT and further diluted to the working concentration (0.1 mM) with 0.1% phenol red or diluted with 2× phenol red dye (0.1 M KCl, 20 mM HEPES, pH 7.4 and 0.01% phenol red) at a 1:4 dilution for injection. Final concentrations of MOs were ~2 ng/nl or 0.25 mM. A volume of 1 and 2 nl of MOs (4.2 and 8.4 ng per embryo), which was backloaded into microinjection needles, was injected into 1-cell stage embryos using a Picospritzer III (General Valve Corporation, Fairfield, NJ) attached to a broken capillary tube. For rescue experiments, 100 pg of *zlyve1* cDNA, which was purified using Qiaquick spin columns (Qiagen), or pCMV-HA vector was co-injected with MO2 or control MO into zebrafish embryos.

2.10. RT-PCR Analysis

Total RNAs from 5 embryos at designated stages were isolated by the use of RNA STAT-60™ reagent (Tel-Test, Inc, USA) or TRIZOL reagent (Invitrogen). First-strand cDNA was synthesized in a 20 µl RT reaction from 1 µg of total RNA by using Super Script III First-Strand synthesis system (Invitrogen), according to the manufacturer's instructions. *zlyve1* cDNA fragments of 308 bp and 461 bp were amplified with the following two-pair primers: 5'-G GCCTGTTCATGTCTGGTCTG-3' (forward)/5'-TCTGCACGCCACACCACAACC-3' (reverse) and 5'-ATG ACA CGA GTC TGC ATG GG-3' (forward)/5'-ATC GGC TTT CCG GTT GTC ATT TGG-3' (reverse), respectively. Amplification was carried out in a thermal cycler (2720 thermal cycler, Applied Biosystems, USA) by 30 s denaturation at 94°C, 30 s annealing at 55°C and 30 s extension at 72°C. A total of 35 cycles were carried out. The *β*-actin and 18 s rRNA RT-PCR products (433 and 241 base pairs, respectively) in total RNAs isolated were used as internal controls.

2.11. FITC-Dextran Microinjection

Using a conventional microinjection setup, 5 - 7 dpf embryos were embedded in 2% low-melting point agarose, and 2.5 ng/nl FITC-dextran (average M.W. 2,000,000) was injected into the tail muscles of zebrafish embryos. Embryos were then imaged every 2 - 3 minutes for evaluation of movement of the large fluorescent molecule. The rate (segments/min) of FITC-dextran fluorescence transit along the thoracic duct was measured between 5 and 20 min after injection of FITC-dextran. The FITC-dextran transit time from injection site to the visible end of the thoracic duct was approximately 20 min.

2.12. Whole-Mount *in Situ* Hybridization

A cDNA fragment of *zlyve1* or *zlyve1l* was cloned into a cloning vector as the DNA template for synthesizing an antisense RNA probe. The T3 polymerase promoter sequence was added to the 3' end of the cDNA by insertion into the 5' end of the reverse primer. Digoxigenin (DIG)-labeled antisense RNA probes were synthesized from the purified PCR products according to the instructions in the MEGAscript Kit (Ambion, Austin, TX) and DIG RNA labeling mix (Roche). Zebrafish larvae at 30 hpf, 3 dpf, 4 dpf and 5 dpf were fixed with 4% paraformaldehyde overnight at 4°C and dehydrated in methanol at −20°C. Whole mount *in situ* hybridization was performed according to the protocol described by the Thisse Lab (https://wiki.zfin.org/display/prot/Thisse+Lab+-+In+Situ +Hybridization+Protocol+-+2010+update) with some modifications. Briefly, zebrafish embryos were rehydrated gradually with PBST (1× PBS, 0.1% Tween 20) and treated with proteinase K for 30 min. After being prehybridized with HYB$^-$ (50% formamide, 5× SSC, and 0.1% Tween 20), zebrafish were incubated at 65°C overnight in HYB$^+$ (HYB$^-$ plus 50 µg/ml heparin, 500 µg/ml wheat germ tRNA) containing 1 µg/µl of the DIG-labeled antisense RNA probe. They were then washed with 75%, 50% and 25% HYB$^-$/2× SSC for 15 min, 2× SSC for 15 min, and twice with 0.2× SSC for 30 min at 70°C and transferred gradually to maleic acid buffer (100 mM maleic acid, 150 mM NaCl, 0.1% Tween 20) and blocked in 1× blocking buffer [1% Blocking Reagent (Roche) in maleic acid buffer] for at least 2 h. Zebrafish were then incubated in 1:5000 anti-DIG-AP (Roche) in 1× blocking buffer overnight at 4°C. After six washes with PBST for 15 min and three washes with alkaline Tris buffer for 5 min, bound antibody was detected using BCIP/NBT (Roche). After staining, the labeled larvae were mounted in 2% methylcellulose. The images were captured using ZEISS ImagerA1 and ZEISS AxioImagerA1 and processed with their associated software.

2.13. Statistical Analysis

The values are presented as mean ± S.D. Two-tailed unpaired Student's *t*-test was used to determine the significance of differences between groups. $P < 0.05$ was considered significant. Comparisons between the two groups were conducted with the Mann-Whitney test.

3. Results

3.1. Cloning of *zlyve1* cDNA

Before cloning the *zlyve1* cDNA, we performed a homology search in the zebrafish gene bank using mammalian LYVE 1. We found two genes, one of which was recently identified by Flores *et al.* [16] and one other, which we named zLyve-1. Because of the protein structure homology found among mammalian LYVE-1, including salmon LYVE-1 (sLYVE-1) and zebrafish Lyve-

1 (zLyve-1), we believe that zLyve-1 is the ortholog of mammalian LYVE-1 in zebrafish. We therefore cloned *zlyve1* cDNA using the primers generated based on the putative *zlyve1* cDNA sequence. The cDNA (*zlyve1*) of the zebrafish LYVE-1 ortholog (zLyve-1) was cloned from zebrafish RNA at 2 - 5 days post fertilization (dpf) using RT-PCR. We obtained the cDNA nucleotide sequence of the cloned *zlyve1* (**Figure 1(A)**). The *zlyve1* cDNA encodes a 328-amino-acid type I membrane glycoprotein containing a 10-amino-acid putative signal peptide, a 187-amino-acid cell-surface domain which includes a 77-amino-acid Link module [17], a 23-amino-acid transmembrane domain and a 93-amino-acid cytoplasmic domain (**Figure 1(B)**). Comparison of the nucleotide and deduced amino acid sequences of zLyve-1 with those of other vertebrate homologs revealed that zLyve-1 has 22% - 45% amino acid and nucleotide sequence identity with those of the salmon, chicken, rat, mouse, cow, dog, chimpanzee and human homologs (**Figure 1(B)**). zLyve-1 and other known vertebrate LYVE-1 homologs share important protein structural features (**Figure 1(C)**, panels a, b and c). It is important to note that all vertebrate homologs, including zLyve-1 and sLYVE-1, possess two N-glycosylation sites on both sides of the Link module (**Figures 1(B)** and **(C)**, panels a, b and c). A zebrafish LYVE-1 ortholog, which was re-

ATGACACGAGTCTGCATGGGGATGATGATGATGATGATTAAGATGGTGCTTTTCAGTGGC

M T R V C M G M M M M M I K M V L F S G 20

CTGTTCATGTCTGGTCTGGCGTTTGACATGCAGCTTGTAAAAGTGCACCCTAAACAAGCC

L F M S G L A F D M Q L V K V H P K Q A 40

ATCTCCGGGGTATCAGAAGCTTCCATTGGGAATCAATATGCTTTAAACGCATCATCAGCC

I S G V S E A S I G N Q Y A L N A S S A 60

AGAGATCTTTGTGAGCATCTCGGATTGACGATTGCAAACAAAGCACAGCTAGCAGAGGCT

R D L C E H L G L T I A N K A Q L A E A 80

CAGAAACACGGCCTGGAGACGTGCAGGTTCGGGTGGATCGATGAGCAGATCGCTGTTGTT

Q K H G L E T C R F G W I D E Q I A V V 100

CCTCGAGTCAAGGTTAACCCCAACTGCGGCAATGGCAAGACTGGGGTTGTGGTGTGGCGT

P R V K V N P N C G N G K T G V V V W R 120

GCAGATCCCAGCAAACAATTTGATGTCTTTTGCTTCAATGTAACCGATTTTGAGACACAG

A D P S K Q F D V F C F N V T D F E T Q 140

GCTTCCATAAAGGTGCACCAAATGACAACCGGAAAGCCGATGACGACACGTTCATCTGTG

A S I K V H Q M T T G K P M T T R S S V 160

GCTCCTACTGCTGGAGTTCATCTGAGGGGAAGCCCATTATCTAAACTCCCTTCCCATTGG

A P T A G V H L R G S P L S K L P S H W 180

TCCAGTGTTCCTCGTTCAGCGTCTGCAGGTCCCTCTGTGGTCCACAGGGCTGATGCCAAA

S S V P R S A S A G P S V V H R A D A K 200

CATCTTGCCCTGAGCAGTGGCTCCACTGAAGCTGTTCCTGCCGCTTTACTGATCACCGTC

H L A L S S G S T E A V P A A L L I T V 220

ACCTTTGCAGTCATGATTGCTGTGTTTCTGGCACTTTATTATGTCAGAACGAAAAGACCC

T F A V M I A V F L A L Y Y V R T K R P 240

TGCAAGGCCCAGTGTGATGTAGAGCAGCAGAAGGAGTACATCGAGACTGAGGTTTGGGAA

C K A Q C D V E Q Q K E Y I E T E V W E 260

CACCATACTAAAGAAGACCTGCAGAAAACACAAGAGGAACAGGTGGAGGACAAAAACCAG

H H T K E D L Q K T Q E E Q V E D K N Q 280

GAGGAGCAGATGGAGGAAAAGAACCAGGAAGAGCTGGATGAGAATCACCAGGAGGAGCAG

E E Q M E E K N Q E E L D E N H Q E E Q 300

ATGGAGGAGAGACGCCAAGAAGAGCACATGAAGAAAAACCAGGAGGAGCAGGTGGAGGAA

M E E R R Q E E H M K K N Q E E Q V E E 320

AACGACAGTAGCTCTGCAGAACAATGA

N D S S S A E Q 328

(A)

```
Human LYVE-1       1  MARCFS--LVLLLTSIWTTRLLVQG-SLRAEELSIQVSCRIMGITLVSKKA-----NQQLNFTEAKEACRLLGLSLAGKD  72
chimpanzee LYVE-1  1  MARCFS--LVLLLTSIWTTRLLVQG-SLRAEELSIQVSCRIMGITLVSKKA-----NQQLNFTEAKEACRLLGLSLAGKD  72
Dog LYVE-1         1  MAKCFS--LVLLLASIWTTRLLVHA-TLRVEELSISGPCRIVGVTLVNKKT-----IQQLNFTEAQEACRLMGLTLASKD  72
Cow LYVE-1         1  MAKFFS--LGLLLASIWTTRLLVQG-SLRSEELSILGPCRIMGVTLVTKKT-----QPLLNFTEAQEACRLVGLTLASQD  72
Mouse LYVE-1       1  MLQHTS--LVLLLASIWTTRHPVQG-ADLVQDLSIS-TCRIMGVALVGRNK-----NPQMNFTEANEACKMLGLTLASRD  71
Rat LYVE-1         1  MLQHCS--LVLLLASLWTTRHPVHG-TVQVQDLSIS-PCRIMGVALVGRNA-----DPQMNFTEAKEVCKVLGLTLASRN  71
Chicken LYVE-1     1  MTTYFG--VTSAVLSVWVMTFMAQN-YFITG--STLSPCRITGVGLYL--------GHKVNFSEASNVCNRLNLQLASKE  67
Salmon LYVE-1      1  MMQVWI------LSLLLPLTLSFSGLHVDPSKIHAFPERHIAGVFLVSYTKDLNQFAYAFNASEAREVCWSLGVTMASNS  74
Zebrafish Lyve-1   1  MTRVCMGMMMMMIKMVLFSGLFMSGLAFDMQLVKVHPKQAISGVSEASIGN-----QYALNASSARDLCEHLGLTIANKA  75

Human LYVE-1      73  QVETALKASFETCSYGWVGDGFVVISRISPNPKCGKNGVGVLIWKVPVSRQFAAYCYNSSDTWTNSCIP-EIITTKDPIF  151
chimpanzee LYVE-1 73  QVETALKASFETCSYGWVGDGFVVISRISPNPKCGKNGVGVLIWKVPVNRQFAAYCYNSSDTWTNSCIP-EIITTKDPIF  151
Dog LYVE-1        73  QVEAARRSGFETCSYGWVADKLLVIPRILPNPKCGKNGIGVLVWRNSLSKKFLAYCYNSSDTWINSCIP-EIITTTDPIF  151
Cow LYVE-1        73  QVEEARKFGFETCSYGWVKNQFVVIPRIISNPKCGKSGVGVVIWRSSLSSRHRSYCHNSSDIWINSCLP-EIITTDDPLF  151
Mouse LYVE-1      72  QVESAQKSGFETCSYGWVGEQFSVIPRIFSNPRCGKNGKGVLIWNAPSSQKFKAYCHNSSDTWVNSCIP-EIVTTFYPVL  150
Rat LYVE-1        72  QVESAQKSGFETCSYGWVGERFSVIPRISPNPKCGKNGKGVLIWNASPSQKFRVYCHNSSDTWANSCFP-EITTTF----  146
Chicken LYVE-1    68  QVEKALNHGFETCSSGWIKDGSVAIPRITSNMKCGKGSVGLVNWRPNHAHKFTVYCFNSSDVQINSCKP-DPTTTIIP--  144
Salmon LYVE-1     75  QVEEAQRLGLETCRFGWIDEHFAVIPRIEASKTCGQNQTGVIKWRASVTKLFDVFCFNASGV---ASFP----------  140
Zebrafish Lyve-1  76  QLAEAQKHGLETCRFGWIDEQIAVVPRVKVNPNCGNGKTGVVVWRADPSKQFDVFCFNVTDFETQASIKVHQMTTGKPMT  155

Human LYVE-1      152  NTQTATQTTEFIVSDSTYSVASPYSTIPAPTTTPPAP--ASTSIPRRKKLICVTEVFMETS-TMSTETEPFVENKAAFKN  228
chimpanzee LYVE-1 152  NTQTATHTTEFIVSDSTYSVASPYSTIPAPTTTPPAP--ASTSIPRRKKLICVTEVFMETS-TMSTETEPFVENKAAFKN  228
Dog LYVE-1        152  NIQETIYTTEMIVSDSTYSASSTDAPYSATPTLAPTSALASTSTRRKRKLICVTEAFVETS-TVSTETESYIENKAAFKN  230
Cow LYVE-1        152  NTETATYTTKLMVSDSTHSELSTDGPDYVTTTVAPP--LASTSTPRKRKLICITEAFMDTS-AVATERESDIQNRPAFKN  228
Mouse LYVE-1      151  DTQTP--ATEFSVSSSAYLASSPDSTTPVSATTRAPP---LTSMARKTKKICITEVYTEPI-TMATETEAFVASGAAFKN  224
Rat LYVE-1        147  NTQTP--AAEFSVSSDTYSASSSDSTTSASATTRAPP---LTTMARKTKMICITEVYTEPI-TMDAETEASVESGAAFKN  220
Chicken LYVE-1    145  ---SSSVPTDLTAYSSS---DLTGNTTAVPTAAEPEQ----TLKNVKFRIICITETILPTEETTTTMPEESYTYPAAFRN  214
Salmon LYVE-1     141  -STSLS-PAIIHLVPSTQS-------------TRPTS-------HSRSSLLSLSSFSDNPE-------EVEVELPQSMSS  191
Zebrafish Lyve-1  156  TRSSVAPTAGVHLRGSPLS-KLPSHWSSVPRSASAGP---SVVHRADAKHLALSSGSTEA--------------------  211

Human LYVE-1      229  EAAGFGGVPTALLVLALLFFGAAAGLGFCYV--KRYVKA-FPFTNKNQQKEMIET-KVVKEEKANDSNPNEESKKTDKNP  304
chimpanzee LYVE-1 229  EAAGFGGVPTALLVLALLFFGAAAGLGFCYV--KRYVKA-FPFTNKNQQKEMIET-IVVKEEKANDSNPNEELKKTDKNP  304
Dog LYVE-1        231  EVVGFGGVPTALLVLALLFFAAAAGLAVCYV--KRYVKT-FPFTNKNQQKEMIET-KVVKEEKADDSNPNEESKKTDKKP  306
Cow LYVE-1        229  EAVGFGGVPTALLVLALLFFAAAAGLAVCYV--KRYVKA-FPFTNKNQQKEMIET-KVVKEEKADDSNPNEESKKMNKTP  304
Mouse LYVE-1      225  EAAGFGGVPTALLVLALLFFGAAAVLAVCYV--KRYVKA-FPFTTKNQQKEMIET-KVVKEEKADDVNANEESKKTIKNP  300
Rat LYVE-1        221  EAAGFGGVPTTLLVLALFFFGAAAVLGVCYV--KRYVKA-FPFTNKNQQKEMIET-KVVKEEKADDVNANEESKKMVKNS  296
Chicken LYVE-1    215  DGVVFGGIPTALLVLAIIFFISSVLLAVCYI--KKYKKC-LLFSRKNQKKEVVET-TALKDTASNDKILEKETKNNGKMV  290
Salmon LYVE-1     192  AKSSIGAVPKALLITSAIVLLLTAMAILLYFRANNGLKTIFPCWDLEQQKEYSETVECAAHTCMKDTKEVQTEGKAKAEP  271
Zebrafish Lyve-1  212  -------VPAALLITVTFAVMIAVFLALYYVRTKRPCKA--QC-DVEQQKEYIET-EVWEHHTKEDLQKTQEEQVEDKNQ  280

Human LYVE-1      305  EE---SK------------SPSKTTVRCLE------------------AEV  322
chimpanzee LYVE-1 305  EE---PK------------SPSKTTVRCLE------------------AEV  322
Dog LYVE-1        307  EE---PK------------SPTKTTVRCME------------------AEV  324
Cow LYVE-1        305  EE---PK------------SPPKTTVRCLE------------------AEV  322
Mouse LYVE-1      301  EE---AK------------SPPKTTVRCLE------------------AEV  318
Rat LYVE-1        297  EE---PK------------SPPKTTVRCLE------------------AEV  314
Chicken LYVE-1    291  EE---AT------------AKPETTVKCLE------------------VEV  308
Salmon LYVE-1     272  EE---DVCKETANDISVNISDETNTDSASE------------------TKP  301
Zebrafish Lyve-1  281  EEQMEEKNQE---ELDENHQEEQMEERRQEEHMKKNQEEQVEENDSSSAEQ  328
```

(B)

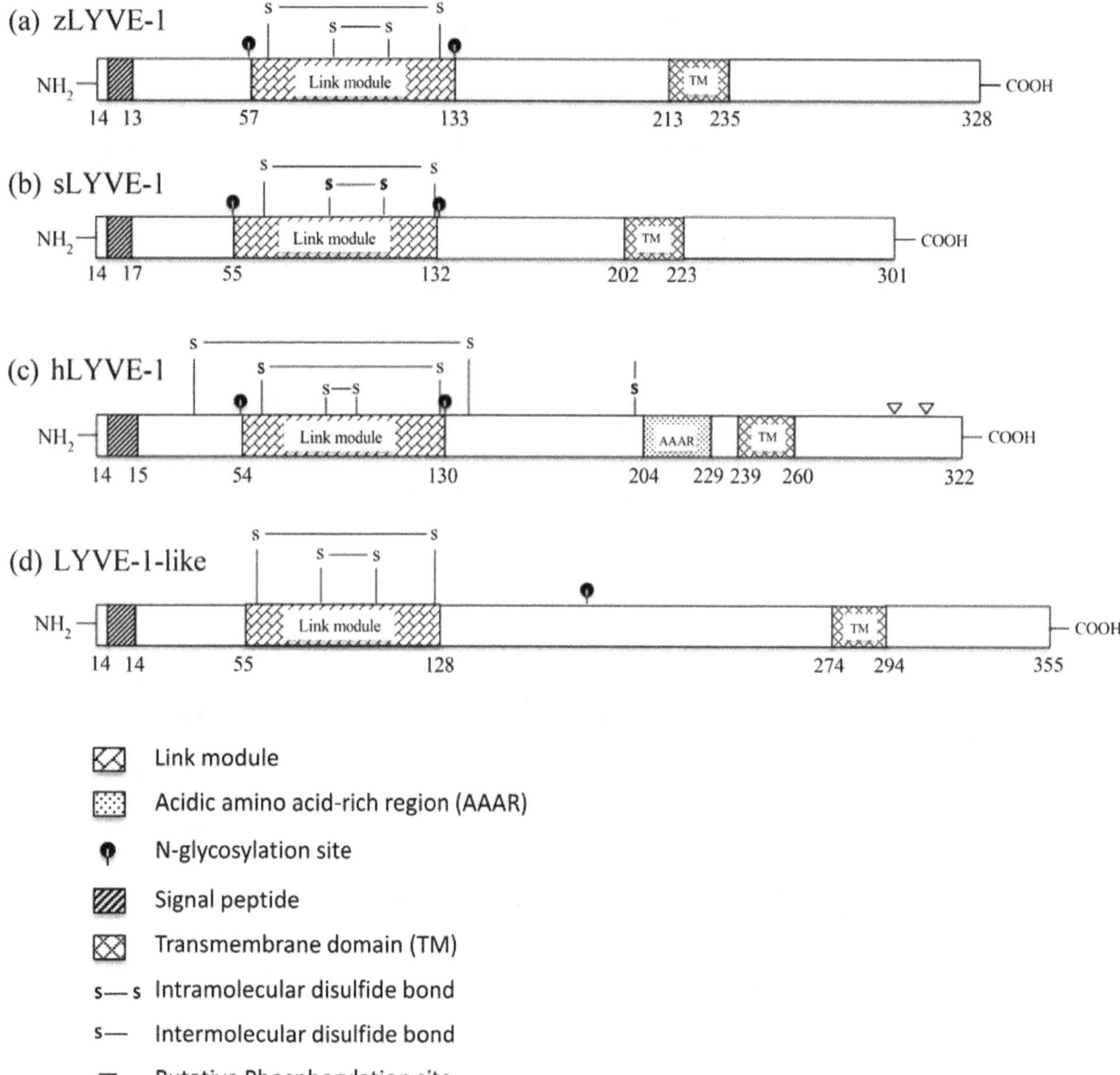

(C)

Figure 1. zLyve-1 (or zLYVE-1) exhibits amino-acid-sequence ((A), (B)) and structural domain (C) homology with those of other vertebrate homologs. (A) The nucleotide and deduced amino-acid sequences of *zlyve1* cDNA cloned from zebrafish were determined as described in the text. Two potential *N*-linked glycosylation sites are *boxed.* The half-cystine residues are marked by *circles.* The boundaries of the exons of *zlyve1* cDNA are distinguished by the shaded and unshaded sequences; (B) The deduced amino acid sequence of zLyve-1 is aligned with those of other vertebrate homologs (human, chimpanzee, dog, cow, mouse, rat, horse, chicken and salmon). The arrow indicates the identified N-terminal amino acid residues in human and cow LYVE-1 (hLYVE-1 and bLYVE-1, respectively) (3). Identical amino acid residues in zebrafish and other vertebrate homologs are marked by shaded boxes. The Link module and transmembrane domain are underlined. The AAAR domain present in other vertebrat homologs is over-lined and is absent in the zebrafish and salmon orthologs (zLyve-1 and sLYVE-1). Identical half-cystine residues in all vertebrate LYVE-1 species are indicated by ▲. Half-cystine residues, which are only present in other vertebrate homologs, are indicated by Δ. Two potential N-linked glycosylation sites are marked by *. The *bar* indicates the absence of the corresponding amino acid residues in the zebrafish ortholog, salmon ortholog and other vertebrate homologs; (C) zLyve-1 (panel a) and sLYVE-1 (panel b) lack the AAAR domain and contain 2 versus 3 disulfide bonds compared to hLYVE-1 (panel c) or other mammalian homologs. The 2 disulfide bonds are present in the Link domain in all vertebrate LYVE-1 homologs. The third disulfide bond seen in hLYVE-1 or other vertebrate LYVE-1 homologs is localized outside the Link module. The two N-linked glycosylation sites are localized at both sides of the Link module in all vertebrate LYVE-1 homologs. In contrast, the zLyve-1-like (panel d) lacks the N-linked glycosylation sites in the Link module.

cently identified by Flores *et al.* [16], does not have this important structural feature and is termed Lyve-1-like in this communication (Supplementary **Figure S1** and **Figure 1(C)**, panel d). The close protein structure-feature similarity of the identified zLyve-1 and other known vertebrate LYVE-1 homologs suggests that the cloned *zlyve1* is the authentic ortholog of LYVE-1 in zebrafish.

In contrast to all other known vertebrate homologs (e.g., human LYVE-1), zLyve-1 and the other fish homolog, salmon LYVE-1 (sLYVE-1) do not contain an acidic-amino-acid-rich domain (AAAR) which is separated from the Link module and close to the transmembrane domain (**Figures 1(B)** and **(C)**, panels a and b vs. panel c). We hypothesize that the AAAR domain orients its acidic side chains of Glu residues toward the basic side chains in the molecule-surface clustered basic amino acid residues of its cognate ligands, including PDGF-BB, VEGF-A^{165}, IGFBP-3 and FGF2 [1-3,7,18]. zLyve-1 and sLYVE-1 do not possess this AAAR domain and are expected to lack growth factor/cytokine (with surface clustered basic amino acid residues in the molecule) binding activity.

3.2. Genomic Structure and Ligand Binding Activity of zLyve-1

The *zlyve1* allele (12.2 kb) resides in chromosome 7 and consists of 6 exons and 5 introns. The genomic structure of the *zlyve1* is quite similar to those of other vertebrate homologs (Supplementary **Figure S2**). A phylogenetic tree analysis indicates that zLyve-1-like protein (*zlyve1l*) is evolutionarily distantly related to all known vertebrate LYVE-1 homologs (Supplementary **Figure S3**). This supports the notion that zLyve-1-like protein (or *zlyve1l*) is not the authentic ortholog of LYVE-1 in zebrafish.

H1299 cells, which express very little endogenous LYVE-1, have been used to analyze the ligand binding activity of bovine CRSBP-1 (bCRSBP-1 or bLYVE-1) after stable transfection of these cells with its cDNA in pCEP4 vector [3]. To define the ligand (hyaluronic acid) binding activity of zLyve-1, H1299cells were transiently transfected with pCMV-*zlyve1*-HA and termed H1299/*zlyve1* cells. As shown in **Figure 2(A)**, zLyve-1 expressed in H1299/*zlyve1* cells was identified as a 48-kDa protein by 7.5% SDS-PAGE followed by Western blot analysis using anti-HA (hemagglutinin epitope) antibody (lane 2). The hyaluronic acid-binding activity of zLyve-1 expressed in H1299/*zlyve1* cells was determined as described previously [3] by the following procedure: Triton X-100 extracts (100 μg protein) of H1299 cells transfected with expression vector containing *zlyve1* cDNA (H1299/*zlyve1* cells) or with empty vector (H1299/vector cells) were incubated with several concentrations of hyaluronic acid (M.W. ~2,000,000). The hyaluronicacid-zLyve-1 complex in the Triton X-100 extracts was then precipitated by cetylpyridinium chloride (CPC) followed by Western blot analysis using anti-HA (hemagglutinin epitope) antibody as described [3]. The relative amounts of zLyve-1 on the blots, which represented those of zLyve-1-hyaluronic acid complexes precipitated by CPC, were estimated. As shown in **Figure 2(B)**, hyaluronic acid formed complexes with zLyve-1 in a concentration-dependent manner with a half-maximal concentration of 30 ± 5 μg/ml (n = 4) (panels a and b). This is equivalent to the apparent K_d (~15 nM) of hyaluronic acid binding to other vertebrate LYVE-1 homologs [3]. This suggests that zLyve-1 and mammalian LYVE-1 homologs bind hyaluronic acid with the similar affinity.

To determine the growth factor binding activity of zLyve-1, we performed direct binding analysis of I^{125}-labeled VEGF peptide (I^{125}-VEGF peptide) in H1299 cells transiently transfected with pCMV-*zlyve1*-HA (termed H1299/*zlyve1* cells). VEGF peptide is a 25-mer peptide containing the cell-surface retention sequence (CRS) motif of VEGF-A^{165} [1-3,7]. We reasoned that, if zLyve-1 possesses I^{125}-VEGF peptide binding activity, H1299/*zlyve1* cells should exhibit more specific binding (estimated in the presence and absence of 100-fold excess of unlabeled VEGF peptide) of I^{125}-VEGF peptide than that found in H1299/vector cells which were H1299 cells stably transfected with vector only [3]. As shown in **Figure 2(C)**, H1299/*zlyve1* cells exhibited less specific binding of I^{125}-VEGF peptide than that found in H1299/vector cells, rather than more specific binding. The apparent I^{125}-VEGF peptide specific binding found in these cells (H1299/vector and H1299/*zlyve1* cells) appeared to be mediated by other membrane proteins (e.g., low-affinity binding proteins such as proteoglycans) [1]. We speculate that the lower low-affinity specific binding in H1299/*zlyve1* cells may be due to lower expression of proteoglycans in these cells. In a positive control experiment, H1299/bLYVE-1 cells (which stably express bovine LYVE-1 (bLYVE-1) that is known to have growth factor binding activity) [3] exhibited more specific binding (bLYVE-1-mediated) of I^{125}-VEGF peptide than those of H1299/vector and H1299/*zlyve1* cells. This suggests that zLyve-1 may not possess I^{125}-VEGF peptide binding activity. To further test for I^{125}-VEGF peptide binding activity of zLyve-1, we performed I^{125}-VEGF peptide-affinity labeling in H1299/*zlyve1* cells and H1299/bLYVE-1 cells (as a positive control). This labeling technique identifies mammalian LYVE-1 (also termed CRSBP-1) in cultured cells and tissue extracts [1-3,7]. The cells were incubated with I^{125}-VEGF peptide and cross-linked with 1 ethyl-3-(3-dimethylaminopropyl) carbodiimide HCl (EDAC) as described previously [1-3,7]. As shown in **Figure 2(D)**, H1299/bLYVE-1cells exhibit a ~60-kDa I^{125}-VEGF peptide-affinity-labeled bLYVE-1 and its ~45-kDa proteolytic product (lane 2),

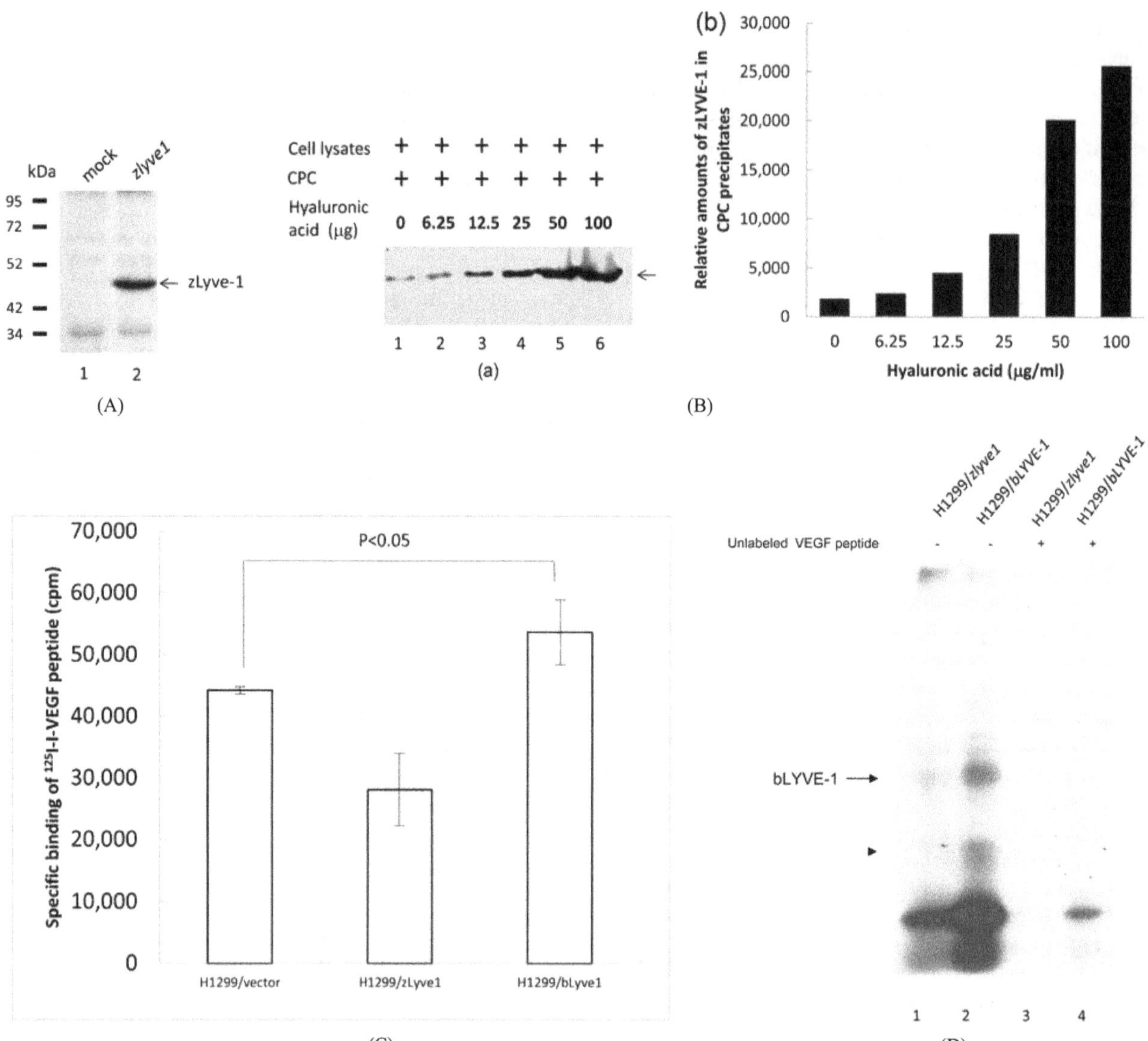

Figure 2. zLyve-1 expressed in H1299/*zlyve1* cells is a 48-kDa protein (A) and exhibits hyaluronic acid binding activity (B) but lacks ^{125}I-VEGF peptide-binding activity ((C), (D)). (A) Cell lysates of H1299/*zlyve1* and H1299/vector cells were analyzed by 7.5% SDS-PAGE followed by Western blot analysis using anti-HA antibody. A 48-kDa protein was found in the cell lysates from H1299/*zlyve1* cells (lane 2) but not in those of H1299/vector cells (lane 1). This suggests that zLyve-1 is a protein with M.W. 48 kDa; (B) The hyaluronic acid binding activity of zLyve-1 was determined by incubation of the Triton X-100 extracts of H1299/*zlyve1* cells with several concentrations of hyaluronic acid followed by CPC precipitation and 7.5% SDS-PAGE/ Western blot analysis using anti-HA antibody (panel a). The arrow indicates the location of zLyve-1. The relative amounts of zLyve-1 in the CPC precipitates were quantified (panel b); (C) The ^{125}I-VEGF peptide-binding activity of zLyve-1 was determined by direct binding analysis in H1299/vector, H1299/*zlyve1* and H1299/bLYVE-1 (as a positive control) cells as described in the text. The specific binding of ^{125}I-VEGF peptide in these cells was determined by subtracting non-specific binding (which was estimated in the presence of 100-fold excess of unlabeled VEGF peptide) from total binding. H1299/*zlyve1* cells did not exhibit more specific binding (zLyve-1-mediated) of ^{125}I-VEGF peptide when compared with that found in H1299/ vector cells. In the control experiment, H1299/bLYVE-1 cells showed more specific binding of ^{125}I-VEGF peptide than H1299/vector cells, as previously described (3); (D) The ^{125}I-VEGF peptide-binding activity of zLyve-1 was also determined by ^{125}I-VEGF peptide-affinity labeling. H1299/*zlyve1* and H1299/bLYVE-1 cells were incubated with ^{125}I-VEGF peptide in the absence (−) and presence (+) of 100-fold excess of unlabeled VEGF peptide (lanes 1, 2 and lanes 3, 4, respectively) at 4˚C for 2.5 h. ^{125}I-VEGF peptide-affinity labeling was performed using EDAC. The ^{125}I-VEGF peptide-affinity labeled cell lysates were analyzed by 7.5% SDS-PAGE followed by autoradiography. A ~60-kDa ^{125}I-VEGF peptide-affinity labeled protein (bLYVE-1) and its ~45-kDa proteolytic product, as indicated by an arrow and arrowhead, respectively, were found in the cell lysates of H1299/bLYVE-1 cells (lane 2) as described (3, 7) but not in those of H1299/*zlyve1* cells (lane 1). The ^{125}I-VEGF peptide-affinity labeling was blocked in the presence of 100-fold excess of unlabeled VEGF peptide (lane 4). This result suggests that, although bLYVE-1 exhibits detectable ^{125}I-VEGF peptide-binding activity, zLyve-1 does not.

as described previously [1-3,7]. The I^{125}-VEGF peptide-affinity labeling of bLYVE-1 was blocked in the presence of 100-fold excess of unlabeled VEGF peptide (lane 4). In contrast, H1299/*zlyve1* cells did not show I^{125}-VEGF peptide-affinity-labeled zLyve-1 (lane 1). The predicted molecular weight of I^{125}-VEGF peptide-affinity-labeled zLyve-1 would be ~50 kDa if zebrafish Lyve-1 was capable of binding I^{125}-VEGF peptide. H1299 cells expressed a very low level of endogenous LYVE-1 (lane 1). Taken together, these results suggest that zLyve-1 does not have I^{125}-VEGF peptide-binding activity, as bLYVE 1 does. These results are also consistent with the observation that zLyve-1 lacks the AAAR domain which is believed to mediate binding of peptides and growth regulators containing CRS motifs [1-3,7].

3.3. Requirement of zLyve-1 for Thoracic Duct (TD) Formation

In zebrafish, the TD is the first perfused lymphatic organ and is located between the dorsal aorta (DA) and posterior cardinal vein (PCV). It has been reported that formation of the TD in zebrafish requires the functions of *prox*1 and *vegfc* genes [19,20]. However, the important role of prox1 in TD formation is currently unclear [21]. Like prox1, mammalian LYVE-1 is primarily expressed in LECs and has been used as a specific marker for these cells. zLyve-1 may also be required for lymphatic vessel formation in zebrafish. To test this possibility, we determined the role of zLyve-1 in the development of lymphatic vessels in wild-type zebrafish by knocking down zLyve-1 levels using embryo microinjection with antisense morpholino oligonucleotides (MOs). We designed antisense MOs which targeted 5' UTR (MO1) and the boundaries between exon 1 and intron 1 (MO2) and intron 1 and exon 2 (MO3), and mis-paired MO2, mis-paired MO3 and control MO (**Table 1**). MO2 was the most effective of the MOs used to inhibit translation of zLyve-1 and to generate a TD defect in the embryos which received MO2 injection (**Table 1**). To exclude off-target effects of MO2, we also co-injected MO2 with pCMV-*zlyve1*-HA into zebrafish embryos and examined the zLyve-1 levels and phenotypes of zebrafish embryos. Rescue of the phenotype by cDNA-based overexpression of the targeted transcript has been used to confirm the specificity of MO targeting in zebrafish [22,23]. As shown in **Figure 3(A)** (panel a), injection of MO2 alone at the one-cell stage greatly reduced the level of zLyve-1 in zebrafish embryos, as determined by RT-PCR which yielded a 308-base pair product. MO2 completely knocked down zLyve-1 expression at 1 and 2 dpf (lanes 2 and 4 vs. lanes 1 and 3). Low levels (<20%) of zLyve-1 were observed at 3 - 5 dpf (lanes 6, 8 and 10 vs. lanes 5, 7 and 9). It is important to note that, in zebrafish embryos

Table 1. Effect of embryo injection with and without MOs on TD formation in *Tg(fli1:EGFP)* zebrafish.

Injection[a]	Embryos (at 0 dpf)		Embryos (at 5 dpf)	
			Total	TD defect[b]
MO1	4.2 ng	200	166 (83%)e	0[f]
	8.4 ng	200	158 (79%)e	0[f]
MO2	4.2 ng	200	155 (77%)e	32
	8.4 ng	200	148 (74%)e	80
MO3	4.2 ng	200	162 (81%)e	0[f]
	8.4 ng	200	154 (77%)e	10
Mis-paired MO2	8.4 ng	200	161 (80%)e	0[f]
Mis-paired MO3	8.4 ng	200	159 (79%)e	0[f]
Control MO	8.4 ng	200	168 (84%)e	0[f]
Mock[c]	8.4 ng	200	165 (82%)e	0[f]
No injection[d]		200	170 (85%)e	0[f]

[a]At 0 dpf, Tg(*fli1*:*EGFP*) zebrafish embryos (200 per experimental group) were injected with 4.2 and 8.4 ng of M01, M02, M03, mis-paired M02 and mis-paired M03, and with mock. At 5 dpf, zebrafish were examined. [b]The embryos (survived at 5 dpf) with abnormal phenotypes (pericardial edema and bent tail) were examined for the absence of the TD under a confocal fluorescence microscope. All embryos with a TD defect (no thoracic duct) had pericardialedema. [c]Phenol red injection. [d]Embryos without injection exhibited a 85% survival rate under the experimental conditions. [e]Survival rate at 5 dpf. [f]No detectable TD defect.

not injected with MO2, zLyve-1 was expressed at the highest level at 3 dpf and gradually decreased afterward (**Figure 3(A)**, panel a, lane 5). Co-injection with pCMV-*zlyve1*-HA reversed MO2-injection-induced knockdown of zLyve-1 levels, as determined by RT-PCR, using a different pair of primers (forward + reverse). This yielded a 461-base pair product (**Figure 3(A)**, panel b, lane 3 vs. lanes 1, 2). Injection with MO2 caused edema in the pericardial sac, bent tail and a decreased survival rate (**Figure 3(B)**, panel b vs. panel a and **Table 1**). Under our experimental conditions, the survival rate of zebrafish at 5 dpf with control MO andd MO1 was 79% - 84% (**Table 1**). The survival rates of zebrafish injected with 8.4 ng MO2 and 8.4 ng MO3 were less (74% - 77%). The MO2-injection-induced phenotypes (e.g., edema) and decreased survival rate were reversed by co-injection with pCMV-*zlyve1*-HA (**Figure 3(B)**, panel c and **Table 1**). These results suggest that MO2-injected embryos may have defects in the formation and/or function of lymphatic vessels, including TD. In zebrafish, loss of the TD has been shown to result in the phenotype of pericardial edema [24]. These results also suggest that the phenotypes of pericardial edema and decreased survival rate in MO2-injected embryos are not due to off-target effects [14] but are specifically caused by the loss of zLyve-1 because the phenotypes can be rescued by co-injection

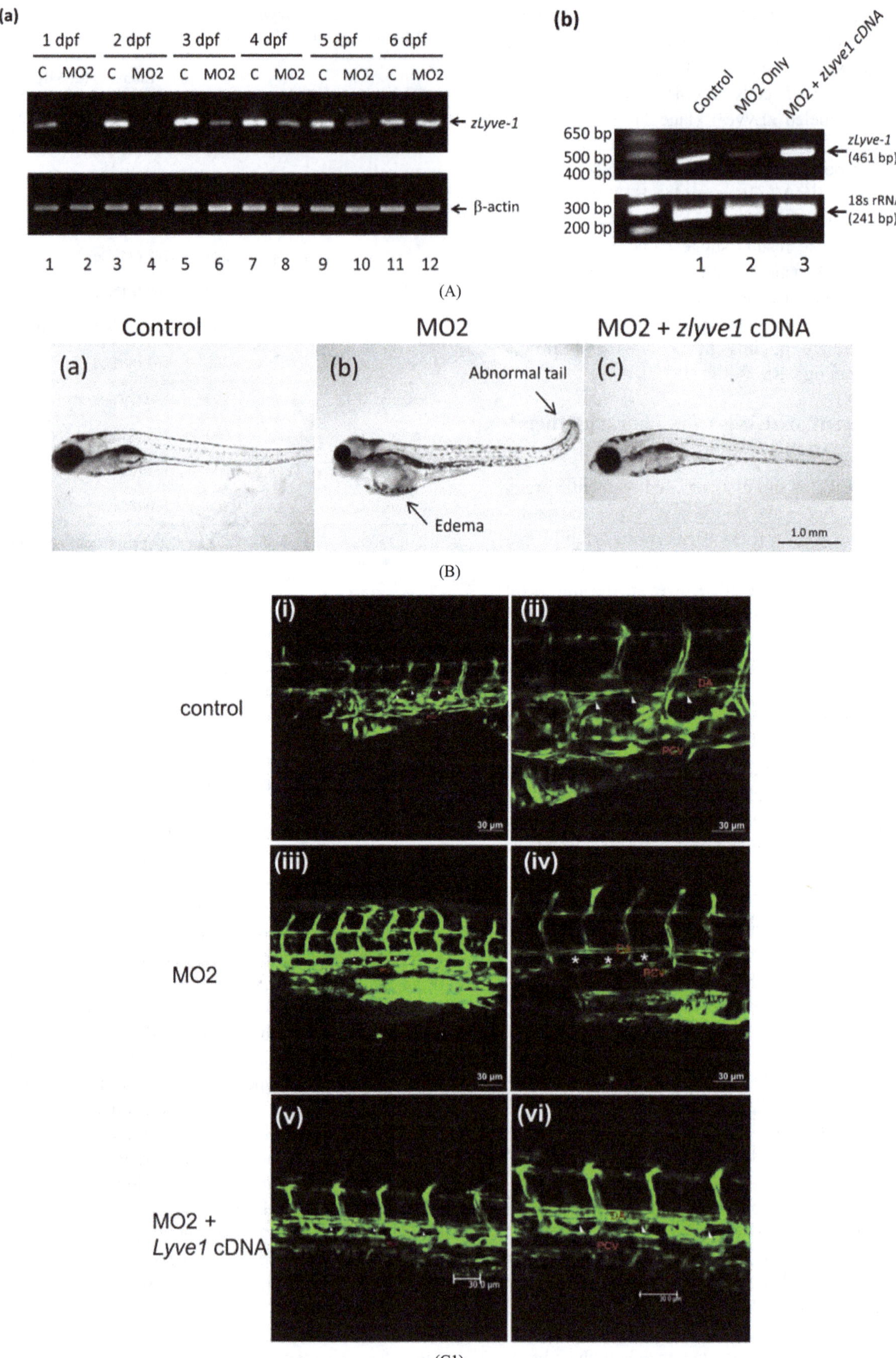
(a)
1 dpf 2 dpf 3 dpf 4 dpf 5 dpf 6 dpf
C MO2 C MO2 C MO2 C MO2 C MO2 C MO2
zLyve-1
β-actin
1 2 3 4 5 6 7 8 9 10 11 12
(b)
Control
MO2 Only
MO2 + zLyve1 cDNA
650 bp
500 bp
400 bp
300 bp
200 bp
zLyve-1 (461 bp)
18s rRNA (241 bp)
1 2 3
(A)
Control
MO2
MO2 + zlyve1 cDNA
(a)
(b)
(c)
Abnormal tail
Edema
1.0 mm
(B)
(i)
(ii)
(iii)
(iv)
(v)
(vi)
control
MO2
MO2 + Lyve1 cDNA
30 µm
(C1)

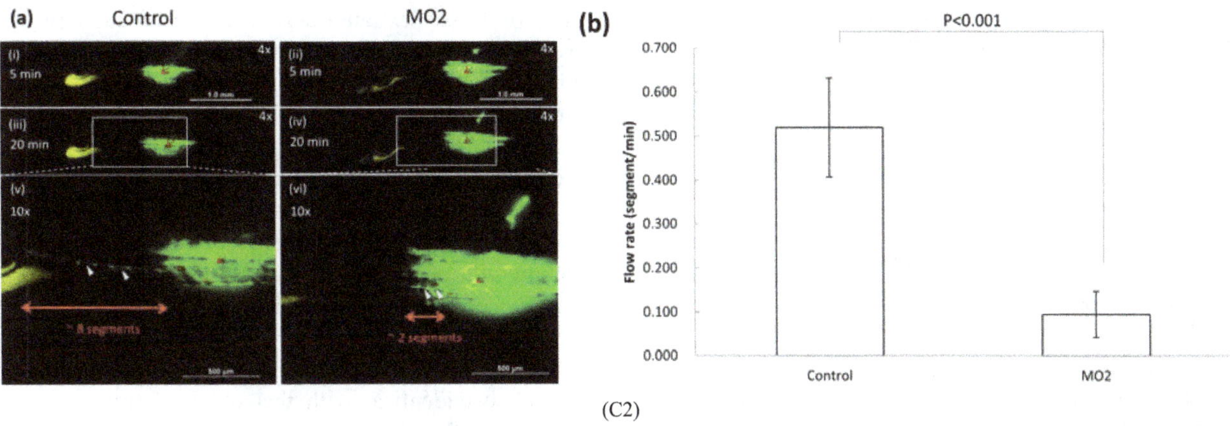

(C2)

Figure 3. Embryo microinjection with MO2 causes decreased levels of zLyve-1 (A), edema formation (B), defects of TD formation (C1), and dysfunction of TD transit of FITC-dextran (C2), that can be reversed by co-microinjection with *zlyve1* cDNA, in zebrafish. (A) (panel a) The levels of *zlyve1* mRNA in zebrafish embryos (at 1 - 6 dpf) injected with control MO (C, lanes 1, 3, 5, 7, 9, 11) and MO2 (lanes 2, 4, 6, 8, 10, 12), were determined by measuring the level of a 308-base pair product (as indicated by the arrow) with RT-PCR analysis using the following primers: 5'-G GCCTGTTCATGTCTGGTCTG-3' (forward) and 5'-TCTGCACGCCACACCACAACC-3' (reverse). The *β*-actin RT-PCR product (433 base pair) was determined an internal control. The zLyve-1 mRNA levels in zebrafish embryos injected with MO2 increased at 3 - 5 dpf but remained at <20% of those seen in zebrafish derived from embryos injected with control MO (lanes 6, 8, 10 vs. lanes 5, 7, 9); (A) (panel b) The levels of zLyve-1 mRNA in zebrafish embryos injected with control MO (C, lanes 1), MO2 (lanes 2) and MO2 + *zlyve1* cDNA (in pCMV-*zlyve1*-HA) (lane 3) were determined by measuring the level of a 461-base pair product (as indicated by an arrow) with RT-PCR analysis using the following primers: 5'-ATG ACA CGA GTC TGC ATG GG–3' (forward) and 5'-ATC GGC TTT CCG GTT GTC ATT TGG-3' (reverse). The RT-PCR product (241 base pairs) of 18 s rRNA in these embryos was determined as an internal control. Embryo injection with MO2 resulted in complete knockdown of zLyve-1 mRNA levels at 2 dpf (lane 2 vs. lane 1). Co-injection of MO2 with *zlyve1* cDNA (in pCMV-*zlyve1*-HA) reversed the effect of MO2 knockdown on zLyve-1 levels in zebrafish embryos at 5 dpf (lane 3 vs. lane 2); (B) Wild-type zebrafish embryos of 5 dpf derived from MO2-injected embryos clearly exhibited tail abnormality and pericardial edema as indicated by the arrow (panel b) compared to control MO-injected zebrafish embryos (panel a). These phenotypes were reversed by co-injection with *zlyve1* cDNA (in pCMV-*zlyve1*-HA) (panel c). Scale bar = 1 mm; (C1) Fluorescent images (lateral view) of the trunk vasculature (head-left and tail-right orientation) from 5-dpf *Tg(fli1:EGFP)* zebrafish embryos derived from control MO-injected embryos (panels i and ii), MO2-injected embryos (panels iii and iv) and MO2 + *zlyve1* cDNA (in pCMV-*zlyve1*-HA)-co-injected embryos (panels v and vi) are shown. The images were taken by confocal fluorescence microscopy. Locations of DA and PCV are indicated. The TD is present (as indicated by arrowheads) in the zebrafish derived from control MO-injected embryos (panels i and ii) and MO2 + *zlyve1* cDNA (in pCMV-*zlyve1*-HA)-coinjected embryos (panels v and vi). The TD is absent between DA and PCV (as indicated by*) in the zebrafish derived from embryos injected with MO2 (panels iii and iv). No fluorescent cells (e.g., PLs) were found in the horizontal midline; (C2) FITC-dextran was injected intramuscularly into 5-dpf wild-type zebrafish embryos derived from embryos injected with control MO (panels a-i, -iii and -v) and MO2 (panels a-ii, -iv and -vi). Epifluorescent images (head-left and tail-right orientation) obtained at 5 min (panels a-i and -ii; 4× magnification) and 20 min (panels a-iii and -iv; 4× magnification and panels a-v and -vi; 10× magnification) after injection with FITC-dextran. The red arrow indicates the injection site; white arrowheads indicate the location of the segment in the TD. The transit speed of FITC-dextran along the segment of TD in zebrafish embryos at 5 dpf (which were derived from embryos injected with control MO or MO2) was quantified in a time period of 20 min from a total of 6 independent experiments (panel b). The maximum transit distance of FITC-dextran after injection in zebrafish embryos derived from embryos injected with MO2 was ~2 segments. This appeared to be due to the defect in TD formation beyond the ~2 segments toward the head. Locations of DA and PCV are indicated.

with pCMV-*zlyve1*-HA but not with pCMV-HA (vector) (**Table 2**).

To test the possibilitythat zLyve-1 is required for lymphatic vessel formation and function, we examined the TDin *Tg* (*fli*1:*EGFP*) zebrafish embryos at 5 dpf after embryo microinjection of *Tg* (*fli*1:*EGFP*) zebrafish with control MO or MO2. The TD is the main conduit of the lymphatic system. It runs through the space between DA and PCV longitudinally, inthe trunk of fluorescent transgenic *Tg* (*fli*1:*EGFP*) zebrafish [10-12,25]. *Tg* (*fli*1:*EGFP*) zebrafish, which bear a *egfp* transgene driven by the zebrafish fli1 promoter *Tg* (*fli*1:*EGFP*), exhibit fluorescence in both blood and lymphatic vascular structures. This is used to examine the formation of blood and lymphatic vessels during development [11,25]. The TD in *Tg* (*fli*1:*EGFP*) zebrafish derived from embryos microinjected with control MO was analyzed using confocal fluorescence microscopy and was found to be present and not altered (**Figure 3(C1)**, panels i and ii). It is not different from that seen in *Tg* (*fli*1:*EGFP*) zebrafish de-

Table 2. Embryo coinjection with MO2 and pCMV-*zlyve1*-HA reverses MO2-induced TD defect in *Tg(fli1:EGFP)* zebrafish.

Injection[a]	Embryos (at 0 dpf)	Embryo (at 5 dpf)	
		Total	TD defect[b]
MO2 + pCMV-HA (vector)	200	142 (71%)[c]	84
MO2 + pCMV-*zlyve1*-HA	200	160 (80%)[c]	5

[a]At 0 dpf, *Tg(fli:EGFP)* zebrafish embryo (200 per experimental group) were injected with 8.4 ng of MO2 ± 0.1 ng of pCMV-*zlyve1*-HA or pCMV-HA. At 5 dpf, zebrafish were examined. [b]The embryos with abnormal phenotypes (pericardial edema and bent tail) were examined for the absence of the TD under a confocal fluorence microscope. All embryo with a TD defect (no TD) had pericardial edema. [c]Survival rate at 5 dpf.

rived from embryos without injection. Injection with MO2 caused defects in TD formation but had no significant effects on the formation of the blood vascular system (e.g., DA, PCV and intersegmental blood vessels) in this animal. As shown in **Figure 3(C)** (panels iii and iv), in embryos injected with MO2, because of the loss of TD and absence of detectable fluorescent cells (e.g., parachordal lymphangioblasts) [12,25], the area in the horizontal midline had no fluorescence (as indicated by *). This suggests that embryos injected with MO2 do not develop the TD. We hypothesize that knockdown of zLyve-1 levels by the MO2 injection in the PCV causes the defects in secondary lymphangiogenic sprouts from PCV that give rise to precursors (parachordal lymphangioblasts) of LECs forming the TD [11,25]. As observed with the phenotype of pericardial edema, the TD defect phenotype in zebrafish derived from embryos injected with MO2 was reversed by embryo co-injection with pCMV-*zlyve1*-HA (**Figure 3(C1)**, panels v and vi and **Table 2**). To determine the effect of MO2 injection on lymphatic function, we injected FITC-dextran intramuscularly into wild-type zebrafish derived from embryos injected with MO2 ± pCMV-*zlyve1*-HA or control MO. We then monitored the 20-min transit of the FITC-dextran along the TD in the trunk from the injection site toward the head in these animals. The FITC-dextran transit time from the injection site to the visible end of the TD was approximately 20 min. As shown in **Figure 3(C2)** and Supplementary **Figure S4** (movie), FITC-dextran fluorescence moved along the TD segment by segment with a speed of 0.52 ± 0.12 segment/min (mean ± S.D., n = 6) in zebrafish derived from embryos injected with a control MO (**Figure 3(C2)**, panel a: Control i, iii, v and Supplementary **Figure S4(a)**). Zebrafish derived from embryos injected with MO2 exhibited minimal transit (~2 segments) of FITC-dextran fluorescence along the TD (**Figure 3(C2)**, panel a: ii, iv vi) with a speed of 0.1 ± 0.05 segment/min (mean ± S.D., n = 6) (**Figure 3(C2)**, panel a: MO2, ii, iv, vi and Supplementary **Figure S4(b)**). The ~2 segments were the maximal traveling distance (in >20 min) of FITC-dextran fluorescence. This supports our observation that defective TD formation is caused by injection with MO2 (**Figure 3(C1)**, panels iii and iv). To rule out the possibility of "off-target" effects of MO2, we co-injected zebrafish embryos with MO2 and pCMV-*zlyve1*-HA or pCMV-HA (vector). The FITC-dextran transit alteration phenotype in zebrafish derived from embryos injected with MO2 was reversed by embryo co-injection with pCMV-*zlyve1*-HA but not with pCMV-HA. The speed of FITC-dextran transit in zebrafish co-injected with MO2 and pCMV-*zlyve1*-HA was estimated to be 0.54 ± 0.16 segment/min (mean ± S.D., n = 6) which is almost identical with that obtained from embryos injected with a control MO (**Figure 3(C2)**, panel a: i, iii, v and panel b). These results indicate that zLyve-1 is required for TD formation.

3.4. zLyve-1 Is Highly Expressed in the Posterior Cardinal Vein (PCV)

The finding that zLyve-1 is required for TD formation in zebrafish raises the question of why LYVE-1 is not required for lymphatic development in mammalian species. Null-mutation of LYVE-1 has been shown not to affect lymphatic development in mice [26,27], although LYVE-1-null mice exhibit altered morphology and function of lymphatic capillary vessels [6,7]. In zebrafish, the TD is known to arise from parachordal lymphangioblasts, which are derived from secondary lymphangiogenic sprouts from PCV [11,25]. A simple possibility is that zLyve-1 is highly expressed in PCV endothelial cells. zLyve-1 may be required for the activation of these cells to become LEC precursors (parachordal lymphangioblasts), which form TD endothelial cells, in zebrafish lymphatic development. To test this possibility, we performed *in situ* hybridization of zLyve-1 (**Figure 4(A)**) and zLyve-1-like protein (**Figure 4(B)**) in zebrafish larvae at 30 hpf (hours post fertilization), 3 dpf, 4 dpf and 5 dpf. As shown in **Figures 4(A)** and **(B)**, zLyve-1 and zLyve-1-like protein were expressed in the PCV in zebrafish embryosat 30 hpf (panels a, b, c, and d), 3 dpf (panels e, f, g and h), 4 dpf (panels i, j, k and l) and 5 dpf (panels m, n, o and p). At 30 hpf prior to venous sprouting (which would be 36 hpf) [25], zLyve-1 and zLyve-1 like protein were expressed in the PCV from the caudal to the anterior parts (panels a to d). This suggests potential involvement of zLyve-1 and zLyve-1-like protein in venous sprouting and/or subsequent events. However, at 3 dpf, zLyve-1 was expressed at the highest level in the PCV (panels e and f). This result is consistent with that obtained from RT-PCR analysis (**Figure 2(A)**, lane 5). In contrast, zLyve-1-like protein was expressed at the highest level in PCVat 30 hpf or earlier than 3 dpf (panels c and d). This indicates that the temporal expression profiles of zLyve-1 and zLyve-1-like protein in PCV are dif-

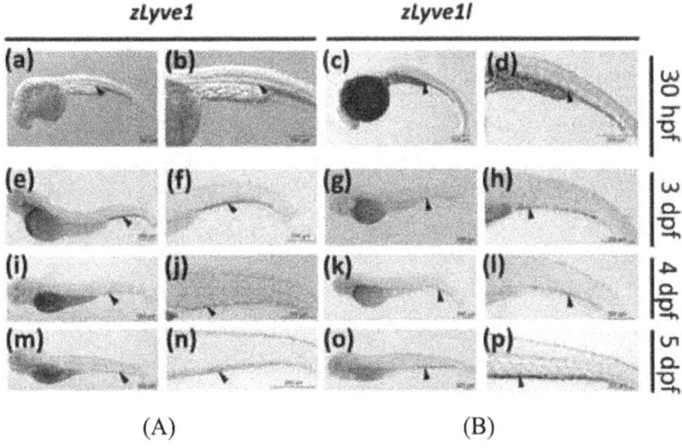

Figure 4. zLyve-1 (A) and zLyve-1-like (B) are expressed in PCV in zebrafish embryos. *In situ* RNA hybridization was performed on zebrafish embryos at 30 hpf (panels a-d), 3 dpf (panels e-h), 4 dpf (panels i-l), and 5 dpf (panels m-p), using digoxigenin-labeled antisense RNA probes for zLyve-1 (panels a, b, e, f, i, j, m, n) and zLyve-1-like (panels c, d, g, h, k, l, o, p). Panels a and b, e and f, I and j, m and n, c and d, g and h, k and l, or o and p represent the pictures of the same animal but with different magnifications (20 and 200 µm, respectively). Hybridization was performed as described in the text. Arrowheads indicate the localization of zLyve-1 or zLyve-1-like in the PCV (panels a-p) in zebrafish embryos. The expression patterns in PCV of zLyve-1 and zLyve-1-like were similar. However, zLyve-1 was expressed at the highest level in PCV at 3 dpf (panels e and f) whereas zLyve-1-like was expressed at the highest level in PCV at 30 hpf or earlier than 3 dpf. As development progresses, the expression levels in PCV of zLyve-1 and zLyve-1-like after 3-dpf or 30 hpf decreased gradually (panels e and f, and panels c and d, respectively). This 3 dpf time point appeared to coincide with the formation of TD, which occurred at 3.5-dpf [25].

ferent. Interestingly, at 3 dpf, the highest expression of zLyve-1 coincides with the time point (3.5 dpf) seen in the formation of lymphatic vessels such as the TD, intersegmental lymphatic vessels (ISLV) and dorsal longitudinal lymphatic vessels (DLLV) in zebrafish development [25]. Both zLyve-1 and zLyve-1-like protein were specifically expressed at the vessel wall of PCV and intersegmental vessels connected to PCV (data not shown).

4. Discussion

Several lines of evidence indicate that zLyve-1 is the ortholog of LYVE-1 in zebrafish. These include: 1) zLyve-1 shares important protein structural features with all other known vertebrate LYVE-1 homologs. Like all other known vertebrate LYVE-1 congeners, including mammalian and sLYVE-1, zLyve-1 is a membrane glycoprotein possessing two N-glycosylation sites which are localized on both sides of the Link module. In contrast, zLyve-1like protein does not have such glycosylation sites on both sides of the Link module. 2) All of known vertebrate LYVE-1 orthologs contain 301 to 328 amino acid residues whereas zLyve-1-like protein is larger than these known LYVE-1 vertebrate orthologs and has 355 amino acid residues. 3) The sizes of the genes of all other known vertebrate LYVE-1 homologs, including *zlyve1* and sLYVE-1 (9.4 - 18.7 kb) are larger than *zlyve1l* which is small (4.2 kb). 4) A phylogenetic tree analysis indicates that zLyve-1-like protein (*zlyve1l*) is evolutionarily distantly related to all known vertebrate LYVE-1 homologs, including zLyve-1 and sLYVE-1.

zLyve-1 and sLYVE-1 exhibit amino acid sequence homology with other mammalian homologs but lack the acidic-amino-acid-rich (AAAR) domain which is believed to mediate its binding to growth factors/cytokines with CRS motifs [1-3,7,18]. This suggests that growth factor (e.g., VEGF-A, PDGF-BB and FGF2) binding activity of LYVE-1 is not important for lymphatic development in zebrafish and salmon. We demonstrate that VEGF-A and PDGF-BB from zebrafish lack the CRS motifs in their molecules (Supplementary **Figures S5(a)** and **(b)**, respectively) in contrast to their mammalian counterparts. FGF2 from zebrafish also does not have a typical CRS motif in the molecule (data not shown). It appears that the LYVE-1 gene acquires the AAAR domain (which confers growth factor binding activity) at evolutionary stages beyond fish. We speculate that the development of the LYVE-1 gene with an AAAR domain in evolution may be associated with the acquisition of certain biological processes (e.g., adaptive immunity) involving growth factors and cytokines that possess CRS motifs. For example, the growth factor/cytokine-stimulated transit of immune cells from the interstitial space to lymph nodes is believed to be involved in the adaptive immune response in mammals [7,8]. LYVE-1 ligands (e.g., PDGF-BB, VEGF-A^{165}, CCL19) have been shown to be required for the transit of dendritic cells from the interstitial space to lymph nodes during immune responses in mammals such as mice [7,28]. Fish is known to have a very limited development of adaptive immunity [29]. This may be due to the lack of the AAAR domain in the LYVE-1 molecule and the CRS motifs in these growth factors/cytokines (PDGF-BB and VEGF-A) in fish.

In this communication, we demonstrate that embryo MO2 knockdown of zLyve-1 results in disruption of TD formation and impaired TD flow in zebrafish. Because zLyve-1-like protein-knock down zebrafish are phenotypically normal (16, data not shown), it appears that zLyve-1, but not zLyve-1-like protein, expressed in PCV plays an important role in the formation of the TD. zLyve-1-like protein does not appear to compensate for the loss of zLyve-1 in TD formation in zebrafish injected with MO2. The exact role of zLyve-1 in the secondary lymphangiogenic sprouts from PCV (that give rise to precursors of LECs forming the TD) [11] is unknown. We hypothesize that zLyve-1 may mediate signaling which leads to cellular activation of PCV endothelial cells to become LEC precursors (parachordal lymphangioblasts) after interaction with hyaluronic acid or unknown ligands. This hypothesis is based on our recent finding that mammalian LYVE-1 mediates signaling via its partner PDGF β-type receptor after interaction with its ligands, including hyaluronic acid [7].

5. Acknowledgements

We thank Frank E. Johnson, M.D. for critically reviewing the manuscript and Rachael Sheridan for zebrafish husbandry. This work was supported by NIH grants NS 060074 to MMV, AR052578 to SSH and AA 019223 to JSH.

REFERENCES

[1] C. Boensch, M. D. Kuo, D. T. Connolly, S. S. Huang and J. S. Huang, "Identification, Purificationand Characterization of Cell-Surface Retention Sequence-Binding Proteins from Human SK-Hep Cells and Bovine Liver Plasma Membranes," *Journal of Biological Chemistry*, Vol. 270, No. 4, 1995, pp. 1807-1816. http://dx.doi.org/10.1074/jbc.270.4.1807

[2] C. Boensch, S. S. Huang, D. T. Connolly and J. S. Huang, "Cell Surface Retention Sequence Binding Protein-1 Interacts with the V-Sis Gene Product and Platelet-Derived Growth Factor Beta-Type Receptor in Simian Sarcoma Virus-Transformed Cells," *Journal of Biological Chemistry*, Vol. 274, 1999, pp. 10582-10589. http://dx.doi.org/10.1074/jbc.274.15.10582

[3] S. S. Huang, F.-M. Tan, Y.-H. Huang, S.-C. Hsu, S.-T. Chen and J. S. Huang, "Cloning, Expression, Characterization, and Role in Autocrine Cell Growth of Cell Surface Retention Sequence Binding Protein-1," *Journal of Biological Chemistry*, Vol. 278, No. 44, 2003, pp. 43855-43869. http://dx.doi.org/10.1074/jbc.M306411200

[4] S. Banerji, J. Ni, S. X. Wang, S. Clasper, J. Su, R. Tammi, M. Jones and D. G. Jackson, "LYVE-1, a New Homologue of the CD44 Glycoprotein, Is a Lymph-Specific Receptor for Hyaluronan," *Journal of Cell Biology*, Vol. 144, No. 4, 1999, pp. 789-801. http://dx.doi.org/10.1083/jcb.144.4.789

[5] R. Prevo, S. Banerji, D. J. Ferguson, S. Clasper and D. G. Jackson, "Mouse LYVE-1 Is an Endocytic Receptor for Hyaluronan in Lymphatic Endothelium," *Journal of Biological Chemistry*, Vol. 276, No. 37, 2001, pp. 19420-19430. http://dx.doi.org/10.1074/jbc.M011004200

[6] S. S. Huang, I.-H. Liu, T. Smith, M. R. Shah, F. E. Johnson and J. S. Huang, "CRSBP-1/LYVE-1-Null Mice Exhibited Identifiable Morphological and Functional Alterations of Lymphatic Capillary Vessels," *FEBS Letters*, Vol. 580, No. 26, 2006, pp. 6259-6268. http://dx.doi.org/10.1016/j.febslet.2006.10.028

[7] W.-H. Hou, I.-H. Liu, C. C. Tsai, F. E. Johnson, S. S. Huang and J. S. Huang, "CRSBP-1/LYVE-1 Ligands Induce Disruption of Lymphatic Intercellular Adhesion by Inducing Tyrosinephosphorylation and Internalization of VE-Cadherin," *Journal of Cell Science*, Vol. 124, 2011, pp. 1231-1244. http://dx.doi.org/10.1242/jcs.078154

[8] W.-H. Hou, I.-H. Liu, S. S. Huang and J. S. Huang, "CRSBP-1/LYVE-1 Ligands Stimulate Contraction of the CRSBP-1-Associated ER Network in Lymphatic Endothelial Cells," *FEBS Letters*, Vol. 586, No. 10, 2012, pp. 1480- 487. http://dx.doi.org/10.1016/j.febslet.2012.04.001

[9] A. M. Kuchler, E. Gjini, J. Peterson-Maduro, B. Cancilla, H. Wolburg and S. Schulte-Merker, "Development of the Zebrafish Lymphatic System Requires VEGF-C Signaling," *Current Biology*, Vol. 16, No. 12, 2006, pp. 1244-1248. http://dx.doi.org/10.1016/j.cub.2006.05.026

[10] K. Yaniv, S. Isogai, D. Castranova, L. Dye, J. Hitomi and B. M. Weinstein, "Live Imaging of Lymphatic Development in the Zebrafish," *Nature Medicine*, Vol. 12, No. 6, 2006, pp. 711-716. http://dx.doi.org/10.1038/nm1427

[11] B. M. Hogan, L. Frank, F. L. Bos, J. Bussman, M. Witte, N. C. Chi, H. J. Duckers and S. Schulte-Merker, "Ccbe1 Is Required for Embryonic Lymphangiogenesis and Venous Sprouting," *Nature Genetics*, Vol. 41, No. 4, 2009, pp. 396-398. http://dx.doi.org/10.1038/ng.321

[12] B. M. Hogan, R. Herpers, M. Witte, H. Heloterä, K. Alitalo, H. J. Duckers and S. Schulte-Merker, "Vegfc/Flt4 Signalling Is Suppressed by Dll4 in Developing Zebrafish Intersegmental Arteries," *Development*, Vol. 136, No. 4, 2009, pp. 4001-4009.

[13] J. S. Eisen and J. C. Smith, "Controlling Morpholino Experiments: Don't Stop Making Antisense," *Development*, Vol. 135, No. 10, 2001, pp. 1735-1743. http://dx.doi.org/10.1242/dev.001115

[14] S. Isogai, J. Hitomi, K. Yaniv and B. M. Weinstein, "Zebrafish as a New Animal Model to Study Lymphangiogenesis," *Anatomical Science International*, Vol. 84, No. 3, 2009, pp. 102-111. http://dx.doi.org/10.1007/s12565-009-0024-3

[15] M. C. McKinney and B. M. Weinstein, "Chapter 4. Using the Zebrafish to Study Vessel Formation," *Methods in Enzymology*, Vol. 444, 2008, pp. 65-97. http://dx.doi.org/10.1016/S0076-6879(08)02804-8

[16] M. V. Flores, C. J. Hall, K. E. Crosier and P. S. Crosier, "Visualization of Embryonic Lymphangiogenesis Advances the Use of the Zebrafish Model for Research in Cancer and Lymphatic Pathologies," *Developmental Dynamics*, Vol. 239, No. 7, 2010, pp. 2128-2135. http://dx.doi.org/10.1002/dvdy.22328

[17] A. J. Day and G. D. Prestwich, "Hyaluronan-Binding Proteins: Tying Up the Giant," *Journal of Biological Chemistry*, Vol. 277, No. 7, 2002, pp. 4585-4588. http://dx.doi.org/10.1074/jbc.R100036200

[18] N. Platonova, G. Miquel, B. Regenfuss, S. Taouji, C. Cursiefen, E. Chevet and A. Bikfalvi, "Evidence for the Interaction of Fibroblast Growth Factor-2 with the Lymphaticendothelial Cell Marker LYVE-1," *Blood*, Vol. 121, No. 7, 2013, pp. 1229-1237. http://dx.doi.org/10.1182/blood-2012-08-450502

[19] L. Del Giacco, A. Pistocchi and A. Ghilardi, "Prox1b Activity Is Essential in Zebrafish Lymphangiogenesis,"

PLoS One, Vol. 5, No. 10, 2010, p. e13170. http://dx.doi.org/10.1371/journal.pone.0013170

[20] A. M. Küchler, E. Gjini, J. Peterson-Maduro, B. Cancilla, H. Wolburg and S. Schulte-Merker, "Development of the Zebrafish Lymphatic System Requires VEGFC Signaling," *Current Biology*, Vol. 16, No. 12, 2006, pp. 1244-1248. http://dx.doi.org/10.1016/j.cub.2006.05.026

[21] S. Tao, M. Witte, R. J. Bryson-Richardson, P. D. Currie, B. M. Hogan and S. Schulte-Merker, "Zebrafish Prox1b Mutants Develop a Lymphatic Vasculature, and Prox1b Does Not Specifically Mark Lymphatic Endothelial Cells," *PLoS One*, Vol. 6, No. 12, 2011, p. e28934. http://dx.doi.org/10.1371/journal.pone.0028934

[22] C. E. Eckfeldt, E. M. Mendenhall, C. M. Flynn, T. F. Wang, M. A. Pickart, S. M. Grindle, S. C. Ekker and C. M. Verfaillie, "Functional Analysis of Human Hematopoietic Stem Cell Gene Expression Using Zebrafish," *PLoS Biology*, Vol. 3, No. 8, 2005, p. e254. http://dx.doi.org/10.1371/journal.pbio.0030254

[23] D. Sheng, D. Qu, K. H. Kwok, S. S. Ng, A. Y. Lim, S. S. Aw, C. W. Lee, W. K. Sung, E. K. Tan, T. Lufkin, S. Jesuthasan, M. Sinnakaruppan and J. Liu, "Deletion of the WD40 Domain of LRRK2 in Zebrafish Causes Parkinsonism-Like Loss of Neurons and Locomotive Defect," *PLoS Genetics*, Vol. 6, No. 4, 2010, p. e1000914. http://dx.doi.org/10.1371/journal.pgen.1000914

[24] S.-J. Lee, T.-H. Chan, T.-C. Chen, B.-K. Liao, P.-P. Hwang and H. Lee, "LPA1 Is Essential for Lymphatic Vessel Development in Zebrafish," *FASEB Journal*, Vol. 22, No. 19, 2008, pp. 3706-3715. http://dx.doi.org/10.1096/fj.08-106088

[25] J. Bussmann, F. L. Bos, A. Urasaki, K. Kawakami, H. J. Duckers and S. Schulte-Merker, "Arteries Provide Essential Guidance Cues for Lymphatic Endothelial Cells in the Zebrafish Trunk," *Development*, Vol. 137, No. 16, 2010, pp. 2653-2657. http://dx.doi.org/10.1242/dev.048207

[26] N. W. Gale, R. Prevo, J. Espinosa, D. J. Ferguson, M. G. Dominguez, G. D. Yancopoulos, G. Thurston and D. G. Jackson, "Normal Lymphatic Development and Function in Mice Deficient for the Lymphatic Hyaluronan Receptor LYVE-1," *Molecular and Cellular Biology*, Vol. 27, No. 2, 2007, pp. 595-604. http://dx.doi.org/10.1128/MCB.01503-06

[27] M. X. Luong, J. Tam, Q. Lin, J. Hagendoorn, K. J. Moore, T. P. Padera, B. Seed, D. Fukumura, R. Kucherlapati and R. K. Jain, "Lack of Lymphatic Vessel Phenotype in LYVE-1/CD44 Double Knockout Mice," *Journal of Cellular Physiology*, Vol. 219, No. 2, 2009, pp. 430-437. http://dx.doi.org/10.1002/jcp.21686

[28] D. F. Robbiani, R. A. Finch, D. Jager, W. A. Muller, A. C. Sartorelli and G. J. Randolph, "The Leukotriene C(4) Transporter MRP1 Regulates CCL19 (MIP-3*β*, ELC)-Dependent Ilization of Dendritic Cells to Lymph Nodes," *Cell*, Vol. 103, No. 5, 2000, pp. 757-768. http://dx.doi.org/10.1016/S0092-8674(00)00179-3

[29] L. Tort, J. C. Balasch and S. Mackenzie, "Fish Immune System. A Crossroads between Innate and Adaptive Responses," *Inmunologia*, Vol. 22, No. 3, 2003, pp. 277-286.

Supplementary

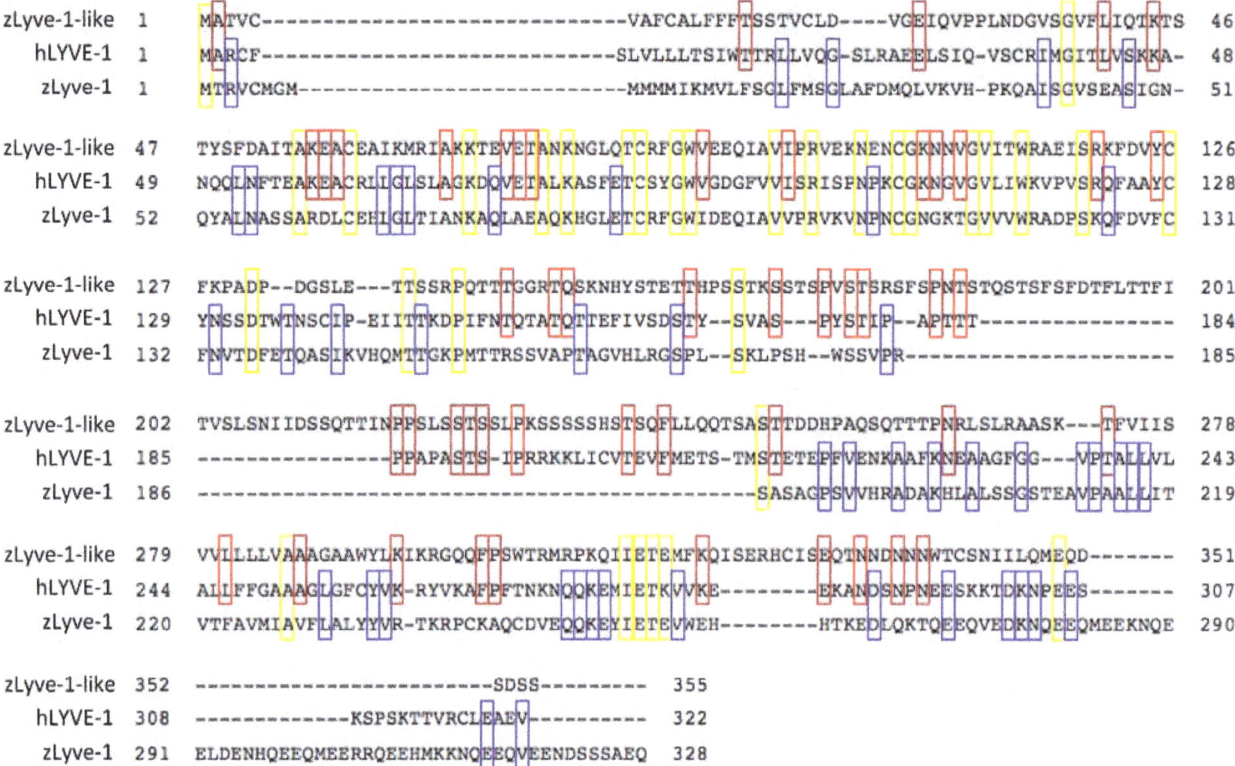

Figure S1. Comparison of the deduced amino acid sequence of zLyve-l with those of zLyve-1-like and human LYVE-1 (hLYVE-1). The deduced amino acid sequences of zLyve-1-like, zLyve-1 and hLYVE-1 are compared. Identical amino acid residues are boxed.

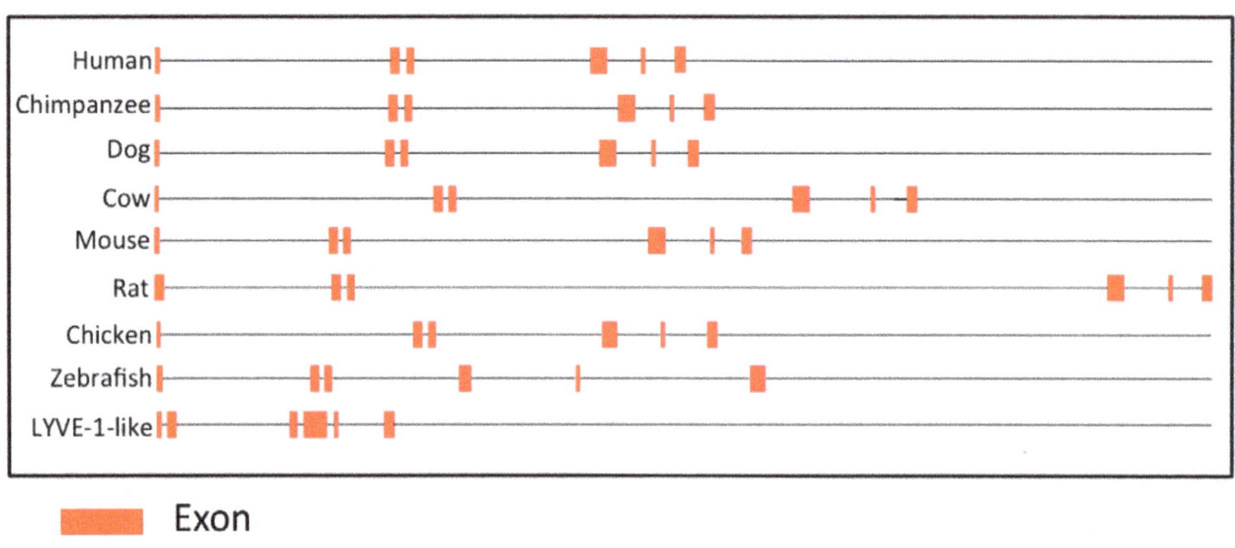

Figure S2. Genomic structure of *zlyve1*, but not *zlyve1l*, is closely similar to those of other vertebrate homologs. Genomic structure analysis was performed using the NCBI sequence viewer for the 9 compared species of the LYVE-1 family (zebrafish, salmon, chicken, rat, mouse, cow, dog, chimpanzee and human) and *zlyve1l.* All these vertebrate *zlyve1* homologs and *zlyve1l* possess 6 exons and 5 introns. However, the sizes of the LYVE-1 family genes are 9.4 - 18.7 kb (human: 9.4 kb; chimpanzee: 9.9 kb; dog: 9.6 kb; mouse: 10.6 kb; rat: 18.7 kb; chicken: 10.0 kb; zebrafish: 10.8 kb) whereas the size of the *zlyve1l* gene is 4.2 kb. The length of introns 1, 3 and 4 in the *zlyve1l* gene are shorter than those of the LYVE-1 family homologs.

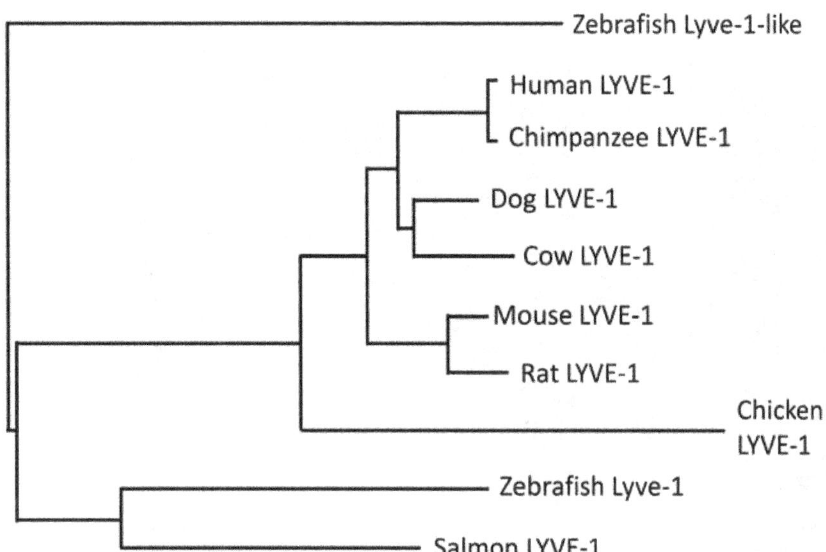

Figure S3. The *zlyve1* gene is evolutionarily related to other vertebrate LYVE-1 homologs. Phylogenetic analysis was performed using the NCBI multiple alignment program. The horizontal branch lengths are proportional to the estimated time from divergence of the gene from the related homologs.

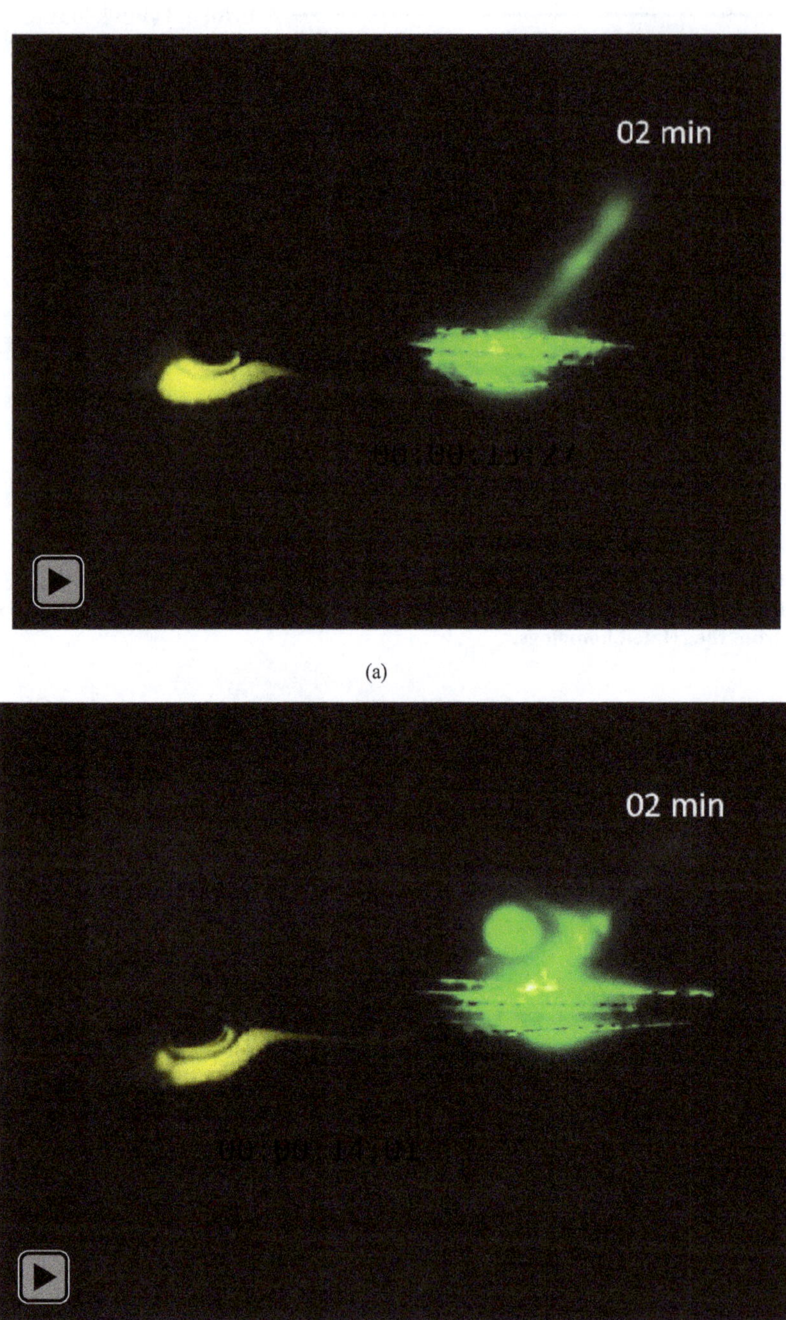

(b)

Figure S4. (movie) Movement of intramuscularly injected FITC-dextran along the TD in wild-type zebrafish derived from embryos injected with control MO (a) or MO2 (b). FITC-dextran was injected intramuscularly into 5-dpf wild-type zebrafish derived from embryos injected with control MO (a) or MO2 (b). The progression of FITC-dextran fluorescence movement along the TD (segment by segment) in the trunk was monitored by taking pictures from 5 to 20 min after FITC-dextran injection.

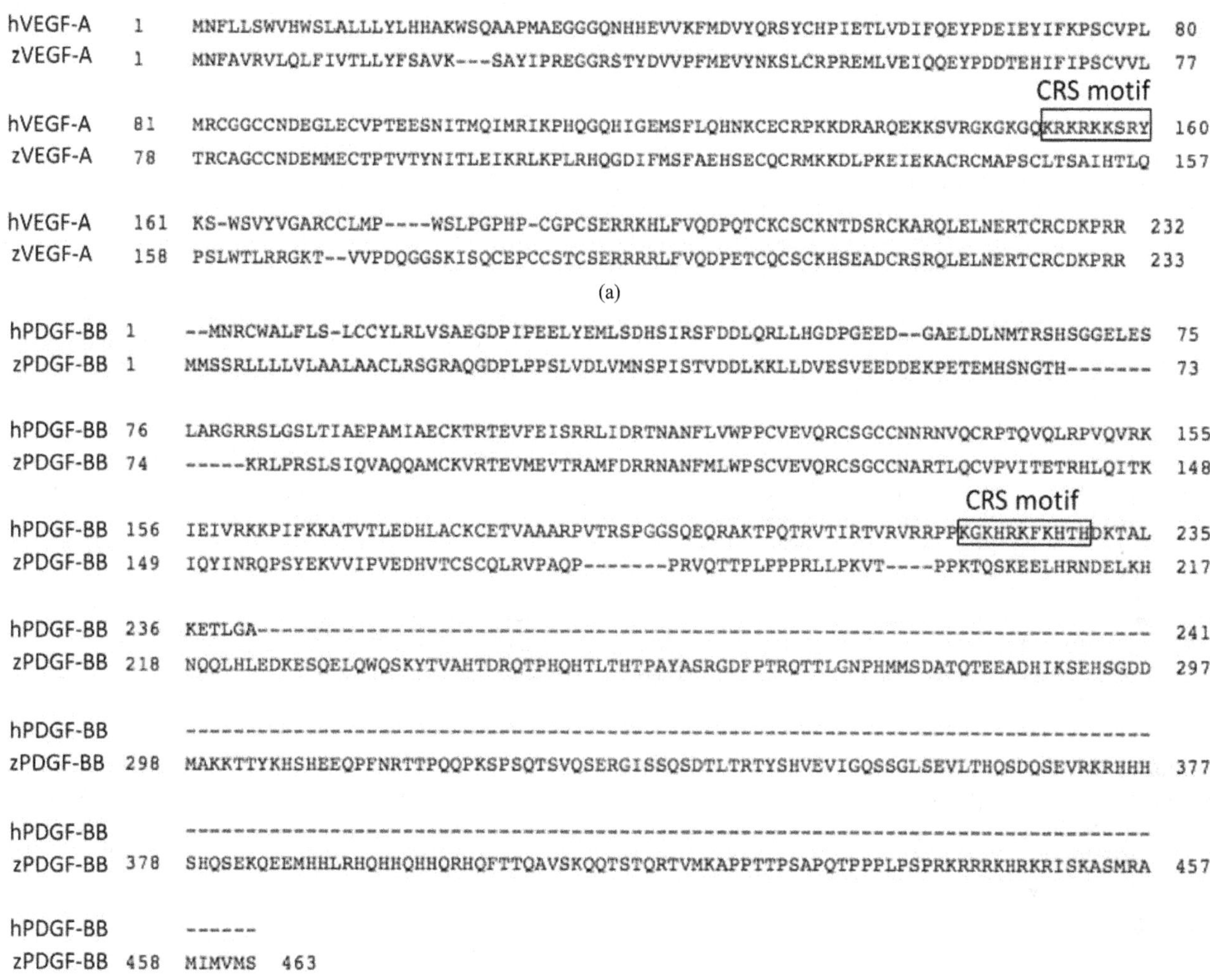

Figure S5. Comparison of deduced amino acid sequences of zebrafish VEGF-A (a) and PDGF-BB (b) with those of their human counterparts. The CRS motifs in the deduced amino acid sequences of human VEGF-A and PDGF-BB are boxed.

Permissions

All chapters in this book were first published in CellBio, by Scientific Research Publishing; hereby published with permission under the Creative Commons Attribution License or equivalent. Every chapter published in this book has been scrutinized by our experts. Their significance has been extensively debated. The topics covered herein carry significant findings which will fuel the growth of the discipline. They may even be implemented as practical applications or may be referred to as a beginning point for another development.

The contributors of this book come from diverse backgrounds, making this book a truly international effort. This book will bring forth new frontiers with its revolutionizing research information and detailed analysis of the nascent developments around the world.

We would like to thank all the contributing authors for lending their expertise to make the book truly unique. They have played a crucial role in the development of this book. Without their invaluable contributions this book wouldn't have been possible. They have made vital efforts to compile up to date information on the varied aspects of this subject to make this book a valuable addition to the collection of many professionals and students.

This book was conceptualized with the vision of imparting up-to-date information and advanced data in this field. To ensure the same, a matchless editorial board was set up. Every individual on the board went through rigorous rounds of assessment to prove their worth. After which they invested a large part of their time researching and compiling the most relevant data for our readers.

The editorial board has been involved in producing this book since its inception. They have spent rigorous hours researching and exploring the diverse topics which have resulted in the successful publishing of this book. They have passed on their knowledge of decades through this book. To expedite this challenging task, the publisher supported the team at every step. A small team of assistant editors was also appointed to further simplify the editing procedure and attain best results for the readers.

Apart from the editorial board, the designing team has also invested a significant amount of their time in understanding the subject and creating the most relevant covers. They scrutinized every image to scout for the most suitable representation of the subject and create an appropriate cover for the book.

The publishing team has been an ardent support to the editorial, designing and production team. Their endless efforts to recruit the best for this project, has resulted in the accomplishment of this book. They are a veteran in the field of academics and their pool of knowledge is as vast as their experience in printing. Their expertise and guidance has proved useful at every step. Their uncompromising quality standards have made this book an exceptional effort. Their encouragement from time to time has been an inspiration for everyone.

The publisher and the editorial board hope that this book will prove to be a valuable piece of knowledge for researchers, students, practitioners and scholars across the globe.

List of Contributors

Sadaki Yokota and Yuko Onohara
Section of Functional Morphology, Faculty of Pharmaceutical Sciences, Nagasaki International University, Nagasaki, Japan

Bor Luen Tang
Department of Biochemistry, Yong Loo Lin School of Medicine, National University Health System, NUS Graduate School for Integrative Sciences and Engineering, National University of Singapore, Singapore, Singapore

Alexei A. Stortchevoi
Department of Pathology, Yale University, New Haven, USA

Magdalena Labieniec-Watala
Department of Thermobiology, Faculty of Biology and Environmental Protection, University of Lodz, Lodz, Poland

Karolina Siewiera
Department of Haemostasis and Haemostatic Disorders, Medical University of Lodz, University Clinical Hospital No. 2, Lodz, Poland

Darina A. Sokolova, Galina S. Vengzhen and Alexandra P. Kravets
Department of Plant Biophysics and Radiobiology, Institute of Cell Biology and Genetic Engineering, National Academy of Science of Ukraine, Kiev, Ukraine

Matthew K. Ball, Kelly Ezell, Jessica B. Henley and Wade A. Grow
Department of Anatomy, Arizona College of Osteopathic Medicine, Midwestern University, Glendale, USA

David H. Campbell and Paul R. Standley
Department of Basic Medical Sciences, University of Arizona College of Medicine, Phoenix, USA

Rashmi Saini, Savita Verma, Abhinav Singh and Manju Lata Gupta
Institute of Nuclear Medicine and Allied Sciences (INMAS), Defence Research and Development Organization (DRDO), New Delhi, India

Shreyasee Chakraborty, Bibiana Sandoval-Bernal and James Kumi-Diaka
Department of Biological Sciences, College of Sciences, Florida Atlantic University at Davie, Davie, USA

Manaka Akashi, Sadaki Yokota and Hideaki Fujita
Section of Functional Morphology, Faculty of Pharmaceutical Sciences, Nagasaki International University, Nagasaki, Japan

Vivek K. Vishnudas
Department of Biology, University of Vermont, Burlington, USA
Berg Biosystems, Framingham, USA

Shawna S. Guillemette
Department of Biology, University of Vermont, Burlington, USA
Department of Cancer Biology, University of Massachusetts Medical School, Worcester, USA

Panagiotis Lekkas
Department of Biology, University of Vermont, Burlington, USA

David W. Maughan
Department of Molecular Physiology and Biophysics, University of Vermont, Burlington, USA

Jim O. Vigoreaux
Department of Biology, University of Vermont, Burlington, USA
Department of Molecular Physiology and Biophysics, University of Vermont, Burlington, USA

Lauren Deady
Department of Biology, Truman State University, Kirksville, USA

James L. Cox
Department of Biochemistry, AT Still University, Kirksville, USA

Vasyl' Chekhun, Natalia Lukianova, Dmytro Demash, Tetiana Borikun, Svyatoslav Chekhun and Yulia Shvets
Department of Mechanisms of Antitumor Therapy, R.E. Kavetsky Institute of Experimental Pathology, Oncology and Radiobiology, NAS of Ukraine, Kyiv, Ukraine

Mercedes Leonor Sánchez
National Council for Scientific Research (CONICET), Buenos Aires, Argentina
School of Medicine, University of Buenos Aires, Buenos Aires, Argentina

Melina María Belén Martínez
Laboratory of Molecular Microbiology, National University of Quilmes, Buenos Aires, Argentina

Paulo César Maffia
National Council for Scientific Research (CONICET), Buenos Aires, Argentina
Laboratory of Molecular Microbiology, National University of Quilmes, Buenos Aires, Argentina

Sadanand Fulzele, Monte Hunter, Rajnikumar Sangani, Norman Chutkan and Carlos Isale
Department of Orthopaedic Surgery, Medical College of Georgia, Augusta, USA

Mark W. Hamrick
Department of Cellular Biology and Anatomy, Medical College of Georgia, Augusta, USA

Lucie Abrouk-Vérot, Claire Brun and Jean-Marie Exbrayat
Université de Lyon, UMRS 449, Biologie Générale, Université Catholique de Lyon, Reproduction et Développement Comparé, Ecole Pratique des Hautes Etudes, Lyon, France

Bruna Amorin, Ana Paula Alegretti, Vanessa de Souza Valim, Annelise Martins Pezzi da Silva, Maria Aparecida Lima da Silva and Lucia Silla
Cell Technology Center, Experimental Research Center, Hospital de Clínicas de Porto Alegre-Porto Alegre, Rio Grande do Sul, Brazil
Graduate Program in Medicine: Medical Sciences, Universidade Federal do Rio Grande do Sul-Porto Alegre, Rio Grande do Sul, Brazil

Felipe Sehn
Cell Technology Center, Experimental Research Center, Hospital de Clínicas de Porto Alegre-Porto Alegre, Rio Grande do Sul, Brazil

Maryam Moslehi and Razieh Yazdanparast
Institute of Biochemistry and Biophysics, University of Tehran, Tehran, Iran

Ayano Harata, Takashi Matsuzaki, Koichi Ozaki and Setsunosuke Ihara
Faculty of Life and Environmental Science, Department of Biological Science, Shimane University, Shimane, Japan

Jin Kawata
Department of New Frontier Sciences, Graduate School of Science and Technology, Kumamoto University, Kumamoto, Japan
Kumamoto Health Science University, Kumamoto, Japan

Makoto Kikuchi
Kumamoto Health Science University, Kumamoto, Japan

Hisato Saitoh
Department of New Frontier Sciences, Graduate School of Science and Technology, Kumamoto University, Kumamoto, Japan

Dmytro Minchenko
Department of Molecular Biology, Palladin Institute of Biochemistry, National Academy of Sciences of Ukraine, Kyiv, Ukraine
Department of Pediatrics, National Bogomolets Medical University, Kyiv, Ukraine

Oksana Ratushna, Yulia Bashta, Ruslana Herasymenko and Oleksandr Minchenko
Department of Molecular Biology, Palladin Institute of Biochemistry, National Academy of Sciences of Ukraine, Kyiv, Ukraine

Sparkle D. Williams and Benny Washington
Department of Biological Sciences, Tennessee State University, Nashville, USA

Rie Irie-Maezono
Department of Gene Therapy and Regenerative Medicine, Kagoshima University, Graduate School of Medical and Dental Sciences, Kagoshima, Japan

Shinichiro Tsuyama
Laboratory for Neuroanatomy, Kagoshima University, Graduate School of Medical and Dental Sciences, Kagoshima, Japan

Kelly Lokka and Spiridon Kintzios
Laboratory of Enzyme Technology, Faculty of Biotechnology, Agricultural University of Athens, Athens, Greece

Panagiotis Skandamis
Laboratory of Food Quality Control and Hygiene, Faculty of Food Science & Technology, Agricultural University of Athens, Athens, Greece

Zaven A. Karalyan, Hranush R. Avagyan, Hovakim S. Zakaryan, Liana O. Abroyan, Lina H. Hakobyan, Aida S. Avetisyan and Elena M. Karalova
Institute of Molecular Biology of the National Academy of Sciences of the Republic of Armenia, Laboratory of Cell Biology, Yerevan, Armenia

Mary Arai, Takashi Matsuzaki and Setsunosuke Ihara
Department of Biological Science, Faculty of Life and Environmental Science, Shimane University, Matsue, Japan

Wen-Han Chen, Jung San Huang and Andrew Schreiner
Department of Biochemistry and Molecular Biology, Saint Louis University School of Medicine,

Wen-Fang Tseng and Chiou-Hwa Yuh
Division of Molecular and Genomic Medicine, National Health Research Institutes, Chunan, Taiwan

Gen-Hwa Lin and Jen-Leih Wu
Institute of Cellular and Organismic Biology, Academia Sinica, Taipei, Taiwan

Hsiao-Rong Chen
Institute of Systems Biology and Bioinformatics, National Central University, Jhongli, Taiwan

Mark M. Voigt
Department of Pharmacology and Physiological Science, Saint Louis University School of Medicine, St. Louis, USA

Shuan Shian Huang
Auxagen Inc., St. Louis, USA

www.ingramcontent.com/pod-product-compliance
Lightning Source LLC
Chambersburg PA
CBHW080502200326
41458CB00012B/4056

* 9 7 8 1 6 3 2 3 9 7 5 0 8 *